Design Solutions and Innovations in Temporary Structures

Robert Beale
Independent Researcher, UK

João André
Portuguese National Laboratory for Civil Engineering, Portugal

A volume in the Advances in Civil and Industrial Engineering (ACIE) Book Series

IGI GLOBAL
DISSEMINATOR OF KNOWLEDGE

www.igi-global.com

Published in the United States of America by
IGI Global
Engineering Science Reference (an imprint of IGI Global)
701 E. Chocolate Avenue
Hershey PA, USA 17033
Tel: 717-533-8845
Fax: 717-533-8661
E-mail: cust@igi-global.com
Web site: http://www.igi-global.com

Copyright © 2017 by IGI Global. All rights reserved. No part of this publication may be reproduced, stored or distributed in any form or by any means, electronic or mechanical, including photocopying, without written permission from the publisher. Product or company names used in this set are for identification purposes only. Inclusion of the names of the products or companies does not indicate a claim of ownership by IGI Global of the trademark or registered trademark.

Library of Congress Cataloging-in-Publication Data

Names: Beale, Robert, 1947- author. | Andrâe, Joäao, 1981- author.
Title: Design solutions and innovations in temporary structures / by Robert Beale and Joäao Andrâe.
Description: Hershey, PA : Engineering Science Reference, [2017] | Includes bibliographical references.
Identifiers: LCCN 2016055143| ISBN 9781522521990 (H/C) | ISBN 9781522522003 (eISBN)
Subjects: LCSH: Temporary structures (Building)--Design and construction. | Temporary structures (Building)--Safety measures.
Classification: LCC TH5280 .B43 2017 | DDC 624.1--dc23
LC record available at https://lccn.loc.gov/2016055143

This book is published in the IGI Global book series Advances in Civil and Industrial Engineering (ACIE) (ISSN: 2326-6139; eISSN: 2326-6155)

British Cataloguing in Publication Data
A Cataloguing in Publication record for this book is available from the British Library.

All work contributed to this book is new, previously-unpublished material. The views expressed in this book are those of the authors, but not necessarily of the publisher.

For electronic access to this publication, please contact: eresources@igi-global.com.

Advances in Civil and Industrial Engineering (ACIE) Book Series

Ioan Constantin Dima
University Valahia of Târgovişte, Romania

ISSN:2326-6139
EISSN:2326-6155

Mission

Private and public sector infrastructures begin to age, or require change in the face of developing technologies, the fields of civil and industrial engineering have become increasingly important as a method to mitigate and manage these changes. As governments and the public at large begin to grapple with climate change and growing populations, civil engineering has become more interdisciplinary and the need for publications that discuss the rapid changes and advancements in the field have become more in-demand. Additionally, private corporations and companies are facing similar changes and challenges, with the pressure for new and innovative methods being placed on those involved in industrial engineering.

The **Advances in Civil and Industrial Engineering (ACIE) Book Series** aims to present research and methodology that will provide solutions and discussions to meet such needs. The latest methodologies, applications, tools, and analysis will be published through the books included in **ACIE** in order to keep the available research in civil and industrial engineering as current and timely as possible.

Coverage

- Hydraulic Engineering
- Optimization Techniques
- Ergonomics
- Earthquake engineering
- Engineering Economics
- Operations Research
- Productivity
- Production Planning and Control
- Construction Engineering
- Materials Management

IGI Global is currently accepting manuscripts for publication within this series. To submit a proposal for a volume in this series, please contact our Acquisition Editors at Acquisitions@igi-global.com or visit: http://www.igi-global.com/publish/.

The Advances in Civil and Industrial Engineering (ACIE) Book Series (ISSN 2326-6139) is published by IGI Global, 701 E. Chocolate Avenue, Hershey, PA 17033-1240, USA, www.igi-global.com. This series is composed of titles available for purchase individually; each title is edited to be contextually exclusive from any other title within the series. For pricing and ordering information please visit http://www.igi-global.com/book-series/advances-civil-industrial-engineering/73673. Postmaster: Send all address changes to above address. ©© 2017 IGI Global. All rights, including translation in other languages reserved by the publisher. No part of this series may be reproduced or used in any form or by any means – graphics, electronic, or mechanical, including photocopying, recording, taping, or information and retrieval systems – without written permission from the publisher, except for non commercial, educational use, including classroom teaching purposes. The views expressed in this series are those of the authors, but not necessarily of IGI Global.

Titles in this Series

For a list of additional titles in this series, please visit:
http://www.igi-global.com/book-series/advances-civil-industrial-engineering/73673

Modeling and Simulation Techniques in Structural Engineering
Pijush Samui (National Institute of Technology Patna, India) Subrata Chakraborty (Indian Institute of Engineering Science and Technology (IIEST), Shibpur, India) and Dookie Kim (Kunsan National University, Soth Korea)
Engineering Science Reference • ©2017 • 524pp • H/C (ISBN: 9781522505884) • US $220.00

Computational Modeling of Masonry Structures Using the Discrete Element Method
Vasilis Sarhosis (Newcastle University, UK) Katalin Bagi (Budapest University of Technology and Economics, Hungary) José V. Lemos (National Laboratory for Civil Engineering, Portugal) and Gabriele Milani (Technical University in Miln, Italy)
Engineering Science Reference • ©2016 • 505pp • H/C (ISBN: 9781522502319) • US $210.00

Handbook of Research on Emerging Innovations in Rail Transportation Engineering
B. Umesh Rai (Chennai Metro Rail Limited, India)
Engineering Science Reference • ©2016 • 664pp • H/C (ISBN: 9781522500841) • US $235.00

Advanced Research on Nanotechnology for Civil Engineering Applications
Anwar Khitab (Mirpur University of Science and Technology, Pakistan) and Waqas Anwar (Mirpur University of Science and Technology, Pakistan)
Engineering Science Reference • ©2016 • 339pp • H/C (ISBN: 9781522503446) • US $195.00

Emerging Challenges and Opportunities of High Speed Rail Development on Business and Society
Raj Selladurai (Indiana University Northwest, USA) Peggy Daniels Lee (Kelley School of Business, Indiana University, USA) and George VandeWerken (Providence Bank, USA)
Engineering Science Reference • ©2016 • 289pp • H/C (ISBN: 9781522501022) • US $210.00

Advanced Manufacturing Techniques Using Laser Material Processing
Esther Titilayo Akinlabi (Univeristy of Johannesburg, South Africa) Rasheedat Modupe Mahamood (University of Johannesburg, South Africa & University of Ilorin, Nigeria) and Stephen Akinwale Akinlabi (University of Johannesburg, South Africa)
Engineering Science Reference • ©2016 • 288pp • H/C (ISBN: 9781522503293) • US $175.00

For an enitre list of titles in this series, please visit:
http://www.igi-global.com/book-series/advances-civil-industrial-engineering/73673

IGI GLOBAL
DISSEMINATOR OF KNOWLEDGE
www.igi-global.com

701 East Chocolate Avenue, Hershey, PA 17033, USA
Tel: 717-533-8845 x100 • Fax: 717-533-8661
E-Mail: cust@igi-global.com • www.igi-global.com

Table of Contents

Foreword

Collapse of temporary structures such as scaffolds, grandstands, shoring etc. is common in both developed and developing countries. Safety of temporary structures consequently has been a big issue to many places as they, possibly with the exception of massive permanent structures during earthquakes, attracted a high collapse rate even greater than the permanent structures. It is actually sad for us as structural engineers to see the frequent collapse leading to casualties which not only take lives of involved workers, but also damage their families as many construction workers are family supports. Authored by world renounced researchers and engineers, this book is devoted comprehensively to various aspects of safe temporary structures namely as actions, analysis, structural modeling, safety design standards and codes, collapse analysis and investigation and quality control, risk management and training for personnel including inspection, coordination, maintenance and so on. Innovation in improving the safety and efficiency is also detailed in this book.

There have not been many documents and guidebooks covering various aspects of temporary structures, especially their safety. Many design codes related to temporary structures have been considered by practitioners as too complex to use by engineers, including lack of explanations and practical considerations. This book fills this gap of producing a comprehensive and authoritative guidance to safe construction and design of temporary structures which include design, analysis, construction, inspection, supervision and productions. The authors have embarked on research and engineering design of scaffolds and temporary structures over the past decades and the experience gained by them is invaluable for the advancement in technology of temporary structures. A number of innovative ideas in improving the quality of temporary structures are also discussed in this book.

The authors are to be congratulated on their contribution to the field of safety of temporary structures through design and control. I have little doubt that this book will be of immense benefit, not only to the designer for enhancement of safety level under the theory of structural stability of temporary structures, but also all parties involved including the public in norm in improving the safety and reliability of temporary structures in delivering their functions of supporters against loads.

S. L. Chan
The Hong Kong Polytechnic University, China

S. L. Chan, *the Chair Professor in Computational Structural Engineering at the Hong Kong Polytechnic University, published extensively in the field of steel structures and computational methods. He is the chief and founding editor of the SCI international journals "Advanced Steel Construction", "Steel and Composite Structures (2002-2005)" and the regional editor of "Interna-*

tional Journal of Applied Mechanics and Engineering" and serves as a member of editorial boards in several international journals, and of ad-hoc committees in drafting guides for design of steel, scaffolding and glass structures. He is also the past chairman of the Structural Division of the Hong Kong Institution of Engineers (2014-2015), incumbent President of the Hong Kong Institute of Steel Construction (HKISC) and adjunct professor at the Southeast University in Nanjing, Harbin Institute of Technology in Harbin and Tongji University in Shanghai. In conjunction with a research team of the Tongji University, Professor Chan was given the first class award for research in steel structures by the Education Ministry in the Mainland China. Professor Chan carried out extensive applied research and, with his PhD students, he received excellence awards by Geotechnical and Structural Divisions of the Hong Kong Institution of Engineers in the same year of 2016 for their research papers in nonlinear analysis and numerous research and consultancy awards from his university.

Preface

Temporary structures are defined as any structure that is used during the construction, rehabilitation and retrofit of buildings and bridges, or any other type of permanent structure. Temporary structures can be divided into various types, being falsework, formwork, scaffolds and shoring systems the most common. Bridge construction equipment (BCE) is a special type of temporary structures.

A scaffold is that structure erected for the maintenance or alterations of an existing building or bridge. It is usually attached or tied to the permanent structure and for this reason is often called an access scaffold. Formwork is used to provide a containing shape into which concrete is poured to obtain the required shape and strength. Falsework structures are spatial framed systems that are used to support buildings and bridges. BCE are highly specialised systems and as the name implies are only used during the construction of bridges. This book will primarily deal with scaffolds, falsework and BCEs with only brief mentions of shoring and formwork.

Temporary structures have a major role in the execution time, cost, quality, durability, safety, efficiency, utility and aesthetics of any construction project. Past reviews conducted by several authors regarding the performance of temporary structures have shown that although these structures are in common use they frequently fail due to poor design or management. A correct choice, good planning, designing and operation of temporary structures are keys for the success of every construction project. This book will enable graduate students, educators, researchers, designers, producers, contractors, buildings and bridge owners, and managers of temporary structures to become aware of the breadth of the subject and the latest methods of analysis, design and management.

The book commences with an overview of the research into temporary structures since 1970 as modern computer technology has enabled the design of the structures to change from traditional designs based on effective lengths and elastic analysis procedures to designs based on advanced nonlinear numerical analyses and methods.

However, temporary structures design is often considered secondary to the design of permanent works and simplifications are often made without proper justification with regard to design actions and their effects on the structure, but also with respect to the interaction of various elements in the structural system. Often, to properly account for the latter factors more accurate analyses are necessary resulting in alternative design solutions. In addition, the requirements concerning the foundations upon which the structure is to be erected are commonly ignored. Past forensic studies have found recurrent deficiencies about foundations, lateral stability, design errors, design details and materials, to which can be added planning errors (e.g. insufficient inspection plans).

When assisting in developing the design manual for the UK National Access and Scaffolding Confederation (NASC) there was considerable opposition from practitioners who found it hard to understand that their previous assumption in an access scaffold of considering that the effective length of the columns were equal to the lift height was in many cases incorrect and unconservative. This and other common misconceptions are an important reason for this book which details the most recent and relevant research into temporary structures emphasising modern knowledge of structural behaviour.

In addition, many designers of permanent works do not always consider the problems arising in the erection or maintenance of their structures, leaving the problems to the contractors and sub-contractors employed to manage the construction.

The prime objective of this book is to enable all practitioners in the temporary structures industry to gain an understanding of safe design and practice, and modern methods of analysis. The authors are based in Europe but since they are aware that engineers may work in many countries during their lives, have looked at practices worldwide in order to make this book relevant to as many people as possible . For example, when design codes are discussed and compared in detail in Chapter 6, not only European codes are presented but also codes developed in the USA, Canada, Australia and Hong Kong for a range of materials from steel and aluminium to bamboo.

Sadly, many changes in design procedures are only made after major disasters occur. For example, the collapse of a falsework structure for a bridge over the river Loddon, near Reading, UK, in 1972 with the loss of three lives led to the UK Government forming an advisory committee on falsework which in 1975 published the Bragg report. This report led to the suggestion that supervisors called Temporary Works Coordinators be appointed to preside over temporary structures projects. It was first codified into the UK code BS 5975 in 1982 and only referred to falsework. However, later revisions included all temporary structures. The lessons that were documented in the Bragg report about poor site control and management still occur today as documented by many research studies. Further information on the causes of temporary structures failures can be found in Chapter 7, where collapses and failures are discussed and in Chapter 8, where detailed analyses of the management of temporary structures projects are presented with strong recommendations for the correct procedures to reduce and potentially eliminate the majority of failures.

A second important objective of this book is to enable new researchers into temporary structures and advanced practitioners to gain knowledge of the latest methods developed for analysis. The book therefore presents the reader with knowledge of new methods of analysis which enable the force distributions, material's stresses and strains to be calculated with improved accuracies. In particular the results of experiments conducted by André which led to improved models for column-to-column connections in Cuplok® systems (a typical proprietary scaffold and falsework system) and detailed models of extensible props by Ms Feng in 1994 are presented for the first time. Models developed by Beale for access scaffolds which have proved efficient and which were used to develop load tables for the UK NASC are also presented. Improved models for the moment-rotation curves of connections between vertical and horizontal scaffold/falsework members are discussed with recommendations on the best model presented. Modern risk analysis procedures applied to these structures and new ideas developed by André are also presented for the first time.

A common assumption made in temporary structures design is that design factors used for permanent structures can be reduced because temporary structures are only in use for a limited period before being dismantled. However, Fyall, for example, when discussing the Milton Keynes scaffold collapse argued that wind distributions should be assumed to be taken as those that occur using a 100-year return cycle as opposed to the conventional limited 5-year or 25-year return periods specified in codes as the structures are more likely to have lower factors of structural stability and hence additional wind will cause failures with potential disproportionate consequences. Therefore, in this book a whole Chapter is devoted to risk analysis and its implication to the structural safety of temporary structures, including an application example.

As will be discussed in the final two Chapters of the book, the management of a temporary structure project is crucial to the safe and successful outcome of any construction project. A significant percentage of the causes of failures and collapses can be shown to have occurred due to inadequate supervision and overview of the design and construction process of temporary structures. When analysing the causes of the Milton Keynes collapse, it was found that the scaffold structure was not built according to its design, changes were made to the scaffold without reference to the design drawings which were not on site, the scaffold was overloaded and crucially no inspections had been made by the supervisor in charge of the project for several months. These inspections would have made the likelihood of collapse smaller and the risk of collapse acceptable.

The book is organised into eight Chapters. Each Chapter is self-contained with only occasional references to other Chapters. This means that each Chapter starts with an introduction which contains some common material to other Chapters, particularly information presented in the historical survey so that the reader does not have to cross reference too often when studying the information provided in the book. The objectives of each chapter are:

CHAPTER 1: INTRODUCTION

This chapter briefly summarises the types of temporary structures, presents the motivation of the book as well as its objectives and significance.

CHAPTER 2: HISTORICAL SURVEY

This chapter briefly overviews the research into temporary structures developed from 1970 up to 2016. The chapter introduces the distinction between scaffolding and falsework, scaffolding being mainly concerned with temporary structures used along the façades of buildings for construction and maintenance, and falsework for those temporary structures used to support formwork for the placing of concrete in bridges and buildings.

The historical survey into scaffold structures covers tubular scaffolds, the most common temporary structure in the UK, proprietary and modular scaffolds which are used in many other countries. The survey shows that traditional designs of falsework structures often lead to underestimates of the safety of the structures by the use of simple design procedures based on effective lengths. The survey also emphasises the importance of site management in ensuring the safety of scaffolds.

The survey into bridge falsework shows that these structures can be sub-divided into two types: firstly fixed falsework for bridges normally less than 30 m in height and typically less than 60 m in span to support concrete formwork; and secondly repeatedly using sections of falsework as the bridge is extended over large spans. In addition, specialist temporary structures called Bridge Construction Equipment (BCE), are used for more complex bridges and contexts to achieve the maximum efficiency in construction.

The materials used for the structures are shown to be metallic (steel or aluminium), bamboo or timber as the latter two materials are still common in Asia and India. It is notable that to a large extent the same analysis procedures are used for these different materials with the exception that the differences in material strengths and elastic stiffnesses must be taken into account.

The chapter then surveys the actions acting on temporary structures, namely permanent and imposed loads, wind loads and notes that limited research has been carried out on earthquake provisions for temporary structures.

Finally, the chapter gives an introduction to failures of falsework and formwork structures which are analysed in more detail in Chapter 7.

CHAPTER 3: ACTIONS

This chapter presents a description of the actions applied to temporary structures. In particular, a classification of actions is presented.

Actions are classified into permanent actions such as self-weight, lateral loads by soil or water; and variable actions such as imposed loads and wind loads. Comparisons are made between design code provisions for loads as specified by European, USA and Australian codes.

The dynamic effects of pouring concrete onto falsework are analysed and a formula is given to enable the analyst to predict the correct loads. The chapter also discusses the load distribution into falsework supports due to the post-tensioning of cables.

The latest research into wind on temporary structures is a significant part of this chapter with its implications to the correct wind forces acting on temporary structures when turbulence and orography are taken into account. The wind section presents extensive useful information to determine the wind action and its effects on the structures and discusses the effects of using different time periods to determine the forces, with a proviso that longer return periods produce safer structures.

The chapter continues with a presentation of the effects of ground settlement, human induced dynamic forces, and accidental actions such earthquakes and vehicle impacts. The chapter ends with a discussion concerning possible notional actions to take into account the effect of human errors during design, execution and operation of temporary structures.

CHAPTER 4: STRUCTURAL ANALYSIS

The analysis procedures for temporary structures have been extensively enhanced during the last two decades. This chapter develops the components required for successful modelling of temporary structures.

Firstly, the fundamentals of structural analysis using the Finite Element Method (FEM) are provided. In particular, material models of steel, aluminium and bamboo are presented with an emphasis on linear and multilinear models for steel and the Ramberg-Osgood model for aluminium.

In order, to accurately model falsework and scaffold structures the various components making up the structure must be correctly modelled. The chapter therefore presents models for beam-to-column connections, top connections, base connections and column-to-column connections based on the latest theoretical and experimental procedures developed by the authors and co-workers. In particular, multilinear and nonlinear models are presented for the rotational moment-curvature relations of the connections, including joint looseness which has traditionally been ignored but can significantly reduce the rotational stiffness of the connections.

Analyses models for components such as braces and props are presented together with the experimental verification of the models.

Two- and three-dimensional analysis models are then developed for access scaffolds, bridge falsework and bamboo scaffolds.

Finally, the chapter presents information on the effects of ground modelling and wind engineering on the determination of parameters such as wind pressure needed for the correct determination of action effects applied to temporary structures.

CHAPTER 5: STRUCTURAL SAFETY

Chapter 5 presents innovative concepts in safety, starting with definitions of reliability, fragility and risk, giving a new definition of structural robustness which enables advanced calculations to be undertaken.

Following this introduction, uncertainties are discussed and a risk management framework for structural design is proposed. A probabilistic structural design philosophy is presented detailing a new methodology for analysing structural fragility and the robustness of structures against failure. An example is presented determining the robustness of a falsework structure against collapse.

The chapter then presents strategies to enhance structural robustness and risk and the importance of using risk-informed decisions giving reasons why these should be applied to temporary structures as well as to traditional structures.

Next, the chapter proposes a design philosophy for temporary structures, giving details on how to proceed with the design and quality management for different levels of consequences of failure. The chapter also presents an application example of the design philosophy.

The chapter concludes with a statistical background behind design with respect to wind action and gives reasons why the traditional approach of reducing the partial factor temporary structures subjected to wind can lead to an underestimate of the influence of wind on temporary structures, suggesting that wind return periods should be in excess of 80 years and not the common five to ten years.

CHAPTER 6: DESIGN CODES

This chapter starts by over viewing the philosophies behind design codes with particular reference to the use of modern limit state design and points out that few codes now use allowable stress design. Comments are made on the design life of temporary structures which vary considerably from Europe, Australia and the USA.

Design codes of the USA, Europe and Australia/New Zealand (common code for Australia and New Zealand) for temporary structures are compared with particular reference to the values of loads and of the partial factors for loads. It is noted that whilst the European design codes do not specify how construction, use and disassembly of the temporary structures is to be executed, the USA code for scaffolding includes such specification and also gives tables of allowable assemblies. The Hong Kong code for bamboo scaffolds is described showing the similarities and differences between bamboo and metal scaffolds.

Examples are then presented of the use of the codes to determine elements of bridge falsework, prop design, bearing design, temporary structures such as grandstands, and the design of Bridge Construction Equipment.

CHAPTER 7: ANALYSIS OF COLLAPSES

In some ways, this chapter is one of the most important in the book as it describes falsework and scaffolding collapses and gives strong recommendations in the processes to be undertaken in the design and operation of the structures to reduce the chances of future collapses occurring. It is notable that the cost of a temporary structure collapse can be over 100 times the cost of preventing it, and in many cases collapses can lead to companies going into administration.

The chapter starts by extending the review into collapses described in Chapter 2 by investigating in detail the causes of collapse in temporary structures encompassing scaffolds, falsework, temporary structures such as grandstands and providing a comprehensive lists of faults which can occur during design, erection, use and disassembly.

As bridge falsework collapses are more commonly reported and usually greater financial implications and possibly greater life loss occur, a survey, conducted by André, is summarised showing that these collapses occur regularly throughout the world with the same faults as itemised in the first part of the chapter.

The third section of the chapter provides a forensic analysis of a scaffold collapse and a bridge falsework collapse.

CHAPTER 8: QUALITY MANAGEMENT

Ensuring that temporary structures projects are managed well so that budgets are satisfied and that safety is preserved throughout the project is the objective of this chapter.

The chapter starts by discussing the importance of having clear project management procedures involving collaboration between the client, designer and the construction engineer. This leads to the importance of an overall project supervisor, sometimes called a Temporary Works Controller or a Temporary Works Supervisor, who has the ultimate authority for the safe execution of the project.

The chapter also gives recommendations to improve planning, design, assembly, operation, maintenance, disassembly, inspection and supervision, documentation and competence of those involved with temporary structures.

Acknowledgment

NOTE

This book is published on the understanding that the authors are solely responsible for the statements made and opinions expressed in it and that its publication does not necessarily imply that such statements and/or opinions are or reflects the views or opinions of the publishers. While every effort has been made to ensure that the statements made and the opinions expressed in this publication provide a safe and accurate guide, no liability or responsibility can be accepted in this respect by the authors or publishers.

It is with grateful thanks that Robert Beale acknowledges all the colleagues with whom he has worked for many years. In particular, mention must be made of Dr. Mike Godley with whom he developed an extremely helpful and collaborative partnership and without whose assistance he would not have been able to do so much research.

He acknowledges the patience and encouragement of his wife, Hilary, during all the years he has been engaged on research.

It is also with grateful thanks that João André thanks Robert Beale for being his PhD supervisor and for mentoring throughout his doctorate studies. He also thanks Dr. António Baptista, who, in over twelve years working at "Laboratório Nacional de Engenharia Civil" (LNEC), Lisbon, Portugal, has been teaching and mentoring him.

João would wish to express his gratitude to Ana for supporting him throughout this enduring challenge. Finally, but foremost, he would like to thank his parents, Mariana and José, for being the best parents in every moment of his life.

Chapter 1
Introduction

ABSTRACT

Chapter 1 is an introduction to the book and provides an overview of the areas in which temporary structures are used, namely the construction and repair of buildings and bridges. A description of the different types of temporary structures is given together with an overview of the problems which may arise in temporary structures projects. The differences between temporary structures projects and projects for permanent structures are highlighted. An introduction to the particularities of the design, assembly, maintenance and operation of temporary structures is presented in this chapter. It is also emphasised that the book compares the design codes used in the USA, Europe, Australia and Hong Kong. Finally, the chapter concludes with an overview of the remaining chapters of the book.

1.1 PROBLEM STATEMENT AND MOTIVATION

The present book concerns temporary structures, in particular the most commonly used temporary structures: scaffolds, shoring, bridge falsework systems made of slender vertical and horizontal steel tubes connected by special couplers. The design of scaffold structures made of bamboo is also discussed, as well as specialised equipment used in modern bridge construction, often named bridge construction equipment (BCE).

Failures involving these structures are amongst the most common types of accidents in civil and construction engineering, and often lead to disproportionate consequences. This reality calls for a paradigm change regarding the design and use of temporary structures systems.

The book will present and explore the most recent advances in research and innovation in the field of civil, construction and structural engineering applied to temporary structures.

There are various types of structural systems available in the market: from towered systems made of steel or aluminium built-up members, frame systems of steel beams and columns with structural profiled sections, to proprietary modular 3-D frame systems of metallic elements connected by special couplers. There are many applications of these structural systems ranging from the construction, rehabilitation to the retrofit of bridge and buildings structures. Figure 1 illustrates some examples of temporary structures systems.

DOI: 10.4018/978-1-5225-2199-0.ch001

Copyright © 2017, IGI Global. Copying or distributing in print or electronic forms without written permission of IGI Global is prohibited.

Figure 1. Examples of temporary structures
**©2016 Brand Energy & Infrastructure Services. Used with permission*
***©2016 RMD Kwikform. Used with permission*
****©2016 NRS AS, Norway. Used with permission*

(a) Access scaffold*
(b) Bridge falsework
(c) Temporary seating and staging**
(d) Shoring system*
(e) Precast span system***
(f) Balanced cantilever system***

There are several stakeholders directly or indirectly concerned with temporary structures: researchers, designers, producers, clients, consultants, insurers, contractors, sub-contractors and workers. In this context, the assemblage, use and dismantling of temporary structures systems is usually done by a specialised sub-contractor, in accordance with a standard design project or with a specially developed design project, depending on the work complexity.

Since the industrial revolution, the construction industry and in particular temporary structures have been experiencing new challenges and some fundamental changes. The International Federation for Structural Concrete (fib) has stated that through time the role of temporary structures in the cost, construction rate, safety, quality, durability, efficiency, utility and aesthetics of any engineering project has increased in a consistent fashion (fib, 2009). Therefore, it is not surprising that a correct choice of temporary structure, good planning, design and operation of the temporary structures are keys for the success of every engineering project. In particular, it is vital that synchronised planning and continuous knowledge exchange exists between the structural designer, the contractor, the temporary structures designer, the temporary structures contractor and others.

Unfortunately, this is not always a reality. The framework of engineering construction consists of complex interactions between all the above mentioned stakeholders who have different backgrounds and can have different priorities, perceptions and goals, some of which can even be contradictory (fib, 2009). Despite the construction phase being the most critical stage of a structures' lifetime – most failures occur during construction rather than after projects have been completed, see Ratay (2009) and Scheer (2010) for examples – some stakeholders still do not recognise the importance of these systems: they are "temporary" and, therefore, their role is considered to be minor compared to that of the permanent structures. Consequently, the design and use of temporary structures are not usually treated as carefully as in the case of permanent structures. Furthermore, they do not receive the same level of research attention and research funding as occurs in permanent structures.

This is clearly evidenced by the number and the state-of-the-art level of existing standards and guidance documents concerning permanent structures as opposed to those relating to temporary structures. Until recently, national and international design codes/standards and/or guidance documents concerning temporary structures were based on simple design procedures. For example: the columns' effective lengths of temporary structures were only governed by the vertical spacing of horizontal members, not considering the system's overall stability.

It should be noted that the use of the effective length concept as a design procedure, although simple, is often not accurate since it is based on an element level safety check and it assumes that the element's deformed shape is very similar to its first global elastic buckling mode. It is important for a structural analyst to be aware that in practical structures, geometrical imperfections in the elements and horizontal loads, such as those associated with wind, may make the common assumption that failure always occurs in an amplification of the first buckling mode incorrect. These additional forces, not considered in eigenvalue analyses, may tend to cause overload in other deflection modes. For example, under wind load if the highest elements of a scaffold are fully sheeted and not tied at the top to a façade, then the top elements may fail under a wind load lower than the anticipated design resistance, as a plastic hinge under bending about the top working level. Therefore, in this book the use of full second-order nonlinear analysis and design procedures is recommended and appropriate models for structural analysis are presented, some for the first time in print.

Traditionally, most of temporary structures are usually designed using safe load tables developed by the producers of the proprietary equipment. Normally they are general based on existing standards or on in-house developed design methods. Often, these tables do not provide information regarding

1. Quality requirements (e.g. the specification of design tolerances), or
2. Risk assessment for specific applications (e.g. special loading and foundation conditions).

Additionally, the design rules applied to temporary structures are not uniform and therefore the actual reliability levels are usually smaller and exhibit a greater variation than the corresponding reliability levels of permanent structures.

To counter this well rooted reality, and under an increasing pressure from public opinion, there has been an effort in some countries such as the UK, beginning with the Bragg report following the River Loddon accident (Bragg, 1975), and continuing with other documents (BSI, 2011a), and more lately at a

European level (BSI, 2011b) to publish standards and guidance documents prepared by special technical commissions. Still, reference is missing to the design working life of these structural components and in Europe especially, to the management of temporary structures, the codes being primarily design codes. In the development of the European Design codes (the structural Eurocodes), a statement is made that structures designed for use and after disassembly, reuse on several occasions should not be considered as temporary structures but designed as permanent structures. Despite the recent research investigations, the design of temporary structures is still frequently associated with high uncertainty levels, due to insufficient information about their real behaviour at the construction site. In particular, little information is available about the size of the actual geometric imperfections and load eccentricities, or the influence of foundation settlements on their resistance, reliability and robustness. Note that the generic term for European codes is "Eurocodes". However, a single code is called a "Euronorm" and abbreviated EN.

It must be acknowledged that most of the problems not dealt with during the planning and design phases will have to be handled on the site. However, the lack of expertise in the field and tight project deadlines have a tendency to make construction workers behave unsafely, take unnecessary chances, and endanger both themselves and the structures (both temporary and permanent). Long sub-contractor chains lead inevitably to loss of communication between the various agents and to loss of responsibility for the supervision, inspection and dismounting procedures.

It should also be stressed that the design and use of temporary structures places very complex and different challenges from the ones associated with permanent structures, such as:

1. BCE systems have the capacity of moving. This is different from common permanent structures which are normally considered as static.
2. Generally, the design of most temporary structures is usually controlled by construction loads, for example, the self-weight of the resulting permanent structure or the materials loaded onto a scaffold. Note that wind loading is often only considered as an equivalent side load of fixed magnitude and dynamic effects ignored. As a result, most temporary structures are subject to load values close to, and sometimes above, the assumed design values during some of their entire service period, whereas the design of permanent structures is often controlled by load cases that will only occur for a brief period of time, or that have a small probability of occurring, during their design working lifetime.
3. Temporary structures are typically used for short periods of time, although due to multiple re-use cycles of the structure, or the material used in the structure such as tubes, their design working life can sum up to 15 years or more. Based on the temporary nature of the use of these structures, some design philosophies specify smaller design values for the actions than the ones used in the design of permanent structures, which may lead to unsafe structures. Indeed, recommendations have been made by engineers and consultants such as Fyall (2012). In a seminar discussing a scaffold collapse, Fyall suggested that due to the nature of the components used and their tendency to elastic failure that enhanced design factors such as a 100 year wind load be used. Furthermore, since the ratio between their cost and the cost associated with their collapse is much lower than for permanent structures, existing standard methodology needs to be reconsidered using a risk informed approach.
4. Temporary structures are assembled, (re)used for short periods and dismantled in general for several times in repetitive cycles. As a result, flaws in erection, inspection and maintenance procedures are likely to occur, leading to construction errors with potentially severe consequences. Additionally, the cooperation between the various stakeholders involved in their design and operation is not

always an appropriate one. These facts can multiply by several orders of magnitude the risk associated with these structures, since their design often does not account for human errors in assembly and operation. All of the above represent possible critical hazard scenarios, and their number is far greater than any permanent structure needs to be designed against. Furthermore, permanent structures are generally assembled only once and are used for large periods of time and exhibit a much higher degree of inherent robustness against human errors.

5. Finally, temporary structures, due to their purpose are commonly made of slender elements, and therefore their performance is more sensitive than permanent structures to errors during their erection and operation. The use of damaged and incorrectly assembled elements is often observed as inadequate maintenance and poor quality control are prevalent. Site control is essential and the appointment of supervisors with experience in such structures who are available throughout the use of the structure including decommission should be implemented.

The framework outlined above contributes strongly to the high number of incidents and accidents involving the use of temporary structures, which frequently cause human casualties and severe injuries, work inefficiency and partial, or total, structural damage of the infrastructure. For example, since 1970, falsework collapses have been reported worldwide, with a growing trend in the developing world like China, India and Dubai where a boom in construction has taken place. One of the most important parts of this book is a review of the causes of temporary structures collapses and failures. The book proposes procedures to reduce the number of collapses by highlighting the causes and presenting methods of controlling temporary structures to minimise the risks. This is undertaken in the second and seventh Chapters where reviews conducted by researchers are described.

Xie and Wang showed that in China, 27 collapses of bridge falsework systems occurred during 2005-2009 period, killing 100 workers and causing a higher, although unspecified, number of injuries (Xie & Wang, 2009). Similarly, in 1976, Matousek and Schneider analysed 800 cases of damage to structures, looking for their causes (Matousek & Schneider, 1976). They found that damage commonly occurred during the execution phase, with temporary structures being responsible for 9% of the collapse cases, 11% of the resulting economic costs and 22% of all casualties. The principal cause was human errors (errors, lapses or omissions) related to deficient planning, design and execution. Hadipriono and Wang have also studied the causes of temporary structures collapses during construction and concluded that almost all triggering and enabling events stemmed from procedural causes such as inadequate review of falsework design/erection and inadequate falsework/formwork inspection during concreting operations (Hadipriono & Wang, 1987). Additionally, they found that 74% of temporary structures collapses occurred during concrete pouring operations.

A study developed in 2004 by the UK's Health and Safety Executive (HSE) (Bennett, 2004) found that approximately one out of six accidents with temporary structures could have been prevented from happening if the original designer had done something to enhance safety, but failed to take that opportunity. This can be justified by the findings of a survey (Pallett, Burrow, Clark, & Ward, 2001) also commissioned by the HSE where a sample of persons directly related to falsework design and procurement were interviewed to assess the level of awareness of the structural behaviour of falsework. The findings show that there is (Pallett et al., 2001):

1. "A lack of understanding of the fundamentals of stability of falsework and the basic principles involved. This shortfall occurs at all levels.

2. Wind load is rarely considered.
3. There is a lack of clarity in terms of design brief and coverage of key aspects such as ground conditions.
4. The lateral restraint assumptions made by designers were often ignored/misunderstood by those on site.
5. There is a lack of adequate checking and a worrying lack of design expertise.
6. Erection accuracy leaves much to be desired".

In 2011, the UK's HSE conducted an extensive study on what are the major hazard events in construction (HSE, 2011). It was found out that failure to recognise hazardous scenarios and influencing factors, poor teamwork and lack of experience and competence were the main causal factors to accidents. Particularly, regarding the design of temporary structures the highlighted causal factors consisted of inadequate design or (late) design changes, underlying lack of robustness and incorrect as-built drawings and information.

Beyond human losses and injuries, these accidents may cause considerable economic, financial, environmental and political costs as well as damage to reputations and increased insurance premiums. Yet, despite their importance and extensive practical use, the existing research concerning temporary structures is very limited, as given by Beale in 2014 concerning scaffolding and falsework (Beale, 2014).

1.2 OBJECTIVES

The present book contributes to a better knowledge about the management, structural design (including reliability and robustness) and operation of temporary structures. In particular, the reader will gain an understanding about:

1. The effects of the most important actions imposed to temporary structures;
2. The implications of short usage periods but large working life spans in determining the value of environmental loads;
3. The different types of human errors that can occur and how they can be controlled and reduced;
4. The development and use of advanced numerical models of temporary structures;
5. The methods of determining the reliability and safety of temporary structures;
6. Existing design codes of the USA, Europe, Australia and Hong Kong applicable to temporary structures;
7. Examples of advanced structural analyses and of design case studies;
8. The causes of collapses and the lessons to be learnt from them;
9. The management of temporary structures projects.

The book is designed to be used by practitioners (designers, developers, contractors, and workers), academic staff, students, researchers, public bodies and private companies and will enhance the knowledge of all readers.

The book contents are diverse and rich, useful for both the less experienced user and the more advanced user.

This book concerns the most commonly used temporary structures. As mentioned, these structures are extremely important to the successful construction and maintenance of buildings and bridges. However, the published works concerning this topic are limited in number and are outdated: they do not account the significant improvements in the state-of-the-art, for example.

This book deals with fundamental, difficult and complex subjects involving temporary structures, applying the most recent research and innovation in the civil and structural engineering fields.

Current design codes were traditionally calibrated to provide an appropriate reliability only at the individual element level. The implied assumption that the adequate resistance of the structure is guaranteed by the resistance of its elements is generally not valid. Therefore, as the global resistance is not directly accounted for, the design efficiency and the global target reliability may not be achieved in practice.

In addition, existing design codes were only calibrated with respect to structures where significant past experience exists, which is clearly not the case in temporary structures. The present basis for design does not assure optimal design in terms of resources allocation and risk acceptance. As a result, the traditional standards-based approach is becoming increasingly inadequate to handle the allocation of limited resources for structures design, operation, repair or improvement, in a climate of growing public scrutiny.

Code limitations are thoroughly discussed in the book and a recently developed risk informed decision-making methodology is presented and applied to temporary structures.

The traditional structural analysis methods are critically assessed, giving examples of how to correctly account for action effects and structural modelling. The book will contain information that will assist the reader in overcoming existing limitations of present design analysis methods.

1.3 SIGNIFICANCE OF THE BOOK

The book concerns the most common types of temporary structures, most of which are made of slender (prone to buckling) elements (tubular members or thin-walled members). The book also includes bridge construction equipment (BCE), such as launching gantries and form travellers. These systems are highly specialised structures that have to handle heavy loads on long spans with adequate safety, but at the same time are light and flexible, so to avoid applying large loads to the bridge structure during the construction phase and to be able to adapt to different projects. As a result, the design and operational requirements are very severe and conflicting. An additional complication is the almost complete absence of design codes and specialised literature concerning BCEs.

In practice the design of scaffolding and falsework is usually an oversimplified process, based on a comparison of the design forces with reference resistance values given by system producers, without knowing their fundamentals, which may lead to their misuse (Baptista & Silva, 2002). This is particularly common in the process of selecting the system bracing configuration, which often suffers from lack of appropriate studies and thus can constitute an enabling cause of collapses.

Various factors that have a decisive influence on the behaviour, resistance and performance of temporary structures are not usually directly accounted for in the design. Examples of these factors are foundation settlements, load redistributions due to asymmetrical concreting, system stiffness variations, system imperfections, joint deformation capacity, use of damaged components such as couplers and tube. Other factors that originate from the interaction between the evolving permanent structure under construction and the temporary structure are also not considered. They are often expected to be covered by the safety margins adopted by the temporary structures producers, but these may be insufficient to withstand the global coupled effect of the above mentioned factors.

Concerning BCEs, their design and operation are subject to large uncertainties despite the significant evolution in structural engineering knowledge brought by the ever-increasing capacity of computational methods, by advances in experimental investigation and by the development of more informative structural monitoring and control methods.

The natural consequence of uncertainty is risk. A risk free structure is a naive, uneconomical objective: risk cannot be eliminated; rather it must be managed rationally through a risk informed decision-making process.

The severe consequences of all the accidents involving temporary structures clearly justify research needs for a holistic approach of temporary structures risk management. The present book provides for an improved understanding of the structural behaviour, robustness and risk of these structures, so that adequate margins against failure may be maintained throughout the whole design/construction/operation process.

Robustness has been present in a more direct or indirect way in several structural codes throughout the last thirty years. Robustness is defined in ISO 2394 (ISO, 2015) as the "ability of a structure not to be damaged by events like fire, explosions, impact or consequences of human errors, to an extent disproportionate to the original cause".

In this way robustness can be seen as a measure of the sensitivity of a given structure to disproportionate collapse. However, to date there is not one document that specifies a general purpose design method for robustness in a consistent manner. Moreover, there is a complete absence in codes about rules, design requirements or procedures to evaluate robustness of temporary structures.

Scaffolding and falsework structures typically exhibit low robustness because:

1. They are made of elements with a similar resistance distributed in a uniform mesh and
2. The critical load case is usually linked with the weight of the permanent structure or to materials stored on the structure and lateral loads such as wind or out-of-plumb.

The system is designed to reach a uniform safety margin for each element. Therefore, a significant number of elements are critical to the global stability of the system, and if one fails it is likely others will also fail leading to an unexpected, sudden and extensive disproportionate collapse of the system. This means that the robustness of the structure to minor failures is low.

BCEs owing to their purpose often consist in isostatic systems, designed for fast assembling, operation and adaptation to project requirements. Contrary to other temporary structures, rigid body instability of BCEs under service loading is a critical design situation that must be verified for static and dynamic conditions. As a result, triggering of failures usually does not involve severe damage of the main elements of the BCEs and therefore they too typically exhibit low structural robustness.

Additionally, factors such as lack of competence in design, absence of rigorous quality control, poor site supervision will also contribute to decrease the robustness of the systems. These factors will also have a negative effect on the reliability of the system and on the levels of uncertainty associated to the risks of collapse of these structures.

A coherent and consistent framework to assess the structural behaviour, robustness and risk of temporary structures is presented, detailed and illustrated in this book. Newly developed structural robustness and structural fragility indices are detailed with advantages over existing analysis methods. The former index can be used as a design option to reduce the structural risk and the latter index is an analysis tool that should be used to assess the structural risk of failure.

This book provides a number of contributions to knowledge. One of the key contributions being the study of the structural behaviour of commonly used temporary structures in a variety of hazard scenarios, and the analysis of the influence of certain decisive factors on the risk of these structures failing, such as the nature of the applied actions, the choice of the structural system and the type of quality management. New analysis procedures for the modelling of temporary structures backed up by experimental verification are presented. The book presents a proposal to evaluate the risk of temporary structures and discuss possible solutions to reduce risks involving temporary structures.

The book aims to fill the gap between research/innovation and practice in temporary structures. Contrary to permanent structures, such as buildings and bridges, temporary structures practice has followed behind the most recent state-of-the-art. The book explains in a clear but detailed way the challenges that set temporary structures apart from permanent structures. It presents new methodologies that can guide the practitioners and other relevant stakeholders, public bodies with responsibilities for public safety, academics and researchers so that they understand the importance of temporary structures and are therefore capable of doing their job properly when it involves temporary structures.

The book differentiates itself from other books starting from the topics it covers. The book covers topics relevant to temporary structures, focusing on the important challenges associated with temporary structures from design office to construction site, pinpointing insufficiencies that can lead to catastrophic failures, identifying aspects that are not covered by existing codes and presenting state-of-the-art methods that help to overcome these limitations. The book covers topics which are not dealt in existing books concerning temporary structures as it includes the latest research. There is significant added value present in the book to stakeholders directly involved with the most commonly used temporary structures, when compared with existing books. Additionally, many existing books concerning temporary structures are limited in number and are outdated and often only written to cover the case of a single country or region.

1.4 CONCLUSION

The present book contributes to a better knowledge about the management, structural design (including reliability, robustness and risk) and operation of temporary structures. In particular, the reader will gain an understanding of:

1. The effects of the most important actions imposed on temporary structures;
2. The implications of short usage periods but large working life spans in determining the value of environmental loads;
3. The different types of human errors that can occur and how they can be controlled and reduced;
4. The development and use of advanced numerical models of temporary structures;
5. The methods of determining the reliability, robustness and risk of temporary structures;
6. Existing design codes of the USA, Europe, Australia and Hong Kong applicable to temporary structures;
7. Examples of advanced structural analyses and of design case studies;
8. The causes of collapses and the lessons to be learnt from them;
9. The management of temporary structures projects.

The book is designed to be used by practitioners (designers, developers, contractors, and workers), academic staff, students, researchers, public bodies and private companies and will enhance the knowledge of all readers.

The book contents are diverse and rich, useful for both the less experienced user and the more advanced user.

This book concerns the most commonly used temporary structures. These structures are extremely important to the successful construction and maintenance of buildings and bridges. However, published works concerning this topic are limited in number and are outdated; they do not account the significant improvements in the state-of-the-art, for example. Fundamental, difficult and complex subjects involving temporary structures, applying the most recent research and innovation in the civil and structural engineering fields are considered and evaluated.

Current design codes were traditionally calibrated to provide an appropriate reliability only at the individual element level. The implied assumption that the adequate resistance of the structure is guaranteed by the resistance of its elements is generally not valid. Therefore, as the global resistance is not directly accounted for, the design efficiency and the global target reliability may not be achieved in practice.

Existing structural design codes were only calibrated with respect to structures where significant past experience exists, which is clearly not the case in temporary structures. The present basis for design does not assure optimal design in terms of resources allocation and risk acceptance. As a result, the traditional standards-based approach is becoming increasingly inadequate to handle the allocation of limited resources for structures design, operation, repair or improvement, in a climate of growing public scrutiny.

Code limitations are thoroughly discussed in the book and a recently developed risk informed decision-making methodology is presented and applied to temporary structures.

The traditional structural analysis methods are critically assessed, giving examples of how to correctly account for action effects and joint modelling. The book contains information that will assist the reader in overcoming existing limitations of present design analysis methods.

REFERENCES

Baptista, A., & Silva, S. (2002). Promotion of quality in the management of construction temporary structures (in Portuguese). In *Proceedings of Portuguese National Conference of Structural Engineering* (pp. 583–594). Lisbon, Portugal: Portuguese National Laboratory for Civil Engineering (LNEC).

Beale, R. (2014). Scaffold research — A review. *Journal of Constructional Steel Research*, *98*, 188–200. doi:10.1016/j.jcsr.2014.01.016

Bennett, L. (2004). Peer review of analysis of specialist group reports on causes of construction accidents (No. 218). London, UK: Health and Safety Executive (HSE).

Bragg, S. (1975). Final report of the advisory committee on falsework. London: Her Majesty's Stationery Office (HMSO).

BSI. (2011a). BS 5975:2008+A1:2011 – Code of practice for temporary works procedures and the permissible stress design of falsework (3rd ed.). London, UK: British Standards Institution (BSI).

BSI. (2011b).*Falsework. Performance requirements and general design*. London, UK: British Standards Institution.

fib. (2009). *Formwork and falsework for heavy construction. Guide to good practice*. Lausanne, Switzerland: International Federation for Structural Concrete (fib), Bulletin No. 48.

Fyall, I. (2012). *Milton Keynes scaffolding collapse*. Retrieved from accessindustryforurum.org.uk (2016, June 27).

Hadipriono, F., & Wang, H. (1987). Causes of falsework collapses during construction. *Structural Safety*, *4*(3), 179–195. doi:10.1016/0167-4730(87)90012-9

HSE. (2011). Preventing catastrophic events in construction. London, UK: The Health and Safety Executive (HSE).

ISO. (2015). ISO 2394 :2015 – General principles on reliability for structures (4th ed.). Geneva, Switzerland: International Organization for Standardization (ISO).

Matousek, M., & Schneider, J. (1976). Untersuchungen zur struktur des sicherheitsproblems bei bauwerken. Zurich, Switzerland: Swiss Federal Institute of Technology Zurich (ETHZ). (in German)

Pallett, P., Burrow, M., Clark, L., & Ward, R. (2001). *Investigation into aspects of falsework (No. crr01394)*. Birmingham, UK: The University of Birmingham.

Ratay, R. (2009). *Forensic structural engineering handbook*. McGraw Hill Professional.

Scheer, J. (2010). *Failed bridges: Case studies, causes and consequences*. London, UK: Wiley VCH. doi:10.1002/9783433600634

Xie, N., & Wang, G. (2009). Test analysis on hidden defect in high falsework and its effect on structural reliability. In *Proceedings of the 8th International Conference on Reliability, Maintainability and Safety, ICRMS 2009* (pp. 1077–1080). Chengdu, China: Institute of Electrical and Electronics Engineers (IEEE). doi:10.1109/ICRMS.2009.5270000

Chapter 2
Historical Survey

ABSTRACT

Chapter 2 overviews the historical development and past research into temporary structures from 1970 up to 2016 and describe the various problems that have occurred necessitating changes to traditional design and construction techniques. The survey covers tubular, proprietary and modular scaffolds, bridge falsework as well as bridge construction equipment. Particular areas emphasised are the changes introduced by the use of advanced structural analysis techniques and the need for changes in procedures following the analyses of collapses of temporary structures. An overview of various solutions is presented, including the use of different materials (steel, aluminium, timber and bamboo). The chapter shows that same analysis procedures are used for these different materials. The chapter then overviews the actions acting on temporary structures such as permanent loads and variable construction loads and finishes with an introduction to failures of falsework and scaffolding structures.

2.1 INTRODUCTION

This Chapter presents a brief review into the history of temporary structures and overviews the development of new systems.

There are many types of temporary structures; the long list includes scaffolding, falsework, and bridge construction equipment. These are the most commonly used temporary structures and the bulk of this book concerns them. This Chapter will also review the management of temporary structures projects but more detail of this is covered later in the book, particularly in Chapter 8.

The Chapter first describes the types of temporary structures commonly used, from tubular scaffolds, through proprietary and modular scaffolds to bamboo and timber structures to bridge construction equipment. During the overview, developments in modelling from the use of effective lengths to full nonlinear finite element models are described. Of particular importance to structural analysis is the incorporation within the models of the material's elastoplastic behaviour, of accurate connection models and of various types of imperfections, such as connection looseness.

DOI: 10.4018/978-1-5225-2199-0.ch002

Copyright © 2017, IGI Global. Copying or distributing in print or electronic forms without written permission of IGI Global is prohibited.

The Chapter also describes the different types of actions that can be applied to temporary structures: from permanent actions due to self-weight and imposed vertical loads, to variable actions such as wind loads. Design recommendations to improve the safety of temporary structures are also introduced.

Finally, the Chapter summarises the research in collapses leading to an introduction to Chapter 7 that will present a forensic analysis into temporary structures collapses with methods to avoid failure.

On the basis of this Chapter it is expected that the reader will acquire knowledge on the following topics:

1. Different types of temporary structures and their main characteristics.
2. Historical evolution of temporary structures, their design and their role in construction.
3. How temporary structures fit in the construction project.
4. Type of actions to which temporary structures are exposed.
5. Examples of temporary structures collapses.

2.2 TYPES OF TEMPORARY STRUCTURES

2.2.1 Scaffolds

2.2.1.1 General

BS 5975:2008 (BSI, 2011) defines scaffolds as:

A temporary structure which provides access, or on or from which persons work, or that is used to support materials, plant or equipment.

Therefore, scaffolds do not have a main structural role; they are mainly used to give access to various levels of the permanent structure during construction related activities.

Different types of scaffolds exist, including the use of different materials such as steel, aluminium or even bamboo. Scaffolds are generally light structures, and each scaffold is unique, because their design varies with the site where they will be used, and with their role in the building process.

As for falsework structures, steel and aluminium scaffolds are based on vertical (standards) and horizontal (ledgers and transoms) tubular elements linked together with couplers to form a three-dimensional structure.

Scaffolds systems can range from proprietary or modular, ready-to-use structures, to structures that can have arbitrary shapes. Scaffolds are generally composed of several bays on which work platform units are placed at different levels, with brace and tie elements arranged in a specific configuration to achieve a better structural integrity and lateral resistance.

A comment must be made about terminology. In Europe, the vertical tubular members are called standards whilst in the USA they are called poles or posts; the horizontal members parallel to the façade are called ledgers in Europe and runners in the USA and the smaller horizontal members are called transoms in the Europe and bearer in the USA. In Europe, a distinction is made between diagonal bracing parallel to the façade and diagonal bracing normal to the façade which are called façade bracing and ledger bracing, respectively, whereas in the USA the term diagonal bracing is used for both types.

Another difference in terminology is that in Europe the different vertical levels are called lifts and reference is made to boarded lifts whereas in the USA the term levels (or storeys) is used together with "decking levels" to imply that a given level is boarded. Also in Europe the distances in plan between adjacent columns are termed bay length and bay width, whereas in the USA the equivalent terms rows and bays are used, respectively.

2.2.1.2 Tubular Scaffolds

Tubular scaffolds are the commonest form of steel scaffold used in the UK. The generic name for these scaffolds in the UK is tube-and-fitting scaffolds and in the USA tube-and-coupler scaffolds. They are often used to enable the sides of buildings to be safely worked on during construction, rehabilitation or retrofit works. See Figure 1 for an example of a small domestic structure.

Traditionally in the UK the design of the structure was determined using effective lengths. These lengths were assumed to be given by the spacing between different layers of horizontal members called transoms and ledgers and hand calculations were used prior to 1970. This often meant that scaffolds were inadequately tied to supporting structures. For example, the UK code BS 5975 (BSI, 1996) used to allow access scaffolds to have one tie in every 40 m^2 of the structure without specifying the spatial arrangement. As a result, in theory, one could have columns 1 m apart and 40 m high only tied at the top and bottom. Unfortunately, there are still scaffold designers and erectors who do not see the need for more accurate calculations.

Figure 1. Example of a five lift domestic tubular scaffold

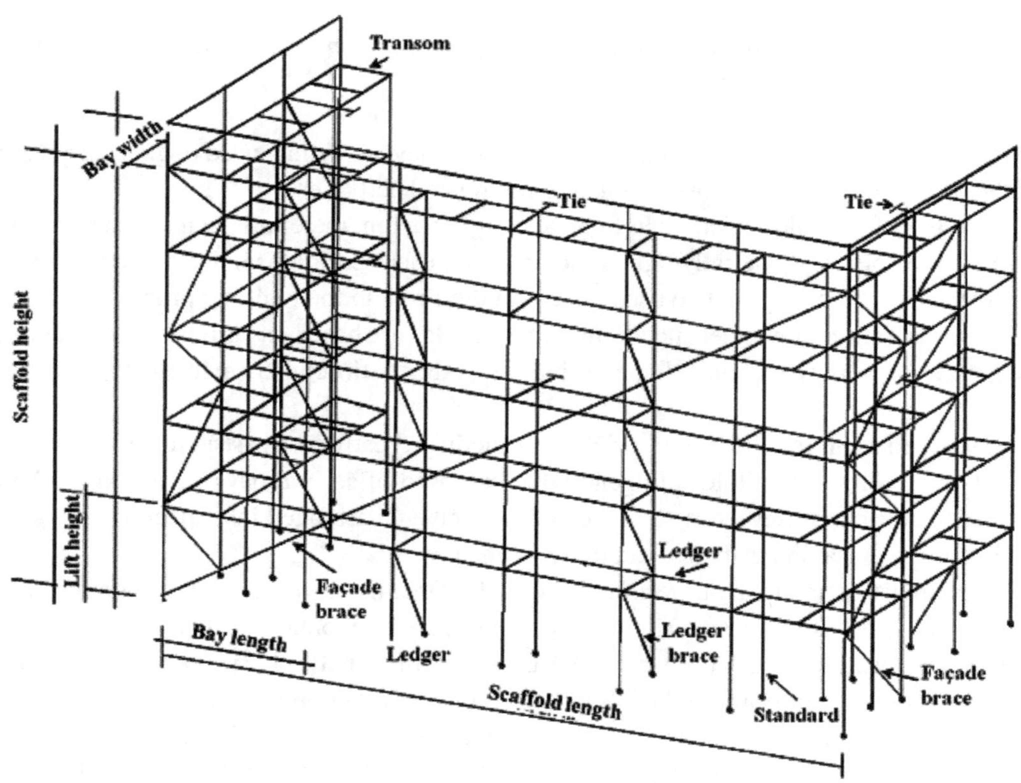

Design results were often presented in textbooks, such as those by Wilshere (1983) and Brand (1975). In addition, tables of ledger spacing were often provided by Institutions such as the American Concrete Institute (ACI, 2004; Hurd, 1995) and UK The Concrete Society (1986, 1995); and by manufacturers, e.g. Prefabricated Access Suppliers' and Manufacturers' Association (PASMA, 2000). Updated versions of the above documents have been published since then: ACI (2014), Johnston (2014) and The Concrete Society (2012).

In the 1960s several falsework and scaffold structures failed in the UK which led to the Institution of Civil Engineers (ICE) and the Concrete Society commissioning a report into falsework procedures (The Concrete Society, 1971). At the same time, the UK government set up an advisory committee into formwork, which in 1975 produced the Bragg Report (Bragg, 1975). A full review of existing research can be found in Beale (2014).

Simultaneously, the UK Science Research Council commissioned research into scaffold structures under Professor Lightfoot's chairmanship at Oxford University. This was published in 1975 and 1977 (Harung, Lightfoot, & Duggan, 1975; Lightfoot & Bhula, 1977a, 1977b; Lightfoot & Lemessurier, 1975; Lightfoot, Olivetto, & Merchant, 1977). Harung et al. (1975) constructed a model of a single story tubular scaffold tower which was loaded at the top by imposed loads. The model was analysed by Lightfoot & Lemessurier (1975) and Lightfoot et al. (1977) using a finite element program where the beam and column elements were modelled using stability functions. These functions enable a single beam element to model the geometric nonlinear interaction between applied force and displacement when large displacements occur. To simplify the analyses all the joints were modelled as pinned or fixed connections and no eccentricity of the joints was included. In all cases, the models failed by buckling with the theoretical buckling loads between 10% and 15% higher than the experimental values. A three-storey model was also tested which had similar differences between the numerical model and the experiment. The discrepancy between experiment and theory is attributed to the fact that the theoretical models at that time were not able to include imperfections in the geometry and to model the nonlinear behaviour of materials and joints due to the limited capacity of the computers available.

The researchers came to the conclusion that the effective lengths of the columns were greater than 1.0 which had been assumed in previous design rules. In cases where columns are made of multiple vertical elements, called standards, connected together by splices, named spigot joints, it was also concluded that the spigot could be considered rigid. This conclusion will be shown to be inaccurate for some structures in Chapter 4 where more accurate models are developed.

Lightfoot & Bhula (1977a, 1977b) constructed a semi-rigid model of a typical scaffold joint where the connection between vertical and horizontal elements was modelled as a small beam with semi-rigid connections at each end. To use this model, a series of six tests on the moment-rotation characteristics together with the translational stiffnesses of the joint had to be conducted, using a special test rig specifically developed for this purpose. However, they soon discovered that the translational extensions of the joints were insignificant and thus it was only required to determine the moment-rotation characteristics in the three rotational axes of the joint. In tubular scaffolds, the two tubes being joined by the connection have an eccentricity of approximately 50 mm. This was explicitly included in the model of Harung et al. (1975), but was found by Milojkovic, Beale, & Godley (1996) to be unnecessary since the accuracy of the models is unaffected by it.

The development of European design codes, led to the investigation of the properties of several types of connection, called couplers, and their effects on structural analysis and design. The traditional couplers in tubular scaffolds are called right-angled, putlog and swivel. Examples are shown in Figure 2.

Figure 2. Three common types of tubular scaffold couplers

(a) Right-angled *(b) Swivel* *(c) Putlog*

Right-angled couplers are typically used to connect ledgers and transoms to the standards. They possess rotational stiffness in at least one axis. Putlog couplers are often used to provide intermediate supports to transoms supporting decking but they have a weak rotation capacity and slip resistance. Swivel couplers are used to connect diagonal bracing to either standards or ledgers/transoms. They have no rotational stiffness and the connections can be considered to be pinned. Full details of the determination of these stiffnesses and their inclusion in analyses will be given in Chapter 4.

As an example, Abdel-Jaber, Beale, Godley, & Abdel-Jaber (2009) undertook a series of tests on putlog and right-angled couplers according to the European Standard BS EN 12811-3 (BSI, 2002) to determine the rotational strength of new and used couplers. Little difference was found between new and used couplers but the authors found there were ambiguities in the code and they recommended changes to remove them. It was noted that both types of coupler tested exhibited significant looseness when subjected to cyclic loading. The implication of this is that many tests on scaffold performance, such as those conducted by Chandrangsu & Rasmussen (2009, 2011a), which load monotonically to failure, ignore the reduced rotational stiffnesses of scaffold connections which have been subjected to varying cyclic loads, for example, due to wind loads or variable vertical loads, and may therefore over-predict the maximum loads of the structures under on-site conditions.

Measurements made on the spacing of adjacent vertical tubes used in domestic constructions conducted by Son & Park (2010) showed that they were typically placed within the range of allowed standard spacing but that the torques applied to the right-angled couplers varied considerably, and were often less than 65% of expected design torque values.

In 2001, a report was published by the UK Health and Safety Executive (HSE) which described the results of an investigation into the faults found in erected scaffolding (Pallett, Burrow, Clark, & Ward, 2001). The report summarised work previously carried out by Birch, Booth, & Walker (1971) and Birch, Walker, & Lee (1977) into props, which showed that 16% were erected with an "out-of-plumb" of 1.5°. The authors commented that the research by Burrows (1989) in his doctoral thesis showed that there was little control on sites into correct erection procedures. Measurements were made at 11 UK sites and the results showed that on most sites significant percentages of the scaffolds were erected with components outside the allowable tolerance limits of both the UK and European codes. Indeed, on one site, 50% of the legs were outside the UK design code, BS 5975 (BSI, 1996). The authors commented that in their belief the reason why collapses were relatively rare was due to the under-utilisation of scaffold capacity. This belief was confirmed by Milojkovic (1999) who in her doctoral thesis showed that a typical domestic access scaffold correctly erected and maintained normally had a factor of safety against collapse of around ten and combinations of errors were required to make the scaffold liable to collapse. Further details are found in Chapter 7 where the findings of her research are presented fully. Misunderstandings of the loads acting on scaffolds in combination with inadequate tying requirements can lead to failure as occurred in the collapse of an access scaffold at Milton-Keynes in 2006 in the UK (Andresen, 2012) – see Chapter 7 for more details on this collapse.

The UK design codes BS 5973 (BSI, 1993) and BS 5975 (BSI, 1996) used effective lengths for the design of standards (Note that a revised edition of BS 5975 (BSI, 2011a) is still in use in 2016 because it describes the erection, maintenance and site supervision of falsework as well as the design of the structure). However, effective lengths for falsework structures are difficult to determine as horizontal ledgers, transoms (see Figure 1) and ties elastically restrain scaffolds at discrete points. The assumption made in the early codes was that the effective length in design was the ledger/transom vertical spacing and it ignored the effects of tie spacing. The introduction of new European design codes led the UK National Access and Scaffolding Confederation (NASC) to commission Oxford Brookes University (formerly Oxford Polytechnic) to produce a new design guide so that scaffolds using tubular scaffolds could be safely erected. During the development of the design guide, many computer models were produced and analysed. Beale & Godley (2006) developed simple two-dimensional (2-D) models which were validated against nonlinear finite element three-dimensional (3-D) analyses. The simple models were evaluated using Excel and agreed to within 10% with the nonlinear models, and enabled load tables to be produced. Full details of both the simplified models and the nonlinear finite element models are given in Chapter 4.

Prabhakaran and co-workers at Oxford Brookes University (Prabhakaran, 2009; Prabhakaran, Beale, & Godley, 2011) developed a computer program including joint looseness and tested several theoretical models. They could show that the effects of simulating joint looseness in the models were insignificant for the behaviour and resistance of diagonally braced frames, but for unbraced frames differences in response were obtained between applying a proportional initial out-of-plumb member imperfection or an equivalent lateral load (as is commonly assumed in today's structural analyses). A result of the analyses was that the inclusion of joint looseness reduced the ultimate load of the modelled unbraced frame by approximately 8% from that of an unbraced frame without joint looseness.

In her doctoral thesis, Prabhakaran also investigated alternative joint models: from a full regression moment-rotation curve following the experimental data, to a tri-linear approximation as defined in the European standard BS EN 12811-3 (BSI, 2002) and a bilinear approximation as defined in the pallet racking standard BS EN 15512 (BSI, 2009). Note that the "BS" simply refers to the English Language version of the European standards produced by the British Standards Institution (BSI). All three models gave similar results which imply that the simple bi-linear model available in most finite element programs is sufficient for analysis.

Theoretical and experimental studies conducted by Liu, Zhao, et al. (2010) and Liu, Chen, Wang, & Zhou (2010) on high scaffolds without diagonal bracing showed that the most important factors for structural safety were the length of exposed U-head at the top of the scaffold and the rotational capacity of the joints. Using the approach presented by Beale & Godley (2006), they derived approximate formulae for the design of these scaffolds.

Recent research in China has been undertaken on theoretical and experimental models of tube-and-fitting scaffolds. Hu, Ge, & Jing (2011) undertook experiments on scaffold assemblies and showed that when the imperfections in the scaffold were correctly modelled, a correspondence between the models could be established. Gao, Gao, Li, & Huo (2013) measured the imperfections occurring in scaffold tubes and developed parametric stochastic models of the scaffolds. The authors applied the models to scaffold structures where they demonstrated that member imperfections significantly reduced the capacity of the scaffolds and recommended stricter quality control measures on allowable imperfections to ensure safe scaffolds. This result reinforces the conclusions on control discussed by Milojkovic (1999).

2.2.1.3 Proprietary Scaffolds

In order to speed up the erection of scaffold structures, proprietary and modular metal systems are often employed. In these structures the standards usually have permanent connections welded onto their ends. The ledgers and transoms are often connected to the standards by wedges knocked in. The rotational stiffness of the connection is affected by the number of knocks, being stiffer after a minimum number that is often not undertaken (André, Beale, & Baptista, 2013a). Tests in Australia (Chandrangsu, 2010; Chandrangsu & Rasmussen, 2011a) showed that the Cuplok® Proprietary Scaffold required at least three knocks using and adequate equipment to get a consistent result.

As part of the process for the development of European design codes for scaffolding, experiments were undertaken at Oxford Brookes University (Godley, 1990) and Stuttgart University (Voelkel, 1990) into the properties of proprietary scaffolds. The rotational properties of the connections about an axis at right angles to the standard were obtained by the cantilever test. Figure 3 shows a cantilever test and a typical moment-rotation curve. The increase of joint looseness as the cyclic loading progresses is clearly visible.

A prototype proprietary scaffold seen in Figure 4 was tested at Stuttgart during the development of the European design code. The prototype structure was analysed by Godley & Beale (1997). They showed that 2-D and 3-D nonlinear analyses ignoring spigot looseness give results where the maximum deflection was only half that observed in the tests, although failure loads were predicted accurately. A good correlation of displacements could only be obtained between theory and experiment when contact elements at spigot joints were added to the analysis. The maximum loads determined in all the analyses were approximately the same as in the experimental tests. The authors showed that the behaviour

Figure 3. Cantilever test and sample moment rotation curve

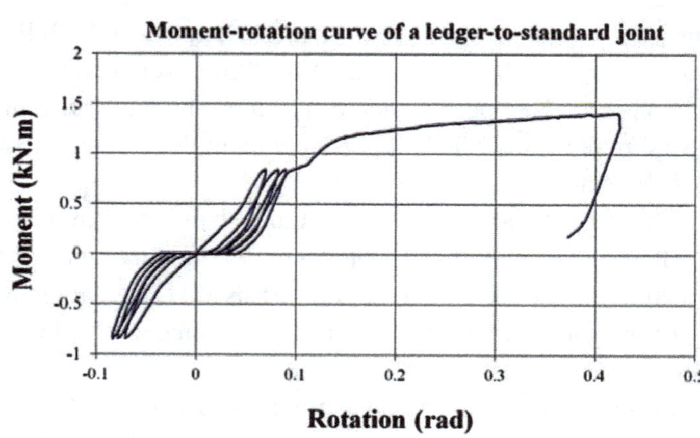

of large proprietary scaffolds could be predicted using 2-D models, as there was little 3-D interaction between failure modes which were predominantly either normal to the façade or parallel to the façade. The authors also showed that 2-D finite element models of a large 3-D scaffold which was based on the standard design given in BS 1139 (BSI, 1990a) could be developed. As the analysis procedures available to the authors at that time did not allow full geometrical nonlinear elastoplastic analyses, the authors undertook a nonlinear geometric elastic analysis. They assumed that failure occurred when the maximum resistance of the columns or connections were exceeded and these were determined in accordance with the current British Standards at that time (BSI, 1993).

Note that in using 2-D models, eccentric diagonal bracing, treated as a bar element, can be modelled as being concentric if the bracing area is reduced using Eq. 1.

$$A_{red} = \frac{L \cdot k \cdot E}{k + 2 \cdot A \cdot E} \tag{1}$$

where A_{red} is the reduced area, L is the length of the brace, A is the area of the bracing element, k is a linear-elastic axial stiffness determined from experiments and E is Young's Modulus of elasticity of the brace element.

A 3-D analysis of the scaffold modelled by Godley & Beale (1997) was also undertaken by Chan, Dymiotis-Wellington, & Zhou (2002) using stability functions who obtained the same buckling load. The failure modes were shown to consist in primarily column buckling normal to the façade in the lower elements only. Similar comparisons between 2-D and 3-D models have been made by other authors, see Beale (2014) for a full account.

In order to obtain data for reliability analyses Chandrangsu under the supervision of Rasmussen undertook bending tests on Cuplok® scaffold/falsework systems, with varying numbers of horizontal members (from 1 to 4) connected to the vertical standard using cantilever tests (Chandrangsu, 2010; Chandrangsu & Rasmussen, 2011a). They produced tri-linear moment-rotation curves. A criticism of these curves is that the tests were conducted to failure without any cycling and hence the looseness of

Figure 4. Schematic of the Stuttgart prototype scaffold

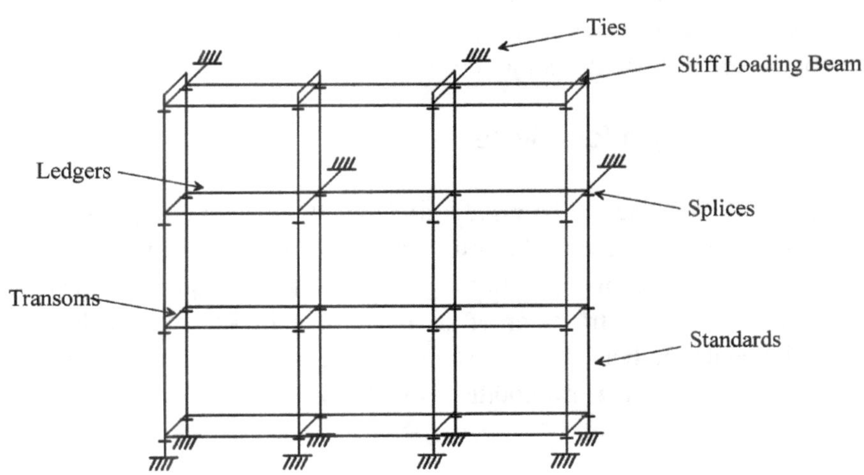

the connection was not determined. The program developed by Prabhakaran (2009) showed that accounting for looseness can yield significantly different analysis results. In part of a continuing research project into bridge falsework reliability (André et al., 2013a) have repeated the tests cycling the loads at early parts of the tests with similar results for loading but also determining some unloading curves. In addition, they reported on tests on forkheads and baseplates and produced reliability results. The results show that it is important to consider joint looseness during the design of slender framed steel temporary work structures.

A numerical finite element analysis of the load bearing capacity of the connection of a modular scaffold with the standards having nodes consisting of a rosette with the ledgers attached by wedges was presented by Pieńko & Błazik-Borowa (2013). The maximum capacity was determined to be three times that which could be determined using a linear stress analysis and 80% higher than that determined using permissible stresses. Unfortunately, there were no experimental results to validate the models.

An analysis of a proprietary scaffold by Błazik-Borowa & Gontarz (2016), similar to that analysed by Godley & Beale (1997), had imperfections of 10 mm, 20 mm and 40 mm applied either normal to façade or parallel to the façade. They showed that the imperfections increase the internal forces transmitted by the loading applied on the top lift but did not give any recommendations about the maximum deflection that could be allowed in scaffolds. Their model assumed that all joints were either rigid or pinned which could lead to an overestimation of axial capacity and an underestimation of rotation capacity of the scaffold.

2.2.1.4 Materials

Different natural materials such as timber and bamboo have been used in the past, given the abundance of available materials, and are still being used in Asia. In the western countries, steel was and still is the primary material option for elements of these systems due to its high strength and wear resistance. Typically, scaffold elements consist in cold-formed or hot-rolled circular hollow steel sections. Recently, following the trend of maximising the efficiency in construction, aluminium is becoming increasingly utilised because of its lighter weight and consequent ease of handling.

Hot-rolled and cold-formed hollow sections must satisfy the requirements specified in BS EN 10210-1 (BSI, 2006a) and BS EN 10219-1 (BSI, 2006b), respectively. However, since many scaffolds have been designed and produced many years ago the grade of the steel used does not follow the existing standards. Nevertheless, it is possible to correlate the old steel grades to the new ones. For example, Table 1 presents a comparison between the current steel grades with the ones specified in BS 4360 (BSI, 1990b).

2.2.1.5 Main Elements of Scaffold Systems

Scaffold systems should be designed to be rapidly assembled, adaptable to particular projects, safely dismantled and the elements to be reused. Various types of scaffold systems exist, each one with different characteristics, therefore complying with the above mentioned requirements at different levels.

Typically, scaffolds are framed structures of an assemblage of vertical and horizontal individual elements braced by diagonal members. These systems require intensive labour work in assembling and dismantling operations, but since their availability is widespread these systems are still the most used solution.

Table 1. Comparison between BS 4360 steel grade and current grades

BS 4360	Current grade (BSI, 2006a, 2006b)
40B	S235JR
40C	S235J0
40D	S235J2
43B	S275JR
43C	S275J0
43D	S275J2
50B	S355JR
50C	S355J0
50D	S355J2
50DD	S355K2

To increase their adaptability and efficiency, vertical (standards), horizontal (ledgers) and brace elements are available in different lengths and threaded universal jacks can be assembled at the bottom and at the top plates (baseplates, headplates or forkheads, respectively) to fine adjust the height of the system. In general, standard, ledger, and brace elements have a circular hollow cross-section, uniform along the length and also with a constant wall thickness. For brace elements there are two possible solutions: one, which is commonly used as façade bracing (i.e. external face bracing), where the length of the element is fixed and a second, which is often used as internal adjustable bracing, consisting of two tubes with different outside diameters so that the inner tube can telescopically slide inside the outer tube to adjust the brace length.

Tubes used in elements often follow the requirements set in BS EN 39:2001 (BSI, 2001). Hence, tubes have a 48.3 mm outside diameter and 3.2 mm (type 3 tube) or 4.0 mm (type 4 tube) wall thickness – additionally non-standard sizes exist such as 60.3×3.2 mm or 48.3×6.0 mm. Tubes can be supplied seamless or welded, and usually have a hot-dip galvanised coating. In Table 2 the characteristics of the elements of a typical scaffold solution are presented.

Joints and foundations

Various types of joints can be found in scaffold systems. Table 3 presents the most common types of joints between elements of these systems.

Joints between two consecutive standards are made by means of a spigot coupler. The spigot has a smaller outside diameter than the one of the standards and can be an individual element or it can be welded to the top section of the lower standard. The length of the spigot should be equal or greater than 150 mm. The spigot can be connected to the upper (and lower) standard(s) by pins, by bolts inserted through centred holes or by locking the upper standard to a special connector welded to the spigot wall. Note that for a standard tubular scaffold, an external sleeve coupler is used to join two coaxial tubes, and a parallel coupler when the tubes are not concentrically aligned – see BS EN 74-1 (BSI, 2005).

Several types of couplers can be used to connect ledgers to standards: from the classical right angle and putlog couplers to the proprietary solutions such as Cuplok® and wedge couplers. The last two types

Table 2. Characteristics of the elements of a typical scaffold system

Element	Length	Cross-section	Typical material	Illustration	
Standard	0.4 m (SL), 0.8 m (SL), 1.0 m (S), 1.3 m (SL), 1.8 m (SL), 2.0 m (S), 2.3 m (SL), 3.0 m (S)	Tube with circular hollow section. Outside diameter×wall thickness: 48.3×3.2 mm	Steel: Grade 50C to BS 4360 (BSI, 1990b), equivalent to S355J0 according to BS 10210-1 (BSI, 2006a) or BS 10219-1 (BSI, 2006b)	Spigotless standard: 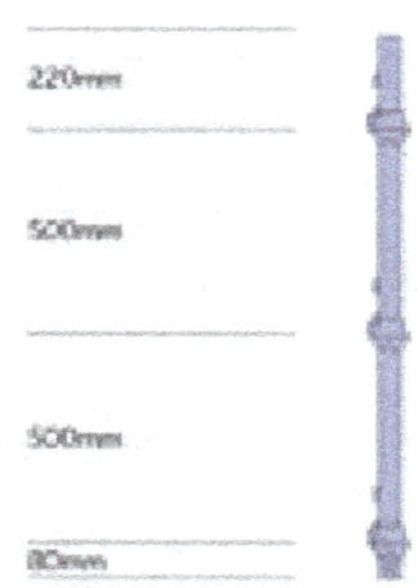	Spigoted standard: 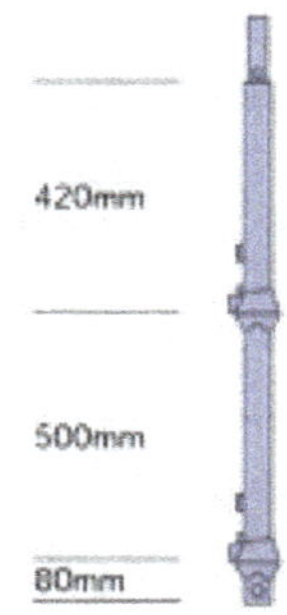
Ledger	0.6 m, 0.9 m, 1.0 m, 1.2 m, 1.3 m, 16 m, 1.8 m, 2.5 m				
Face bracing	(X × Y): 1.8×1.5 m 1.8×2.0 m 2.5×1.5 m 2.5×2.0 m 3.0×2.0 m	Tube with circular hollow section. Outside diameter×wall thickness: 48.3×3.2 mm	Steel: S275 according BS 10210-1 (BSI, 2006a) or BS 10219-1 (BSI, 2006b)	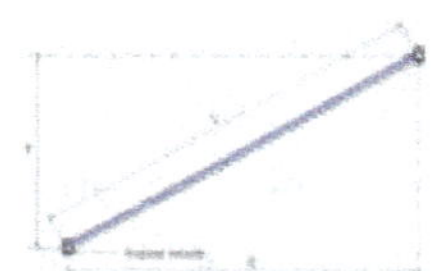	
Internal adjustable bracing	(Bay×Lift): 1.0×1.2 m 1.0×1.3 m 1.0×1.6 m 1.0×1.8 m 1.0×2.5 m 1.5×1.2 m 1.5×1.3 m 1.5×1.6 m 1.5×1.8 m 1.5×2.5 m 2.0×1.3 m 2.0×1.6 m 2.0×1.8 m 2.0×2.5 m	Tubes with circular hollow sections. Outside diameter×wall thickness: Inner tube 38.0×3.2 mm Outer tube 48.3×2.9 mm			

S – Spigoted standards; SL – Spigotless standards
All images ©2016 Brand Energy & Infrastructure Services. Used with permission

of couplers were developed to overcome the limitations of using the first two types, by allowing several ledger elements to connect to one standard element at a single node. For example, in the case of the Cuplok® systems, the standards have steel elements uniformly distributed along its length (in general spaced by 500 mm) consisting in bottom and upper cups – only the former is welded to the standard wall. The ledgers have two steel blades at each end which are introduced within the cups. Finally, the joint is

Table 3. Types of connections between elements of a scaffold system

Connection	Type	Illustration	
Standard-to-standard	Spigot joint	Type A):	Type B):
		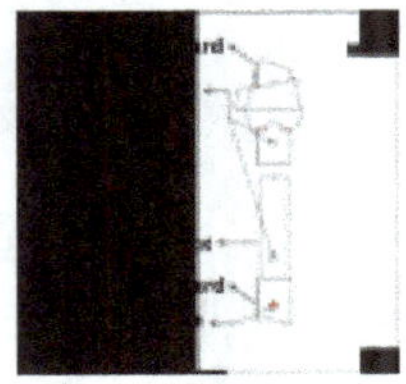	
Ledger-to-standard	Special couplers	Right-angle coupler	Putlog coupler
		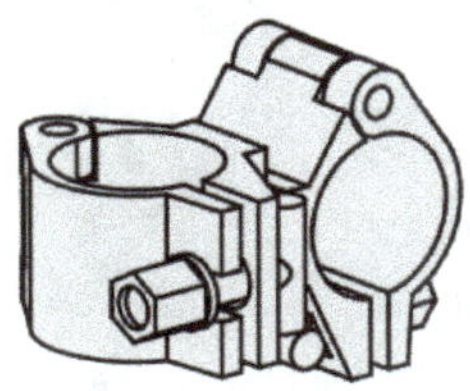	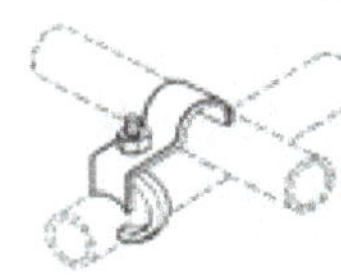
Ledger-to-standard		Cuplok® joint	Wedge joint

Ledger-to-brace	Hook joint or wedge joint	Hook joint:	Wedge joint:
			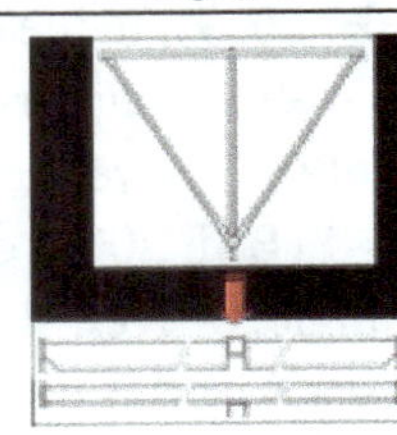

Continued on following page

Table 3. Continued

Connection	Type	Illustration	
Standard-to-brace	Swivel joint	Type A): 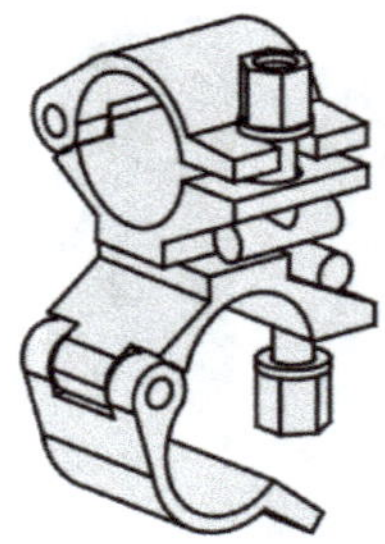	Type B): 
Standard-to-jack Jack-to-baseplate Jack-to-headplate Jack-to-forkhead	Jack joint	  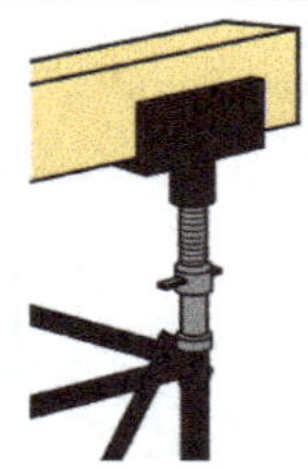	
1) ©2016 Brand Energy & Infrastructure Services. Used with permission 2) ©2016 RMD Kwikform. Used with permission			

locked by striking with two to three hammer blows the upper cup. Each Cuplok® joint can accommodate up to four elements – combination of ledgers and brace elements.

A brace element can be connected to a ledger element by a hook coupler or by a type of wedge coupler. The manufacturers of Cuplok® recommend that the brace is fitted within 150 mm of a Cuplok® node (SGB, 2009). In the case of the connections between a brace element and a standard element, these consist of swivel couplers.

Some of the above mentioned couplers are controlled by specific European standards. For instance, BS EN 74-1 (BSI, 2005) defines design requirements, strength classes (A and B) and testing procedures for right angle, swivel, sleeve and parallel couplers used with tube elements of 48.3 mm external diameter. Classes A and B differ in transmissible internal forces and moments and in values of load bearing capacity and stiffness. For example, a class A swivel coupler has a minimum design axial failure load of 14 kN whereas a class B coupler has a minimum design axial failure load of 20 kN.

BS EN 74-3 (BSI, 2010a) specifies structural requirements for baseplates and geometrical characteristics for spigot couplers. Baseplates made of steel of a minimum grade S235 and with a minimum thickness of 5 mm are deemed to satisfy the structural requirements. Finally, BS 1139-2.2 (BSI, 2015) specifies requirements and test methods for putlog couplers.

2.2.2 Telescopic Props

Telescopic props are temporary structures used to support permanent structures, typically concrete slabs, which lack stiffness or resistance capacity while they are being built, or during rehabilitation or retrofit interventions.

Metal props typically consist of two slender circular hollow tubes with different diameters and connected by means of a length adjustment device consisting of a pin and a collar nut. To reach a desired prop height the pin is inserted in one of the holes of the inner tube, and the collar nut makes it possible to fine tune the extension length. Each tube possesses an endplate, generally square (with rounded corners), although the top endplate could also be a forkhead, to receive the loads from the permanent structure or to transmit them to the foundation. The more commonly used materials for the various components of the props are steel and aluminium, although wooden props are also used.

Among the first references addressing the behaviour and design of such structural elements are two reports prepared in the 1970s for the UK Construction Industry Research and Information Association (CIRIA, 1971, 1977) and the Bragg Report (Bragg, 1975). The former ones, are essentially reports of extensive experimental studies performed in accordance with BS 4074 (BSI, 1982a) and BS 5507-3 (BSI, 1982b). The authors concluded that: for fully closed props, pins can endure excessive bending deformations, for partial and fully extended props a significant curvature of the inner tube could be observed. No evidence was found to conclude that the average strength of used props and new props is significantly different. From these studies safe working loads for adjustable telescopic props were recommended, using a safety factor of two and considering maximum erection tolerances of 25 mm for eccentric load application and a 1.5° for out-of-plumb initial imperfections. These safe loads later became incorporated in the first edition of BS 5975 (BSI, 1982c). The basis of the design consisted either on the allowable stress approach or on the results of experimental tests, together with the use of a global safety factor.

A few years later, an intensive research effort was carried out by W.-F. Chen and his co-workers at Purdue University, then continued by J.L. Peng and others in Taiwan. The research concerned mostly evaluation of loads during construction, namely the estimate of values and distribution of vertical forces transmitted from the concrete slab to the props (El-Sheikh & Chen, 1989a, 1989b), and seismic and wind load considerations (Mohammadi & Heydari, 2008).

Peng (2002), studied the stability behaviour of wood and metal telescopic props and suggested design recommendations for this type of temporary structures. A major conclusion from Peng's paper is that the stiffness of the connection between the two tubes has little influence on the resistance of an individual prop.

The German code DIN 4424 (DIN, 1987), on the design of telescopic props which was incorporated into the Euronorm BS EN 1065 (BSI, 1999) included an analytical model for prop behaviour. Feng derived an exact solution of the differential equations governing the prop model in her doctoral dissertation (Feng, 1994) which unfortunately was not published at the time but is now presented for the first time in this book in Chapter 4.

In 1999, Canisius and Maitra reported the results of 17 adjustable props tests, according to BS 4074 (BSI, 1982a), to study the props strength with and without shear failure at the connection between the two tubes (Canisius & Maitra, 1999). Currently, the design of steel telescopic props in Europe is governed by BS EN 1065 (BSI, 1999) and the design of aluminium telescopic props by BS EN 16031 (BSI, 2012).

A research project was undertaken at the Portuguese National Laboratory for Civil Engineering (LNEC) and finished in 2008 concerning steel telescopic props. Several experimental tests were performed which were used to validate numerical models, see André, Baptista, & Camotim (2007, 2009b). A sensitivity study was performed and the entire set of results were used to develop buckling curves of steel telescopic props, see André (2008) and André, Baptista, & Camotim (2009a). Figure 5 shows an example of telescopic prop being tested.

More recently, Salvadori (2009) derived a theoretical solution of the differential equations governing the telescopic prop's structural model under compression loading. Enright, Harris, & Hancock (2000) developed a simplified model for the connection between the two tubes, later used by Chandrangsu & Rasmussen (2011b) in their numerical modelling of spigot joints in falsework structures. Peng, Wang, Chan, & Huang (2012) and Peng, Wu, Shih, & Yang (2013) carried out experimental tests of steel telescopic props to obtain design curves.

2.2.3 Falsework

2.2.3.1 General

BS 5975:2008 (BSI, 2011a) defines falsework as:

Any temporary structure used to support a permanent structure while it is not self-supporting.

Figure 5. Telescopic prop under test

Therefore, there is a fundamental difference between scaffold structures and falsework structures: the latter is designed to assure the strength, stability and stiffness of the permanent structure during its construction. As a result, the load range applied to each system is completely different with direct implications in their design and safety requirements.

The main role of falsework is to provide structural safety and safety to the workers during the construction of a structure (a bridge, a building, etc.). Falsework consists in temporary structures providing a stable platform upon which the formwork may be built, and giving support for the superstructure until the members being constructed have attained sufficient strength to support themselves and sufficient stiffness to satisfy performance requirements.

Falsework systems can consist of 3-D metal frame structures where "beams" and "columns" are connected to each other by special couplers (see Figure 6), or they can be ready-to-use heavy-duty towers (see Figure 7).

The type of system most commonly used corresponds to proprietary, or modular, units consisting of an assembly of metal (steel or aluminium) tubes, generally constructed in a uniform mesh of vertical and horizontal elements (in both directions of the horizontal plane), connected to each other by special

Figure 6. Falsework frame system. ©2016 Brand Energy & Infrastructure Services. Used with permission

Figure 7. Falsework heavy-duty tower system. ©2016 RMD Kwikform. Used with permission

couplers, braced by diagonal members and placed under the entire formwork. These structural solutions are the same as the ones described for scaffolds.

Additionally to the 3-D frame structures defined above, there are also available in the market other falsework solutions such as heavy-duty towers made of built-up elements. Sometimes, these two different systems are used together in the same construction, see Figure 8.

Falsework includes shoring, a term usually associated with building construction (see Figure 9), and bridge falsework (see Figure 8).

Figure 8. Bridge falsework example. ©2016 RMD Kwikform. Used with permission

Figure 9. Shoring example. ©2016 RMD Kwikform. Used with permission

Bridge falsework systems are mostly used during the construction of concrete bridges using span-by-span in situ casting or span-by-span precast segmental methods, although, falsework elements could be used in the erection of steel bridges. They can also be used in the construction of other types of bridges as support to the main bridge construction equipment (BCE) (see Section 2.2.5).

Bridge falsework systems are stationary temporary structures, i.e. which do not have integrated a mechanical system allowing them to move without having to be dismantled and reassembled in the new location. They were the first type of bridge temporary structures to be developed, and are simultaneously the more basic and most versatile bridge temporary structure. Thus, they had a key role in the construction of major infrastructures around the world. Nowadays with the advances of bridge engineering, novel design and construction techniques have been introduced. Cable stayed bridges, post-tensioned box-girder bridges, composite bridges and precast construction propelled the industry to invent new ways to construct bridges, which led to the development of the BCE which couples civil, mechanical and electrical/electronic engineering.

Nevertheless, bridge falsework systems are still widely used to build low height (9 m to 12 m; 30 m maximum) concrete bridges with small span lengths (up to 60 m (Crémer, 2003)) and not too long (500 m in general, although (Masumoto, Hara, & Yamashita, 1994) referred to a 725 m continuous bridge that was built using this construction method), in wide valleys with good ground conditions, easy access and no major land use. See also Tischer & Kuprenas (2003). This is the most flexible construction method to the designer: since it does not influence the bridge geometry – the bridge deck geometry can vary from span to span and exhibit complex configurations both in plan and in elevation; and it does not control the design of the structure.

Where a bridge crosses waterways or roads, or the soil properties are weak, steel trusses or steel girders can be used to sustain the formwork, transmitting the loads to heavy-duty towers placed at the ends of the span in order to avoid the obstacles. This system can also be used if the height of the bridge piers is high. Additionally, heavy-duty towers can be used as temporary supports in bridge launching or in the construction of arch bridges. Heavy-duty towers can consist in ready-to-use structures made of built-up elements or in an assembly of elements of the 3-D structural systems described previously.

The construction cycle of a cast in situ concrete bridge using this method of construction consists in the following stages: first, the temporary structure is placed underneath the bridge section to be cast; secondly the formwork is assembled; next the concrete is cast and when it has hardened and has achieved sufficient strength, the post-tension cables, if they exist, are at least partially tensioned; and finally the falsework is removed and moved to another section. In multiple span bridges, it is common practice to consider construction joints distancing one fifth of the span length of the bridge piers. In terms of construction rate, a 20 m/week cycle is normally achieved (Crémer, 2003). These systems are usually used in each construction site for only a few weeks, although sometimes they can be continuously used for six months or more. To maximise economic benefits, bridge falsework elements are reused several times in different projects at different locations. Their broad availability, low investment needed together with cheap labour work (less specialised), also contributes to make them a strong competitor to other alternative construction methods such as prefabrication or use of MSS systems.

In 1979, tests on components of a proprietary falsework were conducted by Holmes & Hindson (1979). They then tested a full-scale falsework loaded by applying concrete blocks at the top of the scaffold. The collapse load was compared against the buckling load of the standards obtained using a Perry-Robertson formula. The results were varied with predictions of the theoretical load varying from as low as 40% to as high as 110% of the experimental load. The cases of large discrepancy were attributed to load eccentricity and coupler failure. The numerical procedures described in Chapter 4 would probably yield better results.

Research into modular falsework systems was first reported by Chan and Peng and co-workers. The first papers in 1995 and 1996 (Chan, Zhou, Chen, Peng, & Pan, 1995; Chu et al., 1996) described a finite element model using a polynomial beam element to analyse shoring falsework with between one and three stories. In all their models, the buckling mode was approximately a sine wave normal to the modular sections. Peng and his co-workers in a series of papers included wooden shores on the top of a modular falsework and concluded that these shores significantly reduced the capacity of the structure. They also could obtain a reasonable agreement between tests and numerical models (Peng, Pan, et al., 1996b, 1996a; Peng, Pan, Chen, Yen, & Chan, 1997; Peng, Yen, Pan, Chen, & Chan, 1996). As falsework frequently fails during the construction phase of a building the authors analysed the effect of different placement loads and devised design guidelines. The authors produced simplified methods of analysis and design (Peng, 2002, 2004; Peng, Pan, & Chan, 1998).

Huang. developed simple numerical models of modular components and showed that 2-D models gave accurate results when compared with tests (Huang, Chen, Rosowsky, & Kao, 2000; Huang, Kao, & Rosowsky, 2000).

Weesner & Jones (2001) described tests on three storey modular falsework systems from different manufacturers and compared the results against advanced numerical models with moderate agreement. These tests have since often been used as example results for other researchers.

Xie & Wang (2009) investigated high falsework structures and performed reliability analyses. They showed that incorrect alignment of vertical members significantly reduced the falsework resistance capacity. Yen, Huang, Chen, & Lin (1995) conducted tests on shoring systems up to five stories in height and proposed an empirical equation that could model the experimental results. However, as the experimental frame was only five stories tall, the formula could not be used for higher systems without further work.

A detailed study of the Cuplok® proprietary falsework was undertaken at the University of Sydney Australia starting with an investigation of the spigot joint (Enright et al., 2000) using tests and nonlinear computer models. The spigot joint was modelled by a similar procedure to that used in the European standard for steel props (BSI, 1999). The results showed that if the spigot is concentrically loaded, the capacity of the standard was as high as an equivalent standard without a spigot, but that eccentricity of loading significantly reduced the capacity of the standard. The Sydney research then tested the individual components to determine connection and material properties followed by tests on falsework assemblies. Nonlinear finite element models were constructed and probabilistic analyses undertaken to get reliability data (Chandrangsu, 2010; Chandrangsu & Rasmussen, 2011b; Zhang, Chandrangsu, & Rasmussen, 2010; Zhang & Rasmussen, 2013). The spigot model of Chandrangsu was found not to give consistent results by André in his thesis (André, 2014), see also André et al. (2013a), where an alternative phenomenological model based on statistical analyses was found to be better and André's re-analysis of the Australian tests had improved correlation with experimental results.

It is notable that all the analyses used in the tests summarised above were finite element models using different commercial programs and that typically the first step in the analysis of the structures was to conduct a linear buckling model to get a maximum carrying capacity before conducting a full nonlinear elastoplastic analysis.

2.2.3.2 Materials and Main Elements of Falsework Systems

In general, the materials and main elements are the same as detailed for scaffolds, see Section 2.2.1. The exceptions are the ready-to-use heavy-duty towers that are made of metal built-up elements consisting in four main vertical members (chords) braced by batten or laced elements welded to the chords. These systems allow higher construction rates but their handling requires the use of cranes and their application range is limited.

2.2.3.3 Joints and Foundations

Figure 10 illustrates a schematic representation of a falsework system. The types of joints are usually the same as Presented in Section 2.2.1 for scaffolds.

Regarding falsework foundations, different types of foundations exist for transmitting falsework loads to the supporting ground, see Figure 11. Due to the high concentrated loads and small dimensions of the baseplates high stresses need to be transferred to the ground.

For bridge falsework, considering that the ground over which the foundations rest is often characterised by being soft and weak, there is the need to improve the ground stiffness and resistance and/or to adopt more complex structural solutions than the ones typically used in scaffolds. If the ground is made of soil the top layers of the ground must be removed and in situ testing should be performed in order to characterise the type, depth, lateral and vertical variations of the soil underlying and adjacent to a falsework site. If the slope of the ground exceeds a certain value, BS EN 12812 (BSI, 2011b) suggests 8%, it is recommended that the foundation should be designed accounting for this factor.

Figure 10. Schematic representation of falsework solution. ©2016 Brand Energy & Infrastructure Services. Used with permission

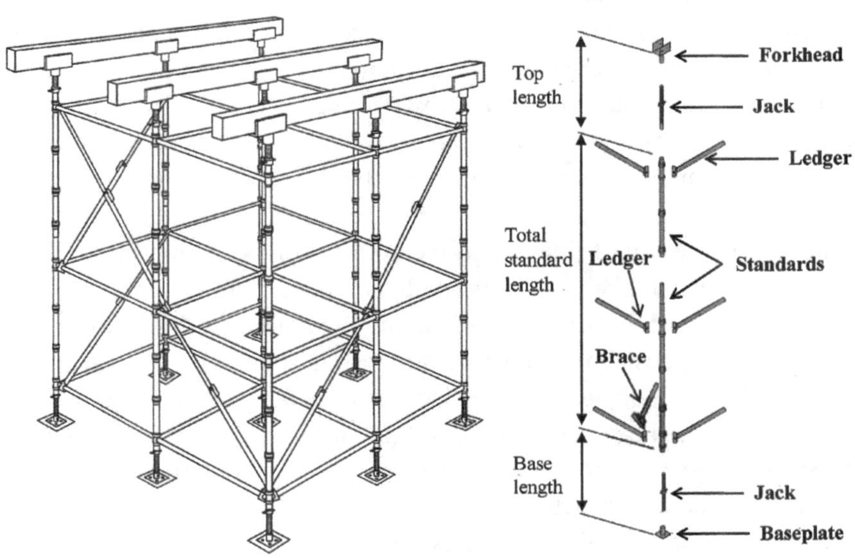

Figure 11. Left: Timber sole plates, Right: concrete footings. ©2016 RMD Kwikform. Used with permission

2.2.4 Timber and Bamboo Temporary Structures

The materials used for most temporary structures in Europe and America are steel and aluminium. However, in Southern Asia, for example Hong Kong, bamboo is still widely used where it forms a significant proportion of the temporary structures. Evidence obtained by the authors in Hong Kong is that practicing scaffolders working there prefer bamboo scaffolds to metal ones as they feel safer.

When a bamboo scaffold shakes more than a certain amount under windy conditions the scaffolders know it is time to get off before collapse. Metal scaffolds, unfortunately, do not move as much before collapse. The structural bamboo used in Hong Kong is of two types – Kao Jue and Mao Jue. Their properties were investigated by K. F. Chung & Yu (2002).

At a conference held in Hong Kong, Chung & Chan (2002) discussed the analysis and design of bamboo scaffolds. In this conference, papers on the design and assessment of bamboo columns were presented. A description was given of the areas where bamboo scaffolding is currently used and design limitations of the scaffolds. It was also stated that the performance of single layer bamboo putlog scaffolds could be improved by using metal tubes as the horizontal putlog connections. This was extended in another paper where steel tubes were used as main standards with bamboo standards at intermediate positions. The joints in bamboo scaffolds are made with rattan or other twine material and are designed as simply supported connections. In the conference, it was also described a double layer bamboo grid system where special PVC joints were used to improve structural performance. Finite element analyses of the joints were presented. Further details on bamboo scaffolds are presented in Chapters 4, 6 and 7.

A comparison between the use of timber and metal falsework systems was undertaken by Poon & Yip (2008) where they showed that the former was often more economical. Peng (2002, 2004) and Peng et al. (1997) analysed and tested timber shoring systems and determined safe spacings for vertical and horizontal members. They suggested that the main supporting standards be placed at between 1.2 – 1.8 m spacings but have additional bracing standards between each main support at 0.5 – 0.6 m spacing. The horizontal ledger spacing should be at between 1.8 – 2.25 m spacing with intermediate ledgers between 0.6 – 0.75 m. The same design and calculation procedures are used for bamboo scaffolds as are used for metal scaffolds as long the reduced structural properties of bamboo poles are included and joint connections are treated as simply supported. The same types of failure that are associated with metal structures occur with bamboo and timber structures.

2.2.5 Bridge Construction Equipment

Bridge construction equipment (BCE), consists in more complex structures than bridge falsework, such as those used in the cantilever form-traveller construction method, where the formwork is incorporated in the equipment, or bridge launching equipment which was developed to facilitate precast construction. Another significant difference between bridge falsework and BCE systems is that the latter have a mechanic hydraulic system that enables the automatic controlled movement of the system without the need for dismantling and reassembling procedures. Figure 12 illustrates different types of BCE. See also André, Beale, & Baptista (2012b).

Therefore, there are several types of bridge temporary structures, each one targeting a special application under certain engineering and economic constraints. Cardwell (2010) presented a possible range of application of bridge temporary structures based on the material of bridge decks: steel, composite and concrete, see Figure 13. A similar classification is presented in Bakhoum (2014) together with detailed and very useful information about step-by-step procedures to be followed for most of the bridge construction methods currently available.

The criteria for choosing the construction method of a bridge are manifold: from the geometrical characteristics of the superstructure, namely the layout of the bridge (plan and elevations), deck type and its material but also the height of the piers, the length of the bridge and of each span and the spans uniformity, the ground properties, the bridge context (deep valleys, crossing a waterway or a road, open field or urban area, ease of access, size of space available, etc.), the labour costs and logistic issues such as availability of materials and equipment, the designer and contractor expertise, etc. In fib (2000) and

Figure 12. Examples of BCE
**©2016 Strukturas AS. Used with permission*
***©2007 VSL. Used with permission*

(a) Self-launching gantry systems*
(b) Movable scaffolding systems (MSS)*
(c) Incremental launching systems**
(d) Form-traveller systems for balanced cantilever construction*

Figure 13. Range of application of bridge temporary structures (based on Cardwell (2010))

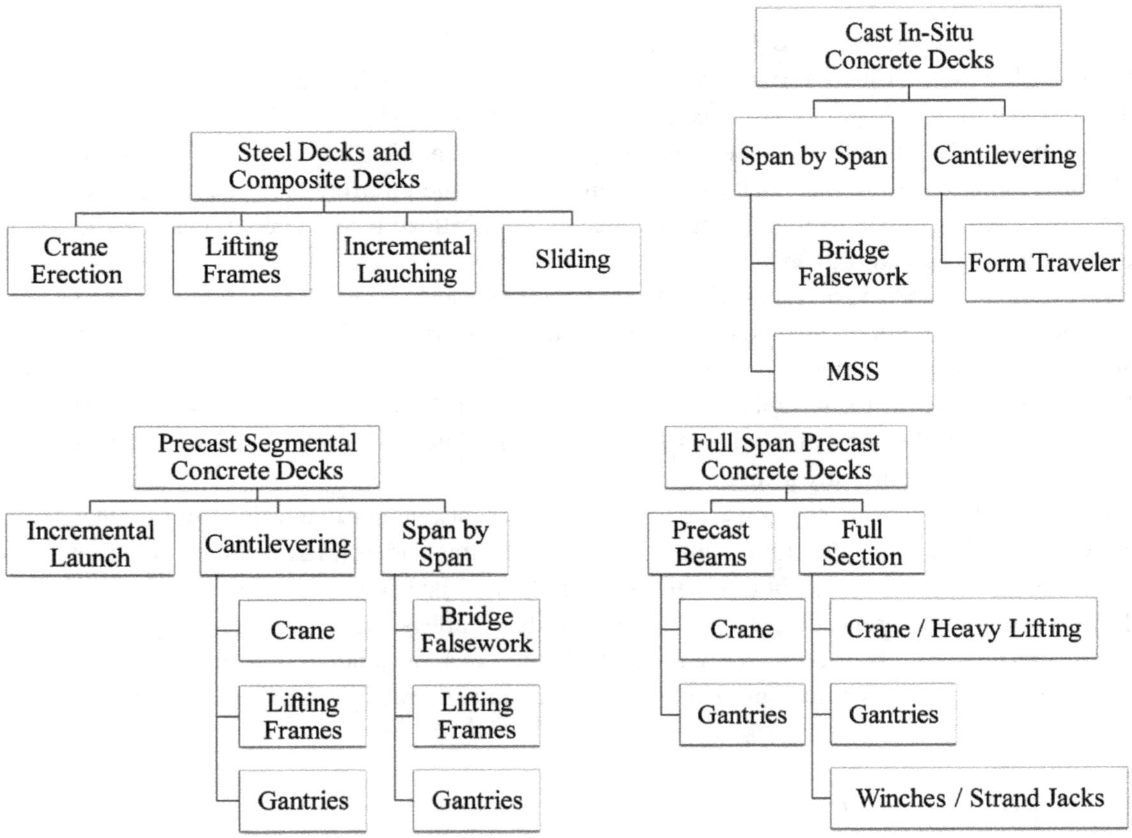

The Concrete Centre (2008) pre-design aids for concrete bridges are available and in Ayaho, Hideyuki, & Hideaki (1997) a system for selecting the erection method for steel bridges is proposed.

Moveable scaffold structures are used in large bridge projects. They are specialized structures which consist of a traditional scaffold hung or supported from systems of supporting beams. These beams must be able to support the weight of the span during casting and being able to be launched from one span to the next. These scaffolds are often used for spans of up to 50 m but new developments may enable spans of 90 m to be accommodated (Lee & Daebritz, 2010; Póvoas, 2012). The advantages quoted for moveable scaffold structures are that they can easily be adapted for different spans and girder weights and that they reduce the manpower required for the scaffold. They enable bridge components to be constructed elsewhere under controlled conditions. Reviews have recently been produced by André et al. (2012b) and André, Beale, & Baptista (2013b) where they gave recommendations that the codes be revised to incorporate a risk management framework for bridge construction as existing codes could lead to unsafe practice. They also commented that the practice of reuse of moveable structures could lead to flaws developing in erection, inspection and maintenance procedures with potentially fatal consequences.

BCE are typically steel structures, usually light twin truss systems or alternatively heavier single or twin girder systems. Depending on the reusability requirements, systems may be modular using pinned or bolted connections for the module splices. A more detailed description of each type of BCE is provided in Chapter 6.

2.3 PROJECT DELIVERY METHODS

There are various strategies which have been developed to management construction projects. They can be classified as: Design-bid-build and Design-build.

- **Design-Bid-Build:** Is the traditional procurement approach for a project. The owner provides the completed plans and specifications and procures the construction services based on the lowest bid. The primary intent of a design-bid-build project is to build the project exactly as the owner specifies. This process limits the ability of the contractor to innovate. In the United States, transportation projects have been traditionally procured through a design-bid-build process. There is considerable interest on the part of transportation agencies in alternative forms of procurement and their benefits.
- Traditionally, consultants design and contractors construct. Both parties are employed by the client through separate contracts. Thus, the designer has a direct relationship with the client and is focused on the client's needs.
- **Design-Build** *:* Or alternatively known as *design and construct (D&C)*, is an approach where the contractor provides both design and construction through a single contract between the owner and the design-build contractor. The owner will prepare a portion of the design, typically between 15 to 35%, before bid. D&C allows the contractor to be innovative during the design phase because the designer and the contractor are on the same team and constructability related issues can be addressed during design. When used with performance/end result specifications, D&C allows the contractor greater innovation. Within the civil engineering industry, D&C is increasingly being adopted as the preferred procurement route (CIRIA, 2000). Involving the constructor, the operator and the eventual end-user in the initial stages of design work can bring the client significant benefits, from single point accountability through to improved operational and maintenance characteristics.

Temporary structures are usually sub-contracted by the main contractor. Depending upon the expertise of the contractor the sub-contractor may design the structure as well as erect and manage its use or be solely responsible for the erection and disassembly of the structure. The designer of a temporary structure must be competent and have experience in temporary structures for safe structures to be erected.

2.4 ACTIONS ON TEMPORARY STRUCTURES

2.4.1 Permanent and Variable Loads

The vertical variable loads applied to temporary structures depend upon whether the structures are used as access scaffolds where they may be applied throughout the scaffold or used as support structures for falsework such as a bridge deck during construction where the load is primarily applied at the top.

When used for access scaffolds, the European standard BS EN 12810 (BSI, 2003a) and its predecessor BS 1139 (BSI, 1990a) stipulated that five top storeys must be considered as boarded although modern European practice is often to board all storeys. The imposed variable load is then applied to the top storey with a reduced load to the storey below. These loading systems induce failure, normally by buckling of the bottom storeys. The buckled modes are either a sway buckling parallel to the façade or by buckling normal to the façade. See for example Figure 14. Note that these buckling modes are only in lower elements of the scaffold as often over 50% of the total load acting on the scaffold is the permanent load of the scaffold structure itself which is distributed uniformly throughout the structure. This is different to some of the modes of failure of bridge falsework where high imposed variable loads on the top of the falsework cause sinusoidal buckling modes that occur throughout the system.

An interesting consequence of this difference in modes is that many tests on falsework structures which are purely loaded through jacks or large imposed loads at the top do not behave in the same way as do access scaffolds. In addition, in these tests, loads are usually imposed monotonically to failure which could give an overestimate of the "true" structures performance as cyclic loads on connections, which could occur due to wind acting horizontally, reduce the stiffness and maximum moment capacity of the connections.

Surveys undertaken by the USA National Bureau of Standards indicated that a significant proportion of failures were attributable to excessive loads applied to the falsework. El-Sheikh & Chen (1989b) undertook a survey of the loads on shoring loads and showed that using the standard simplified design analyses that loads were underestimated by up to 27%. Rosowsky and co-workers (Rosowsky, Huang, Chen, & Yen, 1994; Rosowsky, Huston, Fuhr, & Chen, 1994) measured the loads during placement and recommended that load factors in excess of two should be used to ensure safe design.

Hill (2004) raised the issue as to whether the design loads for temporary structures should be lower than those for permanent structures as is often postulated by some designers as the structures are only in existence for a limited time. However, he argued that this can lead to failures. This topic will be further discussed in Chapter 5 on reliability.

Carlton & Llurba (2007) measured the loads occurring on two sites and compared them with the values determined from a 3D finite element simulation. They discovered that the simulation was unable to predict the experimental results varying from a load underestimation of 20% to an overestimation of 67%. They attributed the discrepancy to props being out-of-plumb and foundations not being as stiff as assumed in the analysis.

Figure 14. Buckling of an access scaffold parallel and normal to the façade

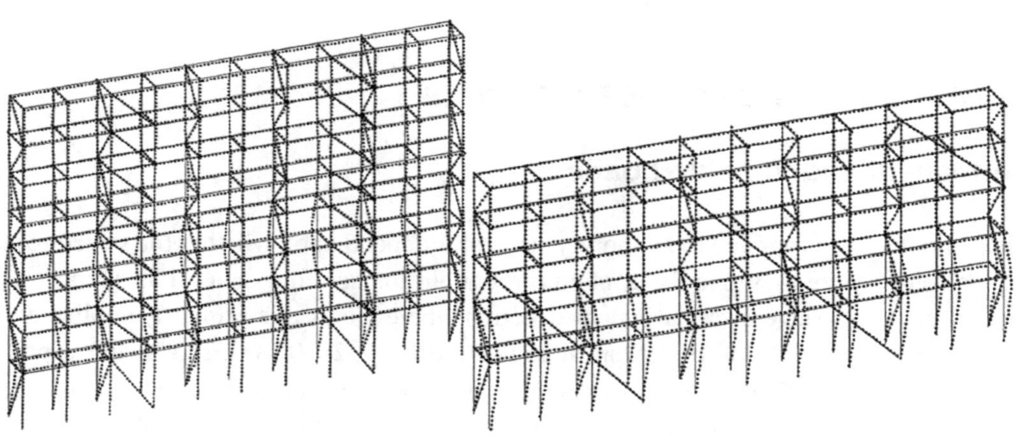

2.4.2 Seismic Loads

Limited research has been reported into the behaviour of temporary structures under seismic conditions.

Blair & Woods (1990) described an analysis of a tubular access scaffold subjected to seismic loads when attached to the Fort Calhoun Nuclear Power Station in Omaha. They found that the friction between the base and the foundation was unable to resist horizontal seismic forces and they recommended that scaffold structures be free to translate under seismic loads. It is interesting to note that this procedure was also recommended for racking structures where the advantage of not restraining the base was that moments induced at the base by the fixity caused failure whereas an unrestrained base was able to "hop" and just settle somewhere else.

Lindley, Pandya, & Khanpour (2001) undertook a series of shaking table tests on 6 foot and 12 foot scaffolds and concluded that damping was high (damping coefficient equal to 0.15), that rubber mats would improve friction resistance and that the fundamental frequency of scaffolds is low (between 3 to 6 Hz). In their tests no structural failures except tie-off wire breaks occurred. They also found that a single degree of freedom models could adequately model the structure.

2.4.3 Wind Loads

Research into wind loading applied to structures has been primarily concerned with determining loads on permanent structures with results codified into National and International standards such as BS EN 1991-1-4 (BSI, 2010b). As far as temporary structures are concerned wind loading can be classified into the effects of wind on clad and unclad structures. Unclad scaffolds are sometimes called bare-pole scaffolds.

In bridge falsework structures which often contain many rows of bare-poles if the outer rows are not clad then the total wind load on the falsework can often be higher than the wind load on the face of the clad row as it cannot be assumed that one row of standards can shield other rows.

CFD analyses of the flow applied to a single bare-pole scaffold tube show that the tube has vortices behind it (Irtaza, 2009; Irtaza, Beale, & Godley, 2007) which could interact with elements in subsequent rows to cause failures similar to that which occurred at Ferrybridge in the UK in1965 when wind caused cooling towers at a power station to collapse (CEGB, 1966).

Lindner & Magnitzke (1990) calculated that wind on scaffolds required tying the scaffold horizontally at 2 m intervals and vertically at 4 m intervals and that the load on the ties of a sheeted scaffold was up to five times the load on an unsheeted scaffold.

A conference was held by the UK Health and Safety Executive (HSE) at Buxton in 1994 (HSE, 1994). Papers on the use of cladding in scaffolds, usually determined by wind tunnel tests or measuring forces on full-scale scaffolds were presented. In this conference, Williams (1994) discussed the dynamic behaviour of fabric sheets suggesting that at that time modelling could only be achieved by wind-tunnel tests. Permeability of debris netting was shown to reduce the total force applied to the netting by over 20%. Hoxey (1994) pointed out that the maximum force applied to a scaffold occurred when the wind was at an angle of approximately 30° – 40° from the plane of the façade.

A fundamental study of the wind loads on porous façade systems were determined experimentally by Gerhardt & Janser (1994) where comparisons were made between full-scale and model experiments.

Between 2000 and 2007 Hino and co-workers published a series of reports and papers on Japanese experiments into wind loads, summarised in Charuvisit, Hino, Ohdo, Maruta, & Kanda (2007). Wind tunnel experiments were undertaken with the scaffold placed around one or two sides of a rectangular building. Reliability analyses of the scaffold systems were also undertaken assuming that the scaffolds could be modelled as a series of series and parallel systems.

Recently, wind tunnel tests on scaffold systems surrounded by cladding with different numbers of storeys and different cladding arrangements have been reported by Wang, Tamura, & Yoshida (2013, 2014) and by Irtaza, Beale, & Godley (2012). It is notable that these wind-tunnel tests are performed on scaffolds which are fully clad with impermeable scaffolds. To conduct wind-tunnel tests on net-clad scaffolds would require the netting to be reduced below sizes which can be currently produced.

CFD analyses into wind loads on scaffolds were first reported by Yue et al. (2005) who analysed the behaviour of integral lift scaffold. A combination of LES and RANS models were used to get pressure distributions acting on fully sheeted and porous clad scaffolds by Irtaza (2009) and Irtaza, Beale, Godley, & Jameel (2013). They found that the wind loads on clad scaffolds on the lee face could be neglected in some cases and that the practice of not cladding the lowest level of a scaffold made negligible differences to the total wind pressure on the scaffold.

This recent research has shown that the wind loads usually allowed in scaffold design, particularly for net-clad structures are often lower than those which actually occur. For example, the European standard BS EN 12811-1 (BSI, 2003b) assumes that the wind load acting on a net clad scaffold is only 40% of the equivalent load acting on a fully impermeable scaffold in a similar situation whereas the CFD simulations show that net-cladding may only reduce the load by approximately 40%.

2.5 TEMPORARY STRUCTURES COLLAPSES

Many researchers have investigated the causes of collapse of temporary structures. Many journals report detailed forensic analyses of particular falsework collapses. For example, the papers by El-Safty, Zinszer, & Morcous (2008) and Pisheh, Shafiei, & Hatambeigi (2009) describe detailed presentations of particular collapses, the former analysing a bridge falsework collapse and the latter a formwork collapse.

Lew (1984) reported that the results of a review of serious construction collapses in the USA showed that common causes were errors in falsework design, lack of communication between designer and builder. The review suggested that design loads for construction and the calculations for falsework loads should be included in any construction plan.

Hadipriono & Wang (1986, 1987) collected data on 85 falsework collapses. They reported that of the known causes of collapse that approximately 40% of the collapses occurred during pouring of concrete and 10% due to improper/premature falsework or falsework removal. Wind loads caused only one collapse. They emphasised that in most cases procedural errors due to inadequate design/construction and/or lack of inspection during concreting caused most failures.

During the six years from 1986-1993 the UK Health and Safety Executive investigated 1091 safety related incidents using the MARCODE HSE Database (Maitra, 1997). Of these there were 471 collapses out of an estimated 7.5 million scaffold erections. The majority of the scaffold accidents were caused by faulty platforms including platform supports, human error, unsafe working procedures and faulty access arrangements. Inadequate guardrails also precipitated 44% of falls from scaffolds. The remaining failures occurred during erection/dismantling scaffolds and climbing up the outside of the scaffolds. A detailed analysis of the failure of scaffolds showed that 28% of the trigger events for scaffold collapse were caused by inadequate tying of the scaffold to the façade and 25% to structural overload. Collapses due to wind occurred on equal numbers of sheeted and unsheeted scaffolds and Milojkovic's research (Milojkovic, 1999) showed that the same faults were still present. The analysis by Maitra (1997) also showed that 13.5% of collapses took place in scaffolds less than 5 m in height, 57% with scaffolds between 5 and 10 m in height, 22% in scaffolds between 10 and 15 m in height and 7.5% in scaffolds in excess of 15 m in height. Whitaker, Graves, James, & McCann (2003) reported on an analysis of over 3000 incidents in the UK by the Health and Safety Executive. They found that common structural causes were the use of defective components, unauthorised modifications to the structure of the scaffold as well as management failures in risk procedures and inadequate training.

Similar problems were reported by Błazik-Borowa & Szer (2014) who stated that the Ministry of Labour in Poland had discovered that, small companies often used old/defective materials as they could not afford to purchase more expensive new components.

In the 1990s domestic access scaffolds, as shown in Figure 1, were collapsing at a rate of approximately one per week in the UK. Oxford Brookes University was commissioned by the HSE to produce a recommendation for HSE Inspectors of scaffolding to enable them to know the major factors which lead to scaffold failures and hence be able to enforce changes to scaffolding. A programme of research was set up which resulted in Milojkovic's doctoral thesis (Milojkovic, 1999), a paper by Milojkovic, Beale, & Godley (2002) and a paper by Beale & Godley (2003) into the causes of scaffold collapses.

As part of her research Milojkovic conducted a survey of 56 scaffolds between 1996 and 1999 and showed that the same faults as found by Maitra (1997) were still present. To investigate the effects of faults on scaffold safety Milojkovic constructed a model of a small domestic scaffold. This scaffold was analysed under various combinations of permanent loads, imposed loads and wind loads to determine the combination producing the lowest overall load factor. Faults were then introduced into the scaffold, typical of those found by Maitra. It was found that inadequate foundations produced a 41% reduction in maximum capacity, excessive curvature of individual standards a 36% reduction, incorrect connections between standards and transoms 24% and inadequate tying 30% reductions. Combinations of faults were then introduced to correspond with poor site controls which showed that these could reduce the capacity to less than 10% of the original design capacity – see Chapter 7 for more details.

Halperin & McCann (2004) surveyed 113 scaffolds in the USA and found that 32% were either near to collapse or missing boards, guardrails or had inadequate access. They recommended improved site safety procedures.

Bridge falsework structures suffer from the same causes of failure as those described above for access scaffolds, namely poor site supervision and use. In addition, Baptista and Siva (2002) also found that the design process was often over-simplified by the use of reference resistance values given in tables by falsework system producers which may not reflect reality.

Xie & Wang (2009) reported 27 collapses of bridge falsework systems in 2005 and 2009 killing 100 workers and injuring many more.

André, Beale, & Baptista (2012a) undertook a comprehensive survey of 73 bridge falsework failures occurring worldwide since 1970. The survey looked into design standards used, context and exposure characteristics, modes of failure and types of bridge falsework systems. Amongst their results they determined that 60% of concrete bridges and viaducts were built using bridge formwork systems, 80% (in developed countries) and 90% (in developing countries) were built after 1970. They determined the individual risk per annum for each country reporting a failure. They observed that in five countries (Andorra, Brazil, India, Portugal and Vietnam) there was an estimated chance of at least 10 in 100,000 of a fatal accident per year for bridge construction compared, for example, with the UK of 2.4 per

100,000. However, the results from the survey were considered conservative because in many countries accidents are under-reported. Unfortunately, in many of the cases surveyed no detailed information on the causes of the accidents was available. The authors subdivided causes into three types – procedural (relating to management and organisational deficiencies), enabling events and triggering events. The main procedural causes were inadequate and/or insufficient design/assembly/operation methods including falsework dismantling; inadequate quality control and quality assurance practices, including design and site practices. The enabling events included inadequate falsework bracing, inadequate falsework main elements and inadequate foundations. The triggering events included construction material loads, effects of improper/premature falsework assembly/removal. An example of a bridge false work collapse is given in Figure 15.

Billings & Routley (1978) measured the axial loads in the standards during the construction of a post-tensioned concrete bridge in New Zealand. They found that there was a considerable variation in the axial loads during concrete pouring of the order of ±25%. Differences were attributed to positional errors in standard placements, out-of-plumb of the standards, stiffness and redundancy of the falsework, incorrect allowances in the falsework design for post-tension and thermal loads and, incorrect tightening of support jacks. They recommended that a minimum factor for design loads should be 2.0.

2.6 CONCLUSION

This Chapter provided an overview of the different types of temporary structures most commonly used during the construction of buildings and bridges: scaffolds, props, falsework and bridge construction equipment. The various elements that constitute the different temporary structures were identified, as well as the materials most often used.

Figure 15. Example of a bridge falsework collapse. ©2016 LUSA. Used with permission

This Chapter also presented the most significant research carried out in the past concerning the structural behaviour and safety of the different temporary structures considered in the book, from numerical analyses to experimental studies.

From the brief historical survey presented here a review has been conducted into the development of more accurate models for temporary structures. Originally, temporary structures design tended to be based on simple models such as column's effective length. The development of finite element models is now a good design practice in many areas of structural engineering, but its implementation in the analysis and design of temporary structures still poses significant challenges. For example, traditionally, structural analyses of scaffolding assumed simple models to simulate connections between horizontal and vertical members. However, as was shown in the description of tubular scaffolds, joint stiffnesses significantly affect the behaviour of this type of temporary structures. The determination of bi-linear and tri-linear models of joint behaviour will be described in Chapter 4.

The survey has also commented on the similarities between bamboo and metal scaffolds and therefore, later in the book, attention will be paid not only to the analysis of metal scaffolds but also to the changes required to analyse timber and bamboo structures.

Following an overview of the actions applied to temporary structures, which is further developed in Chapter 3, the Chapter concludes with a brief survey of scaffold and falsework collapses, emphasising that most collapses occur due to a combination of incorrect design, poor erection procedures and inadequate quality management. A thorough discussion of this topic is given in Chapter 7.

REFERENCES

Abdel-Jaber, M., Beale, R., Godley, M., & Abdel-Jaber, M. (2009). Rotational strength and stiffness of tubular scaffold connectors.*Proceedings of the ICE - Structures and Buildings, 162*(6), 391–403. doi:10.1680/stbu.2009.162.6.391

ACI. (2004). ACI 347–04 – Guide to formwork for concrete. Washington, DC: American Concrete Institute (ACI).

ACI. (2014). ACI 347R–14 – Guide to formwork for concrete. Washington, DC: American Concrete Institute (ACI).

André, J. (2008). *Behaviour and quality of steel telescopic props. Analytical, numerical and experimental investigation* (M.Sc. Thesis). Instituto Superior Técnico, Technical University of Lisbon, Lisbon, Portugal. (in Portuguese)

André, J. (2014). *Determination of the main parameters affecting the performance of bridge falsework systems* (PhD Thesis). Oxford Brookes University, Oxford, UK.

André, J., Baptista, A., & Camotim, D. (2007). Experimental investigation of steel telescopic props (in Portuguese). In *Proceedings of the VI Portuguese National Conference of Steel and Composite Construction* (p. 12). Porto, Portugal: The Portuguese Steelwork Association (cmm).

André, J., Baptista, A., & Camotim, D. (2009a). Design buckling curve for steel telescopic props under compression (in Portuguese). In *Proceedings of the VII Portuguese National Conference of Steel and Composite Construction* (p. 10). Lisbon, Portugal: The Portuguese Steelwork Association (cmm).

André, J., Baptista, A., & Camotim, D. (2009b). Numerical investigation of adjustable telescopic steel props. In *Proceedings of the 12th International Conference on Civil, Structural and Environmental Engineering Computing* (p. 25). Madeira, Portugal: Civil-Comp Press.

André, J., Beale, R., & Baptista, A. (2012a). A survey of failures of bridge falsework systems since 1970. *Proceedings of the ICE - Forensic Engineering, 65*(4), 161–172. doi:10.1680/feng.12.00012

André, J., Beale, R., & Baptista, A. (2012b). Bridge construction equipment: An overview of the existing design guidance. *Structural Engineering International, 22*(3), 15. doi:10.2749/101686612X13363869853419

André, J., Beale, R., & Baptista, A. (2013a). Experimental behaviour of bridge falsework joints. In *Proceedings of the IX Portuguese National Conference of Steel and Composite Construction* (p. 10). Guimarães, Portugal: The Portuguese Steelwork Association (cmm).

André, J., Beale, R., & Baptista, A. (2013b). Recent advances and existing challenges in the design of bridge falsework systems. *Civil Engineering and Environmental Systems, 30*(2), 130–145. doi:10.1080/10286608.2012.733374

Andresen, J. (2012). Investigation of a collapsed scaffold structure. *Proceedings of the Institution of Civil Engineers - Forensic Engineering, 165*(2), 95–104. doi:10.1680/feng.11.00021

Ayaho, M., Hideyuki, K., & Hideaki, N. (1997). Development of a knowledge-based expert system for selection of steel bridge erection method. In *Proceedings of NEW TECH Lisbon 97: New Technologies in Structural Engineering*, 2, 203–211.

Bakhoum, M. (2014). Bridge construction methods. In *Bridge engineering handbook: Construction and maintenance* (2nd ed.; Vol. 2, p. 60). CRC Press.

Beale, R. (2014). Scaffold research — A review. *Journal of Constructional Steel Research, 98*, 188–200. doi:10.1016/j.jcsr.2014.01.016

Beale, R., & Godley, M. (2003). Common causes of failure of metal scaffolds in the UK and stability checks on scaffolds. In *Proceedings of the Technical Seminar on Metal Scaffolding (Falsework) – Design, Construction and Safety* (pp. 62–79). Hong Kong, China: Hong Kong Institute of Steel Construction.

Beale, R., & Godley, M. (2006). Numerical modelling of tube and fitting access scaffold systems. *Advanced Steel Construction, 2*(3), 199–223.

Billings, I., & Routley, R. (1978). Falsework load measurement at Newton No. 1 bridge. *New Zealand Engineering, 33*(6), 127–134.

Birch, N., Booth, J., & Walker, M. (1971). Effect of site factors on the load capacities of adjustable steel props (No. R27). London, UK: Construction Industry Research and Information Association (CIRIA).

Birch, N., Walker, M., & Lee, C. (1977). Safe working loads for adjustable steel props; the influence of prop conditions and site workmanship (No. TN 79). London, UK: Construction Industry Research and Information Association (CIRIA).

Blair, T., & Woods, K. (1990). Seismic scaffolding concerns? Let it 'walk'. *Nuclear Plant Journal*, *8*(3), 76–77.

Błazik-Borowa, E., & Gontarz, J. (2016). The influence of the dimension and configuration of geometric imperfections on the static strength of a typical façade scaffolding. *Archives of Civil and Mechanical Engineering*, *16*(3), 269–281. doi:10.1016/j.acme.2015.11.003

Błazik-Borowa, E., & Szer, J. (2014). The analysis of the stages of scaffolding life with regard to the decrease in the hazard at building works. *Archives of Civil and Mechanical Engineering*, *15*(2), 516–524. doi:10.1016/j.acme.2014.09.009

Bragg, S. (1975). Final report of the advisory committee on falsework. Her Majesty's Stationery Office (HMSO).

Brand, R. (1975). *Falsework and access scaffolds in tubular steel.* London, UK: McGraw-Hill.

BSI. (1982a). BS 4074:1982 – Specification for metal props and struts. London, UK: British Standards Institution (BSI).

BSI. (1982b). BS 5507–3:1982 – Methods of test for falsework equipment – Part 3: Props. London, UK: British Standards Institution (BSI).

BSI. (1982c). BS 5975:1982 – Code of practice for falsework (1st ed.). London, UK: British Standards Institution (BSI).

BSI. (1990a). BS 1139–5:1990, HD 1000:1988 – Metal scaffolding – Part 5: Specification for materials, dimensions, design loads and safety requirements for service and working scaffolds made of prefabricated elements. London, UK: British Standards Institution (BSI).

BSI. (1990b). BS 4360:1990 – Specification for weldable structural steels. London, UK: British Standards Institution (BSI).

BSI. (1993). BS 5973:1993 – Code of practice for access and working scaffolds and special scaffold structures in steel (3rd ed.). London, UK: British Standards Institution (BSI).

BSI. (1996). BS 5975:1996 – Code of practice for falsework (2nd ed.). London, UK: British Standards Institution (BSI).

BSI. (1999). *Adjustable telescopic steel props – Product specifications, design and assessment by calculation and tests.* London, UK: British Standards Institution.

BSI. (2001). *Loose steel tubes for tube and coupler scaffolds – Technical delivery conditions.* London, UK: British Standards Institution.

BSI. (2002). BS EN 12811–3:2002 – Temporary works equipment – Part 3: Load testing. London, UK: British Standards Institution (BSI).

BSI. (2003a). BS EN 12810 – Façade scaffolds made of prefabricated components – Parts 1–2. London, UK: British Standards Institution (BSI).

BSI. (2003b). BS EN 12811–1:2003 – Temporary works equipment – Part 1: Scaffolds – Performance requirements and general design. London, UK: British Standards Institution (BSI).

BSI. (2005). BS EN 74–1:2005 – Couplers, spigot pins and baseplates for use in falsework and scaffolds – Part 1: Couplers for tubes – Requirements and test procedures. London, UK: British Standards Institution (BSI).

BSI. (2006a). BS EN 10210–1:2006 – Hot finished structural hollow sections of non–alloy and fine grain steels – Part 1: Technical delivery conditions. London, UK: British Standards Institution (BSI).

BSI. (2006b). BS EN 10219–1:2006 – Cold formed welded structural hollow sections of non–alloy and fine grain steels – Part 1: Technical delivery requirements. London, UK: British Standards Institution (BSI).

BSI. (2009). *Steel static storage systems – Adjustable pallet racking systems – Principles for structural design*. London, UK: British Standards Institution.

BSI. (2010a). BS EN 74–3:2007 – Couplers, spigot pins and baseplates for use in falsework and scaffolds – Part 3: Plain base plates and spigot pins – Requirements and test procedures. London, UK: British Standards Institution (BSI).

BSI. (2010b). BS EN 1991–1–4:2005+A1:2010 – Actions on structures – General actions – Part 1–4: Wind actions. London, UK: British Standards Institution (BSI).

BSI. (2011a). BS 5975:2008+A1:2011 – Code of practice for temporary works procedures and the permissible stress design of falsework (3rd ed.). London, UK: British Standards Institution (BSI).

BSI. (2011b). *Falsework. Performance requirements and general design*. London, UK: British Standards Institution.

BSI. (2012). *Adjustable telescopic aluminium props – Product specifications, design and assessment by calculation and tests*. London, UK: British Standards Institution.

BSI. (2015). BS 1139–2.2:2009+A1:2015 – Metal scaffolding – Part 2: Couplers – Section 2.2: Aluminium couplers and special couplers in steel. London, UK: British Standards Institution (BSI).

Burrows, B. (1989). *Organisation and quality of falsework construction (PhD)*. Warwick University.

Canisius, G., & Maitra, A. (1999). Strength of adjustable props - effect of shear-pin position.*Proceedings of the ICE - Structures and Buildings, 134*(2), 139–146. doi:10.1680/istbu.1999.31380

Cardwell, S. (2010). *Bridge deck types and construction methods.* Presented at the IABSE WG 6 Seminar: State-of-the-art Bridge Deck Erection; Safe and Efficient use of Special Equipment, Singapore.

Carlton, R., & Llurba, X. (2007). Modelling prop loads during concrete slab construction. *Structures & Buildings*, *160*(4), 221–229.

CEGB. (1966). Report of the Committee of Inquiry into collapse of cooling towers at Ferrybridge, Monday 1 November, 1965. London, UK: Central Electricity Generating Board (CEGB).

Chan, S., Dymiotis-Wellington, C., & Zhou, Z. (2002). Second-order analysis and design of steel scaffold using multiple eigen-imperfection modes. In *Proceedings of the 3rd International Conference on Advances in Steel Structures (ICASS 2002)* (pp. 312–327). Hong Kong, China: Elsevier. doi:10.1016/B978-008044017-0/50037-8

Chan, S., Zhou, Z., Chen, W., Peng, J., & Pan, A. (1995). Stability analysis of semirigid steel scaffolding. *Engineering Structures*, *17*(8), 568–574. doi:10.1016/0141-0296(95)00011-U

Chandrangsu, T. (2010). *Advanced analysis and probabilistic-based design of support scaffold systems* (PhD Thesis). University of Sydney, Sydney, Australia.

Chandrangsu, T., & Rasmussen, K. (2009). Full-scale tests and advanced structural analysis of formwork subassemblies. In *Proceedings of the 6th International Conference on Advances in Steel Structures* (pp. 1083–1090). Hong Kong, China: Hong Kong Institute of Steel Construction.

Chandrangsu, T., & Rasmussen, K. (2011a). Investigation of geometric imperfections and joint stiffness of support scaffold systems. *Journal of Constructional Steel Research*, *67*(4), 576–584. doi:10.1016/j.jcsr.2010.12.004

Chandrangsu, T., & Rasmussen, K. (2011b). Structural modelling of support scaffold systems. *Journal of Constructional Steel Research*, *67*(5), 866–875. doi:10.1016/j.jcsr.2010.12.007

Charuvisit, S., Hino, Y., Ohdo, K., Maruta, E., & Kanda, M. (2007). Wind tunnel experiment on wind pressures acting on the scaffolds in strong winds. *Journal of Wind Engineering*, *32*(1), 1–10. doi:10.5359/jwe.32.1

Chu, A., Zhou, Z., Koon, C., Chan, S., Peng, J., & Pan, A. (1996). Design of steel scaffolding using an integrated design and analysis approach. In *Proceedings of International Conference on Advances in Steel Structures* (pp. 245–250). Hong Kong, China: Pergamon. doi:10.1016/B978-008042830-7/50039-5

Chung, K., & Chan, S. (2002). *Proceedings of the International Seminar 'Bamboo Scaffolds in Building Construction' - An Alternative and Supplement to Metal Scaffolds*. Hong Kong, China: International Network for Bamboo and Rattan (INBAR).

Chung, K. F., & Yu, W. (2002). Mechanical properties of structural bamboo for bamboo scaffoldings. *Engineering Structures*, *24*(4), 429–442. doi:10.1016/S0141-0296(01)00110-9

CIRIA. (1971). Effect of site factors on the Load capacities of adjustable steel props (No. 27). London, UK: Construction Industry Research and Information Association (CIRIA).

CIRIA. (1977). Safe working loads for adjustable steel props: The influence of prop conditions and site workmanship (No. TN79). London, UK: Construction Industry Research and Information Association (CIRIA).

CIRIA. (2000). Civil engineering design and construct - A guide to integrating design into the construction process (No. C534). London, UK: Construction Industry Research and Information Association (CIRIA).

Crémer, J. (2003). *Conception des ponts*. Liège, Belgium: Liège University.

DIN. (1987). *Telescopic steel props – Safety requirements, testing*. Berlin, Germany: Deutsches Institut für Normung.

El-Safty, A., Zinszer, M., & Morcous, G. (2008). Forensic investigation of a bridge construction scaffolding collapse. In *Proceedings of Structures Congress 2008: Crossing Borders* (pp. 1–10). Vancouver, Canada: American Society of Civil Engineers (ASCE). doi:10.1061/41016(314)198

El-Sheikh, M., & Chen, W. (1989a). *Behaviour and strength of telescopic steel shores (No. CE-STR-88-7)*. West Lafayette, IN: Purdue University.

El-Sheikh, M., & Chen, W. (1989b). Maximum probabilistic shore load in multistorey R/C buildings. *Computers & Structures*, *32*(6), 1347–1357. doi:10.1016/0045-7949(89)90311-8

Enright, J., Harris, R., & Hancock, G. (2000). Structural stability of braced scaffolding and formwork with spigot joints. In *Proceedings of 15th International Specialty Conference in Cold-Formed Steel Design and Construction*, 357–376.

Feng, X. (1994). *The influence of semi-rigid connections on the behaviour of slender structures* (PhD Thesis). Oxford Brookes University, Oxford, UK.

fib. (2000). *Guidance for good bridge design*. Lausanne, Switzerland: International Federation for Structural Concrete (fib), Bulletin No. 9.

Gao, H., Gao, W., Li, C., & Huo, X. (2013). Influence of the structural performance of steel tubular scaffold based on measured imperfection. *Applied Mechanics and Materials*, *256–259*, 754–757.

Gerhardt, H., & Janser, F. (1994). Wind loads on permeable façades. *Journal of Wind Engineering and Industrial Aerodynamics*, *53*(1-2), 37–48. doi:10.1016/0167-6105(94)90017-5

Godley, M. (1990). *The performance of SGB Cuplok connectors (No. R221)*. Oxford, UK: Oxford Polytechnic, Department of Civil Engineering, Building & Cartography.

Godley, M., & Beale, R. (1997). Sway stiffness of scaffold structures. *The Structural Engineer*, *75*(1), 4–12.

Hadipriono, F., & Wang, H. (1986). Analysis of causes of falsework failures in concrete structures. *Journal of Construction Engineering and Management*, *112*(1), 112–121. doi:10.1061/(ASCE)0733-9364(1986)112:1(112)

Hadipriono, F., & Wang, H. (1987). Causes of falsework collapses during construction. *Structural Safety*, *4*(3), 179–195. doi:10.1016/0167-4730(87)90012-9

Halperin, K., & McCann, M. (2004). An evaluation of scaffold safety at construction sites. *Journal of Safety Research*, *35*(2), 141–150. doi:10.1016/j.jsr.2003.11.004 PMID:15178232

Harung, H., Lightfoot, E., & Duggan, D. (1975). The strength of scaffold towers under vertical loading. *The Structural Engineer*, *53*(1), 23–30.

Hill, H. (2004). Rational and irrational design loads for temporary' structures. *Practice Periodical on Structural Design and Construction*, *9*(3), 125–129. doi:10.1061/(ASCE)1084-0680(2004)9:3(125)

Holmes, M., & Hindson, D. (1979). Structural behaviour of load bearing falsework. *Proceedings of the ICE*, *67*(3), 721–741.

Hoxey, R. (1994). Full-scale studies of wind-loading on light-weight structures, windbreaks and walls at Silsoe Research Institute. In *Proceedings of Conference into Wind Loading on Temporary Structures* (pp. 83–98). Buxton, UK: Health and Safety Executive (HSE).

HSE. (1994). *Proceedings of the conference on wind loading on temporary structures*. London, UK: Health and Safety Executive (HSE).

Hu, C., Ge, Z., & Jing, W. (2011). Research on ultimate bearing capacity of coupler steel tube falsework with initial defect. *Advanced Materials Research*, *163–167*, 32–35.

Huang, Y., Chen, H., Rosowsky, D., & Kao, Y. (2000). Load carrying capacities and failure modes of scaffold-shoring systems. Part 1: Modelling and experiments. *Structural Engineering & Mechanics*, *10*(1), 53–66. doi:10.12989/sem.2000.10.1.053

Huang, Y., Kao, Y., & Rosowsky, D. (2000). Load carrying capacities and failure modes of scaffold-shoring systems. Part 2: An analytical model and its closed-form solution. *Structural Engineering & Mechanics*, *10*(1), 67–79. doi:10.12989/sem.2000.10.1.067

Hurd, M. (1995). Formwork for concrete (6th ed.). American Concrete Institute (ACI), ACI SP-4.

Irtaza, H. (2009). *Experimental and computational determination of wind loads on netted/sheeted scaffolds* (PhD Thesis). Oxford Brookes University, Oxford, UK.

Irtaza, H., Beale, R., & Godley, M. (2007). Determination of the effects of wind load on bare tube access scaffold structures using computational fluid dynamics. In *Proceedings of the 5th International Conference on Advances in Steel Structures (ICASS 2007)*, 637–642.

Irtaza, H., Beale, R., & Godley, M. (2012). A wind-tunnel investigation into pressure distribution around sheet clad scaffolds. *Journal of Wind Engineering and Industrial Aerodynamics*, *103*, 86–95. doi:10.1016/j.jweia.2012.03.004

Irtaza, H., Beale, R., Godley, M., & Jameel, A. (2013). Comparison of wind pressure measurements on Silsoe experimental building from full-scale observation, wind-tunnel experiments and various CFD techniques. *International Journal of Engineering Science and Technology*, *5*(1), 28–41.

Johnston, D. (2014). Formwork for concrete (8th ed.). American Concrete Institute (ACI), ACI SP-4.

Lee, M., & Daebritz, M. (2010). *Moveable scaffolding systems - MSS*. Presented at the IABSE WG 6 Seminar: State-of-the-art Bridge Deck Erection; Safe and Efficient use of Special Equipment, Singapore.

Lew, H. S. (1984). Construction failures and their lessons. *Batiment International Building Research and Practice*, *12*(5), 272–275. doi:10.1080/09613218408545283

Lightfoot, E., & Bhula, D. (1977a). A test rig for scaffold couplers. *Matériaux et Construction*, *10*(3), 168–173. doi:10.1007/BF02474850

Lightfoot, E., & Bhula, D. (1977b). The idealization of scaffold couplers for performance tests and scaffold analysis. *Matériaux et Construction*, *10*(3), 159–168. doi:10.1007/BF02474849

Lightfoot, E., & Lemessurier, A. (1975). Instability of space frames having elastically connected and offset members. In *Proceedings of the 2nd International Conference on Space Structures*, 143–149.

Lightfoot, E., Olivetto, G., & Merchant, W. (1977). The collapse strength of tubular steel scaffold assemblies. *Proceedings of the ICE*, *63*(2), 311–329.

Lindley, D., Pandya, D., & Khanpour, A. (2001). Estimating seismic response of freestanding scaffolding. In *Transactions of the 16th International Conference on Structural Mechanics in Reactor Technology*. Washington, DC: International Association for Structural Mechanics in Reactor Technology (IASMRT).

Lindner, J., & Magnitzke, P. (1990). Stability of scaffold structures with sheeting[in German]. *Stahlbau*, *59*(2), 39–44.

Liu, H., Chen, Z., Wang, X., & Zhou, T. (2010). Theoretical analysis and experimental research on stability behavior of structural steel tube and coupler falsework with X-bracing. *Advanced Steel Construction*, *6*(4), 949–962.

Liu, H., Zhao, Q., Wang, X., Zhou, T., Wang, D., Liu, J., & Chen, Z. (2010). Experimental and analytical studies on the stability of structural steel tube and coupler scaffolds without X-bracing. *Engineering Structures*, *32*(4), 1003–1015. doi:10.1016/j.engstruct.2009.12.027

Maitra, A. (1997). Accidents associated with scaffolding in the UK. In *Proceedings of Technical Seminar, Design and use of Temporary Structures*. London, UK: Institution of Civil Engineers (ICE).

Masumoto, H., Hara, K., & Yamashita, M. (1994). Design plan of super multi-span continuous Menshin bridge with deck length of 725m. In *Proceedings of the 3rd U.S.- Japan Workshop on Earthquake Protective Systems for Bridges* (Vol. 4, pp. 69–84). U.S. National Center for Earthquake Engineering Research.

Milojkovic, B. (1999). *Factors of safety for standard scaffold structures* (PhD Thesis). Oxford Brookes University, Oxford, UK.

Milojkovic, B., Beale, R., & Godley, M. (1996). Modelling scaffold connections. In *Proceedings of 4th ACME UK Annual Conference* (pp. 85–88). Glasgow, UK: Association of Computational Mechanics in Engineering (ACME).

Milojkovic, B., Beale, R., & Godley, M. (2002). Determination of the factors of safety of standard scaffold structures. In *Proceedings of the 3rd International Conference on Advances in Steel Structures (ICASS 2002)* (pp. 303–310). Hong Kong, China: Elsevier. doi:10.1016/B978-008044017-0/50035-4

Mohammadi, J., & Heydari, A. (2008). Seismic and wind load considerations for temporary structures. *Practice Periodical on Structural Design and Construction*, *13*(3), 128–134. doi:10.1061/(ASCE)1084-0680(2008)13:3(128)

Pallett, P., Burrow, M., Clark, L., & Ward, R. (2001). *Investigation into aspects of falsework (No. crr01394)*. Birmingham, UK: The University of Birmingham.

PASMA. (2000). Operators code of practice for PASMA aluminium alloy towers (7th ed.). Glasgow, UK: Prefabricated Access Suppliers' and Manufacturers' Association (PASMA).

Peng, J. (2002). Stability analyses and design recommendations for practical shoring systems during construction. *Journal of Construction Engineering and Management*, *128*(6), 536–544. doi:10.1061/(ASCE)0733-9364(2002)128:6(536)

Peng, J. (2004). Structural modeling and design considerations for double-layer shoring systems. *Journal of Construction Engineering and Management*, *130*(3), 368–377. doi:10.1061/(ASCE)0733-9364(2004)130:3(368)

Peng, J., Pan, A., & Chan, S. (1998). Simplified models for analysis and design of modular falsework. *Journal of Constructional Steel Research*, *48*(2–3), 189–209. doi:10.1016/S0143-974X(98)00198-9

Peng, J., Pan, A., Chen, W., Yen, T., & Chan, S. (1997). Structural modeling and analysis of modular falsework systems. *Journal of Structural Engineering*, *123*(9), 1245–1251. doi:10.1061/(ASCE)0733-9445(1997)123:9(1245)

Peng, J., Pan, A., Rosowsky, D., Chen, W., Yen, T., & Chan, S. (1996a). High clearance scaffold systems during construction - I. Structural modelling and modes of failure. *Engineering Structures*, *18*(3), 247–257. doi:10.1016/0141-0296(95)00144-1

Peng, J., Pan, A., Rosowsky, D., Chen, W., Yen, T., & Chan, S. (1996b). High clearance scaffold systems during construction - II. Structural analysis and development of design guidelines. *Engineering Structures*, *18*(3), 258–267. doi:10.1016/0141-0296(95)00145-X

Peng, J., Wang, P., Chan, S., & Huang, C. (2012). Load capacities of single-layer shoring systems: An experimental study. *Advances in Structural Engineering*, *15*(8), 1389–1410. doi:10.1260/1369-4332.15.8.1389

Peng, J., Wu, C., Shih, M., & Yang, Y. (2013). Experimental study of load capacities of tubular steel adjustable shores used in construction. *International Journal of Structural Stability and Dynamics*, *13*(04), 32. doi:10.1142/S0219455412500630

Peng, J., Yen, T., Pan, A., Chen, W., & Chan, S. (1996). Structural modeling and analysis of scaffold systems. In *Proceedings of International Conference on Advances in Steel Structures* (pp. 251–256). Hong Kong, China: Pergamon. doi:10.1016/B978-008042830-7/50040-1

Pieńko, M., & Błazik-Borowa, E. (2013). Numerical analysis of load-bearing capacity of modular scaffolding nodes. *Engineering Structures*, *48*, 1–9. doi:10.1016/j.engstruct.2012.08.028

Pisheh, Y., Shafiei, H., & Hatambeigi, M. (2009). A case study of failure due to inappropriate usage of forming scaffold system. In *Proceedings of 5th Forensic Engineering Congress 2009* (pp. 536–545). Washington, DC: American Society of Civil Engineers (ASCE). doi:10.1061/41082(362)55

Poon, C., & Yip, R. (2008). Comparison of timber and metal formwork systems. *Water Resources Management*, *161*(1), 29–36.

Póvoas, A. (2012). The utilization of movable scaffolding systems in large spans. *Structural Engineering International*, *22*(3), 395–400. doi:10.2749/101686612X13363869853374

Prabhakaran, U. (2009). *Nonlinear analysis of scaffolds with semirigid connections* (PhD Thesis). Oxford Brookes University, Oxford, UK.

Prabhakaran, U., Beale, R., & Godley, M. (2011). Analysis of scaffolds with connections containing looseness. *Computers & Structures, 89*(21–22), 1944–1955. doi:10.1016/j.compstruc.2011.03.016

Rosowsky, D., Huang, Y., Chen, W., & Yen, T. (1994). Modeling concrete placement loads during construction. *Structural Engineering Review, 6*(2), 71–84.

Rosowsky, D., Huston, D., Fuhr, P., & Chen, W. (1994). Measuring formwork loads during construction. *Concrete International, 16*(11), 21–25.

Salvadori, A. (2009). Ultimate strength of adjustable telescopic steel props according to standard EN 1065. *Journal of Constructional Steel Research, 65*(10–11), 1964–1970. doi:10.1016/j.jcsr.2009.06.006

SGB. (2009). *Cuplok: Datasheet*. SGB Group (now HARSCO).

The Concrete Centre. (2008). *Concrete bridges: The benefits of concrete in bridge design and construction*. London, UK: The Concrete Centre.

The Concrete Society. (1971). *Falsework - Report of the joint committee, The Concrete Society and the Institution of Structural Engineers (No. TRCS 4)*. London, UK: The Concrete Society.

The Concrete Society. (1986). *Formwork: A guide to good practice* (1st ed.). London, UK: The Concrete Society, Report CS30.

The Concrete Society. (1995). *Formwork: A guide to good practice* (2nd ed.). London, UK: The Concrete Society, Report CS30.

The Concrete Society. (2012). *Formwork: A guide to good practice* (3rd ed.). London, UK: The Concrete Society, Report CS30.

Tischer, T., & Kuprenas, J. (2003). Bridge falsework productivity - Measurement and influences. *Journal of Construction Engineering and Management, 129*(3), 243–250. doi:10.1061/(ASCE)0733-9364(2003)129:3(243)

Voelkel, W. (1990). German assessment for the Cuplok scaffold system within the framework of the comparison of the procedures in different European countries - Part 1: Test report (No. T 2381/1). Berlin, Germany: Institut fur Bautechnik (IfB).

Wang, F., Tamura, Y., & Yoshida, A. (2013). Wind loads on clad scaffolding with different geometries and building opening ratios. *Journal of Wind Engineering and Industrial Aerodynamics, 120*, 37–50. doi:10.1016/j.jweia.2013.06.015

Wang, F., Tamura, Y., & Yoshida, A. (2014). Interference effects of a neighboring building on wind loads on scaffolding. *Journal of Wind Engineering and Industrial Aerodynamics, 125*, 1–12. doi:10.1016/j.jweia.2013.11.009

Weesner, L., & Jones, H. (2001). Experimental and analytical capacity of frame scaffolding. *Engineering Structures, 23*(6), 592–599. doi:10.1016/S0141-0296(00)00087-0

Whitaker, S., Graves, R., James, M., & McCann, P. (2003). Safety with access scaffolds: Development of a prototype decision aid based on accident analysis. *Journal of Safety Research, 34*(3), 249–261. doi:10.1016/S0022-4375(03)00025-2 PMID:12963071

Williams, C. (1994). The structural design of fabric structures to resist wind loading. In *Proceedings of Conference into Wind Loading on Temporary Structures* (pp. 74–82). Buxton, UK: Health and Safety Executive (HSE).

Wilshere, C. (1983). *Falsework. ICE works construction guide*. London, UK: Thomas Telford Ltd.

Xie, N., & Wang, G. (2009). Test analysis on hidden defect in high falsework and its effect on structural reliability. In *Proceedings of the 8th International Conference on Reliability, Maintainability and Safety, ICRMS 2009* (pp. 1077–1080). Chengdu, China: Institute of Electrical and Electronics Engineers (IEEE). doi:10.1109/ICRMS.2009.5270000

Yen, T., Huang, Y., Chen, W., & Lin, Y. (1995). *Design of scaffold systems for concrete buildings during construction (No. CE-STR-95-4)*. West Lafayette, IN: Purdue University.

Yue, F., Yuan, Y., Li, G., Ye, K., Chen, Z., & Wang, Z. (2005). Wind load on integral-lift scaffolds for tall building construction. *Journal of Structural Engineering*, *131*(5), 816–824. doi:10.1061/(ASCE)0733-9445(2005)131:5(816)

Zhang, H., Chandrangsu, T., & Rasmussen, K. (2010). Probabilistic study of the strength of steel scaffold systems. *Structural Safety*, *32*(6), 393–401. doi:10.1016/j.strusafe.2010.02.005

Chapter 3
Actions

ABSTRACT

This chapter presents a general description and discussion of the actions applied to temporary structures such as construction loads, wind loads, impact loads and unidentified hazard events. A classification of actions is presented. Actions are classified into permanent actions such as self-weight, lateral loads by soil or water; and variable actions such as live loads, earthquakes and wind loads. Comparisons are made between design provisions for loads as specified by European, USA and Australian design codes and standards. Methods to estimate the main effects of the actions on temporary structures are presented. The latest research into wind on temporary structures is a significant part of this chapter with its implications to the correct wind forces acting on temporary structures when turbulence and orography are taken into account.

3.1 INTRODUCTION

Every temporary structures project is a unique endeavour, given a particular set of challenges and a specific context. The planning, design, execution and operation processes vary with the application, the site where it will be used and the role of temporary structures in the construction process. Therefore, temporary structures are exposed to a multiplicity of natural and/or man-originated hazardous events.

From a temporary structures' design and operation perspective, there are a large variety of design challenges related to actions, originating from the diversity of the panoply of applications, geography of the site and climate exposure where temporary structures are used. For example, temporary structures can be used in prefabricated (precast) or cast-in-place (in situ) concrete construction, in residential buildings or multi-span bridges, in areas with significant seismic hazards or with challenging geotechnical conditions, in urban or rural areas under possible hurricane wind forces.

Many of the hazards have been appropriately researched and rules have been incorporated in existing codes of practice or guidance documents. However, there are still gaps of knowledge that need to be filled, with emphasis on the risks originating from human interaction, namely human errors during all phases of temporary structures life cycle.

DOI: 10.4018/978-1-5225-2199-0.ch003

Copyright © 2017, IGI Global. Copying or distributing in print or electronic forms without written permission of IGI Global is prohibited.

The aim of this Chapter is to introduce the main types of actions due to external hazard events that are relevant to temporary structures, discuss the challenges associated with their characterisation and quantification, and provide an understanding on how specific actions can affect the performance of different types of temporary structures. The Chapter also identifies aspects that are not covered by existing structural design codes and presents state-of-the-art methods that help to overcome these limitations. Additionally, means to simulate internal hazards due to uncertainties and errors during design, assembly and operation of temporary structures are also analysed.

On the basis of this Chapter it is expected that the reader will acquire knowledge on the following topics:

1. Classification of actions.
2. Typologies of different construction actions and their effects on temporary structures.
3. Assessment of wind actions and their effects on temporary structures.
4. Potential influence of ground characteristics on temporary structures performance.
5. Assessment of human actions and their effects on temporary structures, typically temporary stands and stages.
6. Assessment of accidental actions relevant to temporary structures, such as vehicle impacts and earthquakes.
7. Assessment of notional actions that simulate the effects of unidentified hazard events during design, assembly and use of temporary structures.

3.2 DESIGN CODES

Actions relevant to temporary structures can be determined from the suite of European structural design standards (named the Eurocodes), ASCE 7 (ASCE, 2010) in the USA or Parts 0 to 4 of AS 1170 in Australia and New Zealand, for example. For temporary structures, further guidance is given in BS EN 12811 (BSI, 2004b) for scaffolds, in BS EN 12812 (BSI, 2011b) and BS 5975 (BSI, 2011a) for falsework, and in ASCE/SEI 37 (ASCE, 2014) and BS EN 1991-1-6 (BSI, 2005b) for temporary structures in general. In the USA, the AASHTO bridge code (AASHTO, 2016) and the AASHTO design guide for bridge falsework (AASHTO, 2008), can also be also used to determine design actions. See also Chapter 6.

The ASCE/SEI 37 provides the design loads and load combinations for temporary structures used during construction, as well as for partially completed structures during their construction phases. This standard addresses not only permanent and variable loads due to the construction but also environmental loads, the minimum values of the partial factors and the relevant load combinations to be considered, in accordance with the Limit States design philosophy.

The European standard is BS EN 1991-1-6 (BSI, 2005b) which describes the principles and application rules for the determination of actions to be considered during execution of buildings and civil engineering works.

The Eurocodes have also been adopted in South Africa, and in parts of Asia, such as Singapore and Hong Kong. ASCE 7 is also used outside of the USA, in particular in the Middle East countries.

As the, assumptions, methods and procedures (see Chapters 5 and 6) that form the basis of different design codes may not be the same, one should be careful, perform the necessary analysis and take appropriate precautions before attempting to use interchangeably the rules included in the above standards for the design of temporary structures, in particular with respect to loads and safety factors. For example, the AASHTO bridge code, ASCE/SEI 37 and ASCE 7 are based on Limit States design philosophy whereas the AASHTO design guide for bridge falsework is still based on the Allowable Stress design philosophy. This is not the case of the abovementioned European codes which follow a single and coherent framework. Therefore, before using different codes, it should be demonstrated that the latter will not lead to unconservative designs.

It should not be forgotten that the Eurocodes are only valid if used together with the corresponding National Annexes published by every European Union member state which contain the national choices for the Nationally Determined Parameters (NDPs). In the present book, the UK National Annexes (NAs) will be used as an example.

The versions of the documents reviewed are those current at the time of writing.

The loads presented in the following Sections should always be interpreted as preliminary values. It is crucial that a proper assessment is carried out for each project to determine their values as accurately as possible.

3.3 CLASSIFICATION OF ACTIONS

Actions can be classified by their variation in time as follows (BSI, 2005a; ISO, 2015):

- **Permanent Action (*G*):** An action that is likely to act continuously during a given reference period and for which the variation in space and in magnitude with time is insignificant. Examples are the self-weight of structures, of fixed equipment and of fixed materials (soil for example), and indirect actions caused by post-tensioning and ground settlements.
- **Variable Action (*Q*):** An action that is likely to act during a given reference period but for which the variation in space or in magnitude with time is not insignificant. Examples are construction forces, ordinary operational actions on bridge decks and building slabs, and wind action in general.
- **Accidental Action (*A*):** An action, usually of short duration but of abnormal magnitude, that is unlikely to occur on a given structure during a given reference period. Examples are impact from vehicles or earthquakes in general.

The actual actions classification depends on specific project requirements, including site location, risk management framework, etc. The same action can have different classifications as a function of its magnitude and probability of occurrence, structural effects and ensuing consequences. For example, the use of temporary structures involves transient situations (i.e. short duration periods), such as the construction phase of a permanent structure. In these cases, the self-weight of the permanent structure should be considered a variable action during the construction phase and a permanent action once construction is completed.

For each action, a suitable theoretical/empirical action model must be developed and included in structural analysis of any structure, temporary or permanent. An appropriate action model expresses the action by its fundamental characteristics, such as its origin, nature, magnitude, position, direction, duration and interaction with the exposed structure – in particular for fluid dynamic actions and indirect actions.

Actions may also be classified (BSI, 2005a; ISO, 2015) by their:

1. Origin:
    a. Direct, when originating from a set of loads applied directly to the structure (load is a term frequently used in engineering to mean the external force exerted on a surface or body). The actions' models are independent of the structural properties or the structural response;
    b. Indirect, when originating from a set of imposed deformations or accelerations. The actions' models are dependent of the structural properties or the structural response.
2. Spatial Variation:
    a. Fixed, when the distribution and point of application do not change during a given reference period;
    b. Free, when the above conditions are not fulfilled.
3. Nature and/or the Structural Response:
    a. Static, when inertia effects are irrelevant;
    b. Dynamic, when the above condition is not fulfilled.

In general, the most important actions applied to temporary structures depend of the type of application. For shoring and bridge falsework it may be the pressure from fresh concrete, whereas for scaffolding it may be wind actions. In contrary to permanent structures which only receive their full design force in rare cases (e.g. the design traffic forces on bridges are rarely reached), usually temporary structures are normally subjected for a long period of their design working life to forces whose values are close to their design values. Thus, the actual safety margin of temporary structures is lower than in permanent structures (fib, 2009).

Given the intrinsic specificities of temporary structures projects, complex and not fully resolved challenges exist when defining the actions' models. In particular, regarding the proper way to account for the action-structure interaction and the magnitude of actions: balancing the transient nature of the use of temporary structures and their life time, much larger than each individual period of use, in the considered risk management framework (André, Beale, & Baptista, 2013).

For the construction phase of structures, it is possible to distinguish two categories of actions: construction actions and actions other than construction loads, see Table 1 and Table 2, respectively.

Table 3 to Table 5 provide a summary of the loads for temporary structures used in concrete construction specified in design codes of USA, Europe and Australia. For metal scaffolding see Table 6, and for bamboo scaffolding according to the design guide published by the Labour Department, Hong Kong Government (Hong Kong Labour Department, 2014), see Table 7.

3.4 Permanent Actions

In accordance with BS EN 1991-1-6 (BSI, 2005b), the self-weight of structural and non-structural components should be determined in accordance with BS EN 1991-1-1(BSI, 2002a). The representative value of each permanent action, G_k, should then be taken equal to the mean value of the probability density function of G, provided that the variability of G is small (e.g. coefficient of variation equal to or less than 5%). Table 3.8 contains representative values of the self-weight density of construction materials.

Table 1. Classification of construction loads (BSI, 2005b)

Action (short description)	Classification				Remarks
	Variation in time	Classification / Origin	Spatial variation	Nature (static/ dynamic)	
Personnel and hand tools (e.g. staff and visitors, etc.)	Variable	Direct	Free	Static	
Storage of movable items (e.g. construction materials, equipment, etc.)	Variable	Direct	Free	Static / dynamic	Dynamic in the case of dropped loads
Non-permanent equipment (e.g. formwork panels travelling forms, launching noses, etc.)	Variable	Direct	Fixed / free	Static / dynamic	
Movable heavy machinery and equipment (e.g. cranes, jacks, self-launching gantries, etc.)	Variable	Direct	Free	Static / dynamic	
Accumulation of waste materials (e.g. surplus of construction materials, excavated soil, debris, etc.)	Variable	Direct	Free	Static / dynamic	

Table 2. Classification of actions (other than construction loads) during construction phase (BSI, 2005b)

Action	Classification				Remarks
	Variation in time	Classification / Origin	Spatial variation	Nature (static/ dynamic)	
Self weight	Permanent	Direct	Fixed with tolerance / free	Static	Free during transportation / storage. Dynamic if dropped
Ground settlements	Permanent	Indirect	Free	Static	
Earth pressure	Permanent / variable	Direct	Free	Static	
Post-tensioning	Permanent / variable	Direct	Fixed	Static	Variable for local design (anchorage)
Pre-deformations	Permanent / variable	Indirect	Free	Static	
Temperature	Variable	Indirect	Free	Static	
Shrinkage/hydration effects	Permanent / variable	Indirect	Free	Static	
Wind actions	Variable / accidental	Direct	Fixed/free	Static / dynamic	
Snow loads	Variable / accidental	Direct	Fixed/free	Static / dynamic	
Actions due to water	Permanent / variable / accidental	Direct	Fixed/free	Static / dynamic	Permanent / variable according to project specifications. Dynamic for water currents if relevant
Atmospheric ice loads	Variable	Direct	Free	Static / dynamic	
Accidental	Accidental	Direct/indirect	Free	Static/dynamic	
Seismic	Variable / accidental	Direct	Free	Dynamic	

Table 3. Loads for temporary structures used in concrete construction (USA codes)

<table>
<tr><th rowspan="2">Load type</th><th colspan="2">AASHTO GSBTW-1-M</th><th rowspan="2">Load type</th><th colspan="2">ASCE/SEI 37-02</th></tr>
<tr><th>ID</th><th>Value</th><th>ID</th><th>Value</th></tr>
<tr><td rowspan="4">Permanent loads</td><td>Steel: sections, wires, cables, etc.</td><td>Not provided</td><td>Permanent loads</td><td>Steel (CD): sections, wires, cables, etc.</td><td>Ex: ASCE/SEI 7-10</td></tr>
<tr><td>Concrete: Concrete casting loads</td><td>Min: 25.1 kN/m^3 for normal reinforced concrete</td><td rowspan="7">Construction loads</td><td>Fixed (CFML), e.g. formwork</td><td>Ex: ASCE/SEI 7-10</td></tr>
<tr><td>Wood: formwork</td><td>Not provided</td><td>Variable (CVML), e.g. concrete casting loads, materials</td><td>Analysis dependent ASCE/SEI 7-10</td></tr>
<tr><td>Lateral pressure of concrete (CC)</td><td>ACI 347</td><td>Lateral pressure of concrete (CC)</td><td>ACI 347</td></tr>
<tr><td colspan="2" rowspan="3">Construction loads</td><td rowspan="3">Equipment + 0.96 kN/m^2 (2.16 kN/m^2 if motorized carts are used) + 1.1 kN/m at the outside edges of deck overhangs</td><td>Personnel and equipment (CP)</td><td>Min: 1.1 kN/worker</td></tr>
<tr><td>Equipment reactions (CR)</td><td>See supplier documents for rated equipment</td></tr>
<tr><td>Erection and lifting (CF)</td><td>Analysis dependent</td></tr>
<tr><td colspan="2">Horizontal load</td><td>Min: 2% of vertical load</td><td>Horizontal (CH)</td><td>Max(2% of vertical load; 0.22 kN/person)</td></tr>
<tr><td colspan="2" rowspan="4">Wind load</td><td rowspan="4">Chapter 23, Part II of the Uniform Building Code

The basic wind pressure shall be increased by 240 N/m^2 for falsework members over or adjacent to traffic openings</td><td rowspan="4">Variable loads</td><td>Wind (W)</td><td>ASCE/SEI 7-10 applying reduction factor, see Table 2.8</td></tr>
<tr><td>Thermal (T)</td><td>Analysis dependent</td></tr>
<tr><td>Snow (S)</td><td>ASCE/SEI 7-10 applying reduction factor of 0.8 if construction period is $\leq$ 5 years</td></tr>
<tr><td>Earthquake (E)</td><td>ASCE/SEI 7-10, Category II, using a reduction factor $\geq$ 0.2 and a behaviour factor $\leq$ 2.5</td></tr>
<tr><td>Other loads</td><td>Loads caused by post-tensioning or other actions</td><td>Analysis dependent Foundation settlements should not exceed 25 mm</td><td>Other loads</td><td>Loads caused by post-tensioning or other actions (O)</td><td>Analysis dependent</td></tr>
<tr><td>Accidental loads</td><td>Loads caused by impact, local failure</td><td>Analysis dependent</td><td>Accidental loads</td><td>Loads caused by impact, local failure</td><td>Analysis dependent</td></tr>
</table>

The permanent action of a temporary structure consists of the self-weight of the structure itself and any permanent loads applied to the structure such as kentledge imposed to prevent overturning (BSI, 2011b).

For BCEs the permanent loads are significant. For example, the weight of each form traveller may vary between 0.2 and 1 MN according to the length of the segments and the width of the bridge deck (Hewson, 2003; Sétra, 2007), typically representing 25% to 50% of the weight of the heaviest segment. Launching gantries might weigh up to or more than 10 MN, although MSS systems weight typically less than 2 MN. A typical launching nose weights around 0.8 MN. It is common to calculate the unfactored

Table 4. Loads for temporary structures used in concrete construction (European codes)

Load type	BS EN 12812	
	ID	Value
Permanent loads (Q_1)	Steel: sections, wires, cables, etc.	EN 1991-1-1 + NA
Permanent loads (Q_1)	Wood: formwork	EN 1991-1-1 + NA
Construction loads	Fresh concrete weight, precast units weight (Q_2)	25 kN/m³ for normal reinforced fresh concrete
Construction loads	Concrete casting loads (Q_4)	Additional load for in situ casting (working area 3 m × 3 m): 10% concrete self-weight but ≥0.75 kN/m² and ≤1.75 kN/m² Concrete pressures from CIRIA Report n° 108
Construction loads	Construction loads due to working personnel (Q_2)	Min: 0.75 kN/m²
Construction loads	Horizontal (Q_3)	1% of the Q_2 vertical load
Construction loads	Construction loads due to storage of materials (Q_2)	Min: 1.5 kN/m²
Variable loads	Wind actions (Q_5)	EN 1991-1-4 + NA
Variable loads	Thermal (Q_8)	If $L_{bridge} \geq 60$ m then ±10 K (concrete bridge)
Variable loads	Snow (Q_2)	Consider only if ≥ 0.75 kN/m²
Variable loads	Earthquake (Q_7)	EN 1998-2 + NA
Other loads	Loads caused by post-tensioning or other actions (Q_9)	EN 1990, EN 1992, EN 1997 + NAs
Accidental loads	Loads caused by impact, local failure	EN 1990, EN 1991, EN 1993 + NAs

Load type	Eurocodes	
	ID	Value
Permanent loads	Steel (G): sections, wires, cables, etc.	EN 1991-1-1 + NA
Construction loads (Qc)	Formwork system (Qcc)	See supplier documentation, otherwise: Min: 0.5 kN/m²
Construction loads (Qc)	Concrete casting loads, precast units weight (Qcf)	EN 1991-1-1 + NA 26 kN/m³ for normal reinforced fresh concrete Additional load for in situ casting (working area 3 m × 3 m): 10% concrete self-weight but ≤0.75 kN/m²
Construction loads (Qc)	Construction loads due to working personnel (Qca)	0.75 kN/m² during concrete casting, otherwise 1.0 kN/m²
Construction loads (Qc)	Construction loads due to moveable heavy machinery and equipment, lifting, hoisting (Qcd)	Analysis dependent EN 1991-3 + NA
Construction loads (Qc)	Construction loads due to storage of moveable items (Qcb)	For bridges: Min distributed load: 0.2 kN/m² Min. concentrated load:100 kN
Variable loads	Wind actions (W)	EN 1991-1-4 + NA
Variable loads	Thermal (T)	EN 1991-1-5 + NA
Variable loads	Snow (S)	EN 1991-1-3 + NA
Variable loads	Earthquake (E)	EN 1998-2 + NA
Other loads	Loads caused by post-tensioning or other actions (O)	EN 1990, EN 1992, EN 1997 + NAs
Accidental loads	Loads caused by impact, local failure	EN 1990, EN 1991, EN 1993 + NAs

Table 5. Loads for temporary structures used in concrete construction (Australian codes)

Load type	AS 3610	
	ID	**Value**
Permanent loads (G)	Steel: sections, wires, cables, etc.	AS 1170.1 24 kN/m^3 for normal concrete + $60 \times \frac{\text{volume of reinforcement}}{\text{total volume}} \text{kN} / \text{m}^3$
	Wood: formwork	
	Concrete weight	
Construction loads	Concrete casting loads (Q_c)	Additional load for in situ casting (working area 1.6 m × 1.6 m): 3 kN/m^2
	Personnel and equipment (Q_{uv})	1 kN/m^2
	Lateral pressure of concrete (P)	Section 4.4.5.1 AS 3610
	Load from stacked materials (M)	4 kN/m^2
	Horizontal (Q_{uh})	Min(1 kN/m; 5 kN)
Variable loads	Wind (W_u)	AS 1170.2
	Thermal (T)	Analysis dependent
	Earthquake (E_u)	AS 1170.4 if construction period > 6 months
Other loads	Loads caused by post-tensioning or other actions (X_m)	Analysis dependent
Accidental loads	Loads caused by impact, local failure	Analysis dependent

value of the BCE permanent load by increasing the self-weight of the main elements by up to 40% to account for attached components (Rosignoli, 2010). The exact value should be calculated for each project and depends not only on the geometrical and material properties of each segment but also on the type of system used to balance the construction actions over the supporting pier and adjacent segments.

In analysing the self-weight of a common scaffold and falsework tubular structure, the weight of the connections should also be included. Frequently, to avoid having lots of point loads at the connections the density of the steel or aluminium tube is artificially enhanced so that the connection weight is uniformly distributed throughout the tube length. This assumption is particularly common in the tubes of proprietary scaffolds which have permanent connections such as cups or wedges welded every 300 to 500 mm.

For some permanent loads for which the probabilistic variability cannot be considered small (e.g. coefficient of variation greater than 5%), or for which the positioning is uncertain, or for which the effects of the permanent loads can be both favourable and unfavourable for the stability of the structure, two representative values of G, $G_{k,inf}$ and $G_{k,sup}$, should be considered in design, applied in the most unfavourable way. Typically $G_{k,inf}$ represents the 5% fractile and $G_{k,sup}$ the 95% fractile.

Table 6. Comparison of loads between European standards against the USA ANSI 10.8 standard for access scaffolds

Provision	European standards: ID	European standards: Value
Permanent loads (Q_1)	Self weight of all components including platforms, protective structures, etc.	EN 1991-1-1 + NA
Service loads 6 classes (Q_2)	Uniformly distributed load, Q_2	1 – 0.75 kN/m² 2 – 1.50 kN/m² 3 – 2.00 kN/m² 4 – 3.00 kN/m² 5 – 4.50 kN/m² 6– 6.00 kN/m²
	Concentrated load on area 500 mm × 500 mm	1-3: 1.50 kN 4-6: 3.00 kN
	Concentrated load on area 200 mm × 200 mm	1.00 kN
	Partial area load	1-3: not considered 4: 5.00 kN/m² over an area 0.4* area of bay 5: 7.50 kN/m² over an area 0.4* area of bay 6: 10.00 kN/m² over an area 0.5* area of bay
Variable loads	Wind actions (Q_3)	EN 1991-1-4 + NA Service wind: 0.2 kN/m² (Q_{3a}) Maximum wind: ≥ 0.7×velocity pressure of 50 year return (Q_{3b})
	Snow	Only considered if in national regulations
Loads on side protection	Downward loading	Guardrails or any other side protection must resist point load of 1.25 kN
	Horizontal loading	Guardrails must resist point load of 0.30 kN Toeboards: 0.15 kN
	Upward loading	All fixing except toeboards must resist a point load of 0.30 kN
Dynamic loading	Loads caused by impact	Vertical effect increase load by 20%, horizontally increase load by 10%

Provision	ANSI 10.8: ID	ANSI 10.8: Value
Permanent loads	Self weight of all components including platforms, protective structures, etc.	Not defined
Service loads	Light duty	1.197 kN/m² (= 25 pounds/ square foot)
	Medium duty	2.394 kN/m²(= 25 pounds/ square foot)
	Heavy duty	3.591 kN/m² (= 25 pounds/ square foot)
	No other condition defined	
Variable loads	Wind actions (W)	Nothing defined, except statement that workmen must not work in windy conditions and components must be able to withstand windy conditions
	Snow (S)	Nothing defined, except statement that workmen must not work on scaffolds with snow or ice unless clearing those conditions
Loads on side protection	Downward loading	Guardrails must be able to withstand a downward force of 0.890 kN (200 pounds)
	Horizontal loading	Guardrails must be able to withstand an outward force of 0.890 kN (200 pounds). Toeboards an outward force of 0.225 kN (50 pounds)

Table 7. Load requirements for bamboo scaffolds (Hong Kong code)

Minimum imposed loads			
Duty	**Use of platform**	**Distributed load**	**Concentrated load applied over a square 300 × 300 mm**
Inspection and very light duty	Inspection, painting, access	0.75 kN/m²	2 kN
Light duty	Plastering, glazing, pointing	1.50 kN/m²	2 kN
General purpose	General building work such as brickwork, window fixing	2.00 kN/m²	2 kN
Heavy duty	Blockwork, heavy cladding	2.50 kN/m²	2 kN
Masonry or special duty	Masonry work, concrete blockwork, very heavy cladding	3.00 kN/m²	2 kN

Table 8. Nominal weight densities (kN/m³) of construction materials according to BS EN 1991-1-1 (BSI, 2002a) + UK NA (BSI, 2002b)

Material	kN/m³
Normal weight concrete (see BS EN 206 (BSI, 2013a) for concrete strength classes)	24.0[1), 2)]
Timber strength class C30 (see BS EN 338 (BSI, 2009a) for timber strength classes)	4.6
Softwood plywood	5.0
Birch plywood	7.0
Aluminium	27.0
Steel	77.0
Fresh water	10.0
1) Increase by 1 kN/m³ for normal percentage of reinforcing and pre-stressing steel 2) Increase by 1 kN/m³ for fresh unhardened concrete	

3.5 VARIABLE ACTIONS

3.5.1 Basis

Variable actions are different from permanent actions as they exhibit a marked variability of characteristics with time. Therefore, the models for variable actions have an intrinsic uncertainty, which translates in the structural analysis Chapter 4 in the use of probabilistic models and statistical analyses to simulate the observed values and predict their future evolution (see Chapters 5 and 6).

The involvedness of complete probabilistic structural design makes its application unreasonable to ordinary structures. This led to the need for simplification of procedures and to the development of the sets of rules that constitute the backbone of modern design codes, in what is known as the semi-probabilistic format or the partial factors methods of structural design, see Chapters 5 and 6 for details. This framework has been calibrated based mainly on historical methods and previous experience but also on evolved probabilistic methods (Gulvanessian, Calgaro, & Holický, 2012).

The representative values of actions (termed characteristic values), or of their associated effects, specified in modern design codes, are generally determined based on statistical analyses of historical records and correspond to a prescribed probability of not being exceeded during a chosen base reference period. The latter is chosen taking into account the life time of the structure, the characteristics of the action and the amount of available data. A useful concept is the return period, R; defined as the inverse of the probability of exceedance of a given value, x, of the action within a determined unit observation time, τ, and based on an assumed probabilistic cumulative distribution function of the action values for the period considered, $F_{X[\tau]}$:

$$R = \frac{\tau}{1 - F_{X[\tau]}(x)} = \frac{t_0/n}{1-\left(1-\mathrm{P}_{[t_0]}\right)^{1/n}} \simeq \frac{t_0}{\ln\left(\frac{1}{1-\mathrm{P}_{[t_0]}}\right)} \tag{1}$$

where:

τ represents the unit observation time for which the action maximum values are determined and can be considered to be statistically independent of other maximum action values;

$\mathrm{P}_{[t0]}$ is the probability of the value x not being exceeded during a base reference period t_0;
n is equal to t_0 / τ. When $n = 1.0$:

$$R = \frac{1}{\mathrm{P}_{[t_0]}} \tag{2}$$

In order to obtain a reliable estimate of the action value associated with a low probability of exceedance, it is, in general necessary to have an observational data set several times larger than the base reference period t_0. For example, for wind action it is common to assume a base reference period of one year. Therefore, for a probability P = 0.02 and a base reference period of one year, the return period for the wind action is equal to 50 years.

To account for the influence of uncertainties on the value of the action itself (permanent and variable), and in modelling the effects of actions, modern design codes specify values of partial factors, γ, that must be multiplied by the characteristic values to obtain the corresponding design values of the actions.

3.5.2 Construction Actions

3.5.2.1 General

The sequence of loading of temporary structures can have a major effect on the stresses in individual members of the structure. Important aspects that need proper consideration during planning and design phases include: type of equipment to be used, weight and volume of storage materials, the evolution of the structural system during construction, in particular the interaction between the temporary and the permanent structures, and the method and sequence of loading, including the effect of concrete pours, sequence of post-tensioning, order of removal of elements during disassembling.

A construction load can occur due to execution activities, but ceases to be present once the execution activities are completed. For consistency with this definition, it has been considered that construction loads are classified as variable actions. A construction load may have vertical as well as horizontal components and static as well as dynamic effects. In general, the types of construction loads are characterised by a large diversity, see Table 3.1.

The designer of the temporary structure has to identify the construction loads for the design of an individual project. However, in some types of temporary structures, such as BCE, some heavy loads will only be known after the main contractor is selected, since the latter entity is often responsible for the definition of the construction method and choice of the equipment to be used for each individual project. After the identification of the construction loads, these may be represented either as multiple single variable actions or, where appropriate, grouped and applied as a single variable action.

Construction actions consist of the self-weight of the structural materials of the supported construction (reinforced concrete or structural steel, for example) and if applicable the self-weight of the formwork. The former two become permanent actions once the construction phase is completed, whilst the latter is in general removed at the end of construction. In addition to these, construction loads include local overloads and dynamic effects, the weight of the workers, tools, equipment and stacked materials. The self-weight of formwork and of stacked materials can be obtained from the suppliers. In general, the values of the construction loads associated with the self-weight of the supported construction are considerably larger than the values of the other construction loads.

3.5.2.2 Concrete Casting Actions

For the design of the falsework system, the most critical stage of construction is usually during the pouring of concrete. In the survey presented in Chapter 7 and also in the data reported by Hadipriono & Wang (1987), it was found that over 50% of the falsework collapses occurred during concrete pouring operations. See also André, Beale, & Baptista (2012) and Chapter 7.

For single span concrete bridges, bridge decks when cast in situ can be concreted in a single operation, starting from one end or from the middle of the span. For continuous span concrete bridges, alternative casting methods can be used involving construction joints at one fifth of the span length, see for example (fib, 2000).

Concrete is usually placed either by skips (or buckets) or by pumps. In the former method, a skip filled with concrete is elevated to the required location by a crane, whereas in the latter a special pump is installed on site having a delivery boom through which the concrete flows. The latter method is nowadays the most used procedure for placing concrete. The main relative advantages of using this method are the higher flexibility of the equipment to place concrete in different locations, making it easier to keep a continuous rate of concrete pour with little or no changeover times, but also the less hazardous equipment and construction methods. Regarding skips, although there is no need for special setup and equipment (other than a crane and a skip bucket), the main relative disadvantages are the increased concreting operation times and the risk of impact of heavy loads onto the temporary structures due to pour of concrete from a high bucket level or impact of the crane against the temporary structures.

The self-weight of concrete (except lightweight concrete) is usually assumed at 24 kN/m^3. Most modern design codes assume an additional 1 kN/m^3 to account for the weight of embedded steel, rather than requiring an explicit calculation. The self-weight of the fresh concrete and of the reinforcing steel can be considered equal to 26 kN/m^3, see BSI (2002a).

Figure 1 illustrates the possible local heaping of the concrete during concrete placing (*Left*) and the unfactored load values to account for this variable action suggested in BS EN 12812 (BSI, 2011b) (*Right*). The loads specified in BS EN 12812 only allow for concrete to be dropped by no more than 1 m height and also a heap height not greater than three times the depth of the slab (subject to a maximum imposed load of 1.75 kN/m^2), applied to a maximum area equal to 1 m^2 as shown in Figure 1 (*Left*), (The Concrete Society, 2012). AASHTO bridge temporary structures code (AASHTO, 2008) requires adding to the design load at least 30% of the weight of the material being placed to account for the impact during concrete placement operations, whereas ACI 347 (ACI, 2014) specifies 25%.

The dynamic effects of concrete placing are complex. Ikäheimonen (1997) suggested an approximate method, which is also used in Peurifoy & Oberlender (2010). The maximum dynamic effects can be approximately expressed by a load, *P*, calculated by:

$$P = Q \cdot \sqrt{2 \cdot g \cdot h} \tag{3}$$

where Q is the rate of flow of concrete in kg/s, h is the drop height of concrete in m, and g is the gravitational constant in m/s^2. It can be seen that to reduce the value of the dynamic load it is more efficient to decrease the rate of flow of concrete than reducing the value of drop height of concrete.

By measuring internal forces at falsework columns in different construction sites during concreting, Ikäheimonen (1997) found that when concrete was placed using a pump, the maximum axial force values at every column of the falsework systems monitored occurred at the end of concrete casting operations. In addition, minor to none dynamic effects due to concrete casting were observed. This behaviour was justified owing to the lower rate of flow of concrete (up to 100-150 kg/s) than the ones of using skips (up to 240-480 kg/s). In general, the dynamic load generated by placing concrete using a pump is not greater than 0.5 kN.

However, if concrete is placed using skips, peaks in the column axial force values were observed in a first phase due to the dynamic effects of emptying the concrete from the skip at a given height and later by the accumulation of concrete in a localised area (heaping of concrete) before it could be spread by the workers over the formwork away from the drop area. The magnitude of these short term peak loads could in some cases amount to approximately 30% of the static load value if no dynamic effects were accounted for, but could be much higher depending of the drop height of the concrete and the rate of which the skip was emptied. Based on the values also published in Ikäheimonen (1997) for skips, the dynamic load due to placing concrete using a skip may be typically considered to be in the range of 0.8 to 1.6 kN for rates of flow of concrete between 240 and 480 kg/s and a drop height less than or equal to 0.6 m.

Figure 1. Local heaping of the concrete, adapted from The Concrete Society (2012)

(a) Heaping of concrete (b) In situ concrete construction loads

3.5.2.3 Other Construction Actions

ACI 347 (ACI, 2014) specifies a design value for the construction variable load equal to 2.4 kN/m² for all construction stages of concrete construction. This value should be increased to 2.4 kN/m² if motorised carts are used to level the concrete surface. In the Australian Standard AS 3610 (SAA, 1995), for concrete formwork, the construction variable load varies for different construction stages, being equal to 1.0 kN/m² for the phase of concrete placement.

A minimum vertical load is included in most modern design codes for concrete construction. In the USA, typically, the minimum design value for vertical load is 4.8 kN/m² distributed uniformly over the formwork (ACI, 2014), increased to 6.0 kN/m² if motorised carts are used.

Concerning horizontal loads, BS 5975 (BSI, 2011a) specifies that as a minimum falsework structures should be designed considering the most unfavourable of the two cases:

- A notional horizontal load equivalent to 2.5% of the applied vertical loads at the time under analysis. This minimum load is considered to act horizontally at the top of the falsework.
- Horizontal loads that can arise from wind action, erection tolerances (equivalent horizontal loads taken as 1% of the applied vertical load, provided that the limits of the initial geometrical imperfections are satisfied), concrete pressure loads, dynamic and impact loads, and the loads generated by the permanent works (deformations, post-tensioning, etc.).

Lateral loads can also result from positioning the temporary structures out of vertical, i.e. column elements positioned normal to an inclined formwork soffit. In order to reduce the lateral internal forces generated by the concrete pressures in these cases, the temporary structures is usually installed in the vertical position and the formwork inclination is accommodated using rocking forkheads and wedge elements to secure a concentric transmission of loads to the temporary structures, see report by The Concrete Society (2012) for detailing examples.

BS EN 12812 does not define a minimum notional horizontal load, although it introduces a variable action "Q_3" defined as a horizontal load equal to 1% of the vertical loads. As presented above, BS 5975 also includes allowance for a similar action case, further stating that the horizontal load equal to 1% of the applied vertical loads should be applied in the most unfavourable direction at the point of application of the vertical loads. Again, the specified values are only valid provided the maximum permissible erection tolerances are not exceeded.

BS EN 1991-1-6, concerning actions during execution, does specify minimum horizontal loads for design of falsework and recommends a value equal to 3% of the vertical loads from the most unfavourable combination of actions. For BCE, the design value of the total horizontal friction forces in the bridge longitudinal direction generated during launching should be at least 10% of the vertical loads from the most unfavourable combination of actions.

In the USA, a common requirement is to specify a minimum (design or unfactored) value of the horizontal construction load (e.g. excluding wind action effects) equal to 2% of the total vertical load to be applied at the point under consideration, see ASCE/SEI 37 (ASCE, 2014) for example. The latter code highlights that this minimum load does not need to be applied concurrently with loads not associated with the construction process such as wind or seismic loads, but also should not be considered as a substitute of the latter actions. ACI 347 supplements the requirement by specifying that the minimum design value be the greater of 1.46 kN/m applied along the top formwork edge and 2% of the total vertical load.

For scaffolds, BS EN 12811-1 (BSI, 2003) specifies six load classes with different values for the variable loads and seven width classes of working areas (equal to the sum of the area of the working platform units at each level of the scaffold). The variable loads consist in uniformly distributed loads to be applied in limited parts of the working area or in the entire working area, plus concentrated loads. These loads should be applied separately and not cumulatively. A minimum notional lateral load is also specified, which should not be combined with wind action effects, consisting of 2.5% of the vertical uniformly distributed loads which are applied over the entire working area, or 0.3 kN, whichever is the greater.

For grandstands and stages, minimum notional loads are indicated in Section 3.5.6.

For BCEs, horizontal construction loads may derive additionally from friction over bearing supports, horizontal component of loads due to longitudinal, lateral and plan gradients of the bridge geometry. Guidance is provided in Chapter 6.

For all temporary structures, the effects of the deformed geometry (including the initial imperfections) in the structural behaviour should be taken into account either by simplified design methods based on first-order structural analysis or through results obtained from second-order structural analysis, see Chapter 4.

There are not many studies regarding the adequacy of variable loads specified in design codes simulating the action of concrete pouring during construction of structures. Limited research has been focused mainly on building construction, with less emphasis on bridges and other structures.

In order to analyse the adequacy of variable loads specified in design codes during the construction phase of cast-in place concrete structures using falsework systems, Ikäheimonen (1997) monitored the load on columns at several bridge and residential building sites during concreting. By analysing all the available data registered relative to the maximum values of falsework columns axial loads during concreting he concluded that applying a partial factor of 1.5 to the construction material self-weights (assuming a value for the density of reinforced concrete equal to 24 kN/m^3, and the self-weight of formwork to be equal to 0.4 kN/m^2) is enough to get a safe estimate of the design axial force in each falsework column element. As an alternative method, it could also be possible to use a partial factor equal to 1.3 plus a design load equal to 2.5 kN/m^2.

However, Birch, Booth, & Walker (1971) found evidence of construction load values much higher than the ones presented above. Therefore, it is recommended that the design value of the construction loads specified in design codes, as well as the associated loaded areas, are properly documented and communicated to the interested parties and are adjusted if necessary in a project by project basis so that the necessary safety margins can be ensured throughout the operation. For example, for cast in situ concrete box girder bridges using the balanced cantilever method, Hewson (2003) suggests considering a load equal to 10-20 kN to take into account construction equipment (additional to the form-traveller weight), but for cable stayed bridges using the same method of construction a higher load 30-40 kN is indicated.

Additionally, lateral internal forces could derive from non-balanced, i.e. not auto-equilibrated, concrete pressures applied to the formwork or from discontinuities in the formwork panels (BSI, 2011a; The Concrete Society, 2003, 2012). The lateral internal forces may be equilibrated by structural ties connecting the formwork panels, by structural ties connecting the formwork panels to self-supporting structures such as the permanent structure, or by the falsework system itself in which case must be adequately designed to resist these lateral internal forces. The design of formwork is outside the scope of this book. The interested reader is directed to relevant bibliographic references (ACI, 2014; Johnston, 2014; Peurifoy & Oberlender, 2010; SAA, 1995; The Concrete Society, 2012). Concerning pressures from new concrete types, such as self-compacting concrete, little research has been made. Graubner & Proske (2005, 2010) and T. Proske & Graubner (2007, 2008) are one of the exceptions.

For all types of free variable actions, they should be arranged in the most adverse position in order to get the most unfavourable effects for each individual element during safety verification.

The design of the temporary structures supporting a concrete structure during its construction has to account for secondary effects, including complex creep, shrinkage and temperature movements associated with the segmental and/or time-dependent nature of the construction. From these actions, temperature action (uniform or differential) is usually the most important, in particular for massive concrete or steel structures. The internal forces in the temporary structures induced by temperature actions may be reduced by defining an efficient post-tension cables layout so that the internal forces due to these two actions counteract.

Shrinkage actions can be minimised by using smart concrete casting procedures such as casting first the mid-span and after over the supports for continuous structures.

In all cases, during the construction phase, it is important to carefully monitor the deflections of the temporary structures not only to ensure that the desired shape of the permanent structure is achieved but also to get insight about the behaviour of the temporary structures during the construction process.

An important action that temporary work structures used in post-tension concrete construction is the action effects due to stressing of the post-tension cables. Failure to properly consider and control the post-tensioning and temperature actions may trigger the collapse of the temporary structures, as reported in Huizing, Blakeley, & Ramsay (1977).

For example, Billings & Routley (1978) measured internal forces in falsework standards during the construction of a post-tensioned concrete bridge in New Zealand. The most important observations found by them were:

1. A considerable variation in the axial forces was found in the standards during concreting, of the order of ±25% of the concrete permanent load. These variations were justified by a number of factors: errors in positioning the standards; out of plumb or straightness of some of the standards; stiffness and redundancy of the falsework;
2. The design of falsework supporting major concrete bridges during construction must include allowance for loads caused by post-tensioning and temperature gradients – where large pours are expected the loads induced by the heat of hydration of the concrete may require special investigation;
3. For ground-supported falsework, the base and top jacks should be firmed up as early as possible.

Another study also reports data from monitoring falsework elements during concreting (Quinion, 1984). It was observed that as concrete construction progressed the internal forces in the falsework increased and approached the design values. However, as the concrete matured, particularly after a span was completed, the internal forces dropped in the falsework as some 20% was transferred to the piers. During post-tensioning of the third and last span its extremity rose by 76 mm and the adjacent span, already built, moved 14 mm downwards. Consequently, the internal forces on the measured members of the second span rose by 25% from their previous values. Subsequently, during removal of the falsework the internal forces increased by a further 25% in some of the members which were the last to be fully relieved of loads.

Therefore, it is recommended that for in situ concrete construction, both the structure and falsework (or any type of temporary structures used in post-tensioned concrete construction) need to be integrated in the analysis model to determine how loads are distributed during the application of the post-tension action. When a deck is post-tensioned it normally "lifts up" along the span, reducing any load on the falsework supporting the concrete. However, in continuous multi-span systems, it may be found that if a post-tension load is fully applied without any adjustment of the falsework, internal forces in some elements of the falsework could be significantly increased (as exemplified above). Another example is when the deck is cantilevering over a pier. When the deck over the span is post-tensioned, the end of the cantilever is often deflected downwards and increases the load on the falsework beneath.

In these cases, either the temporary structure is designed to take the increased load (15% to 50%, or more, of the load value due to the casting of concrete), or other measures are implemented. For example, the overloading may be minimised by suitably arranging the layout of the post-tension cables and applying the stressing in stages. In the first stage, only the minimum number of cables required to support the deck permanent load are stressed without causing any significant deflections. This will allow a safe removal the falsework, after which the rest of the post-tensioning load can be applied. Another solution is to incrementally loosen up the temporary structures during the phased application of post-tensioning load, see Metheringham & Townshend (2005) for an application example. These techniques may also be applied for BCEs in the case of in situ concrete construction.

Concerning other construction loads, when using full span method (FSM) gantries, launching gantries for segmental bridge construction or lifting equipment in balanced segmental bridge construction, one must design these BCE for the actions generated during launching and for the actions induced by the lifting operations. Guidance for the determination of these actions including dynamic effects due to operation, braking and acceleration loads is given in BSI (2006a, 2009e). According to AASHTO documents (AASHTO, 2003, 2016) for very gradual and controlled lifting of precast segments, the dynamic effect may be expressed by a dynamic factor taken as 10% of the lifted load. Therefore, the characteristic value of the equivalent static load should be determined by multiplying the characteristic value of the static load with the dynamic factor value. Rosignoli (2013) indicates a higher value for the dynamic factor equal to 20%, the same value recommended in Hewson (2003).

3.5.3 Wind

3.5.3.1 Context

Wind action is one of the most relevant environmental loading for structures. Wind action is always present in any given instant of time with a certain direction and intensity which are both complex to characterise and to predict and subjected to a large uncertainty.

On almost every day of the year an extreme wind event is happening somewhere on the Earth. According to the Centre for Research on the Epidemiology of Disasters (CRED, www.emdat.be/), Munich Re (2015), D. Proske (2008) and UNISDR (2012, 2015), in the period between 1980 and 2014, more than 20 000 natural disasters occurred worldwide, mainly meteorological and hydrological events. More than 4 billion persons have been affected, the cost of damage representing more than US$4 trillion – and increasing each year due among other causes to climate change, from US$250 billion in 2014 to a projected value of US$400 billion in 2030. Extreme wind events are the largest cause of damage costs. The world region most hit by natural disasters is Asia, followed by North America and Europe.

Climate change and natural catastrophes are ranked at the top of the 2015 Global Risks database prepared by the World Economic Forum, and for which less progress has been made in the recent years (WEF, 2015). One needs to remember that most disasters that could happen have not happened yet. Climate change is a reality, not a one-off event which can be addressed on a case-by-case basis (Field et al., 2012). Irrespective of the success of our mitigation efforts, the impact of climate change will increase in the coming decades because of the delayed impacts of past and current greenhouse gas emissions (European Commission, 2013a).

Extreme weather and climate change leave structural infrastructure systems exposed to more extreme and recurrent conditions. Since the available amount of resources is finite, it is highly likely that design thresholds which are built into infrastructure project designs may be breached more frequently in a changing future climate. This may result in threshold failures once considered exceptional but acceptable, becoming unexceptional (i.e. normal) and unacceptable (European Commission, 2013b). In this expected scenario, climate change will greatly increase expected future losses.

The consequences of climate change are increasingly being felt worldwide. For Europe, it is expected that by 2020 climate change effects manifests as a positive variation of average air temperatures of 10%, a decrease in precipitation levels in the Mediterranean countries and a slight increase in the average wind velocities (Rademaekers, Laan, Boeve, & Lise, 2011). Increasing frequency of coastal storms and flooding are also anticipated effects of climate change.

The objective of this Section is to introduce the reader to wind engineering in the context of temporary structures application. It does not replace reference textbooks about wind engineering, to which the interested reader is directed to Holmes (2015) and Simiu & Scanlan (1996) for example.

3.5.3.2 Origin of Wind

Wind is primarily caused by the effect of the sun on earth. Solar radiation is strongest at the equator and this produces temperature differences between the poles and the equator, which in turn produces atmospheric pressure differentials that rise over the surface of the earth. It is intuitively perceived that the rate of change of pressure (over some unit distance) will be proportional to the acceleration applied on a mass of air (Davenport, 1960).

The acceleration produced by these pressure differentials is accompanied by other components of acceleration known as the geostrophic acceleration due to the Coriolis effect and centripetal acceleration.

These accelerations produce large-scale circulation systems of free air in the atmosphere, unaffected by friction near the surface. As a direct result of these circulations, the prevailing wind direction near the equator and the poles tend to be easterly, whereas westerly winds dominate in latitudes in-between. The velocity of the free air is known as the gradient wind velocity, attained only at heights around 300 m to 600 m above the ground depending of terrain roughness.

Closer to the ground, the wind flow is slowed down by frictional forces, which are transmitted through shear between layers of air that form the atmospheric boundary layer (the lowest part of earth's atmosphere), by the obstructions at the surface and by the virtual stresses produced by the vertical exchange of momentum by turbulence (Davenport, 1960). As a result, the wind flow direction is directed towards low-pressure regions. The height of the atmospheric boundary layer normally ranges from a few hundred meters to several kilometres, depending on the wind flow intensity, terrain roughness and latitude (Simiu & Scanlan, 1996).

Turbulence also causes rapid fluctuations in the wind velocity (Figure 2), which is thus usefully expressed in terms of its mean velocity and the deviations from this velocity. Therefore, the wind velocity, V, can be decomposed into a mean wind velocity, V_m, with direction u, and into fluctuating terms, V', along three orthogonal directions u, v, w. The time or distance interval over which the mean wind velocity is averaged depends on the purpose for which the wind velocity is to be used, as a compromise between minimising sampling errors and reducing the effects of non-stationary processes (Cao, 2013).

In cases where the mean wind velocity cannot be assumed as stationary, see Figure 2, sophisticated methods of data processing need to be applied to remove the turbulent part of the signal, such as the Empirical Mode Decomposition (EMD) (Eh Huang & Shen, 2005), or other methods discussed in Kareem & McCullough (2013)

3.5.3.3 Wind Actions

Fundamentals

One of the most important parts of wind engineering is wind action forecasting. There are various methods available for this task, the two main approaches being: numerical weather prediction (NWP) models that use mathematical models of the atmosphere, and statistical predictive models. The choice of using one of the methods is usually made considering the time scale of the desired forecast (X. Wang, Guo, & Huang, 2011):

1. immediate short-term forecasting (up to six hours of look-ahead times);
2. short-term forecasting (day-ahead);
3. long-term forecasting (multiple-days-ahead).

Figure 2. Typical wind velocity variation with time, registered from an anemometer located at Lisbon, Portugal

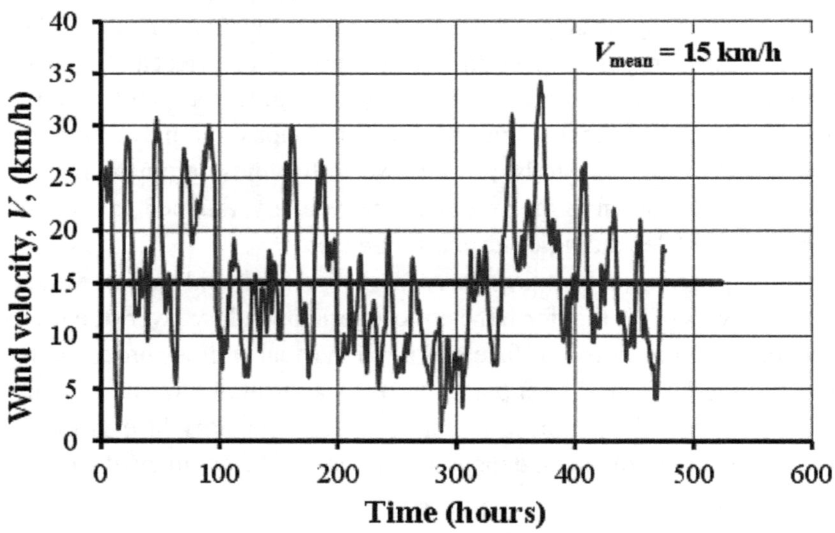

For look-ahead times of more than six hours, NWP models are more accurate (Dowell, Weiss, Hill, & Infield, 2014; Giebel, Brownsword, Kariniotakis, Denhard, & Draxl, 2011). In contrast, for immediate short-term forecasts, the statistical approach is preferred over NWPs because whilst the capabilities of these latter models have improved dramatically over the recent years, some meteorological phenomena are still not resolved, or are poorly forecasted or the time needed to achieve the required accuracy is still prohibitively high.

The most accurate NWP model is the Weather Research and Forecasting (WRF) Model (Done, Davis, & Weisman, 2004), while there are various predictive statistical models that have been successfully used to forecast wind action, see Costa et al. (2008) and Foley, Leahy, Marvuglia, & McKeogh (2012) for examples.

Wind action forecasting depends on the types of wind hazard events, which could be of different nature: from synoptic-scale wind gales produced by large extra-tropical depressions to tropical cyclones (termed hurricanes or typhoons), downbursts and microbursts, tornadoes, gravity winds (katabatic winds), and others, see Cao (2013) and Holmes (2015) for details. The first two types have historically been the focus of research, whereas the latter types, although equally important, only in recent years have reached the research spotlight, see Edwards et al. (2013), Jagger & Elsner (2006), Solari (2016) and L. Wang, McCullough, & Kareem (2013) for example, and are much less well understood. As a result no provisions exist in modern design codes.

The establishment of appropriate design wind velocities for each design situation is a critical first step towards the calculation of design wind effects for structures. The basis for this is the use of probabilistic methods. Classically, for structural collapse situations, the theory of extreme value analysis, e.g. Generalised Extreme Value (GEV) distribution, has been applied. In cases where it is of interest to simulate the complete population of (synoptic-scale) wind velocities, e.g. in the case of fatigue analysis, the Weibull distribution may be used, in particular for the hourly mean wind velocities (Ang & Tang, 2007; Benjamin & Cornell, 1970). All of these methods require the statistical analysis of historical data on recorded wind velocities for different storm types (Castillo, Hadi, Balakrishnan, & Sarabia, 2004; Coles, 2001; Kasperski, 2013; Leadbetter, Bailey, & Rootzén, 1983; Palutikof, Brabson, Lister, & Adcock, 1999; Von Storch & Zwiers, 1999).

Currently, the use of the classical GEV distribution to estimate safe design wind velocities is being replaced by more advanced methods such as the Peaks Over Threshold approach (POT) and more recently by using Markov Chain Monte Carlo methods (MCMC), see Fawcett (2005) and Fawcett & Walshaw (2006, 2008, 2012). A discussion is presented in Chapter 5 where the drawbacks of the GEV against the more recent methods are detailed.

One of the most important factors to be considered in synoptic-scale wind engineering is the increase of the mean wind velocity with height within the boundary layer. Various empirical, semi-empirical and theoretical formulas have been derived to represent the variation of synoptic-scale wind velocity with height. Two of the more familiar forms are the logarithmic law and power law profiles. The former is specified in the modern design codes and is expressed by (BSI, 2010a; ISO, 2009):

$$V(z) = k \cdot \ln\left(\frac{z}{z_0}\right) \cdot V_{ref} \tag{4}$$

where $V(z)$ is the velocity at height z above ground, k is a function parameter, z_0 is the terrain roughness length (see Table 9) and V_{ref} is a reference wind velocity.

Table 9. Terrain roughness length, z_0, for various types of terrain (BSI, 2010a)

Terrain type	Terrain roughness length, m
Sea or coastal area exposed to the open sea	0.003
Lakes or flat and horizontal area with negligible vegetation and without obstacles	0.01
Area with low vegetation such as grass and isolated obstacles (trees, buildings) with separations of at least 20 obstacle heights	0.05
Area with regular cover of vegetation or buildings or with isolated obstacles with separations of maximum 20 obstacle heights (such as villages, suburban terrain, permanent forest)	0.3
Area in which at least 15% of the surface is covered with buildings and their average height exceeds 15 m	1.0

The power law profile is still extensively used because of its simplicity. It can be stated as (Davenport, 1960):

$$V(z) = \left(\frac{z}{z_{\text{ref}}}\right)^{\beta} \cdot V_{\text{ref}} \tag{5}$$

where z_{ref} is a reference height above ground, and β is a function parameter. The value of β is mostly influenced by terrain roughness; β increases when the roughness increases. Other variables that influence, in general less significantly the value of β are the air temperature and the mean wind velocity (Davenport, 1960).

Both laws are applicable only in the atmospheric boundary layer, at the top of which the synoptic wind velocity attains its maximum value, referred to as the gradient velocity. Above this height, the wind velocity may be regarded as constant, flowing along isobars. A more accurate wind velocity profile model is the Deaves & Harris model proposed in a report by CIRIA (1978), later reviewed by Deaves (1981) to extend the model's applicability to heterogeneous terrains. See Drew, Barlow, & Lane (2013) for an application example of the different laws.

It should be emphasized that the term "terrain roughness" refers neither to the shielding due to individual obstacles nor to the orographic effects influencing the airflow in mountain regions, but to the cumulative statistical drag effect of many obstructions on the wind. The terrain roughness, therefore, is characterised by the density, size, and height of the buildings, trees, vegetation, rocks, etc., on the ground, around, and over which the wind must flow; it will be a minimum over the ocean and a maximum in a large city.

Downbursts are transient events produced by thunderstorms, like tornadoes, capable of generating severe winds but for relatively short periods. (Fujita, 1985) defined a downburst as a column of rapidly descending air that when reaches the ground violently bursts out and instantaneously changes direction, producing very high wind velocities near the ground surface. Contrary to synoptic-scale winds, the logarithmic and exponential laws do not apply for downbursts. The wind velocity does not increase with height, instead the maximum wind velocities are likely to happen in the immediate vicinity of the ground and then decrease with height, see Figure 3.

Figure 3. Profile of horizontal wind velocity for synoptic-scale winds and downbursts

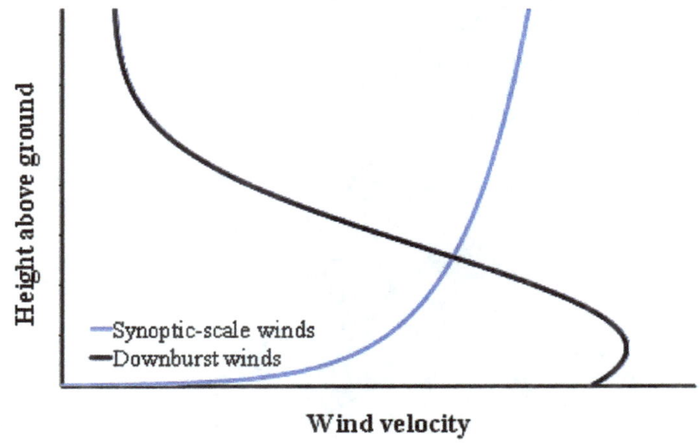

Besides the wind velocity, it is also important to consider the wind direction as it was found that the wind direction relative to a fixed exposed surface with a given rigid shape has a significant influence on the structural effects of the wind action. Several methods have been developed to take proper care of this variable, the most important of these are discussed in Holmes (2015). Note that the wind direction of the mean wind velocity may change with height, although not as noticeably as for wind turbulence.

Turbulence of the wind is expressed by the variation of the wind velocity, for each orthogonal component of the wind velocity: longitudinal wind component (parallel to the mean wind direction) and the two lateral wind components parallel to the unit vectors of the plane normal to the longitudinal wind velocity, see Figure 4. It is usual to express wind turbulence as turbulence intensities, I_t:

$$I_t = \frac{\sigma_V}{V_m} \tag{6}$$

where σ_V is the standard deviation of the wind velocity, and V_m is the mean wind velocity, determined for a given averaging time.

For gales produced by large-scale depression systems, the value of the turbulence intensity can be approximately determined by:

$$I_t \approx \frac{n_d}{\ln\left(z/z_0\right)} \tag{7}$$

where n_d is a constant that depends on the components of the wind velocity. For the longitudinal wind component, I_u, $n_d \approx 1.00$, for the other two components (see Figure 4), I_v and I_w, $n_d \approx 0.88$ and $n_d \approx 0.55$, respectively (Holmes, 2015).

As expected, the turbulence intensity of synoptic winds decreases with height above the ground. For other types of wind events, guidance can be found in Holmes (2015) and ISO (2009).

Figure 4. Wind velocity components

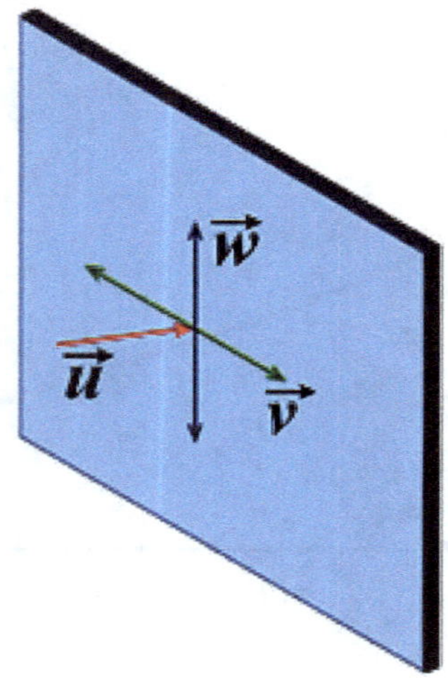

In modern design codes, the evaluation procedures of the effects of the wind action are simplifications of the highly nonlinear and stochastic action-structure interaction. The design wind velocity to be considered is usually derived from an extreme wind velocity value. The latter can be defined by the mean wind velocity, V_{mean}, times a gust factor, G, that for the longitudinal direction of the wind velocity, u, is given by:

$$G_u = 1 + g_u \cdot I_u \tag{8}$$

where g_u is a peak factor that is defined as the ratio of the maximum value of the fluctuating part of the longitudinal wind velocity to the standard deviation of the longitudinal wind velocity fluctuation, σ_u. The values of g depend primarily on the time over which the maximum wind velocity is averaged and on the sample time, being weakly dependent on the height. Values of g for synoptic winds are given in Holmes (2015) and ISO (2009). For other wind regimes, guidance can also be found in Holmes (2015).

A final parameter needed to characterise the wind action, is the impact of the orography of the site. Mean and gust wind velocities can be considerably increased by orography in the form of cliffs, escarpments, hills and ridges. These effects have been already extensively studied and the results have been consolidated in code rules.

As the wind approaches a slope, its velocity near the ground may reduce slightly, but as it moves towards the top of the slope, the wind velocity gradually increases until it reaches a maximum near the top of the slope. The speed-up effects are greatest near the surface, and reduce with height above the ground (see Figure 5). Three dimensional features of the orography may in general be ignored (Holmes, 2015).

The influence of site orography may be accounted for by an orography factor given by (Holmes, 2015):

$$\text{orography factor} = \frac{\text{Wind velocity at height } z \text{ above the feature}}{\text{Wind velocity at height } z \text{ above the flat ground upwind}} \tag{9}$$

Figure 5. Wind velocity over orographic features

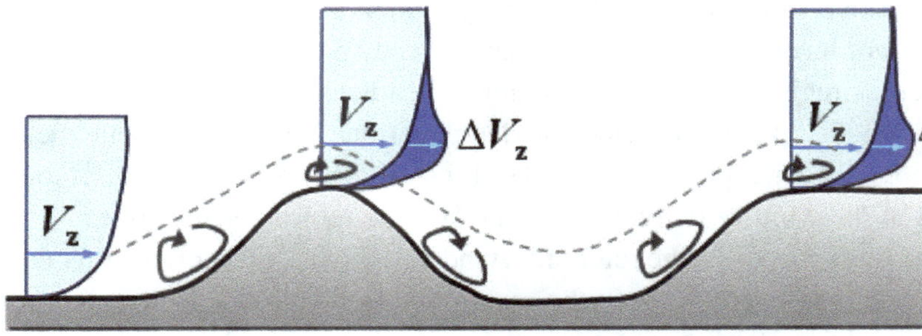

This definition applies to mean, peak gust and standard deviation wind velocities. In general, orography features reduce the standard deviation of the wind turbulence.

The orography factors are best obtained from full-scale or in wind tunnel measurements, or from numerical meso-scale meteorological computer calculations. Several examples of orography factors are given in BSI (2010a) and Holmes (2015).

3.5.3.4 Design Code Provisions

The design wind velocity can be determined from BS EN 1991-1-4 (BSI, 2010a) in Europe, ASCE 7 (ASCE, 2010) in the USA or AS 1170.2 (SAA, 2011) in Australia and New Zealand, for example. For temporary structures, further guidance is given in BS EN 12811 (BSI, 2004b) for scaffolds, in BS EN 12812 (BSI, 2011b) and BS 5975 (BSI, 2011a) for falsework, and in ASCE/SEI 37 (ASCE, 2014) for temporary structures in general. With the exception of the latter, the methods specified in each code are in agreement with the general method specified in BS EN 1991-1-4; therefore, this document is used as a reference throughout this Section. Care should be paid in regions exposed to tropical cyclones, downbursts or tornadoes since most of the rules specified in the mentioned codes do not apply. Guidance can be found in ISO 4354 (ISO, 2009).

It should not be forgotten that the Eurocodes are only valid if used together with the corresponding National Annexes published by every European Union member state which contain the national choices for the Nationally Determined Parameters (NDPs). In the present Section, the UK National Annex to BS EN 1991-1-4 (BSI, 2011c) will be used as an example.

Other code comparisons have been made by Bashor & Kareem (2009), Holmes (2014), Nieto, Hernández, Jurado, & Romera (2010) and Pierre, Kopp, Surry, & Ho (2005).

Wind action is specified in modern design codes as a function of the following features:

1. A specification of a reference wind velocity for various areas, usually expressed by a map with different isopleths, i.e. curves of equal wind velocity values, referenced to a height of 10 m over flat open country terrain (the exposure datum specified by the World Meteorological Organization);
2. Modification factors to take into account the effects of altitude of the site and height of the structure, the roughness of the terrain, the wind direction, the site orography (including the existence of surrounding structures), the averaging time, the value of the exposure period (construction project duration and time of the year), the importance of the temporary structures, and the possible shielding effects.

Caution should be paid when comparing values of reference wind velocity given in different, but corresponding, maps, since values may not be directly comparable as some modification factors, such as those of altitude and/or terrain roughness, may have already been applied. These maps are developed based on wind velocities registered in several national weather stations, nowadays automatic weather stations (AWS), for long observational periods, and application of the probabilistic methods mentioned in the previous Section. For example, in BS EN 1991-1-4, the reference wind velocities correspond to characteristic values of the wind velocity equal to the 50-year return period values.

Typically, for structural purposes, the reference wind velocity is considered either equal to the maximum wind velocity averaged over 3 s, or to the maximum mean wind velocity averaged over 10 min, both referenced to a height of 10 m over flat open country terrain. The latter, is used in the mean wind velocity method that modern European design codes follow. This method, transparently represents the wind action in terms of its mean and fluctuating components, which allows a more accurate description of the wind action (ISO, 2009).

Conversion of maximum mean wind velocities, V, referenced to different averaging times, T, can be done by using an averaging time factor, k_T, (ISO, 2009):

$$V_{\mathrm{T}} = k_{\mathrm{T}} \cdot V_{\mathrm{T}=3600\,\mathrm{s}} = \left(1 + g \cdot I_{\mathrm{t}}\right) \cdot V_{\mathrm{T}=3600\,\mathrm{s}} = \left(1 + g \cdot \frac{\sigma_{V_{\mathrm{T}}}}{V_{\mathrm{m,T}=3600\,\mathrm{s}}}\right) \cdot V_{\mathrm{T}=3600\,\mathrm{s}} \tag{10}$$

For example, the 10 minute mean UK basic wind velocity map was derived from an analysis of the hourly mean wind velocities by applying an averaging time factor equal to 1.06 (BSI, 2008a, 2009c).

Based on some prior information, namely location, altitude and orography of the site, the basic wind velocity is determined by the reference wind velocity. In BS EN 1991-1-4 + UK NA, this translates to:

$$V_{\mathrm{b}} = c_{\mathrm{dir}} \cdot c_{\mathrm{season}} \cdot c_{\mathrm{alt}} \cdot V_{\mathrm{b,0}} \tag{11}$$

where (BSI, 2010a):

$V_{\mathrm{b,0}}$ is the (reference) fundamental basic wind velocity, at 10 m above ground of terrain roughness corresponding to "*area with low vegetation such as grass and isolated obstacles (trees, buildings) with separations of at least 20 obstacle heights*", i.e. terrain category II in the BS EN 1991-1-4 classification;

c_{dir} is a directional factor, with a recommended value equal to 1.0;
c_{season} is a seasonal factor, given in UK NA (BSI, 2011c);
c_{alt} is an altitude factor, given in UK NA (BSI, 2011c), since the $V_{\mathrm{b,0}}$ map reports to the sea level.

The UK NA specifies values of c_{dir} in increments of 30º (interpolation is allowed). When the wind action effects, see the following Section, are simulated using pressure coefficients, the value of c_{dir} to be used should be the maximum value found in a range of ±45º from the considered wind direction, since the latter are only defined in BS EN 1991-1-4 for orthogonal structural axes (for example 0º, 90º, 180º and 270º wind directions).

The seasonal factor can only take values lower than 1.0 for temporary structures and for structures during the construction phase. However, if the structure can be transported and used in different locations, the value of c_{season} must be equal to 1.0. There is an apparent contradiction, since for the same exposures, in nature and in time, different values of c_{season} can be applied depending on whether the structure is permanent or temporary. The UK NA specifies values of c_{season} that can be as low as 0.62 for a one month period corresponding to the month of July, however values lower than 1.0 should only be used if the temporary structures are guaranteed to be used solely during a particular sub-annual period. Due to the uncertain nature of general construction activities, the codes recommend that a value of c_{season} equal to 1.0 be used.

The value of the fundamental basic wind velocity is the 10 minute mean wind velocity having a given probability, p, of annual exceedance. In general, the value of p to be considered is 2%. However, both BS EN 1991-1-4 and BS EN 1991-1-6 (Eurocode part for actions during construction) (BSI, 2005b) and their corresponding UK NAs (BSI, 2008c, 2011c), allow adjusting the considered value of annual exceedance to the exposure period, which may be relevant for temporary structures. Table 3.10 includes the minimum recommended return periods for the determination of the characteristic values of climatic actions given in BS EN 1991-1-6 + UK NA. The actual values to be used should be defined for each individual project.

It can be seen that small values of return periods can be used which may result in unconservative designs in some cases, see Chapter 5 for a full discussion. On the contrary, AS 1170.0 (SAA, 2002) specifies that for common temporary structures, a 100 years return period should be considered for non-cyclonic wind actions and 250 years for cyclonic wind actions. It should be noted that the basic wind velocity in AS 1170.2 (SAA, 2011) is a peak gust wind velocity whereas in BS EN 1991-1-4 it is a mean wind velocity.

In most design codes it is possible to update the basic wind velocity value to account for a reduced exposure period. For instance, using ASCE 7, BS EN 1991-1-4 and AS 1170.2, the basic wind velocity can be derived for different return periods. To this end, the first two documents specify the following formulae:

$$\text{ASCE 7: } V_{b,t} = c_p \cdot V_{b,50} = \left[0.36 + 0.1 \cdot \ln\left(12 \cdot t\right)\right] \cdot V_{b,50} \quad (12)$$

$$\text{BS EN 1991-1-4: } V_{b,t} = c_p \cdot V_{b,50} = \left(\frac{1 - 0.2 \cdot \ln\left(-\ln\left(1 - p\right)\right)}{1 - 0.2 \cdot \ln\left(-\ln\left(0.98\right)\right)}\right)^{0.5} \cdot V_{b,50} \quad (13)$$

Table 10. Minimum recommended return periods for the determination of the characteristic values of climatic actions (BS EN 1991-1-6 + UK NA)

Work duration	Return period (years)	Probability of annual exceedance (%)
≤ 3 days	2	50
≤ 3 months (but > 3 days)	5	20
≤ 1 year (but > 3 months)	10	10
> 1 year	50	2

where $V_{b,t}$ is the basic wind velocity associated with a return period of t years, c_p is a probability factor and p is the probability of annual exceedance.

AS 1170.2 specifies several formulae for different zones and wind regimes. As can be observed in Figure 6, the minimum value of the probability factor to be considered in the calculation of the wind velocity for temporary structures according to the BS EN 1991-1-4 is 0.78, and its values are always larger than the ones obtained using ASCE 7, which returns a minimum value of 0.68, and AS 1170.2 for which a minimum value of 0.73 is obtained for a non-cyclonic wind regime (region A).

In ASCE/SEI 37, the wind velocity applied to temporary structures is a percentage of that applied to a permanent structure, and varies according to the exposure period, see Table 3.11. This Table also shows the return period of the wind velocity considered in ASCE/SEI 37. Comparing Table 3.11 with Table 3.10, it can be observed that the return periods considered in ASCE/SEI 37 are larger than the ones considered in BS EN 1991-1-6 for short exposure periods but smaller for long exposure periods.

The distribution of the mean wind velocity, V_m, with height, z, needs also to be accounted for. This depends on the structure height but also on the site orography and terrain roughness:

Table 11. Reduction factor specified in ASCE/SEI 37 to determine the wind velocity to be used during construction of structures

Construction period	Factor	Return period (years)
less than 6 weeks	0.75	5
6 weeks to 1 year	0.80	7
1 to 2 years	0.85	12
2 to 5 years	0.90	20

Figure 6. Probability factor for wind velocity as function of the return period, according to BS EN 1991-1-4, ASCE 7-10 and AS 1170.2

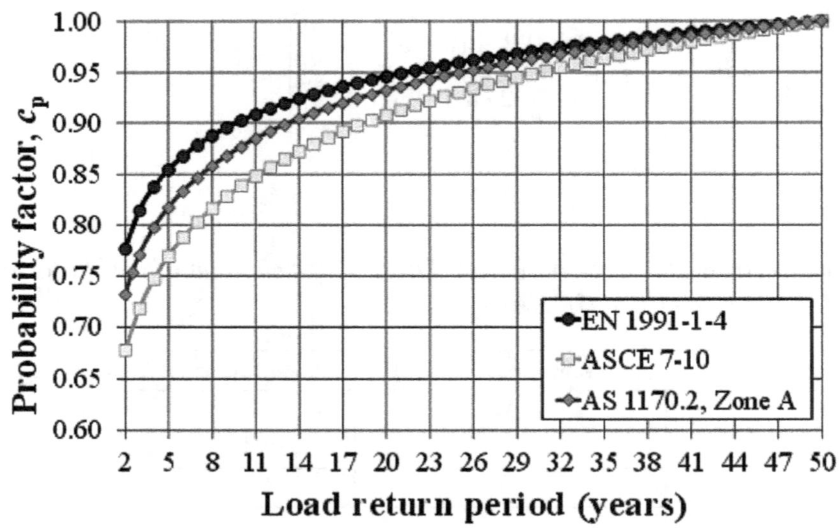

$$V_m(z) = c_r(z) \cdot c_o(z) \cdot V_b \tag{14}$$

where (BSI, 2010a):

c_r is a roughness factor, given in UK NA (BSI, 2011c);
c_o is an orography factor, given in UK NA (BSI, 2011c).

In BS EN 1991-1-4, and AS 1170.2, the simple logarithmic law is used to express the change of mean wind velocity with height, additionally accounting for the effect of terrain roughness. For the former code, it follows:

$$c_r(z) = k_r \cdot \ln\left(\frac{z}{z_0}\right) \text{ for } z_{min} \le z \le z_{max} \quad , \quad c_r(z) = c_r(z_{min}) \text{ for } z \le z_{min} \tag{15}$$

where (BSI, 2010a):

k_r is a terrain factor, depending on the terrain roughness length z_0;
z_{min}, z_{max} are minimum and maximum heights. The former depends on the terrain roughness and the latter is fixed to 200 m and represents an upper limit of the applicability of BS EN 1991-1-4 method.

ASCE 7 use a power-law variation. As mentioned previously, a more accurate model for wind velocity profile with height than these two laws is the Deaves & Harris model. The logarithmic model considerably underestimates the wind velocity for heights larger than 100 m, see Figure 7 and BSI (2008a) and H. Cook (2007). Therefore, the UK NA adopts the Deaves & Harris model and gives charts for the rough-

Figure 7. Models of wind velocity profile with height, considering urban terrain

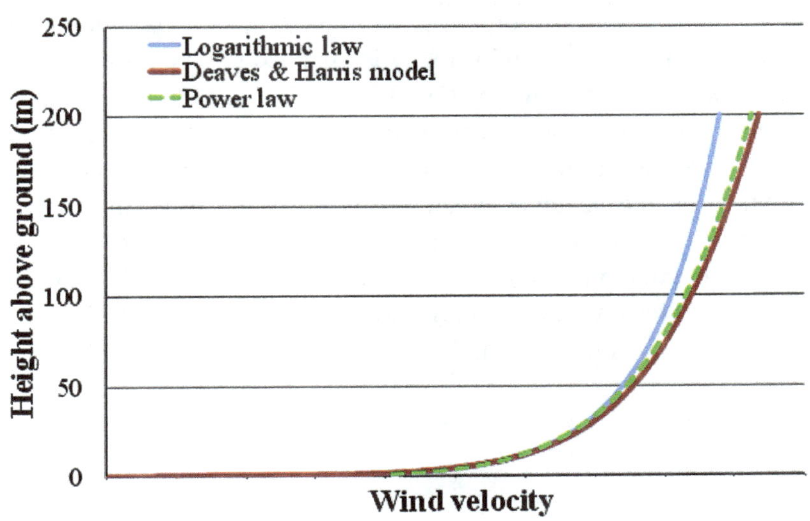

ness factor that depend on three terrain types, instead of five categories suggested in BS EN 1991-1-4 (see Table 9): Sea, Country terrain and Town terrain. When there is choice between two or more terrain types in a given area, then the lowest value of terrain roughness length should be used.

BS EN 1991-1-4 requires orography to be considered when it increases wind velocities by more than 5%. The UK NA reduces the complexities of orography assessment and effects on wind velocity by implementing the altitude factor, c_{alt}. Orography effects must only be evaluated according to the rules specified in Annex A.3 of BS EN 1991-1-4 for sites that fulfil the criteria specified in Figure NA.2 of UK NA. However, the specified method for calculating the orography factor only considers two-dimensional features and does not predict three-dimensional phenomena such as wind funnelling in steep-sided valleys or in urban centres. In these cases, measures should be planned and implemented to determine the local effects on wind action, including seeking expert advice, because site orography could be the most important single factor in the calculation of the wind action.

The effects of interference between neighbouring structures are also addressed in BS EN 1991-1-4. For example, when a structure is located close to another structure that is at least twice its height, it could be exposed to higher wind velocities for certain wind directions. Annex A.4 of BS EN 1991-1-4 gives a conservative method to estimate this effect. Additionally, when groups of structures of similar height are packed closely together, they provide a shielding effect against wind action in a zone extending from the ground to about the average height of the top level of each structure: termed the displacement height, h_{dis}. Therefore, the effective height to be considered for wind action assessment is in this case equal to $z - h_{dis}$. Rules to obtain the value of h_{dis} are given in BS EN 1991-1-4. The use of the latter concept implies that the neighbouring structures can only be removed after the end of the work period of the structure under consideration, and that the shielding effect is uniform in plan.

In BS EN 1991-1-4, the fluctuating component of the wind velocity is simulated by a turbulence intensity parameter, I_v:

$$I_v(z) = \frac{k_I}{c_o(z) \cdot \ln(z/z_0)} \text{ for } z_{min} \leq z \leq z_{max} \quad , \quad I_v(z) = I_v(z_{min}) \text{ for } z \leq z_{min} \tag{16}$$

where k_I is a turbulence factor.

Since the mean wind velocity is also proportional to $\ln(z/z_0)$, see Eqs. -, it can be concluded from Eq., that the standard deviation of the wind velocity is assumed to be constant along the height in BS EN 1991-1-4 (BSI, 2010a). It is shown in BSI (2008a) and H. Cook (2007) that this hypothesis results in conservative values of the wind turbulence for heights lower than 10 m or higher than 100 m, but for values in-between the model returns unconservative values. A better model is presented in the UK NA where the values of the turbulence factor are a function of the height. Several charts are given in the UK NA to determine directly the values of I_v for flat terrains, and correction factors for other types of terrain roughness. For sites where orography is important, see Figure NA.2 of the UK NA, the values of I_v must be divided by the orography factor, c_o, given in Section 4.3.3 of BS EN 1991-1-4.

The two components of the wind, mean and fluctuating, are then combined to obtain the maximum peak gust velocity, $\widehat{V}$:

$$\widehat{V}(z) = \left[1 + g \cdot I_v(z)\right] \cdot V_m(z) \tag{17}$$

In BS EN 1991-1-4 a value of peak factor, g, equal to 3.5 is used for synoptic-scale winds, whereas the UK NA specifies a value equal to 3.0. The definition of a lower value is justified because the basic wind velocity is referenced to 10 minute mean wind velocity values. The value of 3.5 is typically used to determine the peak gust wind velocity from mean wind velocities referenced to 1.0 hour averaging time period (H. Cook, 2007). In the UK NA, the 10 minute mean wind velocities were estimated to be 6% higher than the hourly mean wind velocity values. Based on Holmes, Allsop, & Ginger (2014), the peak gust wind velocity determined using BS EN 1991-1-4 is found to be referenced to a 0.2 s averaging time period.

Therefore, Eq. specified in BS EN 1991-1-4 overestimates the total wind velocity, which partially compensates the errors introduced by the turbulence and height profile models.

The equivalent to Eq. in AS 1170.2 (SAA, 2011) to determine the site wind velocity is, see also Holmes, Kwok, & Ginger (2012):

$$V_{\text{sit},^2} = V_{\text{R}} \cdot M_{\text{d}} \cdot \left(M_{\text{z,cat}} \cdot M_{\text{s}} \cdot M_{\text{t}}\right) \tag{18}$$

where (SAA, 2011):

$V_{\text{sit},\beta}$ is the site wind velocity, where β is the wind direction: N, NE, E, SE for example;
V_{R} is a peak gust wind velocity, measured at 10 m height in open country terrain considering an averaging time of 0.2 s, for a particular region and return period (R years) (SAA, 2011). Values of V_{R} are provided in Section 3 of AS 1170.2;

M are modification coefficients: M_{d} is the wind directional coefficient, $M_{\text{z,cat}}$ is a terrain coefficient, M_{s} is a shielding coefficient, M_{t} is a topographic coefficient. The former is defined in Section 3 of AS 1170.2, while the others in Section 4. M_{s} allows for reductions in the wind velocity to be used when there are structures upwind of greater or similar height.

Sections 26 to 31 of ASCE 7 (ASCE, 2010) concern wind. Section 26 gives "General Requirements" including wind hazard maps, exposure categories, topographic multipliers, gust effects factors and internal pressure coefficients. Sections 27 and 28 relate to buildings, and Section 29 to other structures such as freestanding walls and lattice structures.

There is no expression equivalent to Eq. to determine the site wind velocity. The site wind velocity is given in wind hazard maps. This velocity is a peak gust wind velocity, measured at 10 m height in open country terrain considering an averaging time of 3 s, for a particular region and return period (return period, R, is equal to 300, 700 or 1700 years depending on the type of building, see Table 1.5-1 of ASCE 7). ASCE/SEI 37 (ASCE, 2014) includes temporary structures in Risk Category II). Contrary to BS EN 1991-1-4 and AS 1170.2, the coefficients to account for the effect of terrain roughness and orography, provided in ASCE 7 act on the dynamic pressure (see next Section), rather than on the wind velocity.

In AS 1170.2 the design wind actions are based on the peak gust wind velocity averaged over a 0.2 s time period. To achieve equivalent values using the mean wind velocity method, referenced to a 10 minute mean wind velocity values, a peak factor, g_{u}, equal to 3.4 should be used. As ASCE 7 uses a 3 s averaging time period the corresponding value of g_{u} to be used is 2.5. However, as the wind action analysis method in ASCE 7 is based on mean hourly wind velocities, a g_{u} equal to 2.9 is used (Holmes et al., 2014).

Holmes et al. (2014), demonstrated that if a 3 s peak wind velocity is used instead of 0.2 s, an amplification factor should be added to calculate the wind pressure to be applied to small buildings, or other structures with small exposed areas.

For falsework, a simplified method for determining the wind action is given in BS 5975. The method only applies to:

1. Falsework structures erected less than 100 m high in areas with no significant orography;
2. Falsework structures erected less than 50 m high in areas of significant orography;
3. Falsework structures erected for less than two years;
4. Falsework structures not erected between closely spaced buildings or near to large and considerably higher neighbouring structures.

In cases where the operations during the use of temporary structures require prescribing maximum admissible values of the wind velocity, it is mandatory that an anemometer (wind gauge), for example, is used onsite and monitored continuously, that weather forecasts be reviewed routinely for the period of the operation and some additional time, that an early warning system is implemented and that protection measures be planned beforehand, see BCSA (2005) and IStructE (2007) for guidance.

In general, on most temporary work projects it is usual to limit the operations to a certain limit value of wind velocity, known as the working wind velocity (also termed in-service or operational wind velocity). This is usually set at a Beaufort Scale 6, corresponding to a mean wind velocity of 14 m/s (BCSA, 2005; BSI, 2011a; Newman & Choo, 2003). BS EN 1991-1-6 (BSI, 2005b) gives a minimum recommended value for the basic wind velocity, V_b, equal to 20 m/s for work durations of up to 3 months. Specifically, for falsework, BS EN 12812, and for scaffolds, BS EN 12811-1, stipulate the working wind velocity as a working velocity pressure, assumed to replace q_p, equal to 200 N/m^2. The AASHTO bridge code (AASHTO, 2016) contains a rule for segmental bridges built using the balanced cantilever method in which minimum value of 55 mph (25 m/s) for the wind velocity is specified for erection stability analyses. These values are minimum recommendations; better estimates should be determined by analysis of meteorological records for the area considered. In all cases, reduction factors shall not be applied either to the working wind velocity or to its effects.

3.5.3.5 Wind Effects on Structures

Fundamentals

Having determined the wind velocity, the wind effect on a body (element/structure) can be calculated. Wind action can generate the following effects on exposed structures (ISO, 2009):

1. "excessive forces or instability in the structure or its structural members or elements;
2. excessive deflection or distortion of the structure or its elements;
3. repeated dynamic forces causing fatigue of structural elements;
4. aeroelastic instability, in which motion of the structure in wind produces aerodynamic forces augmenting the motion;
5. excessive dynamic movements causing concern or discomfort to occupants or onlookers;
6. effects of interference from existing and potential future buildings".

The unsteady character of the wind regime, particularly in urban areas, combined with the additional unsteadiness generated by the separated flow after the wind impacts on a structure generates highly fluctuating pressures depending on the flow characteristics and the structure configuration. Indeed, the wind flows around aerodynamic bodies, such as airplane wings, are more easily simulated and characterised because the wind flow streamlines closely follow the wings aerofoil shape (Figure 8a). However, the flow around a bluff body, such as a building or a bridge, is more difficult to model because of separation and reattachment of wind flows that develop around corners and edges, and downstream (wake turbulence) resulting in complex disturbed flows (Figure 8b). These issues are exacerbated in structures with a significant dynamic response to wind action, either due to its shape or due to its structural properties, e.g. structures with low natural frequencies, see ISO (2009).

In general, the wind-induced pressure regime is more complex than the wind flow regime, so its evaluation becomes more cumbersome and analytical techniques fail in most cases. Consequently, boundary layer wind tunnels simulating atmospheric flows have been used and continue to be used extensively for the evaluation of wind effects on buildings. Computational approaches have progressed through the last decade but they are still at a level that hesitation prevails when their results are suggested for use in practical applications (Stathopoulos & Baniotopoulos, 2007).

The simplest analysis is to analyse the along wind structural response, with more complex analyses needed for the across wind and torsional responses (Stathopoulos & Baniotopoulos, 2007).

For the surfaces of a structure exposed to inviscid (zero viscosity) and irrotational (zero vorticity) wind streams, i.e. not affected by fluctuating flows, the wind pressure, p, on the surface of the structure, of area A, is usually expressed in the form of a reference dynamic pressure, q, multiplied by a dimensionless pressure coefficient, C_p:

$$p(z) = q(z) \cdot C_p = \frac{1}{2} \cdot \rho \cdot \left[V(z)\right]^2 \cdot C_p \tag{19}$$

where:

$V(z)$ is the wind velocity at height z;
ρ is the air density, which depends on the altitude, temperature and barometric pressure.

Enclosed structures are subjected to internal pressures generated by the balance of flow through openings and porosities (permeability) in the envelope of the structure, driven by the distribution of

Figure 8. Flow patterns for steady free-stream wind flows

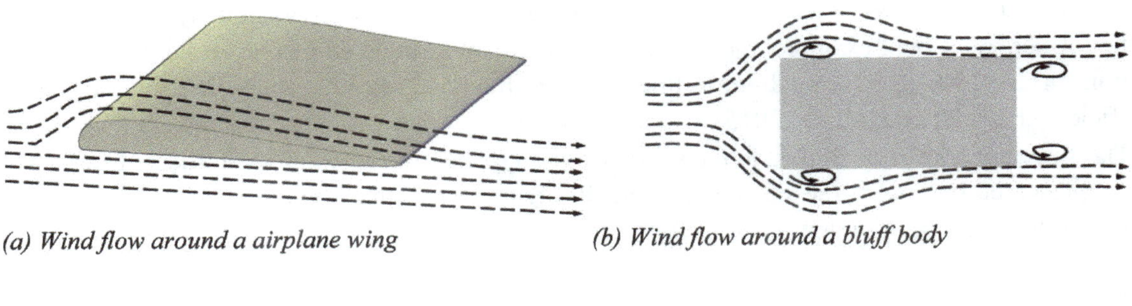

(a) Wind flow around a airplane wing

(b) Wind flow around a bluff body

external pressures over its surface, but also by any pressurisation, mechanical or otherwise. Allowance should be made for these effects by combining pressure coefficients for the external pressures with those for the internal pressures.

Alternatively, the aerodynamic wind-induced effects can be modelled by external forces, F, using a dimensionless force coefficient, C_F:

$$F(z) = q(z) \cdot C_F \cdot A_{ref} \tag{20}$$

where A_{ref} is a reference area, frequently, the projected area of the body in the direction of the mean wind velocity. The force coefficient, C_F, accounts for the total wind effects on the structure, or element, for the considered direction.

Pressure and force coefficients values depend on the shape of the structure, the exposure, the relative wind direction, the windward flow characteristics, and the averaging time (ISO, 2009). It is usual to name the force coefficient along the mean wind velocity direction, i.e. the difference between windward (positive) and leeward (negative) wind effects, as the drag coefficient, C_D, and the force coefficient normal to the mean wind velocity direction as the lift coefficient, C_L. Positive pressure coefficient values indicate pressures acting towards the surface, and negative values indicate pressures acting away from the surface (i.e. suction pressures).

The smallest angle between the mean wind velocity direction and reference axes of the body, for example the principal axes of inertia, is termed angle of attack. This angle can be used to express the mean wind velocity direction in the local coordinate system of the body, or express the relationship between the external forces and force coefficients with respect to the two local coordinate systems, wind and body.

Pressure and force coefficients values are generally derived from wind tunnel tests, and more recently from appropriate computational analyses. Special care is needed in adopting values from different sources to ensure consistency of the methodology. Normally, the aerodynamic force coefficients refer to a mean (time-averaged) wind action, while the aerodynamic pressure coefficients refer to an appropriate fractile value, typically 80%, of the respective extremes (ISO, 2009).

Several studies have been carried out to estimate the pressure and force coefficients for various types of structural forms and wind regimes, see reference list of Chapter 4 of (Holmes, 2015) for various historic references. Modern design codes provide tables with pressure and force coefficients values, see examples in the following Section.

In general, the higher the resistance of the body against the wind flow is, the higher the values of pressure and force coefficients are. The shape of the structure has a significant impact on the resistance to wind action, see Figure 8. The structure porosity also changes the resistance to wind action, which decreases with an increase of the porosity since an increasing amount of air is allowed to flow through the structure, at the same time that a reduction of the leeward pressure occurs. The level of resistance to the wind action is also influenced by the terrain roughness, growing proportionally with the increase of the former. As a final example, the effect of wind stream turbulence increases the drag coefficient for thin surfaces, whereas for medium-to-thick surfaces the wind stream turbulence decreases the drag coefficient (Flay, 2013; Holmes, 2015).

The structure porosity is linked to the solidity ratio, i.e. the ratio between the projected surface area (e.g. represented by area of the shaded elements in Figure 9) and the area enclosed by the boundaries of the surface, e.g. BL in Figure 9.

Figure 9. Exposed surface

The effect of interference between closely spaced bodies is also relevant. However, research is still missing to completely understand the shielding and cluster effects of groups of bodies closely spaced, with mixed findings obtained for groups of thin bodies and thick bodies positioned in series. Holmes (2015) reports that for the former, lower values of the drag coefficient can be obtained for values of spacing equal to 1.5 times the width of equal bodies, while for the latter, evidence shows that drag coefficient may increase up to 15% if equal bodies are spaced half a width.

As was referred before, one of the basis for wind engineering is the assumption of decomposing wind velocity, V, in a mean velocity, V_m, and a fluctuating velocity, V':

$$V(t) = V_m(t) + V'(t) \tag{21}$$

Considering an infinitesimal a small area, A, placed in a turbulent wind flow, the along wind external force, F, i.e. in the direction of the mean wind velocity, experienced by the structure can be derived from Eq. 22:

$$F(t) = \frac{1}{2} \cdot \rho \cdot C_F \cdot A \cdot V^2(t) = \frac{1}{2} \cdot \rho \cdot C_F \cdot A \cdot \left[V_m^2(t) + V'^2(t) + 2 \cdot V_m(t) \cdot V'(t)\right] \tag{22}$$

When linearised, Eq. 23 becomes:

$$F(t) \approx \frac{1}{2} \cdot \rho \cdot C_F \cdot A \cdot \left[V_m^2(t) + 2 \cdot V_m(t) \cdot V'(t)\right] = F_m(t) + F'(t) \tag{23}$$

The total fluctuating external force, F', along the mean wind velocity direction acting on a slender body of finite length, L, subjected to fluctuating wind can be calculated from the fluctuating sectional external forces, knowing the correlation length, l_ρ. The latter is given by the area under the curve of the coefficient of correlation, ρ, between two measurements of the same variable as a function of the distance, τ, at which those two measurements were made. In general, for infinitesimal small values of τ, ρ is equal to 1.0, and as the value of τ increases the value of ρ tends to zero (i.e. there is no statistical relationship between the values). For values of τ smaller than l_ρ, the values are significantly correlated. Conversely, two points separated by a distance greater than l_ρ will be largely uncorrelated. For n pairs of values X_i and X_j of the random variable X measured at two different locations, i and j, the correlation coefficient is given by:

$$\rho_{X_i X_j} = \frac{\mathrm{cov}\left(X_i, X_j\right)}{\sigma_{X_i} \cdot \sigma_{X_j}} \tag{24}$$

where cov(•,•) is the spatial covariance and σ represents the standard deviation. The correlation coefficient between two values of the same variable is also known as autocorrelation (or cross-correlation), and the covariance as autocovariance.

If the variability of X is stationary, i.e. the mean value and the standard deviation of X are constant over the whole domain, the correlation and covariance are only dependent on the distance τ and not on the absolute position of pair of values X_i and X_j. Then, Eq. 25 becomes:

$$\rho_X = \frac{\mathrm{cov}\left(X_i, X_{i+\tau}\right)}{\sigma_X^2} \tag{25}$$

Under the above assumption, the total mean square fluctuating external force along the mean wind velocity direction acting on a slender body of finite length, L, subjected to fluctuating wind can be calculated from (Holmes, 2015):

$$\overline{F'^2} = \int_0^L \int_0^L \overline{f'^2} \cdot \rho_{f_i', f_j'} dX_i dX_j \tag{26}$$

where f' is the fluctuating external force per unit length. In general, the mean square total fluctuating external force is proportional to the correlation length, l_ρ. However, for the case of a perfect (full) correlation between fluctuating sectional external forces:

$$\overline{F'^2} = \overline{f'^2} \cdot L^2 \tag{27}$$

In this case, the fluctuating external forces can be treated as equivalent static external forces.

Fluctuating external forces in the crosswind direction are usually determined from wind tunnel tests (Figure 10) or more recently by computational fluid dynamic analyses (Figure 11), see Holmes (2015) and Simiu & Scanlan (1996) for examples. This is because the phenomena that originate crosswind oscillations are extremely complex and involve action-structure interactions.

Disturbed flows occur when wind flows past a bluff body causing the flow to separate from the surface of the structure rather than to follow the body contour. At relatively low wind velocities, disturbed flows result in spiral vortices that are created periodically and symmetrically on either side of the body. However, for velocities higher than a limiting value, the vortices are shed alternatively, i.e. on one side first then on the other. As a result, alternating low pressure zones are formed on the downstream side of the body and a fluctuating transverse external force is created. This phenomenon is called vortex shedding (Holmes, 2015; Simiu & Scanlan, 1996).

Figure 10. View of a wind tunnel test of a bridge deck. ©2016 LNEC. Used with permission

Figure 11. Results of a CFD analysis of a cladded scaffold. ©2009 H Irtaza. Used with permission

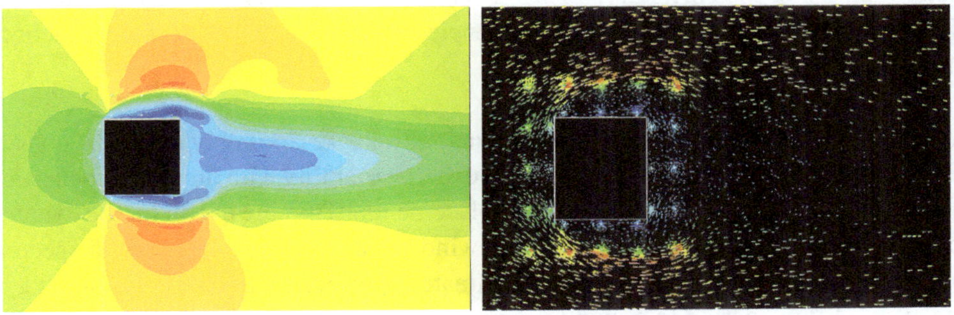

(a) Wind velocity contour *(b) Wind velocity vector*

For a given cross-sectional shape, the frequency of vortex shedding, n_s, is proportional to the approaching wind flow velocity, and inversely proportional to the width of the body. It may be expressed in a non-dimensional form, known as the Strouhal Number, St.

Due to the alternating vortex flows, the oscillations of the transverse (lift) load occur at the vortex shedding frequency and oscillations in drag load occur at twice the vortex shedding frequency.

Galloping is a self-induced phenomenon of flexible bodies and occurs when there are large amplitude lateral or torsional oscillations due to aerodynamic loads that are in-phase with the motion of the body (Holmes, 2015; Simiu & Scanlan, 1996). Lightweight, flexible structures with circular cross sections are more susceptible to this type of phenomenon.

Flutter is also a self-induced phenomenon than occurs when the natural frequencies of the torsional and lateral modes are very similar (Holmes, 2015; Simiu & Scanlan, 1996). A famous example of a structural collapse due to flutter is the Tacoma Narrows Bridge failure of 1940.

For such wind sensitive structures, it is necessary to analyse the resonant structural response. Resonance is a phenomenon that occurs when the frequency of the imposed dynamic action is close to one of the structural system's main natural oscillation frequencies. Resonance translates into a dynamic response of a structural system, excited by a dynamic action, characterised by vibrations with amplitudes significantly larger than the static response. The higher the damping, i.e. dissipation of energy by the system, the lower the resonance response is (Figure 12). However, action-structure interaction may cancel the positive influence of damping and result in self-sustained dynamic instabilities, which can occur during vortex shedding, galloping and flutter for example.

The response of the majority of structures subjected to turbulent wind will not experience significant resonant dynamic response. A well-known rule of thumb is to determine the fundamental natural frequency of the structure, and if this value is below 1 Hz the resonant response may be significant (ASCE, 2010; Holmes, 2015; SAA, 2011).

The analysis of the resonant structural response requires:

1. The development and analysis of a spectral density function, *S*, which describes the distribution of frequency components composing the data series of wind action characteristics, such as wind velocity (Figure 13) and wind turbulence, see Dyrbye & Hansen (1999) and Simiu & Scanlan (1996) for examples of wind action spectral functions. The spectral density function provides an indication of the amount of energy which is present at a given frequency.
2. The determination of the correlation of wind velocity fluctuations at different points of the structure, including their analysis at different frequencies via a cross-spectral density function (Holmes, 2015).

Spectral models are used to perform structural dynamic analyses, possibly including consideration of the uncertainty of actions and of system properties and their variability with time (i.e. stochastic analyses), and consequently of the wind effects (forces and displacements). They are useful for the dynamic analysis of multi-degree-of-freedom systems, especially for wind sensitive structures. A wind spectral model is based upon the assumption that one can compute estimates of response spectral ordinates, e.g. wind load, from the spectrum of wind action (e.g. wind velocity, wind turbulence) and suitable transfer functions that link the spectra of the different variables, see Davenport (1967), Dyrbye & Hansen (1999), Holmes (2015) and Simiu & Scanlan (1996) for a thorough discussion about this topic and application examples.

Figure 12. Illustration of the increase of vibration amplitude of a single degree of freedom system as damping decreases and frequency of the input action approaches resonant frequency of the system

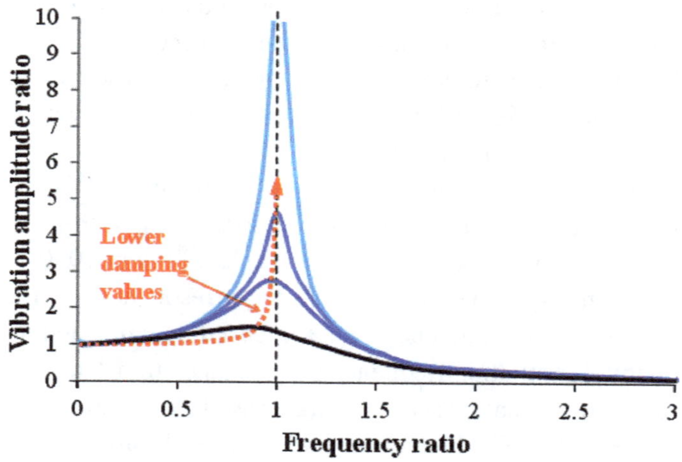

For example, the spectral representation in a frequency domain of the fluctuating longitudinal wind load given by Eq. 23 can be expressed by (Kwok, 2013):

$$S_F(n) = (C_F \cdot A \cdot \rho \cdot V_m)^2 \cdot S_u(n) = 4 \cdot \frac{F_m^2}{V_m^2} \cdot S_u(n) \tag{28}$$

where $S_u(n)$ is the spectrum of the longitudinal wind turbulence and n is the frequency. $S_u(n)$ is usually calculated by the mathematical expression developed by von Karman (Holmes, 2015):

$$S_u(n) = \frac{4 \cdot \left(\frac{n \cdot l_u}{V_m} \right)}{\left[1 + 70.8 \cdot \left(\frac{n \cdot l_u}{V_m} \right)^2 \right]^{5/6}} \cdot \frac{\sigma_u^2}{n} \tag{29}$$

where l_u is the longitudinal wind turbulence integral length scale. Integral scales of turbulence are a measure of the average size of the gusts carried in the mean wind flow. The scales of turbulence represent the length, width, and height of the gust in each of the spatial three dimensions. Design code provisions

The turbulent, fluctuating nature of wind and the complex interaction between wind action and the exposed bodies (e.g. structures, or structural elements/components) generate also highly fluctuating pressures and oscillations on bodies.

The main sources of the fluctuating pressures and loads are (Holmes, 2015):

1. natural random turbulence of the wind stream acting for short durations over the entire or part of the body;
2. fluctuating pressures generated by the body itself, by phenomena such as separations, re-attachments and vortex shedding;
3. fluctuating loads induced by the motion of the body itself, such as galloping and flutter.

Temporary structures have, by their genesis, relatively simple forms and are made by an assembly of a small number of different types of standardised elements. As such, one might think, that the assessment of wind effects on temporary structures is a straightforward analysis. However, temporary structures support, or give access to, permanent structures, and consequently, the assessment of wind effects should take into account the existence of the permanent structure, and the surrounding context. In sites inserted in urban areas, complex aspects of terrain roughness, interference and shielding effects play a key role in the relevance of wind effects on temporary structures. Therefore, assessment of wind effects on most temporary structures is as complicated as it is for permanent structures.

All modern design codes require the effects of wind on a structure to be accounted for either by wind pressures or by wind loads. The former method is specified for structures with similar dimensions in space, or at most two with only one dimension significantly smaller than the other two. Examples are buildings in general, walls, roofs, canopies, cladded temporary structures. The latter method is specified

Figure 13. Example of a spectrum of the horizontal wind velocity (based on Van der Hoven (1957))

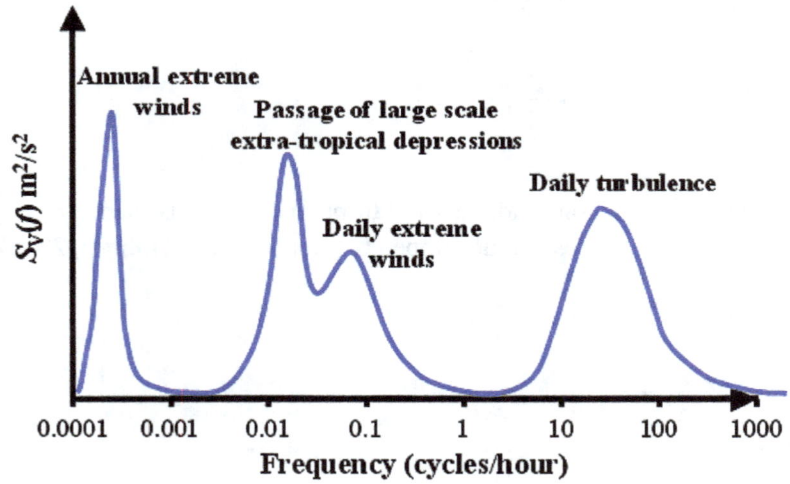

for line-like structures that are characterised for having one dimensional substantially larger than the other two. Examples are lattice structures, bridge decks, bridge piers, masts, chimneys.

Whichever the method used, the effects due to the wind action must be determined, i.e. the combination of the windward and leeward pressures, internal pressures (if applicable), vertical (lift) and crosswind (lateral) pressures. The net pressure on a surface is given by the difference between the wind pressures on either side of the surface. As mentioned in the previous Section, internal pressures do not exist when the structure does not enclose a volume of air, e.g. wind flows on unclad scaffolds, or falsework do not generate internal pressures. According to BS EN 1991-1-4 (BSI, 2010a) when the structure encloses a space but in at least two sides of the structure (façades or roof) the total area of openings in each side is more than 30% of the area of that side, no internal pressures are generated. Force coefficients correspond to the net effect of the external and of the internal pressures on a structure or element.

The complex processes and relationships of wind engineering have been translated to code rules by assuming simplifications, which are understood to result in acceptable conservative designs of common structures. One of the most important assumptions made in most of modern design codes dealing with wind effects is the quasi-static hypothesis, which assumes:

1. the wind turbulence intensity is low;
2. the fluctuations of the effects of the wind action follow the variations of the direction of the mean wind velocity;
3. there is a perfect correlation between fluctuating sectional external forces.

Thus, according to the quasi-static hypothesis, the peak pressures resulting from the mean and turbulent wind action components, along the mean wind velocity direction, can be determined by using the mean values of the pressure coefficient and the peak gust wind velocity. The general expression is given by (ISO, 2009):

$$p(z) = q_{\mathrm{p}}(z) \cdot C_{\mathrm{p,m}} \cdot C_{\mathrm{dyn}} \tag{30}$$

and for wind induced external forces, by (ISO, 2009):

$$F(z) = q_{\mathrm{p}}(z) \cdot C_{\mathrm{F,m}} \cdot C_{\mathrm{dyn}} \cdot A_{\mathrm{ref}} \tag{31}$$

where:

$q_{\mathrm{p}}(z)$ is the peak dynamic pressure at height z, given by Eq. 32;
$C_{\mathrm{p,m}}$ is the mean pressure coefficient;
$C_{\mathrm{F,m}}$ is the mean force coefficient;
C_{dyn} is the peak dynamic response factor

$$q_{\mathrm{p}}(z) = \frac{1}{2} \cdot \rho \cdot \widehat{V}^{2} \tag{32}$$

Inserting Eq. 17 in Eq.32:

$$q_{\mathrm{p}}(z) = \frac{1}{2} \cdot \rho \cdot \left\{\left[1 + g \cdot I_{\mathrm{v}}(\mathrm{z})\right] \cdot V_{\mathrm{m}}(z)\right\}^{2} = \frac{1}{2} \cdot \rho \cdot \left\{1 + \left[g \cdot I_{\mathrm{v}}(\mathrm{z})\right]^{2} + 2 \cdot g \cdot I_{\mathrm{v}}(\mathrm{z})\right\} \cdot \left[V_{\mathrm{m}}(z)\right]^{2} \tag{33}$$

In most modern design codes the above expression has been linearised to (BSI, 2010a):

$$q_{\mathrm{p}}(z) \approx \frac{1}{2} \cdot \rho \cdot \left[1 + 2 \cdot g \cdot I_{\mathrm{v}}(z)\right] \cdot \left[V_{\mathrm{m}}(z)\right]^{2} \tag{34}$$

This is a reasonable simplification when the value of the turbulence intensity I_{v} is small. However, in cases where this condition does not hold, such as near to the ground, it underestimates the peak dynamic pressures. Therefore, in the UK NA to BS EN 1991-1-4 (BSI, 2011c) the complete expression, Eq.33, is used. To ease the calculation, it is customary to find in modern codes design charts with values of the exposure factor, c_{e}, and then determine the peak dynamic pressure by:

$$q_{\mathrm{p}}(z) = c_{\mathrm{e}}(z) \cdot q_{\mathrm{b}} = c_{\mathrm{e}}(z) \cdot \frac{1}{2} \cdot \rho \cdot V_{\mathrm{b}} \tag{35}$$

Thus, the exposure factor combines wind gust, terrain roughness, height profile and orography effects into a single factor, and enables the 10 minute mean wind velocity to be easily converted into a peak gust wind velocity and wind pressure.

In AS 1170.2 (SAA, 2011), as the site wind velocity already accounts for the terrain roughness, height profile and orography effects, Eqs. 30 and 32 are applied (with $C_{p,m}$ replaced by C_{fig}, referred to as the aerodynamic shape factor). In ASCE 7 (ASCE, 2010), the coefficients to account for the effect of terrain roughness and orography, act on the dynamic pressure directly, rather than on the wind velocity:

$$q_z = \frac{1}{2} \cdot \rho \cdot K_z \cdot K_{zt} \cdot K_d \cdot V^2 \quad (36)$$

where (ASCE, 2010):

V is a peak gust wind velocity;
K are modification factors: K_d is the wind directionality factor, K_z is a exposure factor, K_{zt} is a topographic factor. All factors are defined in Section 26 of the code.

The value of the peak dynamic response factor, C_{dyn}, accounts for the dynamic amplification of wind action in the mean wind velocity direction due to wind turbulence and structure interaction. Therefore, the value of C_{dyn} can be taken as one except where the structure is dynamically wind-sensitive (ISO, 2009). Values of C_{dyn} have been derived for various types of wind hazard events, such as gale winds and downbursts.

The main disadvantages of Eqs. 30-31 are:

1. they result in equivalent static wind loads and do not explicitly include dynamic excitations, pressures or external forces;
2. these expressions are only valid in the range of linear elastic structural behaviour;
3. these expressions are not valid when the contribution of nonlinear and unstable wind induced dynamic effects, such as the ones observed in structure sensitive to wind action, cannot be neglected to estimate the overall or local behaviour of the body when expose to wind action. This occurs when the resonant component of the fluctuating wind effects can no longer be considered to be an insignificant percentage of the quasi-static component of the fluctuating wind effects (background component). Annex E of ISO 4354 (ISO, 2009) presents a classification method of structural sensitivity.

BS EN 1991-1-4 (BSI, 2010a), offers in Annex E expressions for determining the equivalent static wind load distribution due to the resonant response in the crosswind direction. These loads should also be accounted for in the structural analysis. It is conservative to simply combine the maximum values of the loads due to the along wind (see Eqs. -) and due to the crosswind. In AS 1170.2 guidance can be found in Sections 6.3 and 6.4. No guidance is given in ASCE 7.

Eqs. 30-31 are given in BS EN 1991-1-4 as:

$$w_e = q_p(z_e) \cdot c_{pe} \quad , \quad w_i = q_p(z_i) \cdot c_{pi} \quad (37)$$

$$F_w = c_s c_d \cdot c_f \cdot q_p(z_e) \cdot A_{ref} \quad , \quad F_w = c_s c_d \cdot \sum_{\text{element } j=1}^{j=n} c_{f,j} \cdot q_p(z_{e,j}) \cdot A_{ref,j} \quad (38)$$

where:

w_e is the wind pressure acting on external surfaces;
w_i is the wind pressure acting on internal surfaces;
z_e, z_i are reference heights defined in Section 7 of the code;
c_{pe} is the external pressure coefficient defined in Section 7 of the code and in the UK NA;
c_{pi} is the internal pressure coefficient defined in Section 7 of the code and in the UK NA;
F_w is the wind external force acting on the whole structure or structural component. It may be given by the sum of the external forces on all elements multiplied by the structural factor;
$c_s c_d$ is the structural factor, which takes into account the fact that the peak velocity does not act simultaneously on a surface, c_s, and the dynamic response of the structure due to wind turbulence, c_d. It is the equivalent to C_{dyn}. The value of $c_s c_d$ can be calculated using the formulas presented in Section 6 and Annex B of the code (Annex C should not be used);
c_f is the force coefficient given in Section 7 of the code and in the UK NA for various types of structure and structural elements.

Similar equations are given in AS 1170.2 and in ASCE 7 (Sections 27, 28 and 29). In the former document, the calculation of the peak dynamic response factor, C_{dyn}, is defined in Section 6. In the latter document, C_{dyn} is termed gust effect factor, *G*, and given in Section 26 of the code. An analytical procedure for the determination of *G* is also presented in the commentary Section of code. The procedure follows closely the one presented in Annex B of BS EN 1991-1-4. In addition, ASCE/SEI 37 specifies that wind loads shall be calculated for each principal structural axis, but should be applied considering that only 50% of the wind load calculated for the perpendicular direction acts simultaneously.

The UK NA recommends that the size factor, c_s, and the dynamic factor, c_d, be calculated separately, providing a table with values of the former factor and figures for the latter factor. Therefore, for static wind effects, instead of considering $c_s c_d$ equal to 1.0 which is allowed in BS EN 1991-1-4, it is most appropriate to consider $c_d = 1.0$ and calculate and to use the actual value of c_s.

For the majority of temporary structures, the dynamic factor, c_d, should be taken as one. However, for very slender and tall structures such as the one shown in Figure 14, or for temporary structures attached to structures vulnerable to dynamic instabilities due to wind action, it may be greater than 1.0, see the guidance given in BSI (2009c, 2010a, 2011c), Holmes (2015, Chapter 11) and SAA (2011), also for vortex shedding phenomena. For lattice structures, the effects of vortex shedding can be ignored for porous structures (SAA, 2011), needing only to be considered when the solidity ratio is greater or equal than 0.6 (BSI, 2006b).

BS EN 1991-1-4 also allows the wind load, F_w, to be calculated from the external and internal pressures:

$$F_w = c_s c_d \cdot \sum_{\text{surface } j=1}^{j=m} w_{e,j} \cdot A_{\text{ref},j} + \sum_{\text{surface } p=1}^{p=s} w_{i,p} \cdot A_{\text{ref},p} \quad (39)$$

For some structures, the second term of Eq. 39, representing the sum of loads due to internal pressures, may cancel out. This should not be considered for surfaces where internal pressures are important (e.g. when they add to the external pressure). Friction loads can also be calculated, using specific pressure or force coefficients, and added to Eq. 39. However, in general their values are small compared with the

Figure 14. Example of a slender and tall temporary structure. ©2016 RMD Kwikform. Used with permission

other loads and can be disregarded when the total area of surfaces parallel to the wind is less than four times the total area of windward and leeward surfaces (BSI, 2010a). In addition, the force coefficients provided in the codes may already account for friction effects: this is the case of BS EN 1991-1-4, with specific exceptions (see Section 7.5 of BS EN 1991-1-4).

The total wind external force, F_w, obtained from Eq. 39 as the sum of the pressures on the windward and leeward surfaces, represents the maximum possible value. However, the values of the windward and leeward pressures are themselves maximum values, which are unlikely to be concomitant (e.g. to occur simultaneously), due to lack of autocorrelation. Since, the value of c_s only accounts for lack of correlation of wind pressures across each individual surface, the value derived from Eq. 39 is conservative. BS EN 1991-1-4 allows the use of a reduction factor to account for this lack of correlation: for structures with ratios height, h, over depth, d, greater or equal to five, the resulting wind external force, F_w, is multiplied by one; for structures with $h/d \leq 1$, F_w is multiplied by 0.85; for all other cases, linear interpolation may be applied. The depth, d, is the length of the surface parallel to the mean wind velocity direction, see Figure 14. This reduction factor shall not be applied to the wind effects on an isolated surface, for instance the external cladded surface of a scaffold attached to a building, but it can be used to determine the total wind load to be applied to the structural elements of a fully cladded lattice structure, such as a falsework with all external surfaces cladded.

Alternatively, the total, F_w, can be obtained from Eq. 39 using net pressure coefficients, $c_{p,net}$, instead of summing pressures coefficients for the windward and leeward surfaces, see the UK NA.

For structures, typically buildings, that are tall and slender, i.e. satisfying $h/d > 5$, BS EN 1991-1-4 allows the use of force coefficients to determine the wind effects, instead of pressure coefficients.

BS EN 1991-1-4 specify an application procedure suitable for each of the three methods to determine wind loads: Eq. 38, Eq. 39 and the latter using c_{pnet} values. When force coefficients or net pressure coefficients are used, a procedure involving the division of the structure in a sufficient number of sections can be applied. The configuration and shape of the elements belonging to a single section should be uniform (or with a small variation), as well as the solidity ratio.

When the pressure coefficients are used, the following division by parts procedure can be used (BSI, 2010a):

1. A structure, typically a building, whose height h is less or equal than its width, b, should be considered to be one part;
2. A structure with $h > b$ but $h \leq 2b$, may be considered to be two parts, comprising: a lower part extending upwards from the ground by a height equal to b and an upper part consisting of the remainder;
3. A structure with $h > 2b$, may be considered to be in multiple parts, comprising: a lower part extending upwards from the ground by a height equal to b; an upper part extending downwards from the top by a height equal to b and a middle region, between the upper and lower parts, which may be divided into horizontal strips with a height h_{strip}.

The above procedure is applicable only to windward surfaces. For the side and leeward surfaces a uniform pressure must be applied along its height.

The wind loads (forces) are calculated for each section (or part) determining the peak dynamic pressure, $q_p(z_e)$, assuming z_e equal to the maximum height of each part. When loads are applied, the point of application is at the middle of each section.

In the following, only rules directly addressing temporary structures will be presented. For other structures, such as buildings, Section 7 of BS EN 1991-1-4, amended and supplemented by the UK NA, gives pressure and force coefficients for most types of structural forms, except bridge decks and bridge piers which may be found in Section 8 of BS EN 1991-1-4. The background document of the UK NA provides further guidance on which pressure and force coefficients values can be used for common structural forms, see BSI (2009c).

Some temporary structures, typically scaffolds, have external surfaces covered with cladding (by netting or sheeting) to prevent construction materials and debris falling from the structure to the area adjacent to it, and to enable work to be carried out in wet conditions. Wide falsework systems may also have external surfaces cladded, in cases where it is possible to reduce the total imposed wind load when compared with the uncladded solution, e.g. when the projected area of the internal elements is larger than the area enclosed by the boundaries of the external surface of the falsework.

For structures with all four external faces fully cladded, Section 7.2.2 of BS EN 1991-1-4, amended and supplemented by UK NA rules, gives guidance on the values and distribution to be considered for the external pressure coefficients, see Table 12 and Figure 14. External pressure coefficients, c_{pe}, are given for the windward (area D), side (areas A, B and C) and leeward (area E) surfaces. Each value represents the most unfavourable value obtained in the range ±45° of the mean wind velocity direction. The UK NA also provides net pressure coefficients that can be used for the determination of total wind load, instead of summing the pressure coefficients for areas D and E.

Table 12. Values of external pressure coefficients for vertical walls (BS EN 1991-1-6 + UK NA)

h/d	Area A (e=min(b,2 h))		Area B (e=min(b,2 h))		Area C (e=min(b,2 h))		Area D	Area E
	c_{pe}	d_A	c_{pe}	d_B	c_{pe}	d_C	c_{pe}	c_{pe}
5	-1.2	d, if $e \geq 5\,d$ e / 5, if $e < 5 \cdot d$	-0.8	0.0, if $e \geq 5\,d$ d - e / 5, if $e \geq d$ 4 e / 5, if $e < d$	-0.5	0.0, if $e \geq d$ d - e, if $e < d$	+0.8	-0.7
1							+0.8	-0.5
≤ 0.25							+0.7	-0.3

Figure 15. Illustration of areas to calculate the external pressure coefficients for vertical walls

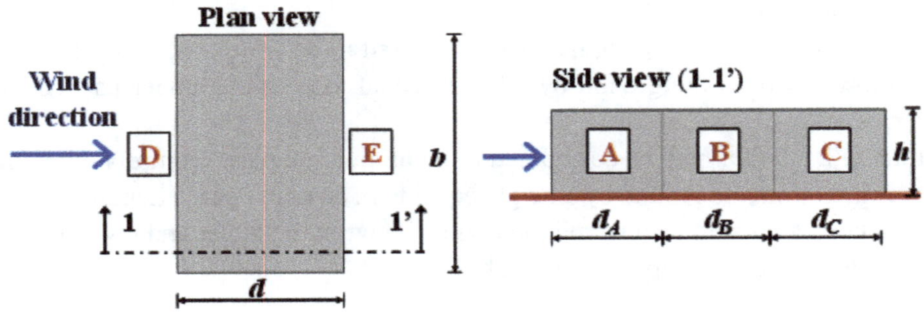

AS 1170.2 (SAA, 2011) provide values which are the same as the ISO standard (ISO, 2009), which are comparable to Table 3.12 for areas D and E, but are smaller for areas A to C. Corresponding values are given in Sections 27 and 28 of ASCE 7 (ASCE, 2010).

For wind acting normal to the plane of a sheeting with no openings, BS EN 12811-1 (BSI, 2003) specifies a single value for the force coefficient equal to 1.3. This value should also be used for netting, unless more accurate values are available obtained from wind tunnel tests. For wind acting parallel to the plane of a sheeting with no openings, values of net pressure coefficients vary, from 0.3 in the case of netting to 0.1 in the case of sheeting.

However, the interaction between temporary structures and permanent structures should not be neglected as occurs in scaffolds (clad or unclad). Rules pertaining to buildings should be considered to assess the wind effects on scaffolds for the windward surfaces, side surfaces as well as for the leeward surfaces. The latter two are critical to establish the number, resistance and layout of the ties to the permanent structure, as well as of the holding down elements to the ground. If not accounted for properly, premature failure may ensue. The vertical ascending (lift) load due to windward flows, needs to be considered to ensure that scaffold working platforms are appropriately secured to the structural members of the scaffolding system.

A scaffold attached to the side face or to the leeward face of a building will experience suction pressures. When the building has no openings or just in one face, or has openings in two or more faces but with an area smaller than 30% of the area of the respective face, the values specified in Table 12 (see Figure 14) for areas A to C and E apply.

The UK NA gives further guidance in cases where the distance between the external wall of the building and the cladded surface of a scaffold is small and susceptible to wind funnelling effects, which increase considerably the external pressure coefficients for areas A to C: -1.6, -0.9, -0.9, respectively.

In both cases above, the resulting suction loads shall be added to the direct wind effects on the scaffold and the total loads will be used to design the temporary structures, in particular to determine the tie resistance and tie spacing to the supporting structures.

Of course, the shielding effect of the building with respect to the wind action to which the temporary structures attached to the leeward façade are exposed to may be also considered. The resulting net effect is complex. Section 6.2.7 and Annex A of BS EN 12811-1 (BSI, 2003) for example, provide values for this shielding effect in terms of a "*site coefficient,* c_s", that should only be used to design the scaffolding with regard to the leeward wind. In these cases, the wind force coefficients applicable to the type of scaffolding (clad or unclad) should be multiplied by c_s.

When the cladding or the building have openings, in at least two faces, with an area greater than 30% of the area of the respective face, the rules given in Section 7.2.2 of BS EN 1991-1-4 do not apply (i.e. the four-sided cladding or the building do not interact, positively or adversely, with the wind action, and do not reduce or increase the wind loads transmitted to the temporary structures. Note however that the provisions specified in BS EN 12811-1 that take into account the leeward shielding effect due to the building are still applicable in these cases for scaffolding (clad or unclad).

In all other cases, the rules of BS EN 1991-1-4 for structures resembling freestanding walls should be used to determine the wind loading acting on cladded temporary structures and on other cases, such as short still girders. In these cases, the corresponding values of net pressure coefficients, for wind acting normal to the plane of the cladding, are given in Section 7.4.1 of BS EN 1991-1-4, Annex D of AS 1170.2 or Section 29 of ASCE 7, for solidity ratios greater or equal than 0.8, i.e. at most 20% of the cladded area corresponds to openings. It is found that upper corners are the most vulnerable parts of these temporary structures to wind action. Values of net pressure coefficients range from 1.2 to 2.3 (near windward edges) for usual temporary structures, but could go higher than 3.0 for long structures ($b/h > 5$).

For surfaces with smaller solidity ratios, the provisions given in Section 7.11 of BS EN 1991-1-4, Annex E of AS 1170.2 and Section 29 of ASCE 7, relative to plane lattice structures should be used instead, or in Sections 7.6 to 7.10 of BS EN 1991-1-4 relative to individual infinitely long elements.

Estimates of the wind loading on lattice structures can be determined by summing the loads on individual members, using the force coefficients provided in Sections 7.6 to 7.10 of BS EN 1991-1-4, or Annex E of AS 1170.2. For rectangular sections the force coefficient values range between 2.4 for wide sections and 0.9 for very long sections, see Section 7.6 of BS EN 1991-1-4, but usually a value equal to 2.0 is used. For circular sections, the value most often used is 1.2, although more accurate values can be obtained from figures that relate the force coefficient with the Reynolds number, Re, see Eq. 41. For angle sections the value most often used is 2.0; it is noteworthy to mention that AS 1170.2 and BS EN 13001-2 (BSI, 2014b) provide tables with several values of force coefficient for various types of angle sections. BS EN 12811-1 indicates that for toeboards a value equal to 1.3 should be used.

The abovementioned procedure can be time-consuming and will give overly conservative values except in cases where the solidity ratio is low. For specific cases of lattice structures, namely those having three or four columns (i.e. three or four faces), such as heavy-duty towers, Section 7.11 of BS EN 1991-1-4, or Annex E of AS 1170.2, provides values of overall (drag) force coefficients which account for group and shielding effects. They should not be applied to the area of a single element but directly to the area equal to the sum of the area of the elements of the most unfavourable exposed face (reference face), projected normal to the face.

The force coefficients provided in Sections 7.6 to 7.11 of BS EN 1991-1-4 correspond to infinitely long bodies (represented with symbol $c_{f,0}$). To account for the reduced wind effects caused by wind flows around the ends of a finite body the code multiplies the force coefficients by an end-effect factor, ψ_λ, which depends on the body slenderness, λ, and solidity ratio, φ, see Figure 16. For elements of typical temporary structures (scaffolds, falsework, shoring) such as vertical and horizontal elements, the slenderness is equal to l/b for bottom lift elements (altered to infinite in the UK NA) and $2 \cdot l/b$ for all other elements (altered to $4 \cdot l/(b \cdot c_{f,0})$ in the UK NA), where l represents the length of the element and b the width of the element.

Care should be paid when determining the value of the end-effect factor: for the force coefficients provided in Sections 7.6 to 7.10 the slenderness and relative position should be determined for the single element, whereas for Section 7.11 the corresponding values should be determined for the reference face.

Figure 16. End-effect factor, ψ_λ, given in BS EN 1991-1-4 (BSI, 2010a)

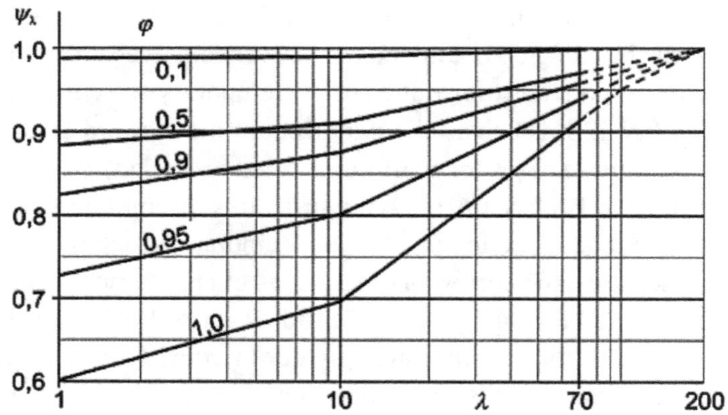

When there are elements with different shapes in a single face, the overall force coefficient could be determined by a weighted average:

$$c_f = c_{f,c} \cdot \frac{A_c}{A_{face}} + c_{f,c,sup} \cdot \frac{A_{c,sup}}{A_{face}} + c_{f,a} \cdot \frac{A_a}{A_{face}} \tag{40}$$

where:

A_c and $c_{f,c}$ represent the exposed area and force coefficient of the elements with circular shape in sub critical flow regimes, respectively;
$A_{c,sup}$ and $c_{f,c,sup}$ represent the exposed area and force coefficient of the elements with circular shape in supercritical flow regimes, respectively;
A_a and $c_{f,a}$ represent the exposed area and force coefficient of the elements with angular shape, respectively.

Different codes provide different values of force coefficients. For example, Figure 17 illustrates the force coefficients provided in BS EN 1991-1-4 (BSI, 2010a), BS EN 13001-2 (BSI, 2014b) and BS EN 1993-3-1 (BSI, 2006b) for three or four faces lattice structures with angle elements. The data for solidity ratios smaller than 0.2 or greater 0.6 are merely indicative and should be used with prudence. Analysing Figure 17, it can be concluded that there is a reasonable agreement between the values given in different codes for solidity ratio values in the range of 0.2 to 0.6.

Comparing the values given in Figure 17 for very low solidity ratios, with the sum of the force coefficients of the single angle elements exposed to the wind, it can be observed that the former values provide a reduction of the drag load to be considered in the design. The force coefficients are always lower than 4.0 which is the value equal to the sum of the force coefficients for two angle elements.

For lattice structures with circular elements, as the behaviour of the wind flow around a circular body depends on the Reynolds number the definition of the value of the force coefficient to be used is more complex. The graphs given in BS EN 13001-2 to calculate the force coefficients of lattice structures with circular elements are more detailed than the ones found in BS EN 1991-1-4, BS EN 1993-3-1 and AS 1170.2.

Figure 17. Force coefficients for three or four faces lattice structures with angle elements

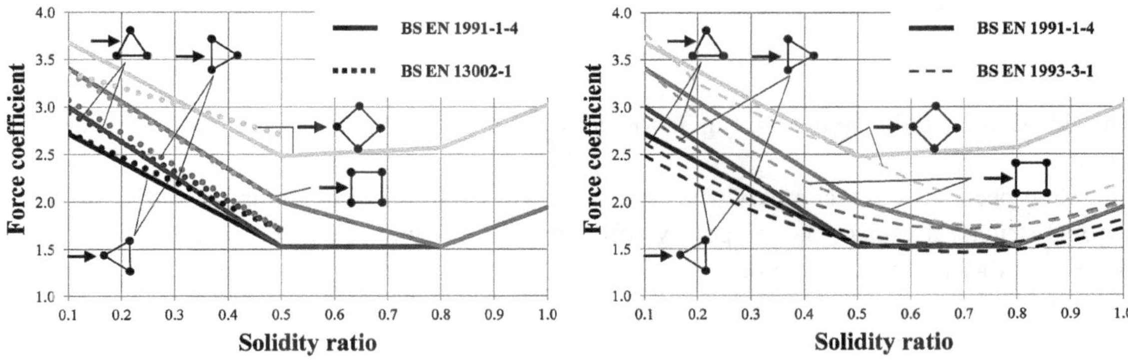

The Reynolds number is a dimensionless quantity that is used to help predict the type of flow regime, turbulent flow or smooth (laminar) flow, over a body in different wind flow situations. The Reynolds number, Re, is defined as the ratio between fluid inertia loads and viscous loads, and can be determined from Eq. 41 for circular elements (where d is the external diameter of the circular element and V is the wind velocity). The viscosity of air, ν, is a relatively low value, 15 10^{-6} m^2/s at 20ºC, but nonetheless, in certain cases, such as circular elements, this small viscosity plays an important role (Simiu & Scanlan, 1996).

$$\mathrm{Re} = \frac{d \cdot V}{\nu} \tag{41}$$

For elements, or structures with elements, of circular shape, it should be evaluated if it is possible that the highest values of wind effects occur for wind velocities smaller than the design wind velocity.

Some temporary structures are assemblies of slender elements. These skeletal arrangements allow wind to pass through but provide a certain degree of internal shielding. This positive effect is accounted for in Annex E.2.3 of AS 1170.2 via a shielding factor, but is not accounted for in BS EN 1991-1-4. As non-contradictory complementary information (NCCI), the general procedure specified in Appendix A.1 of N. Cook (1999) can be used, simplified in Blackmore (2004) and in Annex M of BS 5975 (BSI, 2011a). The shielding factor, η, for unclad temporary structures is expressed by (N. Cook, 1999):

$$\eta = \left(1 - \frac{\varphi \cdot c_{\mathrm{f}}}{2}\right)^2 \tag{42}$$

where the solidity ratio, φ, and force coefficient c_{f} refer to the shielding bay.

For multiple frames, the shielding factor of the nth bay of a series of multiple bays is given by summing over the shielding effect of the upwind n - 1 bays (N. Cook, 1999):

$$\eta = \left(1 - \sum_{j=1}^{n-1} \frac{\eta_j \cdot \varphi_j \cdot c_{f,j}}{2}\right)^2 \tag{43}$$

The shielding factor is applied to Eq. 38 to derive the total wind load on each bay of the temporary structures.

Shielding becomes more complex when the mean wind velocity direction is skewed relative to the orthogonal structural axes of the temporary structures. For common rectangular framed systems, the ratio between the total wind force on a bay for a given angle of attack θ and $\theta = 0.0°$ is given by (N. Cook, 1999):

$$\frac{F_{w,\theta}}{F_{w,\theta=0°}} = \cos(\theta) + 0.12 \cdot \cos(2 \cdot \theta) \cdot \left(\frac{d}{b}\right)^{0.8} \tag{44}$$

In general, for systems of props, shoring, and falsework systems, little internal shielding effect is achieved. For scaffolds, BS EN 12811-1 determines that internal shielding shall not be accounted for. For formwork, Section 17.5.1.15 of BS 5975 gives a possible procedure to determine the wind shielding effect of the leeward edge panels of the formwork before casting of the concrete.

Of course, shielding effects are also relevant in the case of groups of similar structures. In BS EN 1991-1-4, the general guidance is to make use of displacement height, h_{dis}. However, for freestanding walls, BS EN 1991-1-4 gives a shielding factor caused by other upwind walls, provided that these sheltering walls are at least as tall as the wall being considered. The shelter factor should not be applied in the end zones of the wall within a distance of h (height of the wall) measured from the free end of the wall. The assumptions behind the use of this shielding factor are the same as the ones specified for the use of displacement height, h_{dis}.

When analysing the effect of shielding by neighbouring structures, keep in mind that wind may, in principle, blow from any direction. Temporary structures that are totally shielded from wind in one direction may be fully exposed when the wind gust occurs from another likely direction. In these cases, where wind shielding is only partial, i.e. is only effective for a range of upwind directions, guidance has been developed and detailed in Blackmore (2004) and N. Cook (1999, ChapterAnnex A.2) to determine the wind effects. As a rule, it will only be useful to consider shielding when the upwind neighbouring structures provide continuous shelter over a range of at least 120°, including the prevailing wind direction.

In ASCE 7, no shielding, by buildings and other structures or terrain features, is allowed to be accounted for in determining wind effects in temporary structures, except when determined by wind tunnel testing. ASCE/SEI 37 allows taking advantage of the shielding effect for unclad lattice structures, as follows:

1. "The loads on the first three rows of elements along the direction parallel to the wind shall not be reduced for shielding.
2. The loads on the fourth and subsequent rows shall be permitted to be reduced by 15%.".

Some temporary structures consist in isolated towers, or other very flexible systems. In these cases, it may be necessary to laterally stabilise the system using guy wires, i.e. tension cables that connect the falsework to the ground or to a rigid structure. The information included in Annex B of BS EN 1993-3-1

(BSI, 2006b) and in the corresponding UK NA (BSI, 2010b), could be used to derive the wind loads for guyed temporary structures. In the USA, information compatible with ASCE 7 is given in TIA-222-G (TIA, 2016).

Guidance on temporary stage decks can be found in Blackmore & Freathy (2004) or in Section 7.3 of BS EN 1994-1-1 + UK NA, or Annex D of AS 1170.2, which address canopy-like structures.

One should not forget to account for the evolving structural forms of the temporary structures, and possibly of the permanent structures, but also of the structural properties of the temporary structures, when applying the rules given in the design of temporary structures, since the effects of wind action may change significantly. Typically, the most important phases are: during assembly of the temporary structures, in particular for free-standing systems, during the construction of the permanent structure, if applicable, and after the permanent structure has been completed, if applicable. In some cases, the course of the aforementioned phases is concomitant. In particular during the construction of the permanent structure, when the latter is still not capable of resisting loads (e.g. fresh concrete or unconnected steel beam), the mass of the temporary structures increases without the correspondent increase in stiffness, and as result the aerodynamic characteristics of the temporary structures change (i.e. the fundamental natural frequency of the structure decreases) making it more susceptible to resonant dynamic phenomena, such as vortex shedding.

3.5.3.6 Recent Research and Innovation

Research into wind engineering has been primarily concerned with determining accurate design wind velocities and loads on permanent structures. As far as temporary structures are concerned, research has been steadily increasing in recent years and mostly related to complex CFD analyses. Concerning wind tunnel tests, very limited research has been performed owing to the intrinsic difficulties of reproducing the geometry and stiffness of the elements in a reduced scale; in a scale of 1:50 the external diameter of the typical structural element of a scaffold is less than 1 mm. The same limitations apply to modelling permeable cladding, such as netting.

Irtaza (2009) and Irtaza, Beale, Godley, & Jameel (2014) carried out CFD analyses of unclad and clad (sheets and nets) scaffolding under turbulent wind flows. The scaffold model is shown in Figure 18. The CFD analyses were validated by wind tunnel tests (Irtaza, Beale, & Godley, 2012). The results obtained showed that the wind load on net clad scaffolds (with permeabilities varying from $1.0\ 10^{-6}$ to $1.0\ 10^{-10}\ m^2$) could be considered to be 40% of the total wind load on the covered area of the scaffold. It was found that the practice of not cladding the lowest level of a scaffold made negligible differences to the total wind load on the scaffold.

As important, CFD analyses showed that the maximum positive net pressure on the windward surface of clad scaffolds occurred when the direction of wind was perpendicular to it. The maximum negative net pressure on the side surface of clad scaffolds occurred when the direction of wind was parallel to the surface plane. The maximum negative net pressure on the leeward surface of clad scaffolds occurred when the direction of wind was at ±45° from the windward face. Table 13 presents the recommended values. It is possible to observe that in this case, the net pressure coefficient value obtained for the windward surface of net clad scaffold is much smaller than the corresponding value of a sheet clad scaffold, due to the positive effect of the permeability of the net.

Table 13. Average net pressure coefficients on the surfaces of a clad scaffold surrounding the perimeter of a building with h/d = 0.67 and b/h = 1.33, see Figure 18, taken from Irtaza (2009)

Type of cladding	Windward pressures	Side pressures	Leeward pressures
Sheet	+1.3	-0.15	≈0
Net (with permeabilities varying from 5 10^{-9} m^2 to 7 10^{-9} m^2)	+0.4	-0.15	≈0

Figure 18. Overview of building and scaffold assembly analysed by Irtaza (2009). ©2009 H Irtaza. Used with permission

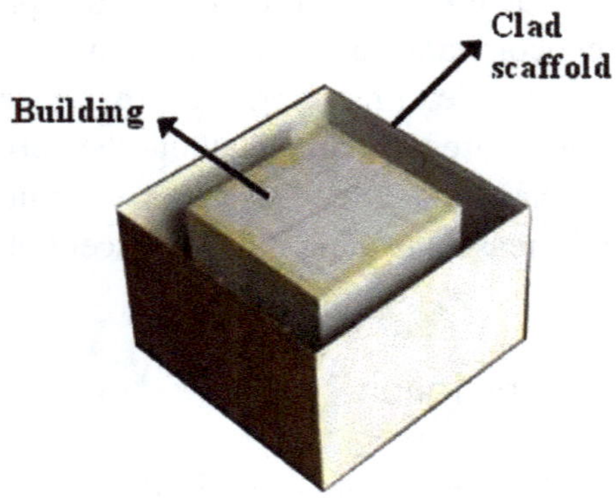

Charuvisit, Hino, Ohdo, Maruta, & Kanda (2007) examined the characteristics of wind pressures acting on surfaces of scaffolds based on wind tunnel test results, only considering non-turbulent wind flows. The results show that the maximum positive wind loads were obtained when the mean wind velocity direction is normal to the scaffold façade and the latter is located windward, while the maximum negative wind loads were obtained when the latter is located leeward or when the mean wind velocity direction is parallel with the scaffold façade. This is in agreement with the findings of Irtaza (2009). The values of the net pressure coefficients are presented in Table 14 ($h/d = 0.5$, $b/h = 1.0$ for wind directions 0° and 180°, and $h/d = 1.0$, $b/h = 2.0$ for wind directions 90° and 270°, see Figure 19a). Comparing the values with Table 13, it is possible to conclude that different scaffolding and building geometries generate different leeward and side pressure coefficients, but similar windward values. In addition, the plan proportions of the building have a large influence on wind pressures acting on the scaffolds: larger exposed surface areas and elongated shapes lead to smaller wind pressure values.

Table 14. Average net pressure coefficients on the surfaces of a sheet clad scaffold, see Figure 19, taken from Charuvisit et al. (2007)

Figure 19	Windward pressures	Side pressures	Leeward pressures
Case 1	+1.5 (0°)	-0.5 (90°)	-0.2 (180°)
Case 2	+1.35 (0°)	-0.4 (90°)	+0.15 (180°)
Case 3	Surface A: +1.5 (0°, A) Surface B: +1.35 (0°, B)	Surface A: -0.7 (90°, A) Surface B: -0.55 (270°, B)	Surface A: -0.2 (180°, A) Surface B: +0.1 (180°, B)

More recently, F. Wang, Tamura, & Yoshida (2013) carried out wind tunnel tests of cladded scaffold systems attached to a building with different numbers of storeys and opening ratios, for different cladding arrangements, scaffolding geometries and wind directions. Mean and peak values of the net pressure coefficients were obtained for the scaffolding surfaces, as well as global force coefficients for the entire scaffolding and local force coefficients for specific areas of the scaffold where wind effects are higher: top or side areas.

The values of the net pressure coefficients for Case 2 of Figure 19, are presented in Table 14 ($h/d = 2.0$, $b/h = 0.8$ for wind directions 0º and 180º, and $h/d = 1.25$, $b/h = 0.5$ for wind directions 90º and 270º, see Figure 19a). The same conclusion as above can be made, that different scaffolding and building geometries generate different leeward and side pressure coefficients, but similar windward values. It was also possible to confirm the results published by Irtaza (2009), Table 13, in which it was found that the smallest values of suction loads are obtained when scaffolds fully surround the building perimeter.

It was also found that when the building opening ratio is higher than 40% the interaction between the building and the scaffold is not significant on average, but could be important for local peak pressure values. In addition, for most geometries considered, the maximum positive wind loads tend to decrease and the maximum negative wind loads tend to increase as the cladded area decreases. For some scaffolding geometries, the wind force coefficients for clad scaffolding are higher than 1.3 but in all cases not more than 1.7. These values are lower than the values specified in BS EN 1991-1-4 for freestanding walls for $b/h < 1.0$ which is the case of the considered scaffolding.

Later, in 2014, these authors analysed the influence of a neighbouring buildings, relative position and height, on the previous studied cases (F. Wang, Tamura, & Yoshida, 2014). It was found that, when the neighbouring building is in front of the scaffolding, the values of the positive (windward) force coefficient decrease significantly, but the values of the negative (leeward) force coefficient increase, relative to the isolated scaffolding case. Changing the relative height of the neighbouring building amplifies these effects. See also Dagnew & Bitsuamlak (2014).

Wind force coefficients on self-climbing scaffold, used for construction of a high-rise buildings, were studied by Okubo, Hongo, & Kondo (2010) through wind tunnel tests of a 1/70 scale model of a typical system. It was found that the up-lift load and the lateral load are as large as the drag load. Therefore, these loads should not be disregarded in the design of self-climbing scaffolds. See also research published by Yue, Li, & Yuan (2012).

3.5.4 Snow and Ice

Snow and ice actions are often disregarded in ordinary weather. However, they may become important actions for temporary structures when located on mountain regions with the chance of snow accumulation and ice formation.

Table 15. Average net pressure coefficients on the surfaces of a sheet clad scaffold, taken from F. Wang et al. (2013)

Figure 19	Windward pressures	Side pressures	Leeward pressures
Case 2	+1.6 (0º)	-0.3 (90º)	-0.2 (180º)

Figure 19. Overview of building and scaffold assemblies analysed by Charuvisit et al. (2007)

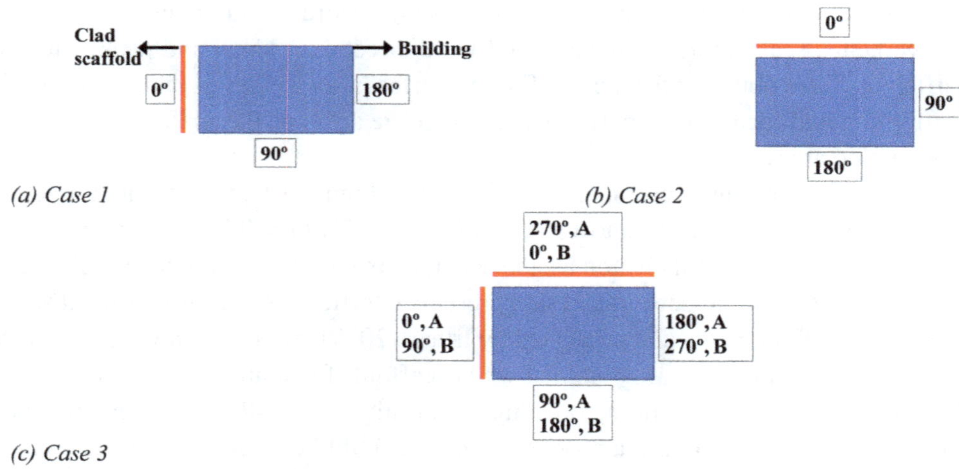

In most design codes, it is possible to update the annual maximum snow load value to account for a reduced exposure period. For instance, using BS EN 1991-1-3 (BSI, 2004a, 2007b):

$$s_{k,t} = c_p \cdot s_{k,50} = \left(\frac{1 - V \cdot 0.78 \cdot \left(\ln\left(-\ln\left(1 - p\right)\right) + 0.57722\right)}{1 + 2.5923 \cdot V} \right) \cdot s_{k,50} \tag{45}$$

where V is the coefficient of variation of the annual maximum snow load, $s_{k,t}$ is the characteristic value of the maximum annual snow load associated with a return period of t years, c_p is a probability factor and p is the probability of annual exceedance which according to the UK NA (BSI, 2007b) must not be considered higher or equal to 0.20. Assuming values of V between 0.1 and 0.4, the results for the values of the probability factor are presented in Figure 20.

BS EN 1991-1-6 limits the permissible reduction of the characteristic value to 25%, but if a planned procedure is enforced to remove of snow every day the recommended value for the probability factor is 0.30.

Snow action can be modelled using a distributed load, s, which value can be determined by (BSI, 2004a, 2007b):

$$s = \mu_i \cdot C_e \cdot s_k \tag{46}$$

where:

μ_i represents the snow load shape coefficient;
C_e represents the exposure coefficient;
s_k represents the characteristic value of snow load on the ground.

Figure 20. Probability factor for annual maximum snow load as function of the return period, according to BS EN 1991-1-3 and UK NA

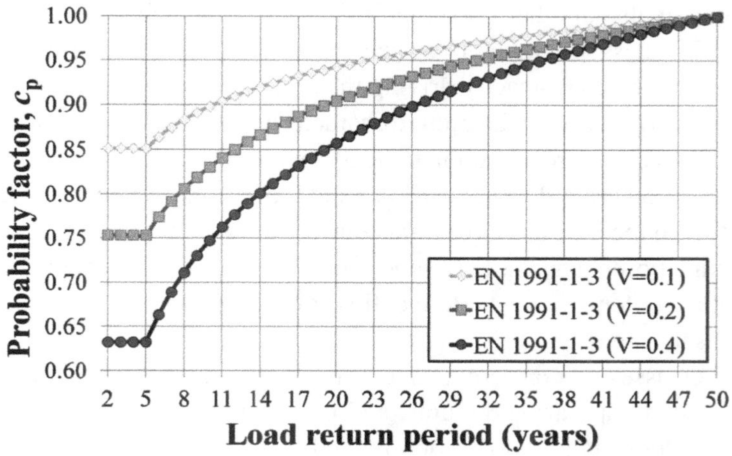

The load should be assumed to act vertically and applied to the plan projection of the area of the structure exposed to snow. The characteristic value of the snow load is usually obtained from snow load maps of the region of interest.

Ice action has in general two relevant structural effects: the increased vertical loads on the iced structure and increased wind drag caused by the increased wind-exposed area and less favourable aerodynamic shape. There are many types of ice and to simplify the analysis, BS 5975 provides a design value for the ice density equal to 920 kg/m^3 (other values, more accurate, are available from ISO 12494 (ISO, 2001)) and indicates a maximum ice thickness surrounding elements not larger than 25 mm.

BS 5975 also provides particular design guidance to take into account snow and ice actions in the analysis and design of falsework. Structures such as masts and towers, together with tensioned steel ropes, cables, mast guys, etc., are sensitive to increased wind drag caused by icing, in particular linear objects with small cross-sectional dimensions.

Wind action on iced structures may be calculated based on the same principles as the action on the ice-free structure. However, both the dimensions of the structural members and their drag coefficients are subject to changes. Force coefficients for different types of elements are provided in ISO 12494 (ISO, 2001).

3.5.5 Geotechnical Actions

Geotechnical actions are diverse but the most important with respect to temporary structures are the pressures imposed by the soil on the structural elements and the potential movements of the supporting ground. Both are classified as permanent actions. Of great importance to define these actions is the correct characterisation of the ground stratigraphy and properties relevant for the expected loading, including the present and potential levels of the ground water table. This information should be gathered during ground investigation.

Soil pressures are imposed in retaining structures as the latter hold the internal equilibrium of the soil pressures. The shape and magnitude of the pressures is a function of the soil properties. Using simplified analysis methods, soil pressures can be modelled by triangular, rectangular or trapezoidal diagrams depending on the soil theory being used, see Chapter 4.

Ground settlements are also another potential critical hazard which deserves an in-depth analysis. Due to the usual low robustness of temporary structures, any imposed internal force redistribution may not find the required redistribution capacity driving the system to collapse.

Ground settlements are a function of the ground characteristics and applied actions. The ground where temporary structures foundations are laid upon typically exhibits poor resistance and rigidity characteristics since they consist of top ground layers, e.g. soft and loose soils. Without proper care, large ground settlements can result from the internal forces transmitted to the temporary structures and from this to the ground via the foundation elements. Differences between displacements of the foundation ground can originate differential settlements at the foundation level of the temporary structures with potential negative structural consequences.

Settlements should be assessed correctly to avoid unwanted and unusual internal force distributions within the elements of the temporary structures, and problems related to the geometry control of the permanent structure. Settlements are most often related to movements in the foundations, but elastic deformations and initial gaps between elements and within the connections can also produce settlements.

Differential settlements translate into unbalanced internal forces and consequently into overloaded main elements and foundations. Furthermore, the occurrence of these settlements may result in the overturning of part of the structure, causing secondary stresses for which the temporary structures was not designed for. This behaviour, if neglected in the design phase may lead to the collapse of the structure.

Free-standing structures where the vertical elements are unbraced or the joints are weak and do not allow the redistribution of the internal forces to adjacent elements, are more sensitive to the effects of differential settlements. However, elements of internally stiff structures, e.g. with many bracing elements, can as well be very sensitive to differential settlements as these introduce additional compression/tension internal forces to neighbour elements.

Ground movements at the foundations of temporary structures should be assessed from the results of ground investigations. Results should provide sufficient information on both absolute and relative values of movements, their time dependency and variability. Guidance is available from AGS (2011), Bowles (1997), BSI (2007a, 2009f, 2011a, 2015a, 2015b) and Dowrick (2009).

For shallow foundation elements, the maximum acceptable relative rotations of *L*/500 is adequate for many structures, where *L* represents the larger of the dimensions of the foundation element, whereas a settlement up to 50 mm is often acceptable (BSI, 2013b, 2014d).

3.5.6 Dynamic Loads Induced by Human Motion

Considerable research has been carried out into the dynamic loads induced by human action. However, little has been undertaken into the effects on temporary structures.

The action of human activity (walking, running, jumping, etc.) on structures introduces dynamic loads that cause the structures to undergo dynamic movements (e.g. vibrations). In most cases, these movements are not perceptible to humans and have no significance in terms of structural integrity. However, in flexible structures like footbridges, stages and grandstands, human activity induced vibrations can have detrimental effects to the functionality and behaviour of the structure, in particular when resonance vibrations occur causing severe discomfort and possibly leading to structural collapse but also through material fatigue when the number of repetitions of the dynamic effects is sufficiently high.

The intense lateral vibration caused by humans when the Millennium Bridge in the UK was opened showed the importance of analysing the human motion dynamic loads (Dallard, Fitzpatrick, & Flint, 2001). The research into human dynamic loads has been summarised by Racic, Pavic, & Brownjohn (2009).

As for the wind action (see Section 3.5.3), in order to assess the effects (e.g. displacements and stresses) induced by human activities, it is necessary to analyse the human motion actions applied to the structure, the dynamic properties of the structure and the action-structure interaction.

The dynamic action produced by human activities depends primarily on the weight of the persons involved in the activity, the density of persons per unit area and on the degree of synchronisation between the persons (ISO, 2007). For actions of human activities such as dancing, running of a group of people, spectator action in halls or in stadiums, the models for the dynamic loads, F, that simulate each of these types of actions are provided in Ellis & Ji (2004), Annex A of ISO 10137 (ISO, 2007) and also in Willford & Young (2006) for footfall human actions. The general model for the dynamic loads, F, in the frequency domain can be expressed by a Fourier series expansion:

$$F\left(t\right) = Q \cdot \left[1 + \sum_{n=1}^{k} \alpha_{n,\mathrm{d}} \cdot \sin\left(2 \cdot \pi \cdot f \cdot t + \phi_{n,\mathrm{d}}\right)\right] \tag{47}$$

where:

Q represents the static load of one person;

$\alpha_{n,\mathrm{d}}$ represents the coefficient corresponding to the nth Fourier term, in the direction d (vertical or horizontal);

f is the frequency component of repetitive loading;

t is the time in seconds;

$\phi_{n,\mathrm{d}}$ is the phase angle of the nth Fourier term, in the direction d (vertical or horizontal);

k is the number of Fourier terms to be considered.

For a group of people, Q is replaced by the load density and distribution, $Q(x,y)$, of the crowd over the area occupied. Since, in general, not every person in a group of people will introduce a dynamic excitation to the structure, the dynamic response of the structure will be smaller when compared to the extreme case of a group of people with perfect synchronisation. For design, it is therefore important to identify and characterise the relevant hazard scenarios. Annex A of ISO 10137 (ISO, 2007) suggests values for a factor that can be used to reduce the dynamic effects for some common human activities determined from Eq. 47. An alternative model for the vertical dynamic load, F_{v}, with improved accuracy against experimental data, is presented in Fernández, Hermanns, Alarcón, & Fraile (2013) consisting of a summation of two square cosines functions:

$$F_{\mathrm{v}}\left(t\right) = A_1 \cdot \cos^2\left(\omega_1 \cdot \left(t - t_1\right)\right) + A_2 \cdot \cos^2\left(\omega_2 \cdot \left(t - t_2\right)\right) \tag{48}$$

where A_1, A_2, ω_1, ω_2, t_1 and t_2 are function parameters. A procedure to determine the values of these parameters is presented in Fernández et al. (2013).

The dynamic action of most of human activities (including dancing, running of a group of people, spectator action in halls or in stadiums), can be assumed to change with time but be stationary in space (i.e. distributed more or less uniformly over a major portion of the structure). In the remaining cases, such as spectators jumping to their feet (for example at a sports event), the dynamic action also changes with time and the dynamic problem become very difficult to solve. In the latter cases, empirical methods can be used based on experimental tests on structures, or it may be possible to simplify the problem representing the action approximately by a series of single pulses, see also Ellis & Ji (2004), Annex A of ISO 10137 (ISO, 2007) and Willford & Young (2006).

Based on d'Alembert's and Hamilton principles, the principle of virtual work for static (equilibrium) problems can be applied to dynamic problems and the resulting equations of motions can be derived from scalar quantities (thus invariant to the applied coordinate system) (Humar, 2012). The equation of motion of a system is given by:

$$\mathrm{M} \cdot \ddot{\mathrm{u}} + \mathrm{C} \cdot \dot{\mathrm{u}} + \mathrm{K} \cdot \mathrm{u} = \mathrm{F} \tag{49}$$

where: **M**, **K** and **C** represent the relevant dynamic scalar quantities of a structural system: the mass, stiffness and damping, respectively; **ü**, **u̇**, **u** represent the accelerations, velocity and displacement of the degrees of freedom of the system, respectively; **F** represents the dynamic loads (external forces).

Setting **F** and **C** equal to zero, and solving Eq. 49 the undamped free-vibration response of the system is obtained which is characterised by the system's natural frequencies and natural modes of vibration. This is called Modal Analysis and is nowadays performed using modern numerical methods such as the Finite Element Method.

The forced vibration response (i.e. when **F** is not equal to zero) has in general two components: a transient component and a steady-state component. The former component is present at the beginning of the response and its effects (magnitude and duration) depend on the type of dynamic external force applied and on the damping of the system. The latter component exists as long as the dynamic external force is applied.

When evaluating the structural response to human activities that are random in space, the characteristics of the unloaded structure should be used in the calculations, since the human activities act simply as loads. For human activities that are approximately stationary in space, part of the mass of the group of people should be added to the mass of the structure, and the system damping value can be increased (Ellis & Ji, 1997) but as pointed out by Sim (2006) only for structures with a fundamental natural frequency higher than 2 Hz.

As previously mentioned, resonance occurs when the frequency of the applied external force is equal to or close to the natural frequency of the system, and the resulting response may significantly exceed the static response. When the frequency of the applied external force is very small compared with the fundamental natural frequency of the system, the response will be close to the static case. Finally, when the frequency of the applied external force is very large compared with the natural frequency of the system, the action occurs so fast that the response of the structure will be insignificant. A useful representation of the response of the system to a range of frequencies of the applied action is to develop a response spectrum of a relevant system quantity (maximum displacement, velocity, acceleration or stress for example). Brownjohn, Racic, & Chen (2016) and Chen, Li, & Racic (2016) derived a response spectrum for human walking derived from experiments involving over 800 people.

The UK National Annex to BS EN 1991-1-1 states that "*resonance of the structure should be avoided by limiting its natural frequencies so that the vertical frequency is greater than 8.4 Hz and the horizontal frequency is greater than 4.0 Hz. These frequencies should be evaluated for the appropriate mode of vibration of an empty structure.*" (BSI, 2002b).

Numerous factors influence the response of humans to vibrations "*including the frequencies, magnitude, duration, variability, form, directions of the vibration and intervals between vibration events, or exposure of the human subjects to the vibration*" (ISO, 2007), for example. For human actions on grandstands Table 16 gives recommended vertical peak accelerations as a function of the gravity acceleration value, $g = 9.8$ m/s^2. These values were confirmed to be safe for use by Browning (2011).

Design codes deal with human loads (workers, spectators, etc.) in different ways. All, however, treat these loads mainly as vertical equivalent static external forces (i.e. vertical static loads multiplied by a dynamic factor). For temporary demountable structures, such as grandstands or stages, guidance is given in a report by IStructE (2007). In this document, minimum notional horizontal loads are also specified, given as a percentage of the vertical imposed loads (ranging from 6% to 10% depending on the potential for synchronised and periodic crowd movement, see Table 17), that implicitly include the effects of different types of human induced motions (therefore, they should only be considered when the structure is in use). For stages, the floor surfaces should be designed to carry a point load of 3.6 kN over an area 50 × 50 mm without causing any damage to the floor and without causing excessive deflection of the floor panels (e.g. deflection of not more than 10 mm) (IStructE, 2007).

Table 16. Recommended peak vertical accelerations for grandstands for a frequency range <10 Hz (Ellis & Ji, 2004)

Reasonable limit	<0.05 g
Disturbing limit	<0.18 g
Unacceptable limit	<0.35 g
Probably causing panic	>0.35 g

Table 17. Notional horizontal loads for design of temporary demountable grandstands (IStructE, 2007)

Category of spectator activity	Notional horizontal load
Category 1 Nominal potential for spectator movement, which excludes synchronised and periodic crowd movement, e.g. most sports events	6%
Category 2 Potential for spectator movement more vigorous than Category 1, e.g. major musical concerts and football matches	7.5%
Category 3 Stands with a potential for synchronised and periodic crowd movement and having vertical and horizontal fundamental frequencies which avoid resonance effects, e.g. most pop concerts	10%

3.6 ACCIDENTAL ACTIONS

BS EN 1991-1-6 (BSI, 2005b, 2008b) states that accidental actions such as impact from construction vehicles, cranes, building equipment, or materials in transit (e.g. skip of fresh concrete), and/or local failure of final or temporary supports, including dynamic effects, that may result in collapse of load-bearing structural members, shall be taken into account, where relevant. It is the responsibility of the designer to select the accidental design situations and the design values of accidental actions during construction, depending on the type of temporary structures.

Very limited guidance is given in design codes regarding accidental actions. Typical cases are those involving impact from materials, equipment or vehicles, but also from local failure of temporary supports and bracing elements. However, it is specified that if a static analysis is performed, the characteristic value of the equivalent static accidental action should be determined by multiplying the characteristic value of the static action value by an appropriate dynamic factor. In BS EN 1991-1-6 (BSI, 2005b), and the corresponding UK NA (BSI, 2008b), but also in ASCE/SEI 37 (ASCE, 2014) and in AASHTO bridge code (AASHTO, 2016), a value equal to 2.0 is recommended applicable to all cases in order to take account the strain energy imposed by the body affected by the fall, subject to better assessment (see also a report by Sétra (2007)). In AASHTO bridge temporary structures code, which is based on the ASD philosophy, a recommended minimum value equal to 1.3 is specified.

More accurate values can be found in other parts of the Eurocodes depending on the nature of the accidental action. For example, the dynamic action generated by the sudden release of the payload being lifted using a crane can be simulated by an upward static load multiplied by a dynamic factor, whose value can be determined from BS EN 1991-3 (BSI, 2006a, 2009e). Another example is the impact action of the crane crab against the buffers for which a minimum value of 1.25 for the dynamic factor is specified in BS EN 1991-3. This standard specifies other accidental load cases relevant for lifting equipment.

For BCE, some of the most important accidental design situations are:

- Loss of stability of a bridge deck during launching due to slip from temporary bearings;
- Fall of equipment (for example a travelling form during its operation), including the dynamic effects;
- Fall of structural elements (for example the fall of a prefabricated segment before the final post-tensioning is applied), including dynamic effects.

Concerning the impact of vehicles or cranes on structural elements of temporary structures, such as bridge falsework, no published information exists about the action effects on the structure. Impact is an interaction phenomenon between a moving object and a structure, in which the kinetic energy of the object is suddenly transformed into energy of deformation. To find the dynamic interaction forces, the mechanical properties of both the object and the structure should be determined. Existing simplified models (BSI, 2014a, 2014c) assume a rigid structure and a deformable colliding body, which is not applicable to the majority of temporary structure. However, it is anticipated that important damage will be inevitable if the event of an accidental action caused by an impact of a body with the temporary structure cannot be prevented. It is not practicable to design the elements directly hit to sustain without significant damage the load values generated by this kind of actions. Therefore, in these cases and in the absence of additional guidance, it is recommended that alternative design strategies such as to prevent and/or to mitigate the effects of such accidental design situations, see Chapter 5, be adopted. However, some design codes provide prescriptive rules for the design of falsework elements located adjacent to roads, such as AASHTO (2008) which indicates that the vertical loads used for the design of these elements shall be increased by not less than 50%, complemented by bracing requirements and the protection of the temporary structures with concrete barriers. A common requirement in the USA is to specify additionally that falsework elements located adjacent to roads should resist a horizontal load equal to 2 000 pounds (≈9 kN) applied at the base.

Earthquakes are also a very important accidental action that needs proper consideration in the design of temporary structures.

An earthquake is a spasm of ground shaking caused by a sudden release of energy in the earth's lithosphere (i.e. the crust plus part of the upper mantle). This energy arises mainly from stresses built up during tectonic processes, which consist of interactions between the crust and the interior of the earth. Almost all earthquake, volcanic and mountain-building activity closely follow the tectonic plate boundaries and are related to movements between them.

The resulting effects of an earthquake are waves of different speed and frequency that expand from the hypocentre of the earthquake and propagate in all directions throughout the earth's crust. The characteristics of the propagation of the seismic waves depends on many factors such as spatial location of the hypocentre, amount of energy released, fault type and configuration, geotechnical stratification (i.e. thickness, configuration and heterogeneity of ground layers) from source to site surface. For example, waves can be amplified or attenuated depending on the type of soils. In order to ascertain the seismic risk for a specific site, a ground investigation may be necessary if sufficient information is not available. Additionally, several types of laboratory tests of ground samples can also be performed to determine ground characteristics relevant to seismic design of structures (e.g. shear-wave velocity, damping, potential for liquefaction). Guidance concerning ground testing is available from AGS (2011), BSI (2007a, 2009f, 2015a, 2015b) and Dowrick (2009), see also Chapter 4.

The acceleration experienced by a structure above ground depends also on the ground-structure interaction and also on the structure dynamic properties (e.g. mass, stiffness, hysteretic behaviour, etc.).

Earthquake structural engineering often starts from consultation of seismic hazard maps which provide the probability distribution of the earthquake induced surface ground motions (e.g. the maximum ground acceleration, also called peak ground acceleration) for a region. These maps may be based on probabilistic studies of historic records of surface ground motions *vs.* time (e.g. accelerograms), generated by past earthquakes. When the latter information is insufficient or inexistent, accelerograms can be numerically simulated taking into account empirical relationships between magnitude and frequency of earthquakes (e.g. the Gutenberg-Richter law (Gutenberg & Richter, 1954)) for the region under consideration and complex attenuation models (also known as ground-motion prediction equations) that describe the propagation of the resulting seismic waves throughout the ground media from source to site surface.

From these maps, it is possible to develop surface ground motion frequency spectra (e.g. peak ground acceleration spectrum), or alternatively a group of accelerograms, for earthquakes with a given annual probability of exceedance (i.e. seismic events with a given return period). With this information it is possible to solve the dynamic equations of motion of the system, see Eq. 50, to assess their structural behaviour and proceed with their design. A thorough presentation and discussion of earthquake structural engineering is provided in Dowrick (2009) and Tesfamariam & Goda (2013).

$$\mathrm{M} \cdot \ddot{\mathrm{u}} + \mathrm{C} \cdot \dot{\mathrm{u}} + \mathrm{K} \cdot \mathrm{u} = \mathrm{F} \tag{50}$$

where: **M**, **K** and **C** represent the relevant dynamic scalar quantities of a structural system: the mass, stiffness and damping, respectively; **ü**, **u̇**, **u** represent the accelerations, velocity and displacement of the degrees of freedom of the system, respectively; **F** represents the dynamic external forces, given by $\mathbf{M}\,\mathbf{a}_g$, where $\mathbf{a}_g$ represents the vector with the ground acceleration that may vary with time.

The stiffness and the damping matrices will be a function of the ground acceleration (time and frequency domain), material and system properties (including boundary conditions).

Therefore, earthquakes are a base excitation rather than a clearly defined load. Structural response is dynamic and often some damage is acceptable to occur in the structure as long as adequate safety margins against global collapse are verified in order to reduce the risk to human lives.

Within the frame of performance-based design philosophy, various seismic hazard levels can be considered associated with equal number of design situations and requirements. A common metric for seismic design for the no-collapse requirement is a seismic event with a 10% exceedance probability in 50 years, equivalent to 475 years return period.

Using the procedure indicated in BS EN 1998-2 (BSI, 2005d) it is possible to determine the probability factor to be applied to the reference peak ground acceleration in order to get the seismic action for reduced work durations. The results are presented in Figure 21. Note that the seismic action during the construction phase does not need to be considered in the UK (BSI, 2009c).

The earthquake is often not considered during the design of falsework and scaffold structures, since the horizontal seismic shear force at the base of the structure must be transferred through the soil to the structure. As typically the foundation elements of these structures consist in simple baseplates not connected to the ground, it is safe to assume that most of the resistance to the lateral load is provided by friction between the surfaces of the ground material and the baseplate in contact. The friction force will be very low and the structures will move like a rigid body with no significant damage. For all other foundation solutions (e.g. piles) and temporary structures, such as BCEs, earthquakes must be considered during design.

In some cases, it is common practice to design the temporary structures against wind and earthquake separately and to insure the system regarding the unlikely event of a simultaneous action of the wind and the earthquake.

Figure 21. Probability factor for the peak ground acceleration as function of the work duration, according to BS EN 1998-2

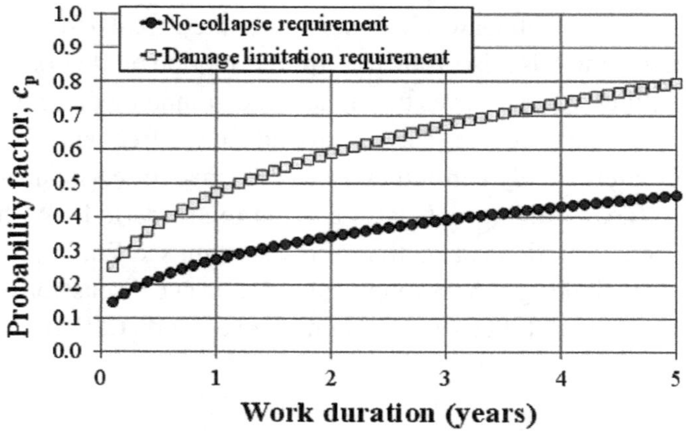

3.7 UNIDENTIFIED HAZARD EVENTS

Temporary structures should be designed with respect to relevant identified and unidentified hazard events. The latter are additional to the identified hazard events that have been explicitly defined in the previous Sections. Unidentified hazard events simulate the effects of uncertainties and errors during the design, execution and operation of a structure (permanent and temporary) that are not accounted for in design codes, or which exceed the limits considered in these documents. Included in these uncertainties and errors during the design, are system effects associated with the global reliability of the structure (e.g. reliability of the structure against disproportionate collapse), that are not usually considered in modern design codes (see Chapter 5).

Typically, unidentified hazard events consist in recommended notional actions, which should be applied as accidental actions in accidental design situations (see Chapters 5 and 6). Note that for some design failure mode scenarios (see definition in Chapter 5, Section 5.6.1), in particular the ones involving brittle failure modes, it may be necessary to apply notional actions as accidental actions in persistent or transient design situations (see Chapters 5 and 6).

For structures for which the expected failure consequences in terms of material damage, risk to human life, and economic, social or environmental are limited to moderate (see Chapter 5), two types of notional actions are usually recommended and defined below. Note that when no guidance is available or where appropriate, the notional actions can be agreed with the client and/or the regulatory authority. For each specific design, one of the two, or both, notional actions should be applied, as relevant.

- Notional removal of elements or joints of the structure, or parts thereof, or limited parts of the structure.

The recommended action for temporary structures is the removal of elements. As a minimum, the element, or elements, to be removed should be the one which has the smallest safety margin with respect to the range of applicable ultimate limit states, when subject to the relevant identified non-accidental hazard events (it may also be relevant to consider the load case where only permanent actions are applied).

Note that the key elements design method defined in Chapter 5 is not applicable to this notional action.

The structural analysis with respect to the removal of elements may involve the consideration of the dynamic effects associated with the actions that can cause the failure of the elements but also the consideration of the dynamic effects associated with the loss of the elements, as relevant.

- Application of a notional load.

There is no guidance available in the present design codes and standards for the recommended load type and value for temporary structures. Therefore, the notional actions may be agreed with the client and/or the regulatory authority.

In BS EN 1991-1-7, for building structures the notional load is defined as a uniformly distributed equivalent static load of 34 kN/m^2. This load was calculated after the collapse of the Ronan Point residential building in 1968 in the UK and is restricted to gas explosions. However, it is now commonly used in building structures as a multi-purpose load case without much justification.

As a minimum, the notional load should be applied to the element, or elements, which has the smallest safety margin with respect to the range of applicable ultimate limit states, when subject to the relevant identified non-accidental hazard events (it may also be relevant to consider the load case where only permanent actions are applied).

For structures for which the expected failure consequences in terms of material damage, risk to human life, and economic, social or environmental are limited to large (see Chapter 5), a risk-based analysis may be used. See Chapter 5.

3.8 CONCLUSION

This Chapter presented the main types of actions due to external and internal hazard events that are relevant to temporary structures. At the start, a classification of different types of actions was presented. Next, the permanent actions were described, followed by the main types of variable actions, namely: the various kinds of construction actions, the wind action, the snow and ice actions, the geotechnical actions, the human motion actions and the accidental actions. Finally, actions that may be used to simulate the effects of uncertainties and errors during design, assembly and operation of temporary works were introduced.

For each type of action, this Chapter provided a discussion about the difficulties associated with the characterisation and quantification of each of them, both in terms of structural engineering thinking and in terms of provisions included in modern design codes. Specifically, the most important rules given in modern design codes for each type of action are presented.

In addition, an understanding on how specific actions can affect the performance of different types of temporary structures was provided.

This Chapter also provided state-of-the-art guidance concerning the analyses of actions and of action effects for which rules specified in modern design codes are not applicable, or are incomplete or do not exist.

REFERENCES

AASHTO. (2003). AASHTO GSCB–2–M – Guide specifications for design and construction of segmental concrete bridges, with 2003 interim revisions (2nd ed.). Washington, DC: American Association of State Highway and Transportation Officials (AASHTO).

AASHTO. (2008). AASHTO GSBTW–1–M – Guide design specifications for bridge temporary works, with 2008 interim revisions (1st ed.). Washington, DC: American Association of State Highway and Transportation Officials (AASHTO).

AASHTO. (2016). AASHTO LRFDUS–7 – LRFD Bridge design specifications, with 2015 and 2016 interim revisions (7th ed.). Washington, DC: American Association of State Highway and Transportation Officials (AASHTO).

ACI. (2014). ACI 347R–14 – Guide to formwork for concrete. Washington, DC: American Concrete Institute (ACI).

AGS. (2011). *UK specification for ground investigation* (2nd ed.). London, UK: ICE Publishing.

André, J., Beale, R., & Baptista, A. (2012). A survey of failures of bridge falsework systems since 1970. *Proceedings of the ICE - Forensic Engineering, 65*(4), 161–172. doi:10.1680/feng.12.00012

André, J., Beale, R., & Baptista, A. (2013). Recent advances and existing challenges in the design of bridge falsework systems. *Civil Engineering and Environmental Systems*, *30*(2), 130–145. doi:10.1080/10286608.2012.733374

Ang, A., & Tang, W. (2007). *Probability concepts in engineering: Emphasis on applications in civil and environmental engineering* (2nd ed.). London, UK: Wiley.

ASCE. (2010). ASCE 7–10 – Minimum design loads for buildings and other structures. American Society of Civil Engineers (ASCE).

ASCE. (2014). ASCE/SEI 37–14 – Design loads on structures during construction. American Society of Civil Engineers (ASCE).

Bashor, R., & Kareem, A. (2009). Comparative study of major international standards.*Proceedings of the 7th Asia-Pacific Conference on Wind Engineering (APCWE-VII)*, 8.

BCSA. (2005). BCSA Guide to steel erection in windy conditions (No. 39/05). London, UK: The British Constructional Steelwork Association Ltd. (BCSA).

Benjamin, J., & Cornell, C. (1970). *Probability, statistics, and decision–making for civil engineers* (3rd ed.). New York: McGraw-Hill.

Billings, I., & Routley, R. (1978). Falsework load measurement at Newton No. 1 bridge. *New Zealand Engineering*, *33*(6), 127–134.

Birch, N., Booth, J., & Walker, M. (1971). Effect of site factors on the load capacities of adjustable steel props (No. R27). London, UK: Construction Industry Research and Information Association (CIRIA).

Blackmore, P. (2004). Wind loads on unclad structures (No. Special Digest 5). Bracknell, UK: Building Research Establishment (BRE).

Blackmore, P., & Freathy, P. (2004). Wind loads on temporary stage decks (No. Special Digest 483). Bracknell, UK: Building Research Establishment (BRE).

Bowles, J. (1997). *Foundation analysis and design* (5th ed.). New York: McGraw-Hill.

Browning, G. (2011). *Human perception of vibrations due to synchronised crowd loading in grandstands* (PhD Thesis). University of Bath, Bath, UK.

Brownjohn, J., Racic, V., & Chen, J. (2016). Universal response spectrum procedure for predicting walking-induced floor vibration. *Mechanical Systems and Signal Processing*, *70–71*, 741–755. doi:10.1016/j.ymssp.2015.09.010

BSI. (2002a). BS EN 1991–1–1:2002 – Actions on structures – Part 1–1: General actions – Densities, self–weight, imposed loads for buildings. London, UK: British Standards Institution (BSI).

BSI. (2002b). UK National annex to BS EN 1991–1–1:2002. London, UK: British Standards Institution (BSI).

BSI. (2003). BS EN 12811–1:2003 – Temporary works equipment – Part 1: Scaffolds – Performance requirements and general design. London, UK: British Standards Institution (BSI).

BSI. (2004a). BS EN 1991–1–3:2003 – Actions on structures – Part 1–3: General actions – Snow loads. London, UK: British Standards Institution (BSI).

BSI. (2004b). BS EN 12811 – Temporary works equipment – Parts 1–3. London, UK: British Standards Institution (BSI).

BSI. (2005a). BS EN 1990:2002+A1:2005 – Basis of structural design. London, UK: British Standards Institution (BSI).

BSI. (2005b). BS EN 1991–1–6:2005 – Actions on structures – Part 1–6: General actions – Actions during execution. London, UK: British Standards Institution (BSI).

BSI. (2006a). BS EN 1991–3:2006 – Actions on structures – Part 3: Actions induced by cranes and machines. London, UK: British Standards Institution (BSI).

BSI. (2006b). BS EN 1993–3–1:2006 – Design of steel structures – Part 3–1: Towers, masts and chimneys – Towers and masts. London, UK: British Standards Institution (BSI).

BSI. (2007a). BS EN 1997–2:2007 – Geotechnical design – Part 2: Ground investigation and testing. London, UK: British Standards Institution (BSI).

BSI. (2007b). UK National annex to BS EN 1991–1–3:2003. London, UK: British Standards Institution (BSI).

BSI. (2008a). Report on the calibration of Eurocode for wind loading (BS EN 1991-1-4) and its UK National Annex against the current UK wind code (BS 6399: part 2). London, UK: British Standards Institution (BSI).

BSI. (2008b). UK National annex to BS EN 1991–1–6:2005. London, UK: British Standards Institution (BSI).

BSI. (2008c). UK National annex to BS EN 1998–6:2005. London, UK: British Standards Institution (BSI).

BSI. (2009a). *Structural timber. Strength classes*. London, UK: British Standards Institution.

BSI. (2009b). BS EN 1998–2:2005+A1:2009 – Design of structures for earthquake resistance – Part 2: Bridges. London, UK: British Standards Institution (BSI).

BSI. (2009c). PD 6688–1–4:2009 – Background information to the national annex to BS EN 1991–1–4 and additional guidance. London, UK: British Standards Institution (BSI).

BSI. (2009d). PD 6698:2009 – Recommendations for the design of structures for earthquake resistance to BS EN 1998. London, UK: British Standards Institution (BSI).

BSI. (2009e). UK National annex to BS EN 1991–3:2006. London, UK: British Standards Institution (BSI).

BSI. (2009f). UK National annex to BS EN 1997–2:2007. London, UK: British Standards Institution (BSI).

BSI. (2010a). BS EN 1991–1–4:2005+A1:2010 – Actions on structures – General actions – Part 1–4: Wind actions. London, UK: British Standards Institution (BSI).

BSI. (2010b). UK National annex to BS EN 1993–3–1:2006. London, UK: British Standards Institution (BSI).

BSI. (2011a). BS 5975:2008+A1:2011 – Code of practice for temporary works procedures and the permissible stress design of falsework (3rd ed.). London, UK: British Standards Institution (BSI).

BSI. (2011b).*Falsework. Performance requirements and general design*. London, UK: British Standards Institution.

BSI. (2011c). UK National annex to BS EN 1991–1–4:2005+A1:2010. London, UK: British Standards Institution (BSI).

BSI. (2013a). *Concrete. Specification, performance, production and conformity*. London, UK: British Standards Institution.

BSI. (2013b). BS EN 1997–1:2004+A1:2013 – Geotechnical design – Part 1: General rules. London, UK: British Standards Institution (BSI).

BSI. (2014a). BS EN 1991–1–7:2006+A1:2014 – Actions on structures – Part 1–7: General actions – Accidental actions. London, UK: British Standards Institution (BSI).

BSI. (2014b). BS EN 13001–2:2014 – Crane safety. General design – Part 2: Load actions. London, UK: British Standards Institution (BSI).

BSI. (2014c). UK National annex to BS EN 1991–1–7:2006+A1:2014. London, UK: British Standards Institution (BSI).

BSI. (2014d). UK National annex to BS EN 1997–1:2004+A1:2013. London, UK: British Standards Institution (BSI).

BSI. (2015a). BS 5930:2015 – Code of practice for ground investigations. London, UK: British Standards Institution (BSI).

BSI. (2015b). BS 8004:2015 – Code of practice for foundations. London, UK: British Standards Institution (BSI).

Cao, S. (2013). Strong winds and their characteristics. In Y. Tamura & A. Kareem (Eds.), *Advanced structural wind engineering* (pp. 1–25). Tokyo, Japan: Springer. doi:10.1007/978-4-431-54337-4_1

Castillo, E., Hadi, A., Balakrishnan, N., & Sarabia, J. (2004). *Extreme value and related models with applications in engineering and science*. Wiley.

Charuvisit, S., Hino, Y., Ohdo, K., Maruta, E., & Kanda, M. (2007). Wind tunnel experiment on wind pressures acting on the scaffolds in strong winds. *Journal of Wind Engineering*, *32*(1), 1–10. doi:10.5359/jwe.32.1

Chen, J., Li, G., & Racic, V. (2016). Acceleration response spectrum for predicting floor vibration due to occupants jumping. *Engineering Structures*, *112*, 71–80. doi:10.1016/j.engstruct.2016.01.013

CIRIA. (1978). A mathematical model of the structure of strong winds (No. 76). London, UK: Construction Industry Research and Information Association (CIRIA).

Coles, S. (2001). *An introduction to statistical modeling of extreme values*. London, UK: Springer-Verlag. doi:10.1007/978-1-4471-3675-0

Cook, H. (2007). Designer's guide to EN 1991–1–4 – Eurocode 1: Actions on structures. Part 1–4: General actions. Wind actions. London, UK: Thomas Telford Ltd.

Cook, N. (1999). *Wind loading: A practical guide to BS 6399–2*. London, UK: Thomas Telford Ltd. doi:10.1680/wl.27558

Costa, A., Crespo, A., Navarro, J., Lizcano, G., Madsen, H., & Feitosa, E. (2008). A review on the young history of the wind power short-term prediction. *Renewable & Sustainable Energy Reviews*, *12*(6), 1725–1744. doi:10.1016/j.rser.2007.01.015

Dagnew, A., & Bitsuamlak, G. (2014). Computational evaluation of wind loads on a standard tall building using LES. *Wind and Structures An International Journal*, *18*(5), 567–598. doi:10.12989/was.2014.18.5.567

Dallard, P., Fitzpatrick, T., & Flint, A. (2001). The Millennium bridge, London: Problems and solutions. *The Structural Engineer*, *79*(8), 15–17.

Davenport, A. (1960). Rationale for determining design wind velocities. *Journal of the Structural Division*, *86*(5), 39–68.

Davenport, A. (1967). Gust loading factors. *Journal of the Structural Division*, *93*(ST3), 11–34.

Deaves, D. (1981). Computations of wind flow over changes in surface roughness. *Journal of Wind Engineering and Industrial Aerodynamics*, *7*(1), 65–94. doi:10.1016/0167-6105(81)90068-4

Done, J., Davis, C., & Weisman, M. (2004). The next generation of NWP: Explicit forecasts of convection using the weather research and forecasting (WRF) model. *Atmospheric Science Letters*, *5*(6), 110–117. doi:10.1002/asl.72

Dowell, J., Weiss, S., Hill, S., & Infield, D. (2014). Short-term spatio-temporal prediction of wind speed and direction. *Wind Energy (Chichester, England)*, *17*(12), 1945–1955. doi:10.1002/we.1682

Dowrick, D. (2009). *Earthquake resistant design and risk reduction* (2nd ed.). John Wiley & Sons. doi:10.1002/9780470747018

Drew, D., Barlow, J., & Lane, S. (2013). Observations of wind speed profiles over Greater London, UK, using a Doppler lidar. *Journal of Wind Engineering and Industrial Aerodynamics*, *121*, 98–105. doi:10.1016/j.jweia.2013.07.019

Dyrbye, C., & Hansen, S. (1999). *Wind loads on structures*. Chichester, UK: John Wiley & Sons.

Edwards, R., LaDue, J., Ferree, J., Scharfenberg, K., Maier, C., & Coulbourne, W. (2013). Tornado intensity estimation: Past, present, and future. *Bulletin of the American Meteorological Society*, *94*(5), 641–653. doi:10.1175/BAMS-D-11-00006.1

Eh Huang, N., & Shen, S. (2005). *Hilbert-Huang transform and its applications* (Vol. 5). Singapore: World Scientific.

Ellis, B., & Ji, T. (1997). Human-structure interaction in vertical vibrations.*Proceedings of the ICE - Structures and Buildings, 122*, 1–9.

Ellis, B., & Ji, T. (2004). The response of structures to dynamic crowd loads (No. Digest 426). Bracknell, UK: Building Research Establishment (BRE).

European Commission. (2013a). An EU Strategy on adaptation to climate change. Brussels, Belgium: European Commision (EC).

European Commission. (2013b). An EU Strategy on adaptation to climate change. Adapting infrastructure to climate change. Brussels, Belgium: European Commision (EC).

Fawcett, L. (2005). *Statistical methodology for the estimation of environmental extremes* (PhD Thesis). University of Newcastle, Newcastle upon Tyne, UK.

Fawcett, L., & Walshaw, D. (2006). Markov chain models for extreme wind speeds. *Environmetrics, 17*(8), 795–809. doi:10.1002/env.794

Fawcett, L., & Walshaw, D. (2008). Bayesian inference for clustered extremes. *Extremes, 11*(3), 217–233. doi:10.1007/s10687-007-0054-y

Fawcett, L., & Walshaw, D. (2012). Estimating return levels from serially dependent extremes. *Environmetrics, 23*(3), 272–283. doi:10.1002/env.2133

Fernández, J., Hermanns, L., Alarcón, E., & Fraile, A. (2013). A review of load models for the prediction of footfall-induced vibration in structures. In *Proceedings of 11th International Conference on Vibration Problems* (p. 9). Lisbon, Portugal: DEC/FCT/UNL and IDMEC/IST.

fib. (2000). *Guidance for good bridge design*. Lausanne, Switzerland: International Federation for Structural Concrete (fib), Bulletin No. 9.

fib. (2009). *Formwork and falsework for heavy construction. Guide to good practice*. Lausanne, Switzerland: International Federation for Structural Concrete (fib), Bulletin No. 48.

Field, C., Barros, V., Stocker, T., Qin, D., Dokken, D., Ebi, K., … Midgley, P. (2012). Managing the risks of extreme events and disasters to advance climate change adaptation. A special report of Working Groups I and II of the Intergovernmental Panel on Climate Change (IPCC). Cambridge University Press, Cambridge, UK: Intergovernmental Panel on Climate Change (IPCC).

Flay, R. (2013). Bluff body aerodynamics. In Y. Tamura & A. Kareem (Eds.), *Advanced structural wind engineering* (pp. 59–84). Tokyo, Japan: Springer Japan. doi:10.1007/978-4-431-54337-4_3

Foley, A., Leahy, P., Marvuglia, A., & McKeogh, E. (2012). Current methods and advances in forecasting of wind power generation. *Renewable Energy, 37*(1), 1–8. doi:10.1016/j.renene.2011.05.033

Fujita, T. (1985). *The downburst: Microburst and macroburst: report of projects NIMROD and JAWS (Satellite and Mesometeorology Research Project)*. Chicago: University of Chicago.

Giebel, G., Brownsword, R., Kariniotakis, G., Denhard, M., & Draxl, C. (2011). The state–of–the–art in short–term prediction of windpower (2nd ed.). Copenhagen, Denmark: ANEMOS.plus.

Graubner, C., & Proske, T. (2005). Formwork pressure - A new concept for the calculation.*Proceedings of the 2nd North American Conference on the Design and Use of Self-Consolidating Concrete and the Fourth RILEM International Symposium on Self-Compacting Concrete*, 605–612.

Graubner, C., & Proske, T. (2010). Formwork pressure of highly workable concretes– Experiments focused on setting and vibration and design approach. In *Proceedings of the 6th RILEM Conference on Self-Compacting Concrete*. Montreal: Springer Verlag.

Gulvanessian, H., Calgaro, J., & Holický, M. (2012). *Designer's guide to Eurocode: Basis of structural design* (2nd ed.). London, UK: Thomas Telford Ltd. doi:10.1680/bsd.41714

Gutenberg, B., & Richter, C. (1954). *Seismicity of the earth and associated phenomena* (2nd ed.). Princeton, NJ: Princeton University Press.

Hadipriono, F., & Wang, H. (1987). Causes of falsework collapses during construction. *Structural Safety*, *4*(3), 179–195. doi:10.1016/0167-4730(87)90012-9

Hewson, N. (2003). *Prestressed concrete bridges: Design and construction*. London, UK: Thomas Telford. doi:10.1680/pcbdac.32231

Holmes, J. (2014). Along- and cross-wind response of a generic tall building: Comparison of wind-tunnel data with codes and standards. *Journal of Wind Engineering and Industrial Aerodynamics*, *132*, 136–141. doi:10.1016/j.jweia.2014.06.022

Holmes, J. (2015). *Wind loading of structures* (3rd ed.). CRC Press.

Holmes, J., Allsop, A., & Ginger, J. (2014). Gust durations, gust factors and gust response factors in wind codes and standards. *Wind and Structures An International Journal*, *19*(3), 339–352. doi:10.12989/was.2014.19.3.339

Holmes, J., Kwok, K., & Ginger, J. (2012). *Wind loading handbook for Australia and New Zealand: Background to AS/NZS1170.2 wind actions*. Sydney, Australia: Australasian Wind Engineering Society.

Hong Kong Labour Department. (2014). *Code of practice for bamboo scaffolding safety* (3rd ed.). Hong Kong, China: Occupational Safety and Health Branch.

Huizing, J., Blakeley, R., & Ramsay, G. (1977). Falsework. *New Zealand Engineering*, *32*(1), 2–9.

Humar, J. (2012). *Dynamics of structures* (3rd ed.). Leiden, The Netherlands: CRC Press.

Ikäheimonen, J. (1997). *Construction loads on shores and stability of horizontal formworks* (PhD Thesis). Royal Institute of Technology (KTH), Stockholm, Sweden.

Irtaza, H. (2009). *Experimental and computational determination of wind loads on netted/sheeted scaffolds* (PhD Thesis). Oxford Brookes University, Oxford, UK.

Irtaza, H., Beale, R., & Godley, M. (2012). A wind-tunnel investigation into pressure distribution around sheet clad scaffolds. *Journal of Wind Engineering and Industrial Aerodynamics*, *103*, 86–95. doi:10.1016/j.jweia.2012.03.004

Irtaza, H., Beale, R., Godley, M., & Jameel, A. (2014). Pressure distribution around net clad scaffolds using computation fluid dynamics techniques. *IOSR Journal of Mechanical and Civil Engineering*, *4*, 54–62.

ISO. (2001). ISO 12494:2001 – Atmospheric icing of structures. Geneva, Switzerland: International Organization for Standardization (ISO).

ISO. (2007). ISO 10137:2007 – Bases for design of structures. Serviceability of buildings and walkways against vibrations (2nd ed.). Geneva, Switzerland: International Organization for Standardization (ISO).

ISO. (2009). ISO 4354:2009 – Wind actions on structures. Geneva, Switzerland: International Organization for Standardization (ISO).

ISO. (2015). ISO 2394 :2015 – General principles on reliability for structures (4th ed.). Geneva, Switzerland: International Organization for Standardization (ISO).

IStructE. (2007). Temporary demountable structures: Guidance on procurement, design and use (3rd ed.). London, UK: The Institution of Structural Engineers.

Jagger, T., & Elsner, J. (2006). Climatology models for extreme hurricane winds near the United States. *Journal of Climate*, *19*(13), 3220–3236. doi:10.1175/JCLI3913.1

Johnston, D. (2014). Formwork for concrete (8th ed.). American Concrete Institute (ACI), ACI SP-4.

Kareem, A., & McCullough, M. (2013). Numerical simulation of wind effects. In Y. Tamura & A. Kareem (Eds.), *Advanced structural wind engineering* (pp. 261–299). Tokyo, Japan: Springer Japan. doi:10.1007/978-4-431-54337-4_10

Kasperski, M. (2013). Estimation of the design wind speed. In Y. Tamura & A. Kareem (Eds.), *Advanced structural wind engineering* (pp. 27–58). Tokyo, Japan: Springer. doi:10.1007/978-4-431-54337-4_2

Kwok, K. (2013). Wind-induced vibrations of structures. In Y. Tamura & A. Kareem (Eds.), *Advanced structural wind engineering* (pp. 121–155). Tokyo, Japan: Springer Japan. doi:10.1007/978-4-431-54337-4_5

Leadbetter, M., Bailey, G., & Rootzén, H. (1983). *Extremes and related properties of random sequences and processes*. New York: Springer-Verlag. doi:10.1007/978-1-4612-5449-2

Metheringham, S., & Townshend, J. (2005). A1 River Tyne bridge, UK-design and construction. *Bridge Engineering*, *158*(2), 45–52.

Munich Re. (2015). *NatCatSERVICE. Loss events worldwide 1980 – 2014*. Munich Re.

Newman, J., & Choo, B. (2003). *Advanced concrete technology: Processes*. Butterworth-Heinemann.

Nieto, F., Hernández, S., Jurado, J., & Romera, L. (2010). Code provisions for wind Loads on short road bridges: Spanish IAP code, UNE-EN 1991-1-4 & 2007 AASHTO LRFD bridge design specifications. In *Proceedings of the 2010 Structures Congress* (pp. 2122–2133). American Society of Civil Engineers (ASCE). doi:10.1061/41130(369)192

Okubo, K., Hongo, T., & Kondo, K. (2010). Evaluation of wind force coefficients of the self-climbing scaffold (in Japanese). In *Proceedings of the 21st National Symposium on Wind Engineering*. Japan Association for Wind Engineering.

Palutikof, J., Brabson, B., Lister, D., & Adcock, S. (1999). A review of methods to calculate extreme wind speeds. *Meteorological Applications*, *6*(2), 119–132. doi:10.1017/S1350482799001103

Peurifoy, R., & Oberlender, G. (2010). *Formwork for Concrete Structures* (4th ed.). New York: McGraw-Hill Education.

Pierre, L., Kopp, G., Surry, D., & Ho, T. (2005). The UWO contribution to the NIST aerodynamic database for wind loads on low buildings: Part 2. Comparison of data with wind load provisions. *Journal of Wind Engineering and Industrial Aerodynamics*, *93*(1), 31–59. doi:10.1016/j.jweia.2004.07.007

Proske, D. (2008). *Catalogue of risks. Natural, technical, social and health risks*. Berlin, Germany: Springer Berlin Heidelberg.

Proske, T., & Graubner, C. (2007). Pressure on formwork using SCC - experimental studies and modelling. In *Proceedings of 5th International RILEM Symposium on Self-Compacting Concrete* (pp. 473–478). Ghent, Belgium: RILEM Publications SARL.

Proske, T., & Graubner, C. (2008). Formwork pressure asserted by highly workable concretes. *Darmstadt Concrete - Annual Journal on Concrete and Concrete Structures, 23*.

Quinion, D. (1984). Load variations in bridge falsework. In *Proceedings of IABSE 12th Congress* (p. 1104). Vancouver, Canada: International Association for Bridge and Structural Engineering (IABSE).

Racic, V., Pavic, A., & Brownjohn, J. (2009). Experimental identification and analytical modelling of human walking forces: Literature review. *Journal of Sound and Vibration*, *326*(1–2), 1–49. doi:10.1016/j.jsv.2009.04.020

Rademaekers, K., Laan, J., Boeve, S., & Lise, W. (2011). Investment needs for future adaptation measures in EU nuclear power plants and other electricity generation technologies due to effects of climate change. Brussels, Belgium: Directorate-General for Energy, European Commision (EC).

Rosignoli, M. (2010). Self-launching erection machines for precast concrete bridges. *PCI Journal*, *55*(1), 36–57. doi:10.15554/pcij.01012010.36.57

Rosignoli, M. (2013). *Bridge construction equipment*. London, UK: ICE Publishing. doi:10.1680/bce.58088

SAA. (1995). *3610:1995 – Formwork for concrete. Standards Association of Australia.*

SAA. (2002). *1170.0:2002 – Structural design actions – Part 0: General principles. Standards Association of Australia.*

SAA. (2011). *1170.2:2011 – Structural design actions – Part 2: Wind actions. Australia: Standards Association of Australia.*

Sétra. (2007). Prestressed concrete bridges built using the cantilever method. Design guide. Paris, France: Service d'études sur les transports, les routes et leurs aménagements (Sétra).

Sim, J. (2006). *Human-structure interaction in cantilever grandstands* (PhD Thesis). University of Oxford, Oxford, UK.

Simiu, E., & Scanlan, R. (1996). *Wind effects on structures: Fundamentals and applications to design* (3rd ed.). New York: John Wiley and Sons.

Solari, G. (2016). Thunderstorm response spectrum technique: Theory and applications. *Engineering Structures*, *108*, 28–46. doi:10.1016/j.engstruct.2015.11.012

Stathopoulos, T., & Baniotopoulos, C. (2007). *Wind effects on buildings and design of wind–sensitive structures* (Vol. 493). Springer. doi:10.1007/978-3-211-73076-8_1

Tesfamariam, S., & Goda, K. (2013). *Handbook of seismic risk analysis and management of civil infrastructure systems*. Cambridge, UK: Woodhead Publishing Limited. doi:10.1533/9780857098986

The Concrete Society. (2003). *Guide to flat slab formwork and falsework*. London, UK: The Concrete Society, Report CS140.

The Concrete Society. (2012). *Formwork: A guide to good practice* (3rd ed.). London, UK: The Concrete Society, Report CS30.

TIA. (2016). TIA–222–G – Structural standard for antenna supporting structures and antennas. Arlington, VA: Telecommunications Industry Association (TIA).

UNISDR. (2012). *Impacts of disasters since the 1992 Rio de Janeiro Earth Summit. United Nations Office for Disaster Risk Reduction*. UNISDR.

UNISDR. (2015). Making development sustainable: The future of disaster risk management. Global assessment report on disaster risk reduction. Geneva, Switzerland: United Nations Office for Disaster Risk Reduction (UNISDR).

Van der Hoven, I. (1957). Power spectrum of horizontal wind speed in the frequency range from 0.0007 to 900 cycles per hour. *Journal of Meteorology*, *14*(2), 160–164. doi:10.1175/1520-0469(1957)014<0160:PSOHWS>2.0.CO;2

Von Storch, H., & Zwiers, F. (1999). *Statistical analysis in climate research*. Cambridge, UK: Cambridge University Press. doi:10.1017/CBO9780511612336

Wang, F., Tamura, Y., & Yoshida, A. (2013). Wind loads on clad scaffolding with different geometries and building opening ratios. *Journal of Wind Engineering and Industrial Aerodynamics*, *120*, 37–50. doi:10.1016/j.jweia.2013.06.015

Wang, F., Tamura, Y., & Yoshida, A. (2014). Interference effects of a neighboring building on wind loads on scaffolding. *Journal of Wind Engineering and Industrial Aerodynamics*, *125*, 1–12. doi:10.1016/j.jweia.2013.11.009

Wang, L., McCullough, M., & Kareem, A. (2013). A data-driven approach for simulation of full-scale downburst wind speeds. *Journal of Wind Engineering and Industrial Aerodynamics, 123,* 171–190.

Wang, X., Guo, P., & Huang, X. (2011). A review of wind power forecasting models. *Energy Procedia, 12*, 770–778. doi:10.1016/j.egypro.2011.10.103

WEF. (2015). *Global risks 2015*. Geneva, Switzerland: World Economic Forum.

Willford, M., & Young, P. (2006). *A design guide for footfall induced vibration of structures (No. CCIP-016)*. Surrey, UK: The Concrete Centre.

Yue, F., Li, G., & Yuan, Y. (2012). Design methods of integral-lift tubular steel scaffolds for high-rise building construction. *Structural Design of Tall and Special Buildings, 21*(1), 46–56. doi:10.1002/tal.635

Chapter 4
Structural Analysis

ABSTRACT

This chapter develops the components required for successful modelling of temporary structures. It presents the principles, methods and the associated limitations that currently are seen as the state-of-the-art in structural analysis using the Finite Element Method. Material models of steel, aluminium and bamboo are presented with an emphasis on linear and multilinear models for steel and the Ramberg-Osgood model for aluminium. Models are presented for braces, props, beam-to-column connections, top connections, base connections and column-to-column connections based on the latest theoretical and experimental procedures developed by the authors and co-workers. Examples of two and three dimensional models are then developed for access scaffolds, bridge falsework and bamboo scaffolds. Finally, the chapter presents information on the effects of ground modelling and on advanced wind engineering using complex numerical methods.

4.1 INTRODUCTION

Structural analysis concerns the assessment of the internal forces and deformations of the structural system for a predefined hazard scenario, establishing the basis for the subsequent design verification compliant with the operational requirements set out in design codes for various design situations.

Structural analysis requires initially the definition of the typology of the structural system, which will then be simulated by an approximate conceptual model based on the theory of structural mechanics which impose the compatibility between material deformations and applied displacements and the equilibrium between internal forces and external actions. The basic input variables of the idealised model are the topology and geometry of the structural system and of its elements, the properties and spatial distribution of the materials used, the boundary conditions with the surrounding systems and the hazard scenario characteristics. The characteristics of the basic variables may change with time. Furthermore, uncertainties are always present due to our incomplete and insufficient knowledge of the real world.

Structural analysis involves the simulation of complex real structural system by approximate conceptual models trying to balance rigour and work feasibility.

DOI: 10.4018/978-1-5225-2199-0.ch004

Copyright © 2017, IGI Global. Copying or distributing in print or electronic forms without written permission of IGI Global is prohibited.

This Chapter develops the numerical models used in design programs with a particular emphasis on the material models used for the temporary structures made of steel, aluminium and bamboo. Sub-models for various types of connections are developed and then applied in two and three dimensional models of access scaffolds and bridge falsework structures. The modelling of soil is addressed and finally the determination of wind pressures on scaffolding is introduced before conclusions in the Chapter.

On the basis of this Chapter it is expected that the reader will acquire knowledge on the following topics:

1. Fundamentals of the Finite Element Method.
2. Differences between types of analysis methods.
3. Models of materials and of different connections of temporary works.
4. Fundamentals of models of soil.
5. Fundamentals of Computational Fluid Dynamics applied to temporary works.

4.2 FINITE ELEMENT METHOD

4.2.1 Basis

Structural analysis methods are based on structural mechanics principles. For statically determinate structures theory of elasticity, strength of materials and simple statics are enough for analysis and design. As structural systems become more complex there is a need for the development of more sophisticated methods which ideally result in more exact analysis, economical and safer designs within feasible time spans. The advance in computational science opened a new era for structural engineering based on numerical methods. Finally, it was possible to take into account in the analysis the geometrical and material nonlinear aspects of the structural behaviour under static or dynamic actions. The most popular numerical method is called the Finite Element Method (FEM) and there is a wide range of finite element analysis computer software programs available. In this Section, a compact overview will be presented about the FEM. Comprehensive presentation and discussion about the methods and procedures of the FEM may be obtained from reference documents (Bathe, 2006; De Borst, Crisfield, Remmers, & Verhoosel, 2012; Zienkiewicz, Taylor, & Fox, 2014; Zienkiewicz, Taylor, & Zhu, 2013).

In the FEM, the equilibrium between external actions and internal forces is approximated by the principle of virtual displacements: the work done by the external forces on a arbitrary (virtual) displacement field is equal to the work done by the internal stresses on the deformation field compatible with the virtual displacement field.

The structural geometry is approximated by discretising it with finite elements. As a result the displacements are only exactly known at the nodes of the finite elements. Displacements within the domain of each element are approximated using the Galerkin method by special interpolation functions (e.g. Legendre polynomials) which enforce compatibility with all kinematic constraints. The strain field, and the conjugate stress field, are obtained from the derivatives of the approximated displacement field. Consequently, the approximation of strains and stresses is at least one order lower than the approximation of the displacements.

The most common interpolation functions are those used in the formulation of isoparametric elements where the functions define both the element's geometric shape and the displacements within the element.

Structural analysis using FEM consists in four main steps:

- Definition of the geometry of the model and finite element mesh generation;
- Calculation of the stiffness matrix in global coordinate system;
- Solving of the equilibrium equations and calculation of the nodal displacements;
- Determination of the stress field.

In general, most of the computational runtime is spent in the second and third steps above, since modern finite element analysis programs have a Computer Aided Design (CAD) module that eases the development of the first step which also automatically optimises the mesh to reduce the runtimes of the following steps.

4.2.2 Solving the Equilibrium Equations

4.2.2.1 Basis

The FEM is the most frequently used numerical method to solve nonlinear equilibrium equations. There are two types of solvers: implicit and explicit solvers. The former solvers are unconditionally stable, and can solve static, quasi-static and dynamic problems, but require the system's stiffness matrix to be inverted at least at every increment. The latter solvers are conditionally stable, designed for short time events and do not require the system's stiffness matrix to be inverted.

4.2.2.2 Implicit Solvers

The most common method to solve nonlinear differential equations is the Newton-Raphson method. Alternative methods exist, such as the Riks arc-length method, but for systems exhibiting smooth equilibrium paths (i.e. with no snap through and snap back phenomena) the convergence rate obtained by using Newton's method is superior.

For static analyses, the problem resumes in solving the following equation through Newton-Raphson iterations (i):

$$^{t+\Delta t}R - {}^{t+\Delta t}F^{(i-1)} = {}^{t+\Delta t}\mathrm{K}^{(i-1)} \cdot \Delta U^{(i)} \quad \text{with} \quad {}^{t+\Delta t}U^{(i)} = {}^{t+\Delta t}U^{(i-1)} + \Delta U^{(i)} \tag{1}$$

where:

R represents the vector of externally applied loads;
F represents the vector of nodal internal forces;
K represents the stiffness matrix;

U and ΔU represent the nodal incremental displacements and the corrections to the nodal incremental displacements, respectively.

Convergence is measured by determining the value at the end of each iteration of the force residuals between the externally applied loads and the nodal internal forces, and also the corrections to the nodal incremental displacements. Convergence at the end of each increment is established if the value of the residuals and corrections is not larger than acceptable, sufficiently small, tolerance values.

For dynamic analyses, solving the nonlinear equilibrium equations is based on the following equation:

$$ {}^{t+\Delta t}R - {}^{t+\Delta t}F^{(i-1)} = \mathrm{M} \cdot {}^{t+\Delta t}\ddot{U}^{(i)} + \mathrm{C} \cdot {}^{t+\Delta t}\dot{U}^{(i)} + {}^{t}\mathrm{K} \cdot \Delta U^{(i)} \quad \text{with} \quad {}^{t+\Delta t}U^{(i)} = {}^{t+\Delta t}U^{(i-1)} + \Delta U^{(i)} \tag{2} $$

where **M** and **C** represent the mass and damping stiffnesses matrices, respectively.

For time integration of the dynamic problem the Hilber-Hughes-Taylor (HHT) method can be used. This method is an extension of the classic Newmark method, introducing the parameter, α, which controls the amount of numerical dissipation that is introduced. For $\alpha = 0$ this method reduces to the Newmark method. For quasi-static analysis the backward Euler method can be used.

4.2.2.3 Explicit solvers

Explicit solvers are a good alternative for pure dynamic problems but require a large number, for some analysis millions, of small time increments to complete the time step. The nonlinear equilibrium equations are solved based on the following equation:

$$ {}^{t}R - {}^{t}F = \mathrm{M} \cdot {}^{t}\ddot{U} + \mathrm{C} \cdot {}^{t}\dot{U} \tag{3} $$

The equations of motion for the body are integrated using the explicit central-difference integration method which is conditionally stable. The undamped stability limit is given in terms of the highest frequency of the finite element model as:

$$ \Delta t \le \frac{2}{\omega_{\max}} \tag{4} $$

The stability limit can be approximately obtained by the smallest time a dilatational wave needs to cross any of the elements in the mesh:

$$ \Delta t \approx \frac{\min\left(L\right)}{c_d} \tag{5} $$

where min(L) is the smallest element dimension and c_d is the material's dilatational wave speed which can be obtained by:

$$ c_d = \sqrt{\frac{\lambda + 2 \cdot \mu}{\rho}} \tag{6} $$

where λ and μ are the Lamé constants. For steel c_d can be taken to be equal to 5000 m/s.

Explicit solvers can also be used in quasi-static analysis but the results should be carefully analysed. It is recommended that the kinetic energy should not exceed 5% of the internal strain energy of the system.

Since explicit solvers can take a long runtime to complete it is possible to artificially boost the speed of time integration by using mass scaling techniques. By increasing the material density the highest frequency of the system is reduced and thus the maximum value of the stable time increment increases. Special attention should be paid to the results obtained.

4.2.3 Finite Element Types

Several types of finite elements have been developed: from the simplest case of beam elements to simulate linear objects, to shell elements to simulate planar objects and solid elements to simulate 3-D objects, see Figure 1. Additionally to elements, multi-point constraints may be used to simulate joints, interfaces and enforce compatibility between different element types.

For all elements, first-order (linear) and second-order (quadratic) interpolation functions can be used. All elements are integrated numerically to calculate their stiffness matrix. Usually the Gaussian Quadrature method is used to select the position of the integration points. The number of integration points in the element may vary. For full integration the number of integration points is the one necessary to integrate exactly the interpolation function. Reduced integration uses less integration points. Both methods have their relative advantages and disadvantages. Reduced integration favours smaller runtimes over accuracy (e.g. hourglass modes may occur in particular in first-order elements). Full integration may in some particular cases lead to overestimation of the stiffness matrix (e.g. in problems involving large localised plastic strains or incompressible materials).

Multi-point constraints (MPC) typically involve two nodes: one slave and one master. They behave by imposing constraints between the degrees of freedom of the slave node to those of the master node. For example, a rotational spring element consists in specifying a relationship for the relative rotation between the slave and the master nodes.

Beam elements are used when the body can be assumed to be adequately simulated by a one dimension element, i.e. when the length is much larger than the other two dimensions. Beam theory is founded in the classical Euler-Bernoulli assumption: plane cross-sections initially normal to the beam's longitudinal axis remain plane, normal to the beam axis, and undistorted. Timoshenko beam theory should be used in cases where the influence of shear strains must be accounted for in the kinematic compatibility conditions.

Figure 1. Example of beam elements (left), shell elements (centre) and solid elements (right)

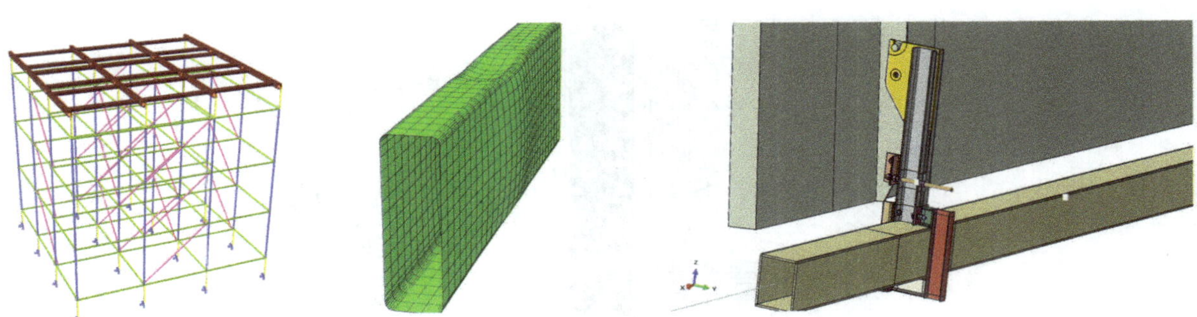

Traditional beam element formulations do not account for warping and local buckling deformation modes. More recent developments such as the Generalised Beam Theory (GBT) do account for these additional deformation modes thereby increasing the range of applicability of the beam elements.

Shell elements are suited for simulating bodies in which one of the dimensions is significantly smaller than the other two. Libraries of shell elements include triangular and quadrilateral elements. Shell elements can be formulated using one of two shell theories: (i) Kirchhoff theory, suited for thin shell elements and (ii) Mindlin theory for thick shell elements where transverse shear flexibility is non-negligible and must be accounted for.

The solid element library includes triangular, tetrahedral, quadrilaterals and hexahedral elements. Contrary to beam and shell elements, the nodal degrees of freedom available in solid elements consist only in three nodal displacements, whereas in the former elements six degrees of freedom are typically available per node.

4.2.4 Finite Element Mesh

The accuracy of the finite element model depends on the type of finite elements used but also on the quality of the finite element mesh. In general, the higher the number of elements of the same type, the better the accuracy of the solution is (*h-convergence*) but with a potential cost of a longer computational runtimes. Therefore, a compromise is generally followed between accuracy and feasibility of the numerical analysis.

However, through a wise choice of types of elements and definition of the mesh it may be possible to optimise both the accuracy and the runtime. For example, it is known that for most frequent analyses, second-order elements perform better than first-order elements and require fewer elements for the same accuracy (*p-convergence*).

A good parameter to check the quality of the finite element mesh is to analyse the distribution of stresses across the boundary between two adjacent elements. In a good quality mesh, the differences between values should be minimal. Areas where the difference is not satisfactory should be refined or the mesh should be changed to have the same direction of the stress gradients, see Figure 2.

The modelling of structures using 1-D elements (beam elements) involves the consideration of several specific aspects due to the simplifications made concerning the simulated geometry of the model (Kindmann & Kraus, 2012). Examples of subjects to consider are the case of non-prismatic elements (e.g. tapered elements) and eccentricities at joints (e.g. beam-to-column joints).

Figure 2. Example of a poor (left) and a good (right) finite element mesh

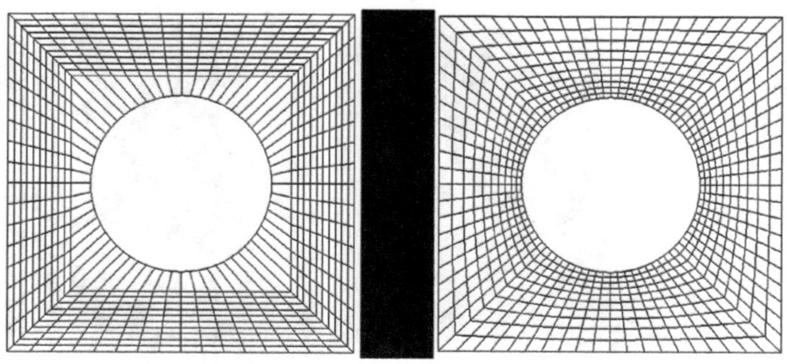

4.2.5 Verification and Validation

Verification and validation (V&V) are essential tools for the successful development of a simulation project (Oberkampf & Roy, 2010; Szabó & Babuška, 2011). Preferably it is beneficial to start the development of the numerical model based on some prior knowledge about the expected structural behaviour. The analysis of the results will also benefit from this pre-existing knowledge. Results are often displayed in terms of relationships between nodal displacements and nodal internal forces or element strains and element stresses.

During the development process of the numerical model, several assumptions and simplifications are often made. Therefore, a number of different conceptual models of the same structural system can be developed. The process of selecting the model most appropriate for the analysis objectives is called model verification. What is achieved is to answer the following question: is this the correct model? Sensitivity analyses are usually performed to assist answering this question supplemented with the use of good simulation practices.

Validation follows after verification and aims to answer the question is this the right model? Typically, validation consists in comparing the results of the numerical model with best known benchmark results from real structures comparable to the one being analysed. The legitimacy of the assumptions made during the simulation can be resolved by verifying if the accuracy achieved complies with the acceptable tolerance limits.

4.3 TYPES OF STRUCTURAL ANALYSES

4.3.1 Global Analyses

4.3.1.1 Basis

There are four distinct types of global analysis methods:

First-order elastic: initial geometry and fully linear material behaviour;

- Second-order elastic: deformed geometry and fully linear material behaviour;
- First-order elastoplastic: initial geometry and nonlinear material behaviour;
- Second-order elastoplastic: deformed geometry and nonlinear material behaviour.

In nonlinear analyses (geometrical and/or material), the superposition principle is not applicable, meaning that it is not possible to combine results from separate analysis models, for example adding internal forces obtained from individual actions (even from the same type of action).

4.3.1.2 First- and Second-Order Analyses

In a first-order analysis (elastic or elastoplastic), equilibrium equations are satisfied in the initial undeformed geometry of the structure. When only first-order results are considered, the influence of local and global deformations in the structural analysis (i.e. second-order effects) is not accounted for and therefore the design verification procedures should include it using conservative methods, such as the amplification method for sway action effects (BSI, 2005b, 2008), usually by incorporating the effect of bow displacements (P-δ effects) and of global sway displacements (P-Δ effects) at an element level. These simplified methods should also make adequate consideration of local and global stability phenomena that the structural elements may exhibit which can reduce considerably the design resistance of the system.

Second-order analysis may be used in all cases, for elastic and elastoplastic material models. Equilibrium equations are satisfied taking into account the influence of the deformation of the structure, and, therefore, reference must be made to the current deformed geometry under load.

For linear structural elements (e.g. beams, columns, braces), local stability may be accounted for analytically by using the concept of effective cross-sections, where the dimensions of the parts of the cross-section under compression stresses are reduced. The properties (area, moment of inertia, etc.) of the new cross-section are determined and used instead of the gross values of the properties. More accurately, local stability can be analysed by numerical methods, e.g. the Finite Element Method (FEM), using either traditional shell finite elements or linear elements using the Generalised Beam Theory (GBT), see Schardt (1994), or including higher-order displacement shape functions in the traditional linear finite elements formulation (Vieira, Virtuoso, & Pereira, 2014). Local stability phenomena are not significant if linear elements with compact sections are used in the structural system.

First-order analysis may be used in structural analysis only if the increase of the relevant internal forces or moments or any other change of structural behaviour caused by deformations can be neglected. The effects of the deformed geometry (second-order effects) should be considered if they increase the action effects significantly or modify considerably the structural behaviour. Second-order analyses often require the use of numerical methods involving iterative procedures.

4.3.1.3 Elastic and Elastoplastic Analysis

Elastic analysis uses theory of elasticity, which assumes a linear relationship between material's stress and strain. Elastic analysis predicts useful results for materials which undergo small reversible deformations and which behaviour is independent of the deformation rate.

However, almost all materials will undergo some permanent irreversible deformation during loading. For example, hot-rolled steels will experience permanent deformations when the stress is higher than the yield stress under a uniaxial stress state. Permanent deformations in materials are caused by plastic flow of deformations. A further difference between elastic and elastoplastic analysis, is that in the latter the numerical methods have to work with plastic flow rules, i.e. constitutive relationships between current stress and current increments of strain, not the models of the accumulated stress and strain constitutive relationships valid only for elastic analyses.

Elastic global analysis may be used in all cases. Elastic analysis is generally used to study the serviceability performance of a structure, and it can also be used to obtain member internal forces for subsequent use in the element design checks, as long as the material's behaviour stay in elastic regime. This analysis method is well accepted, can be shown to lead to safe solutions and has the great advantage that superposition of results may be used when considering different load cases, provided the constitutive law of the material is linear elastic and no significant second-order deformations occur.

Elastic analysis should be based on the assumption that the constitutive relationship of the material is elastic, often linear elastic (e.g. Hooke's law in steel), in the whole range of loading. Once more, no permanent plastic strains are formed.

Under certain strain fields, materials start to deviate from the elastic linear behaviour by beginning to accumulate plastic strains. Typically, this means that the stiffness of the material reduces from its previous linear elastic stiffness. Elastoplastic materials have the potential to withstand loads in excess of those obtained only considering the elastic limit. This is particularly of interest in the case of statically indeterminate structural systems, where internal forces from regions under elastoplastic regime are redistributed to regions where materials are still in the elastic strain range. For example, consider a beam made of an elastoplastic ductile material (i.e. capable of enduring large plastic strains), uniform cross-section, fixed at both ends under a uniformly distributed vertical load. The maximum value of elastic bending moments will be attained at both the end supports. When the yield bending moment is reached, two plastic hinges are formed. Contrary to a brittle material, for a ductile material the latter state does not imply structural failure, and the beam can resist higher loads. As the load increases, the bending moments at the plastic hinges stay constant or can continue to augment if the material exhibits strain hardening, although at a much lower rate since the post-yield stiffness is significantly smaller than the elastic stiffness. Therefore, the rotational restraint at the end supports is reduced. Consequently, the value of the still elastic positive bending moment increases until the yield bending moment is reached, from which a mechanism is attained with the formation of three plastic hinges.

Redistribution of internal forces is allowed by the static theorem of plastic analysis but it is dependent on adequate ductility of the elastoplastic regions. If certain ductility and detailing conditions are met, it may be possible to specify conservative values for the redistribution of internal forces using the results of an elastic linear analysis. In these cases, the internal forces and moments may be calculated according to elastic global analysis even if the resistance of a cross section is based on its plastic resistance.

Due consideration should be paid when performing redistribution of elastic internal forces to serviceability verification, since when the elastic behaviour deviates significantly from the elastoplastic behaviour considerable structural deformations are to be expected. Therefore, elastoplastic global analysis may be used only when the structure fulfils specific requirements (in general specified in design codes, see Chapter 6).

Elastoplastic analysis allows for the effects of material nonlinearity in calculating the action effects of a structural system. The behaviour should be modelled by one of the following methods:

- Plastic hinge method (concentrated plasticity);
- Nonlinear elastoplastic analysis (distributed plasticity).

Simplified elastoplastic analysis methods, such as plastic hinge, imply not only full plastification within the cross-section, but also sufficient internal force redistribution within the structural system in order to develop all the plastic hinges that are needed to give rise to a plastic mechanism. As a result, elements characteristics (cross-section geometry, joint configuration and material properties) must satisfy special requirements by which sufficient deformation capacity can be exploited to enable the required redistributions of internal forces to develop under static and/or dynamic loading:

- Material with adequate ductility by using materials with a minimum strain after fracture and minimum ratio between the ultimate tensile strength and the yield stress;
- No local stability phenomena by using compact sections;
- No global instability phenomena occurs by providing appropriate bracing;
- No brittle failure at joints by correctly designing and detailing the joint components.

Figure 3. Illustration of structural behaviour obtained with different analysis models

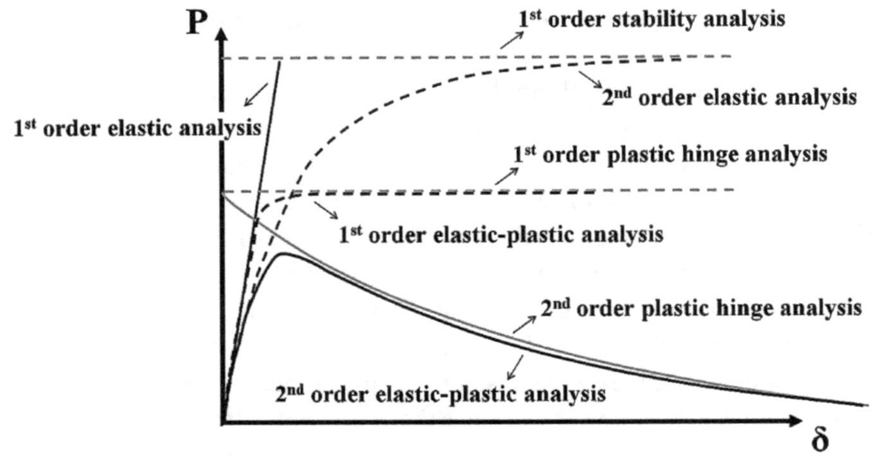

First order plastic hinge method is also known as rigid plastic method.

Figure 3 illustrates an example of different results that can be obtained using the various possible types of structural analyses, assuming instability phenomena does not occur before full plastic resistance is attained.

4.3.2 Imperfections

Appropriate allowances should be incorporated in the structural analysis to cover the effects of initial imperfections, including residual stresses and geometrical imperfections such as lack of verticality, lack of straightness, lack of flatness, lack of roundness, dimples and minor eccentricities present in joints due to fabrication and/or erection operations. The joint effect of the most common types of deviations from the idealised perfect state of the structure is often considered in a simplified manner using equivalent initial geometric imperfections whose values were calibrated to obtain conservative results.

The following equivalent initial geometric imperfections are typically introduced in the analysis:

- Global imperfections expressed as initial sway imperfections of the system;
- Local imperfections such as element out of straightness, joint looseness and load eccentricities.

Often the assumed shape of global initial imperfections and local element initial imperfections are derived from the first elastic instability mode of the structure obtained from a Linear Elastic Stability Analysis. A careful validation analysis should be performed before attempting to use only the first elastic instability mode during design, in particular when equivalent lateral loads are used to simulate the effect of the initial geometric imperfections. This is because the structure might be sensitive to a combination of types of initial geometric imperfections in particular in the elastoplastic phase, but also because the actual critical loading pattern may be very different from the one considered to obtain the first elastic instability mode. Alternatively, the magnitudes and shapes of the admissible initial imperfections may be defined by agreement between the designer and the producer of the elements and the entity responsible for the assembly of the structure. This procedure is especially important for thin-walled steel elements which are very sensitive to imperfections.

4.4 MATERIALS MODELLING

4.4.1 Basis

Material models consist of relations between internal forces or stresses on the one hand and deformations on the other (i.e. flow rules, or constitutive relationships for uniaxial states).

Constitutive relationships are typically based on the results of tests carried out on small size specimens in laboratory conditions. For the convenience of structural analyses and structural design, mathematical representations of the material behaviour are used, for instance in the form of a stress-strain curve under uniaxial loading. Therefore, structural materials should be modelled by functions and parameters (material properties) describing (generalized) stress-strain relationships with a detailing as relevant. The most commonly used variables in such relations are the modulus of elasticity, the yield strength, the ultimate strength, the strain at fracture, etc, under uniaxial loading.

Constitutive relationships for elastomers need only to account for the elastic response (possibly nonlinear, i.e. viscoelastic). For metals, elastoplastic relationships are needed to simulate the behaviour materials after yielding. Concrete and soil materials require relationships which take into account the frictional mechanisms and brittle failure.

Constitutive relationships may be function of deformation rate and of degradation of material properties with damage initiation and accumulation (e.g. stress and strain softening), and eventually include material failure (e.g. tearing or ripping) by removing elements from the mesh.

The material's constitutive relationship may be defined in engineering measures or true measures of stress and strain. The former are determined with respect to the original dimensions of the element (area and length, A_0 and L_0, respectively), see Eq. 7, whereas the latter refer to the current dimensions of the element (A and L), see Eq. 8.

$$\sigma_e = \frac{F}{A_o} \quad , \quad \varepsilon_e = \frac{L - L_o}{L_o} \tag{7}$$

$$\sigma_t = \frac{F}{A} \quad , \quad \varepsilon_t = \int_{L_o}^{L} \frac{1}{L} dL = \ln\left(\frac{L}{L_o}\right) \tag{8}$$

In a multiaxial stress state, the material yield criterion may be achieved before one of the principal stress components equals the material's yield stress. In general, the yield criterion is expressed by a yield surface, f, given by:

$$f\left(\sigma_1, \sigma_2, \sigma_3, \mathrm{n}, \mathrm{k}\right) = 0 \tag{9}$$

where σ_1, σ_2 and σ_3 are the principal stresses of the material, **n** is the unit vector of the directions of the principal stresses with respect to the material orientation, and **k** is a vector with the strain hardening parameters.

Additionally, in a multiaxial stress state, the flow rule that governs the relationship between plastic deformations and stresses needs to be defined. A common hypothesis is to assume that the multiaxial flow rule resembles the uniaxial constitutive relationship, with suitable calculated parameters called effective plastic strains and effective stresses, see Chen & Han (2007).

If the flow rule is a model where the plastic strain increments can be determined directly from the derivative of the yield criterion with respect to the stresses, it is said to be an associated flow rule. When the latter is not true, it is said to be a non-associated flow rule. In associated flow rules, the plastic strain increments vector develops along the normal to the yield surface. Examples of such flow rules are the ones associated with Tresca and von Mises yield criteria. These flow rules are valid for stable materials (i.e. in which a positive work is done for increasing loads). Therefore, they are valid for strain hardening elastoplastic materials, but not in general for strain softening elastoplastic materials (in particular if softening occurs when the material stress-strain state is located along the yield surface). In the latter cases, a non-associated flow rule is required, which may be associated with the Mohr-Coulomb and with the Cam clay yield surface models.

4.4.2 Steel and Aluminium

The materials used in temporary structures are usually steel and aluminium for the load bearing members. However, in Asia bamboo is sometimes used for scaffolding. The formwork used to support concrete whilst being formed is often from timber and plywood.

In Europe, the tubes used in scaffolding and falsework structures are usually made from steel or aluminium and have normally have external diameters of 48.3 mm with thicknesses of either 3.2 mm or 4.0 mm (BSI, 2001). They are usually supplied in lengths of 6.0 m or 6.4 m but shorter lengths can be specially prepared. To ensure satisfactory performance, the allowable tolerance of out-of-straightness must be less than 0.2% of the overall length with no more than a 3 mm out-of-straightness in any 1 m sub-length. The tubes must in addition be subjected to tensile tests in accordance with the international standard ISO 6982-1 (ISO, 2009a), see Figure 4, and flattening tests in accordance with ISO 8492 (ISO, 2013). Figure 5 illustrates a stress-strain curve obtained from a tensile test of hot rolled steel sample. The authors have experience in conducting tests on tubes and found that typical tensile testing machines may not have sufficient force capacity to determine the ultimate tensile strength. In these cases the specimen's thickness may be reduced to a sufficient value by milling the outside diameter.

Figure 4. Tensile test on a tubular sample

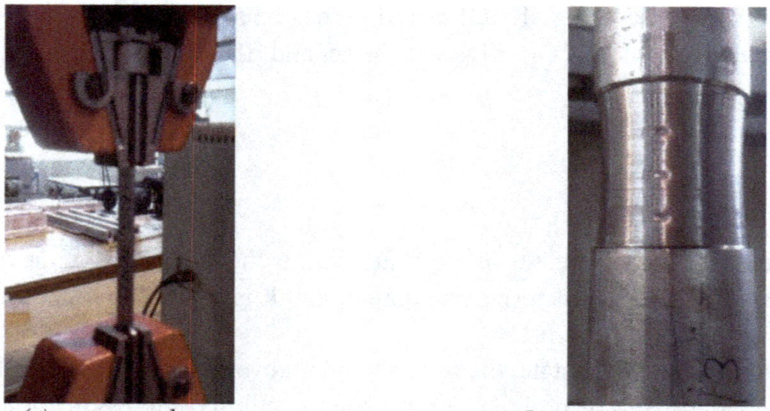

(a) test procedure *(b) sample just before failure*

Figure 5. Schematic stress-strain curve of a hot rolled steel under tensile axial load

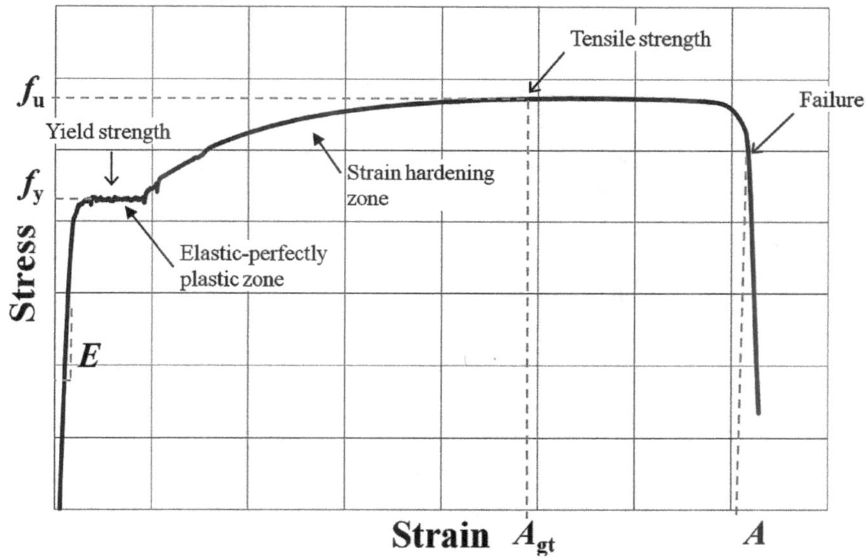

Many components of proprietary scaffold and falsework systems, such as the cups, wedges and tongues for the connections between different standards or to horizontal members can only have their ultimate capacity tested by Hardness tests such as the Brinell test (ISO, 2014) or Vickers test (ISO, 2005) as they are too small or too irregularly shaped to produce the standard "dog bone" tensile specimen. This means that for these components a material model cannot be determined experimentally but standard curves found in material textbooks have to be used.

Concerning BCE, the elements are constituted by large sections made of structural steel, specified in accordance with applicable national product standards, such as BS EN 10025-2 (BSI, 2004) in Europe.

The uniaxial constitutive relationships for steel and aluminium are usually linear, multi-linear or nonlinear. Mild steel can be modelled as a linear elastic material up to its yield point after which it deforms plastically with no increase in stress before going into a strain hardening regime (e.g. stress strengthening induced by plastic deformation).

Many uniaxial constitutive relationships of mild steel only consider the elastic state and the perfectly plastic state until failure occurs (ignoring strain hardening). High strength and stainless steel have nonlinear stress-strain curves which, depending upon the analysis software capabilities, can either be modelled by a polynomial curve found by regression from experimental data, or by a series of multi-linear straight lines, again taken from experimental results, see Figure 6.

Therefore, a range of uniaxial constitutive relationships is available to describe the steel's behaviour. For example (Ottosen & Ristinmaa, 2005):

- Isotropic, linear elastic model;
- Isotropic, elastic-perfectly plastic model which assumes the steel to behave perfectly plastic up to a limiting maximum extension, at which point the material breaks;
- Isotropic, elastoplastic models with isotropic, kinematic or mixed strain hardening.

Figure 6. Stress-strain curves for mild steel and low or high alloy steel, or aluminium

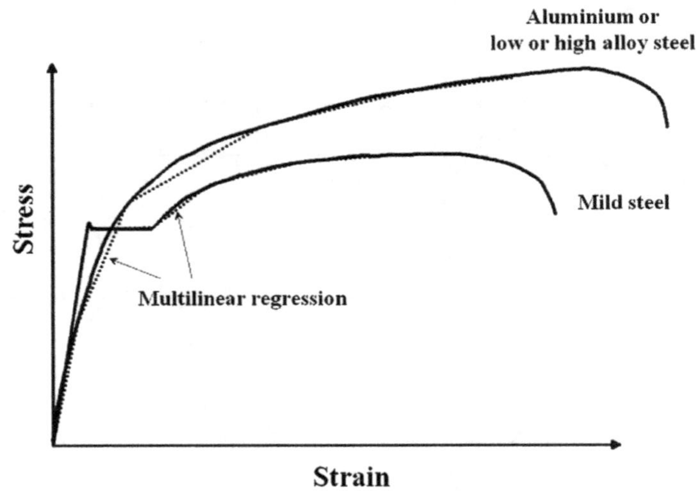

An isotropic material is one in which the mechanical properties (e.g. the uniaxial yield stress) are the same, independent of the stress and material directions. For such a material, the flow rule can be considered to express a proportional relationship between strain increments and deviatoric stress values. Metals may be accurately modelled by an isotropic model for small to moderately large strains.

Isotropic hardening assumes that strain hardening causes the yield surface to expand as stress increases. This approximation is valid for monotonic loading regimes. Kinematic hardening is required when hysteretic behaviour in the plastic strains range is relevant (e.g. cyclic loading). It is known that in these conditions the yield surface of steel no longer remains symmetric due to the Bauschinger effect (e.g. strain-hardening in one direction reduces the yield stress in the opposite direction). Mixed hardening combines the features of the previous two models.

For complex three-dimensional stress states, the most often used yield criteria are the von Mises and the Tresca yield surfaces.

The elastic-perfectly plastic von Mises criterion is given by:

$$\sigma_y = \sqrt{\frac{1}{2}\cdot\left[\left(\sigma_1 - \sigma_2\right)^2 + \left(\sigma_2 - \sigma_3\right)^2 + \left(\sigma_3 - \sigma_1\right)^2\right]} \quad (10)$$

where σ_y represents the uniaxial yield stress of the material and σ_1, σ_2 and σ_3 are the principal stresses of the material.

The elastic-perfectly plastic Tresca criterion states that the shear strength at yielding is given by:

$$\sigma_y = \max\left(\left|\sigma_1 - \sigma_2\right|, \left|\sigma_2 - \sigma_3\right|, \left|\sigma_3 - \sigma_1\right|\right) \quad (11)$$

Both the Tresca and the von Mises yield surfaces do not change shape as a function of the hydrostatic pressure being only dependent of the deviatoric stress tensor. Therefore, the topologic forms of both criteria are cylinders with symmetric cross-sections evolving along an axis defined by a line where each point corresponds to equal values of the three principal stress components.

It is noticeable that in the case of uniaxial stress both criteria yield the same stress for failure but that in a pure shear situation the Tresca shear stress is σ_y / 2 whereas the equivalent von Mises stress is σ_y / $\sqrt{3}$. The von Mises criterion is the commonest in use.

The isotropic elastic properties of steel under uniaxial loading are usually defined by the Young's Modulus of Elasticity, *E*, and the Poisson's coefficient, *υ*, taken as 0.3. In Europe, *E* is taken as 210 N/mm^2 whereas in the USA, it is taken as 200 N/mm^2. Statistical analysis of results of tensile testing of steel coupons showed that the former value is a good estimate of the average value for all grades of steel available in Europe (Simões da Silva et al., 2009). Regarding variables that define the uniaxial plasticity regime, such as yield strength, tensile strength and strain at fracture, results from the previous study are presented in Table 4.1 to Table 3 and in Chapter 5. Values to be used in the design are provided in design codes as exemplified in Chapter 6.

Table 1. All results related to f_y, steel grades according to EN 10025-2 (Simões da Silva et al., 2009)

Steel grade	Thickness	Number of tests	Average value (MPa)	Standard deviation (MPa)	Coefficient of variation (CoV)
S275	≤16	1991	327.93	18.96	0.06
	>16 ≤ 40	2342	306.28	15.63	0.05
	>40 ≤ 63	71	299.23	14.07	0.05
	>63 ≤ 80	21	290.38	9.68	0.03
S355	≤16	733	419.38	20.25	0.05
	>16 ≤ 40	1146	395.82	15.16	0.04
	>40 ≤ 63	77	380.51	10.01	0.03
	>63 ≤ 80	23	361.87	10.25	0.03
S460	>3 ≤ 50	666	474.63	20.29	0.04
	>50 ≤ 100	6	476.00	14.14	0.03

Table 2. All results related to f_u, steel grades according to EN 10025-2 (Simões da Silva et al., 2009)

Steel grade	Number of tests	Average value (MPa)	Standard deviation (MPa)	Coefficient of variation (CoV)
S275	4132	476.10	13.85	0.03
S355	1972	533.44	16.53	0.03
S460	672	632.73	23.18	0.04

Table 3. All results related to ε_u, steel grades according to EN 10025-2 (Simões da Silva et al., 2009)

Steel grade	Number of tests	Average value (MPa)	Standard deviation (MPa)	Coefficient of variation (CoV)
S235	10	31.50	4.82	0.15
S275	12	29.83	6.70	0.22
S355	33	26.45	6.13	0.23
S690	20	16.99	1.55	0.09

The uniaxial stress-strain curve of aluminium has a very small elastic part and the yield stress is usually not clearly defined. Therefore, it is more common to use a nonlinear curve from the start such as the Ramberg-Osgood formula, Eq. 12.

$$\varepsilon = \frac{\sigma}{E} + p \cdot \left(\frac{\sigma}{\sigma_{\mathrm{p}}} \right)^{n} \tag{12}$$

where ε is the strain, σ is the stress, σ_{p} is the proof strength corresponding to the plastic strain p, (typically taken as the 0.2% proof stress) and n is the Ramberg-Osgood parameter (strain hardening coefficient) and is greater than five. An improved version of the Ramberg-Osgood formula is given in Rasmussen (2003). Care should be taken when using this model for cyclic loading.

Simple analysis procedures are to use the initial linear part of the aluminium uniaxial stress-strain curve as the elastic stiffness. To consider the onset of plasticity, the proof stress may be used, determined by placing a straight line, with a slope equal to the elastic stiffness, at 0.2% strain and the point where this straight line intersects the stress-strain curve is defined to be the equivalent to the yield stress of the material.

The Eurocode for aluminium BS EN 1999-1-1 (BSI, 2013), suggests determining the appropriate value of n by Eq. 13:

$$n = \frac{\ln\left(0.002 / \varepsilon_{\mathrm{x}}\right)}{\ln\left(\sigma_{\mathrm{y}} / \sigma_{\mathrm{x}}\right)} \tag{13}$$

where σ_{x} and ε_{x} are the stress and strain at a second point on the experimental curve (the code suggests using the 0.1% proof stress point). The Eurocode gives tables of the parameter n for different aluminium alloys.

4.4.3 Timber and Plywood

Timber and plywood are considered to be orthotropic elastic materials. This means that they have different elastic moduli in three orthogonal directions (longitudinal, radial, and tangential directions). The standard isotropic Hooke's law $\sigma = E{\cdot}\varepsilon$, where σ is the stress, E, Young's Modulus of elasticity and ε the strain, becomes:

$$\begin{bmatrix} \sigma_{\mathrm{x}} \\ \sigma_{\mathrm{y}} \\ \sigma_{\mathrm{z}} \\ \tau_{\mathrm{xy}} \\ \tau_{\mathrm{yz}} \\ \tau_{\mathrm{zx}} \end{bmatrix} = \begin{bmatrix} C_{11} & C_{12} & C_{13} & & & \\ C_{21} & C_{22} & 23 & & & \\ C_{31} & C_{32} & C_{33} & & & \\ & & & C_{44} & & \\ & & & & C_{55} & \\ & & & & & C_{66} \end{bmatrix} \begin{bmatrix} \varepsilon_{\mathrm{x}} \\ \varepsilon_{\mathrm{y}} \\ \varepsilon_{\mathrm{z}} \\ \gamma_{\mathrm{xy}} \\ \gamma_{\mathrm{yz}} \\ \gamma_{\mathrm{zx}} \end{bmatrix} \tag{14}$$

where **C** is the stiffness matrix tensor (symmetric), and C_{11}, C_{22} and C_{33} are moduli of elasticity in the x, y and z directions respectively, $C_{13}=C_{31}$, $C_{23}=C_{32}$ and $C_{12}=C_{21}$ are cross compliances relating extension in one direction to the extension in a direction at right angles. C_{44}, C_{55} and C_{66} relate the shear stresses and strains in the corresponding axes. These terms are all determined by experimental procedures. Orthotropic elastic properties of plywood are given in Gerrand (1987).

4.4.4 Bamboo

Bamboo is used as a structural material in Asia, particularly the Philippines, Hong Kong and Southern China. Figure 7 shows a diagrammatic cross-section of a typical structural bamboo culm. A disadvantage of using bamboo is its propensity to be attacked by fungi and insects. However, if properly treated a bamboo pole can be used for between 10 to 15 years (Chan & Xian, 2004).

The properties of structural bamboo were reported by Chung and co-workers (Chung, Yu, & Chan, 2002; Yu., Chung, & Chan, 2003). They analysed two varieties of structural bamboo – Kao Jue (*Bambusa pervariabilis*) and Mao Jue (*Phyllostachys pubescens*). Compression and bending tests yielded two forms of failure: end bearing and splitting, see Figure 8. For Kao Jue, the authors discovered that the external diameter for the columns was constant at approximately 45 mm but that the internal diameter varied from 4 mm at the top to 8 mm at the bottom. Mao Jue, on the other hand, had external diameters varying from 80 mm at the bottom to 60 mm at the top with thicknesses varying 10 mm to 6 mm. The Young's Modulus varied from about 6 kN/mm^2 to 12 kN/mm^2.

The mechanical properties were found to vary considerably between wet and dry conditions. For example, in dry conditions Kao Jue had a bending and axial compressive strength of over 75 N/mm^2 which reduced in both cases to 35 N/mm^2 in wet conditions. Mao Jue had a dry axial compressive strength of 115 N/mm^2 which reduced to 40 N/mm^2 in wet conditions but its bending strength of 50 N/mm^2 was independent of wet or dry conditions. To obtain the properties, tests can be conducted on compression samples with the length equal to twice the external diameter of the culm in accordance with ISO 8375 (ISO, 2009b) or ASTM D143-14 (ASTM, 2014). In addition, strengths vary from the bottom to the top of culm. Figure 9 shows a stress-strain curve for Kao Jue. Note that D represents the external diameter of the culm and *m.c.* the percentage moisture content. Figure 10 shows the influence of moisture on ultimate stress.

Figure 7. Schematic of bamboo culm

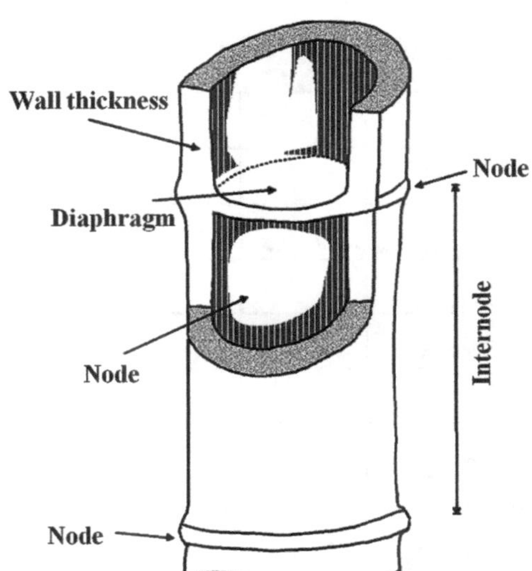

Figure 8. Bamboo failure modes (Yu. & Chung, 2000)

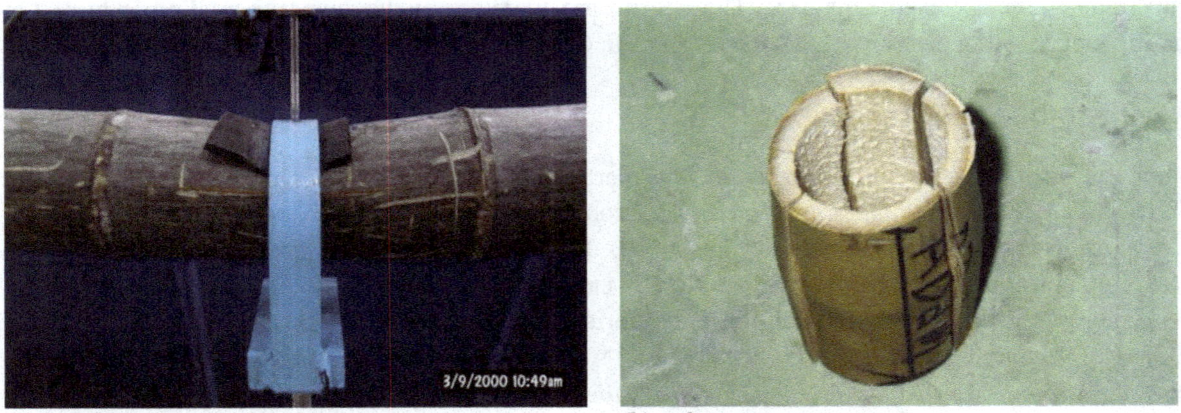

(a) local crushing in bending *(b) splitting in compression*

Bending strength is also highly dependent upon moisture content, see Figure 11.

The shear strength of the Kao Jue and Mao Jue is significantly lower than the bending and compressive strengths, being approximately 10 N/mm^2, being also very sensitive to the moisture content (Chan & Xian, 2004). The shear strength increases with age before six years and also increases with the height of the culm above the ground (Ota, 1950).

The tensile strength of bamboo is relatively high being of the order of 200 – 300 N/mm^2. However, since the transverse compressive strength is low, tensile tests must be made on thin bamboo strips (Chan & Xian, 2004).

The simplest material model of bamboo is to use an isotropic elastic material constitutive relationship. However, due to its fibrous nature more accurate results may be obtained using a composite layered cross-section with orthotropic materials.

Figure 9. Axial compressive stress against strain for Kao Jue (Yu. & Chung, 2000; Yuen, 1994)

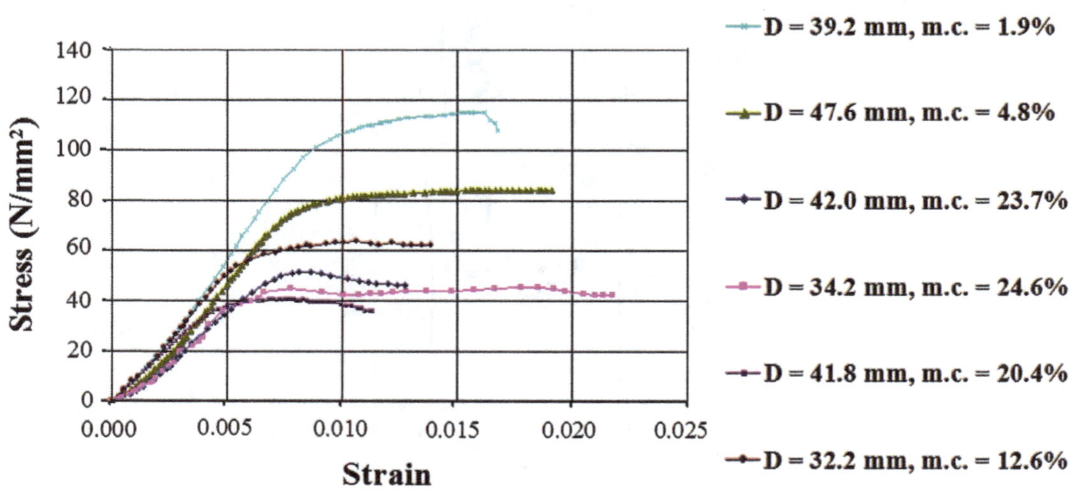

Figure 10. Axial compressive strength vs. moisture content of Mao Jue (Yu. & Chung, 2000)

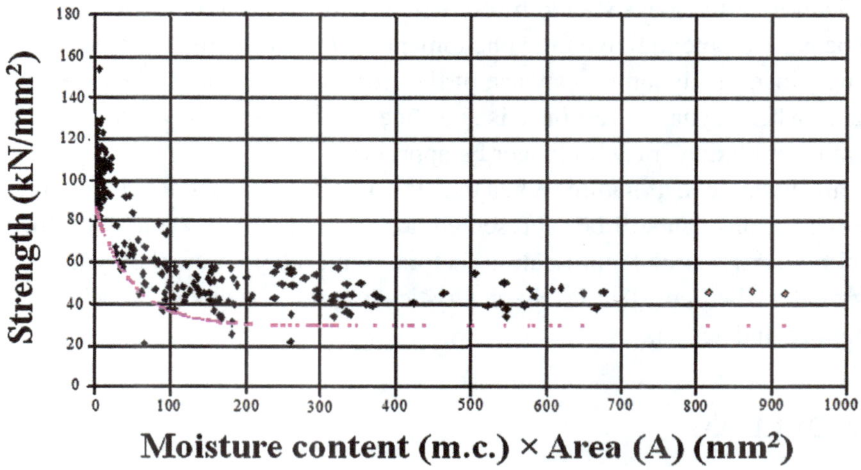

Figure 11. Load – deflection curve for a three-point bending test for a Kao Jue culm (Yuen, 1994)

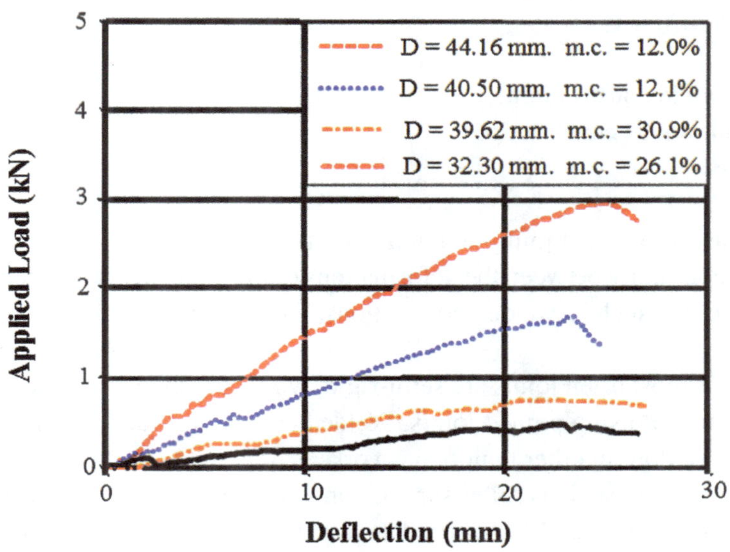

4.5 ACTIONS MODELLING

As was detailed in Chapter 3, the nature of actions that need to be considered during design of structures, in particular temporary works, is diverse and various classification schemes are available (e.g. time dependent or time independent, fixed or free distribution in space, and static or dynamic in nature).

The assumptions included in the analysis model to simulate the actions, or the actions effects, should be appropriate for predicting the structural behaviour with an acceptable level of accuracy and precision.

Static actions may be modelled as point loads, line loads or distributed loads (written in ascending order of accuracy). The choice between each model to use should be based on the nature of the action (e.g. vehicle wheel load and material self-weight) and the sensitivity of the structure to each option. The area of application of the loads should be consistent with the design intended use of the structure.

Care should be paid to potential dynamic phenomena that the structure behaviour may exhibit under continued loading, since the dynamic response of the structure is influenced by the mass which is applied to it. In cases where dynamic response is significant, static models of actions which could result initially in accurate simulations may no longer be appropriate.

Most actions are dynamic in its nature. When the dynamic response of the structure is only moderately nonlinear, dynamic actions can still be represented accurately by equivalent static load models using appropriately calibrated dynamic amplification factors. In all other cases, the dynamic actions need to simulated using models that represent its characteristics (e.g. mass, velocity, acceleration). See Section 4.9 for modelling of wind action.

4.6 JOINT MODELLING

4.6.1 Basis

For the purposes of analysis and design, joints in structural systems can be classified in the following categories in terms of stiffness:

- Pinned joints (simple construction);
- Rigid joints (continuous construction);
- Semi-rigid joints.

Simple connections assume that joints do not transmit bending moments. In semi-rigid connections, there is only a partial continuity between the deformations of the elements connected at the joint. Finally, in continuous connections, such as welded connections, the behaviour of the joint may be modelled as a rigid connection.

In temporary works, most often joints are semi-rigid, which for analysis purposes can in some cases be conservatively simulated using pinned joints. Welded connections are used in Bridge Construction Equipment (BCE) but seldom in other temporary works.

Other types of joints exist, such as roller supports in which the sliding displacements are free.

In terms of resistance, joints can be classified as full strength or partial strength joints, depending if the resistance against a given internal force of the joint is larger or lower than the corresponding resistance of the connecting elements, respectively.

Assumptions on the joint analysis should be translated to the design of joints. For example, a pin joint should have the capacity to rotate under loading with minimum restrictions. Therefore, the maximum allowable rotation capacity at the joint should be verified and compared against the required rotation of the pinned joint.

The effects of the behaviour of the various types of connections, on the distribution of internal forces and deformations within a temporary structure, may be significant.

In order to assess accurately the behaviour of temporary works, the analysis should simulate the joints by suitable models that are able to replicate the potential highly nonlinear performance of the connections under applied loading. These joint models may be determined experimentally, numerically or be based

Figure 12. Sample moment-rotation curve showing regression curve with looseness removed

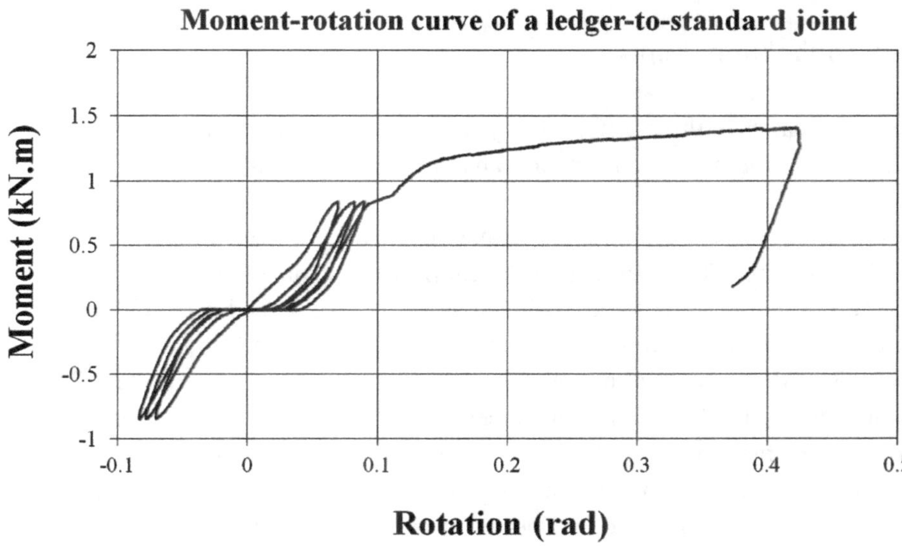

on theoretical conceptual models calibrated using results of the previous methods. Figure 12 shows the results of a typical experiment to determine the moment-rotation characteristic curve of a typical scaffold connection - called the M-θ curve. It can clearly be seen that the curve is inherently highly nonlinear and exhibits considerable looseness when the connection is subjected to cyclic loads, causing rotations in opposite directions as could occur if the structure is subjected to wind or seismic actions.

Regarding falsework and scaffolding, very few studies have been carried out using advanced and complex three-dimensional numerical models of joints. (Pieńko & Błazik-Borowa, 2013) modelled a type of wedge connection of beam-to-column joints using solid 3-D finite elements. However, time and cost involved as well as the uncertainty inherent in the analysis make this method less popular for practical use. Alternatively, mathematical models obtained by curve fitting the experimental data with simple expressions could be used. The latter method is more commonly used due to its simplicity and relative ease of integrating it in the analysis program.

Regarding BCE, the structural system most often resembles solutions already in use for heavy construction of buildings and bridges. Frequently, joints in BCE consist of bolted and welded rigid connection assemblies, although shear-pin connections in splices of modular units are often used for fast site assembly. Extensive past investigations have focused on studying the behaviour of these types of joints and on their correct detailing. Reference is made to the information presented in the following reference documents (Faella, Piluso, & Rizzano, 1999; Ivanyi & Baniotopoulos, 2014; Tamboli, 2009). For this reason, the remainder of the Section concerns only those types of joints most frequently used in other temporary work structures.

4.6.2 Experimental Determination of Joint Behaviour

4.6.2.1 Beam-To-Column Joints

Figure 13 shows schematics of three different ways of determining moment-rotation curves for the joints between standards (columns) and ledgers or transoms (beams) in proprietary and tubular scaffolds/ falsework systems.

The cantilever test is the commonest experimental method to determine the M-θ about an axis at right-angles to the standard. The lever arm of the cantilever is typically between 400 and 600 mm.

Figure 13(a) shows that in order to get the most accurate results the horizontal member is restrained from rotating by a pair of guide columns situated beyond the jack. In addition, to avoid errors due to shear or beam deflections in the horizontal member (see Abdel-Jaber, Beale, & Godley (2006)) the rotation of the connection is measured using two transducers placed on a small lever placed adjacent to the connection. The European standard BS EN 12811-3 (BSI, 2002) requires initial pilot tests monotonically loaded directly to failure in both positive and negative directions to obtain an estimate of the ultimate moment capacity. This is then followed by a series of tests cycled between

$$-0.1 \cdot \frac{R_k^-}{\gamma_R \cdot \gamma_E} \text{ and } +0.1 \cdot \frac{R_k^+}{\gamma_R \cdot \gamma_E}$$

of the ultimate loads in each direction (these can be, and often are, not the same, as many connections have different behaviour when loaded in either direction as wedge connections are unsymmetrical) three times before going to ultimate failure. R_k^+ and R_k^- are respectively the maximum moments in the positive and negative directions obtained from the pilot tests, γ_R and γ_E are respectively the partial factors for resistance and action, normally taken as 1.1 and 1.5. The analysis procedure is then:

1. Plot the experimental results (as seen in Figure 12) and determine a regression curve using the positive part of the third cycle to failure but ignoring that part of the curve showing looseness and the regression curve shifted so that it goes through the origin. Note that the regression curve is only accepted if its correlation coefficient, $r^2 \geq 0.95$. The one shown was better than 0.995.

Figure 13. Schematics of test procedures for rotational connections

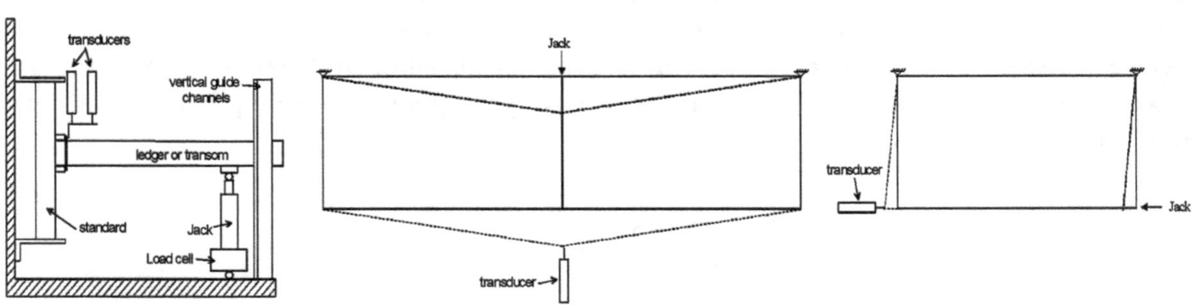

Determine the slope of the unloading curve by either taking measurements on the experimental results or by fitting a regression straight line to the unloading curve. The unloading curve can either be taken from the final unloading curve or if there is too much distortion or failure from the last cyclic unloading curve.

2. Evaluate the areas under the loading and unloading curves, E_{lo} and E_{ul} and determine the dissipation of energy quotient:

$$q_e = \frac{E_{lo}}{E_{ul}} \tag{15}$$

3. If $q_e > 11$ determine the maximum load as that load where the area under the loading curve makes $q_e = 11$. If $q_e < 11$ the maximum load is determined to be the load corresponding to the failure point on the graph.
   Determine

$$\overline{q}_e = \frac{1}{n} \cdot \sum_{i=1}^{n} q_e(i) \tag{16}$$

   by averaging the results of the cyclic tests.
4. The partial factor γ_{R2} is determined by the following equation

$$1.25 \geq \gamma_{R2} = -0.025 \cdot \overline{q}_e + 1.275 \geq 1.00 \tag{17}$$

5. The characteristic value of the moment $R_{k,mon}$ is obtained using the maximum calculated value corresponding to $\overline{q}_e$ and applying a statistical adjustment to obtain the 5% fractile value.
   An adjustment to take account of the actual material mechanical properties of the tested samples is also performed. This is done by determining the ratio of the tensile strength of the materials used in the construction of the connection as compared to their nominal characteristic strengths as supplied by the manufacturer and reducing the resistances accordingly. See example below. Note that the maximum value of the adjustment ratio is 1.00. Also, note that the lowest value of the adjustment ratio is used if different components have different ratios.
6. The nominal value of the maximum moment $R_{k,nom}$ is obtained by dividing the reduced moment by γ_{R2} and the design value by dividing this in turn by $\gamma_E \cdot \gamma_R$.
   The initial stiffness of the connection is the harmonic mean of the line between the origin and the design load or moment (secant stiffness) and the unloading curve on the mean moment-rotation or load-deflection curve. The second stiffness is the slope of the secant between the design moment and the characteristic load or moment.

For example, let us consider the determination of the design and ultimate moments of a ledger/standard connection.

Five tensile tests on ledger tube with a nominal ULS of 255 N/mm^2 had the following show in Table 4.

The characteristic strength is determined from the logarithmic determination of standard deviation using the test with tests at 5% quantile, 75% confidence, which gives a test value of 2.46.

Hence, ln(Characteristic strength) = 5.582 – 0.193 · 2.46 = 5.535.

Therefore, characteristic strength = exp(5.535) = 253.3 N/mm^2.

Adjustment ratio = 255/265.8 = 0.988.

Similar calculations are made for all the tensile tests and the lowest chosen, say 0.952 for example.

q_e is now determined for all the moment rotation tests. This is most simply done by using the regression curve derived for each test and calculating the area under the curve to the maximum moment and dividing by the area under the unloading curve, taken as a straight line. If the value of q_e is above 11.0 then the maximum allowed moment is that moment which produces a ratio of 11.0.

If we assume that $\overline{q}_e$ = 9.318 (found as a normal non-logarithmic average).

$$\gamma_{R2} = -0.025 \cdot \overline{q}_e + 1.275 = -0.025 \cdot 9.318 + 1.275 = 1.042 \tag{18}$$

From Table 5 the mean ln(max moment) = 0.307 with a standard deviation of 0.024. Hence mean moment is 1.359 kN.m.

This must be reduced by the adjustment factor as the tensile strength showed that the tested material was stronger than the characteristic material strength to give $R_{k,b}$ = 1.359 · 0.988 = 1.343.

Therefore, characteristic R_{mom} = 1.343/1.042 = 1.289 kN.m.

This value is taken as the maximum moment that the connection can sustain.

The design moment is $R_{mom}/(\gamma_E \cdot \gamma_R)$ = 1.289/1.042=0.781 kN.m.

The design rotational stiffness of the connection is taken as the slope of the straight line of the regression line from the point where it crosses the x axis to the design moment value.

Above the design moment, a straight line is taken connecting the design moment to the maximum moment and the slope of that line gives the reduced stiffness to the allowed maximum moment. Rotations in excess of that one which gives the maximum moment have a stiffness of zero up until the connection fails.

The above calculations ignore the looseness that occurs in the connection and hysteretic differences between loading and unloading. The looseness is determined in two stages. Firstly, half the difference in straight line tangents determined from the last cycle of loading in the positive and negative directions (the difference between points A and B in Figure 14) and secondly, half the hysteretic difference between loading and unloading in the same direction (points B and C). This looseness is often added into the initial slope to give a reduced initial stiffness of the connection.

Table 4. Tensile test sample calculation

Test Number	ULS (N/mm^2)	ln (ULS)
1	268	5.59099
2	261	5.56452
3	272	5.60580
4	268	5.59099
5	260	5.56068
Mean	265.8	5.58260
Standard deviation	5.12	0.01928

Table 5. Maximum moment sample calculation

Test Number	kN.m	ln (kN.m)
1	1.3224	0.27945
2	1.3388	0.29177
3	1.3576	0.30572
4	1.4084	0.34245
5	1.3696	0.31452
Mean	1.3593	0.30678
Standard deviation	0.0327	0.02401

Figure 14. Connection looseness determination

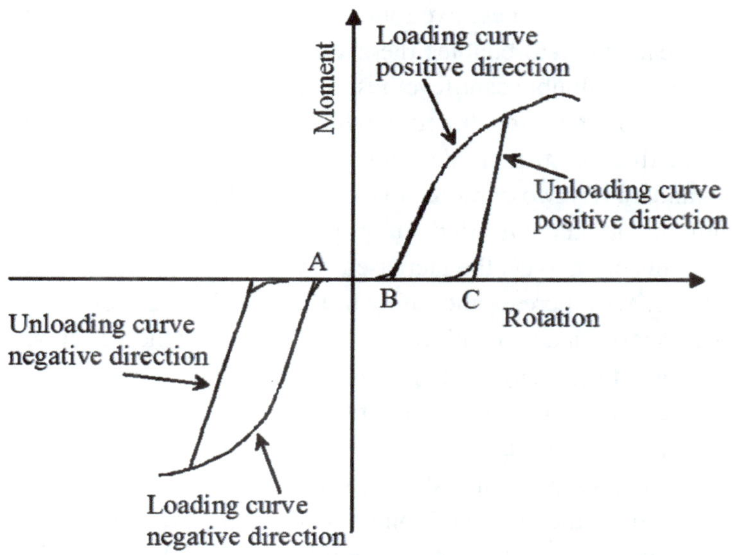

Looseness is important to both the serviceability and ultimate limit states range and justifies the need for carrying out cyclic tests (André 2014). Therefore, testing monotonically to failure could lead to artificially high initial stiffness values.

Under hysteretic cyclic loading, the rotational behaviour is often characterised by stiffness and resistance degradation along with looseness increase, see Figure 15, with failure being attained for a few number or cycles.

Figure 15. Behaviour under cyclic bending loading of a type of beam-to-column joint

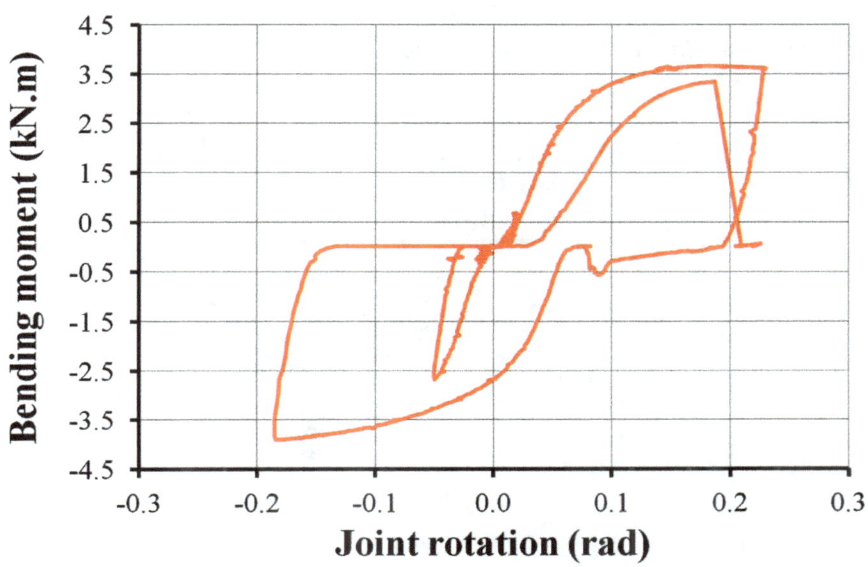

By conducting cantilever tests with the horizontal members in an inverted orientation, rotation stiffness can be found in the alternative, or negative rotation direction. If the differences in stiffness are less than 10% then the European standard enables the average stiffness between the two directions to be used. Some experimenters use a double cantilever test to get average stiffnesses directly but this method assumes that all rotations are in the same direction which is not true under wind or seismic conditions. It is also important to note that the stiffness determined by a simple cantilever test may be smaller than the stiffness when the standard has two or more horizontal members simultaneously attached at the connection as can happen in the nodal connections in proprietary scaffolds/falsework. Small stub-sections can be added to the test connection to get the increased stiffnesses, see André, Beale, & Baptista (2013), but using the lower stiffness will increase the safety of the resulting temporary structure.

The rotation stiffness determined by cantilever tests produces stiffnesses about the primary (strong) axis of rotation, i.e. bending at right angles to the axis of the standard, see Figure 16 in the case of the cuplok joint. However, if the rotation stiffness about the weak axis, bending parallel to the standard, is required then frame tests are often used.

Figure 13(c) show schematics of such tests. A 3-D structure is made up of two arrangements as shown in the Figures: one above the other and joined by the standards. The same procedure for testing as described for the cantilever test is conducted – one pilot followed by a series of cyclic tests.

For example, the rotation stiffness (M/θ) of the connections in

Figure 13(b) is given by Eq. 19:

$$\frac{M}{\theta} = \frac{W \cdot L^2}{8 \cdot \Delta} \tag{19}$$

where W is the applied load, L the ledger length and Δ the displacement of the jack. Note that the looseness determined by these frame tests is often large and jack displacements of 150 mm can occur with zero stiffness.

Figure 16. Illustration of bending axes of the cuplok joint

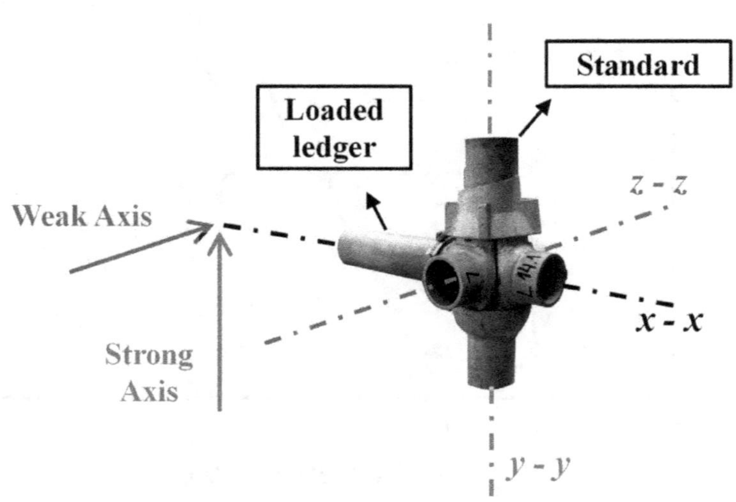

Alternatively, the test setup adopted can be essentially the same as described for the bending tests about the connection strong axis, apart from a 90º rotation of the tested elements about the ledgers axial axis (André et al., 2013).

In addition to bending tests, tensile tests for the connections between standards and ledgers can also be performed (André et al., 2013; Voelkel, 1990). These tests are important since the post-failure behaviour of temporary works, such as bridge falsework systems, may be influenced by the strength and stiffness of these joints due to the development of large pull forces, see André (2014).

The test setup adopted can be the one illustrated in Figure 17. Two 400 mm ledgers are connected to a standard element at diametrically opposed positions and the end extremities clamped to the test machine grips. This test requires the design of special grips. For a maximum axial load of 100 kN, these can consist in two sets (upper and lower grips) of two S275 steel pieces joined by eight preloaded M12 10.9 bolts. A preload force needs to be applied in order to avoid slippage at the grips. Each grip is then connected by a pin connection (consisting of a M24 10.9 bolt) to the testing machine. Figure 18 illustrates the behaviour under monotonic axial loading of a type of beam-to-column joint.

4.6.2.2 Column-To-Column Joints

Column-to-column joints of proprietary and tubular scaffold/falsework systems are often called spigot connections (or spigot joints). Spigot connections in modular and proprietary scaffolds/falsework systems are either permanent inserts welded to one end of each tube or the insert is bolted to one end of the tube. The external dimensions of the spigot (usually an SHS or CHS element) are smaller than the internal

Figure 17. Test setup for the axial tests of the beam-to-column joint

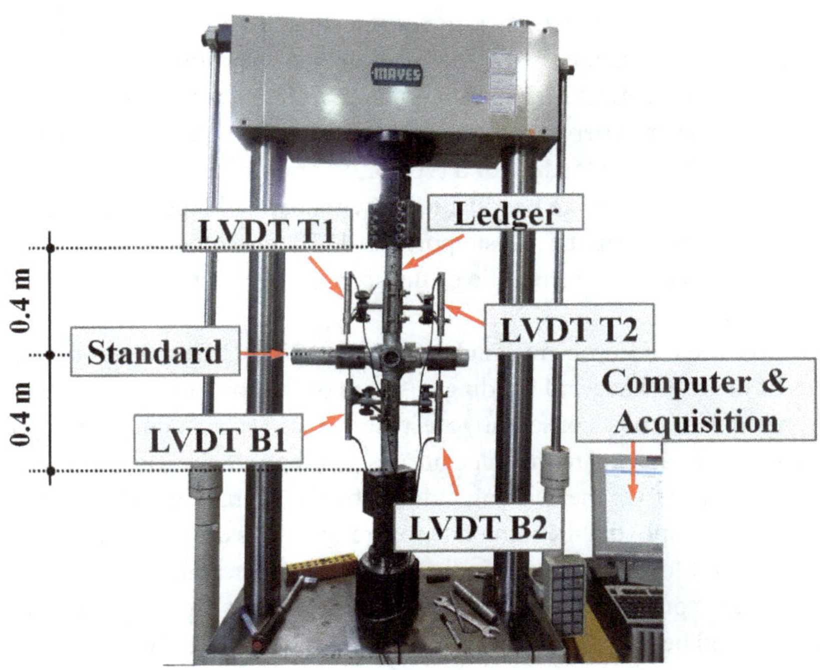

Figure 18. Behaviour under monotonic axial loading of a type of beam-to-column joint

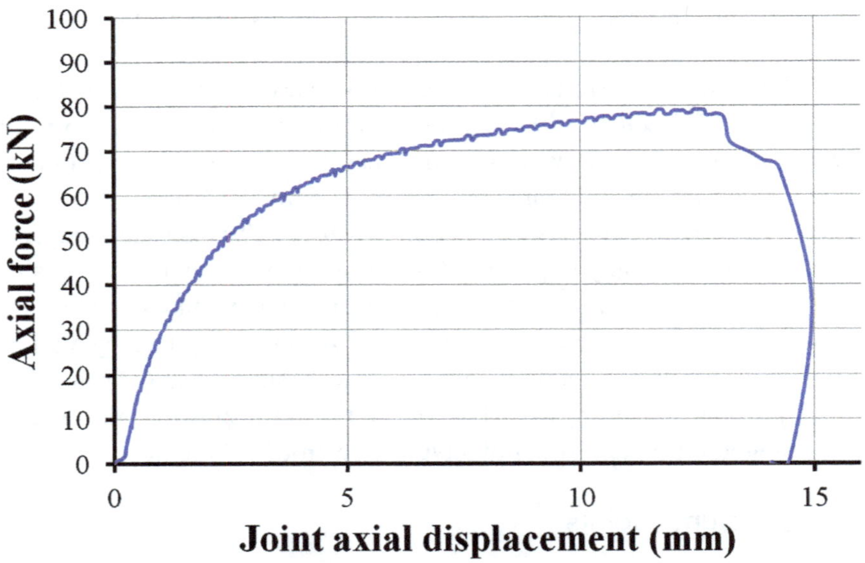

diameter of the standards' CHS. Therefore, an initial play exists. When the standards are bent, this gap will introduce initial member and global geometrical imperfections.

As demonstrated by the full-scale tests carried out in Australia (Chandrangsu & Rasmussen, 2009), the maximum resistance of proprietary and tubular scaffold/falsework systems is often limited by the strength of the spigot joints.

Possible bending test setups for spigot joints (with compressive axial load) are illustrated in Figure 19 (André et al., 2013). In the tests with axial load, the length of each specimen (between the end supports) is 340 mm, small enough to avoid global buckling of the standard elements and large enough to avoid local buckling of the standard elements. The experimental configuration to determine the rotation stiffness of spigot connections is identical to that of baseplates with the exception that there is no need for a block of concrete or other material between two sections of standard. Figure 20 illustrates the behaviour under monotonic axial and bending loading of a type of column-to-column joint.

The test setup adopted for the simple bending tests (without axial load) is a simply supported specimen with a total free length (between the end supports) of 1770 mm (other values can of course be used), where the spigot joint was placed at the middle of the span, subjected to a concentrated transverse load applied at mid-span.

The test method given in the European standard BS EN 15512 (BSI, 2009a) does not always return conservative values of the resistance and bending stiffness of the spigot joint. In fact, as the spigot joint involves a contact problem, the most conservative test method can correspond to the application of a high lateral load to axial load ratio and not to the opposite case, as far as the second order effects produced by the axial load do not dominate over the bending induced by the lateral load. High load ratios (horizontal load divided by axial load) imply that bending due to lateral load is dominant, meaning that the contact area between the upper and lower standards is, in general, smaller than the one for lower load ratios. As the joint stiffness varies proportionally with the bearing contact area, high load ratios imply lower values of the resistance and bending stiffness of the spigot joint (André, 2014).

Figure 19. Test setup for the bending tests of the spigot joint considering axial force

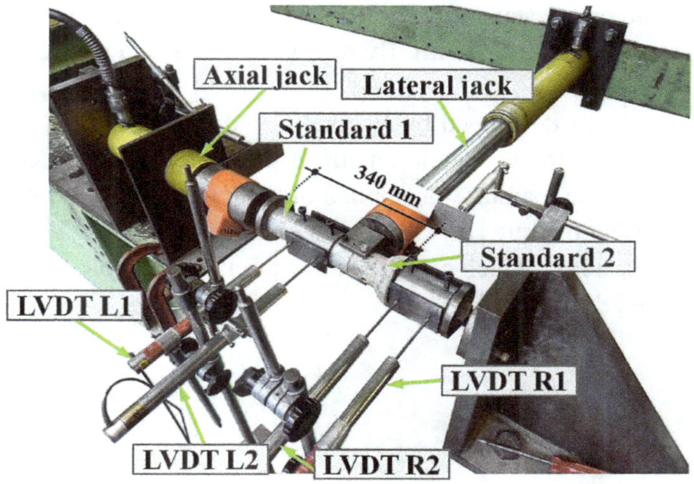

Figure 20. Behaviour under monotonic axial and bending loading of a type of column-to-column joint

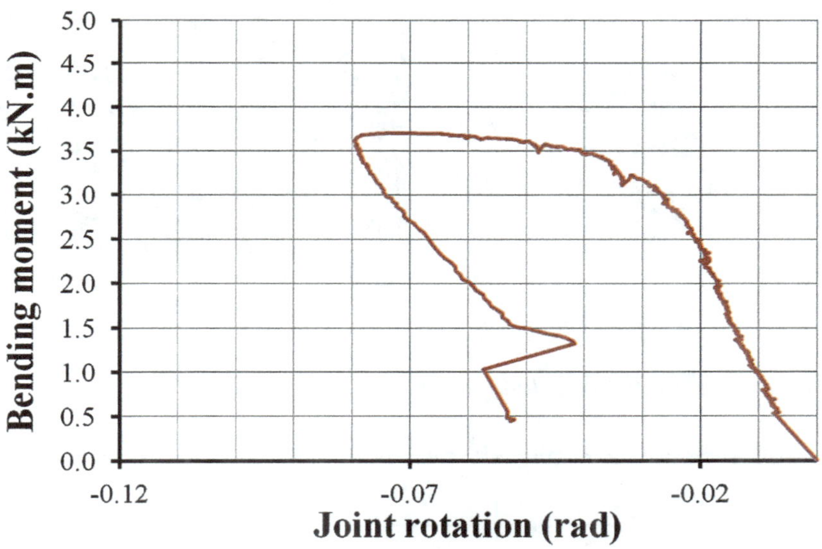

In order to make a good decision regarding the load ratios to be used, a limited number of initial tests should be carried out (André et al., 2013). This problem of different moment-rotation characteristics depending upon the ratio of axial load to lateral bending moment will be developed in the models of these connections (See Section 4.6.5 of this Chapter).

4.6.2.3 Brace Joints

Diagonal braces are inserted into temporary structures to triangulate the frames. A schematic of a brace test and a picture of a test are given in Figure 21. In a brace test the frame consists of two standards connected to a pair of ledgers or transoms. Two braces are then connected, one on either side of the frame, to ensure torsional symmetry. One end of each standard is pinned to a support and a horizontal load is applied to the bottom of one standard. The horizontal displacement of the frame measured at the bottom of the ledger on the opposite standard. By resolving forces and deflections the stiffness of the brace is determined.

An alternative test setup was developed by Voelkel (1990). A standard was rigidly fixed to a test frame with the cuplok joint located at middle length. A brace diagonal was connected to the cuplok joint using a swivel coupler. The loading was applied directly to the free end of the brace element. At the beginning full hysteresis loops were carried through on different loading levels. Finally the load was increased until failure was reached, which occurred most often at the cuplok joint. Figure 22 illustrates the joint behaviour observed during testing.

Figure 21. Brace test

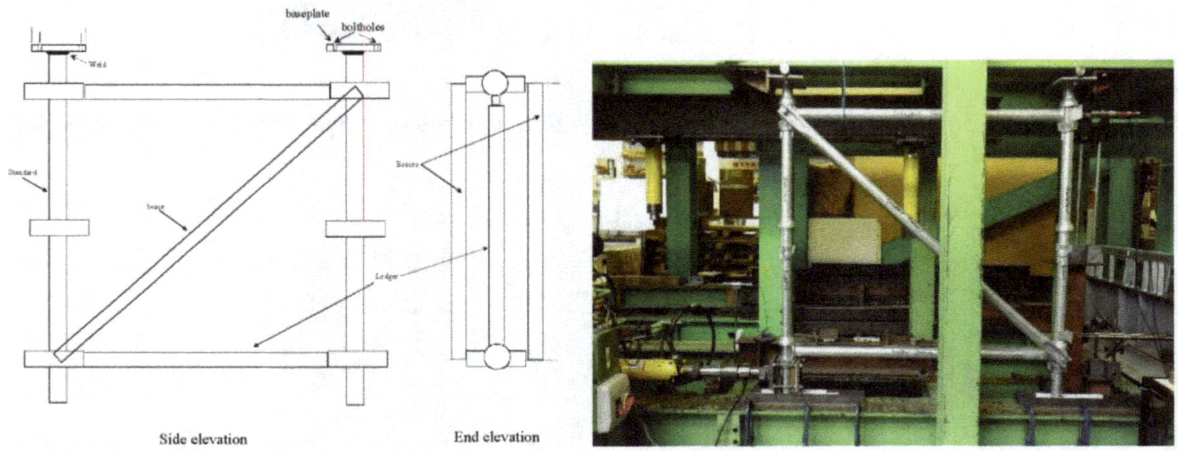

(a) Schematic of test *(b) Brace under test*

Figure 22. Behaviour under cyclic axial loading of a type of brace joint

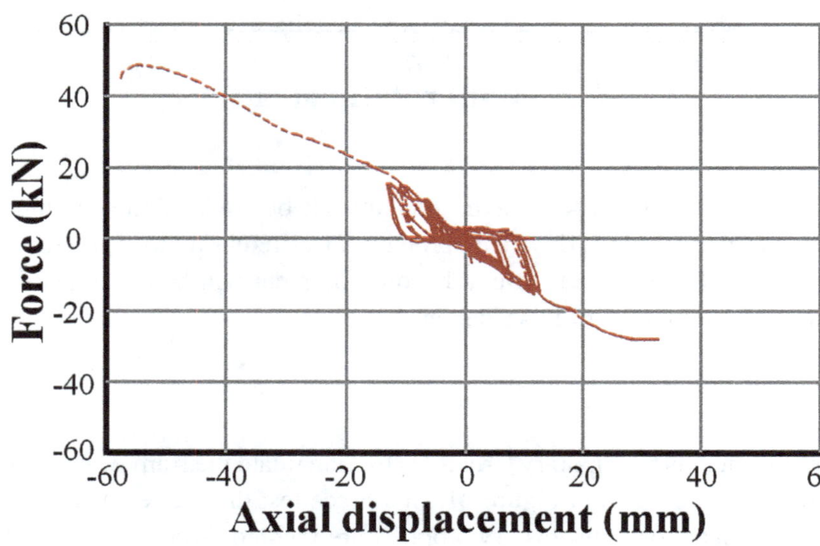

4.6.2.4 Top Joints

The joint between the top of a proprietary and tubular falsework system and the formwork system is in general a grey area. The stiffness and resistance of the joints between the falsework and the formwork depends on the type of the falsework top element (forkhead or baseplate) and on the geometrical, material and stiffness characteristics of the formwork beams and formwork system as a whole. For example, using large width formwork beams may result in higher joint bending stiffness due to larger contact surfaces than when a smaller width beam is used. This difference may be reduced considerably if wedge elements are used to bind the formwork beam to the forkhead side plates as good construction practices recommends but sometimes are not followed.

Despite the large uncertainties associated with this joint, it is important to perform a structural assessment which can be valid for cases where general good practices are followed during planning, design and operation. A setup for bending tests of the joint materialised by the interaction between the forkhead plate and the formwork beam is presented in André et al. (2013).

The forkhead joint has two bending axis, one along an the longitudinal axis of the formwork beam and the other at right angles, see Figure 23 (André, 2014; André et al., 2013). Figure 24 illustrates a possible test setup to perform bending tests of forkhead joints.

For bending tests about axis 1, the joint rotation occurs only due to plastic deformations at the tube segment and no relative rotation of the forkhead was observed. Therefore, the forkhead joint rotational stiffness about axis 1 can be considered as rigid if the tube segment is explicitly simulated in the analysis models. For bending tests about axis 2, the joint rotation occurs mainly due to plastic deformations of the forkhead plates while the tube segment deformed elastically, see Figure 25. In this bending axis, the forkhead joint behaves as illustrated in Figure 26 and can be modelled as a semi-rigid connection.

4.6.2.5 Base Joints

The experimental setup can be very similar to that for column-to-column connections with the difference that instead of the two columns being together, a block of concrete is between them. The test was developed for baseplates in the Pallet Racking Industry but is equally applicable to falsework and scaffold bases. Examples of the tests for racks can be found in BSI (2009a), Feng (1994) and Godley, Beale, & Feng (1998).

Figure 23. Illustration of bending axes of the forkhead joint

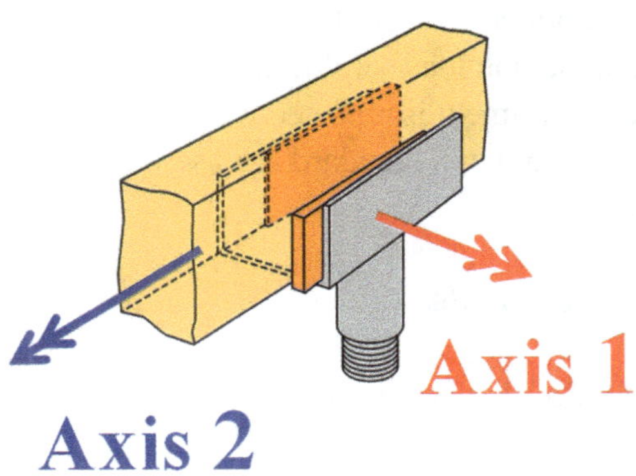

Figure 24. Test setup for the forkhead joint bending tests about axis 2

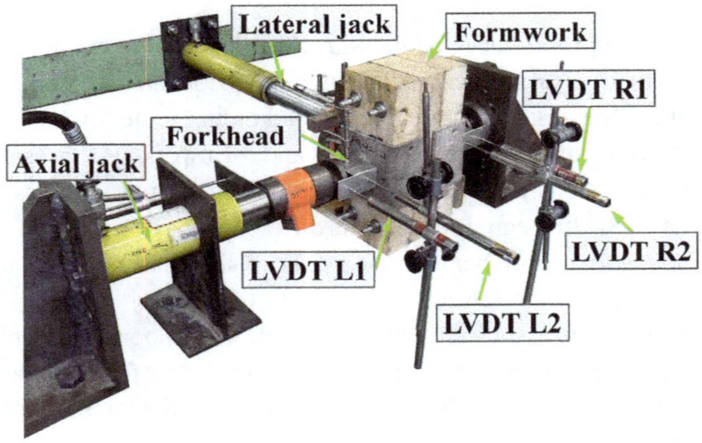

Figure 25. Deformations observed from forkhead bending tests about axis 1 (Left) and axis 2 (Right)

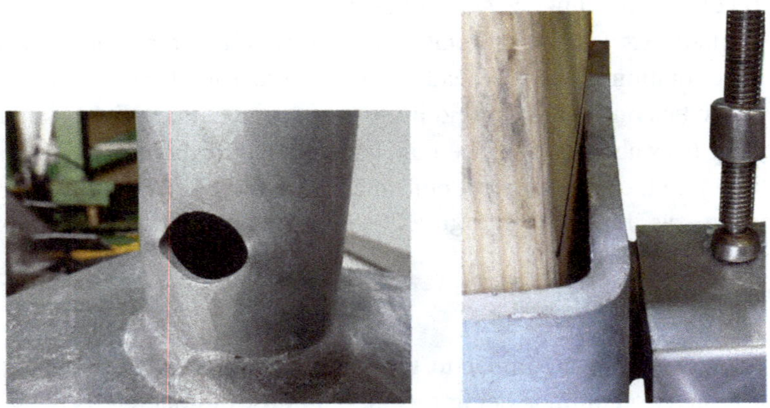

An experimental configuration to determine the baseplate stiffness is shown in Figure 27. Two sections of standard with baseplates attached (typically between 200 and 300 mm long) are placed in line with a jack on one end and a load cell on the other with a sample of the foundation between. In most experiments this would be a block of concrete. A second side jack is attached at right angles to the foundation block. The experiment is conducted by applying a fixed load along the axis, say for example 20 kN, and then applying increments of side load until failure, recording the rotation of the standards and baseplates with transducers. The experiment is repeated with different axial loads so as to determine the full moment-rotation curve. Two sections of standard are used to ensure that torsion of the foundation does not occur.

4.6.3 Theoretical Models for Joints

In the past, several mathematical models have been developed. Some of these are described below:

Figure 26. Behaviour under monotonic axial and bending loading of a type of forkhead joint

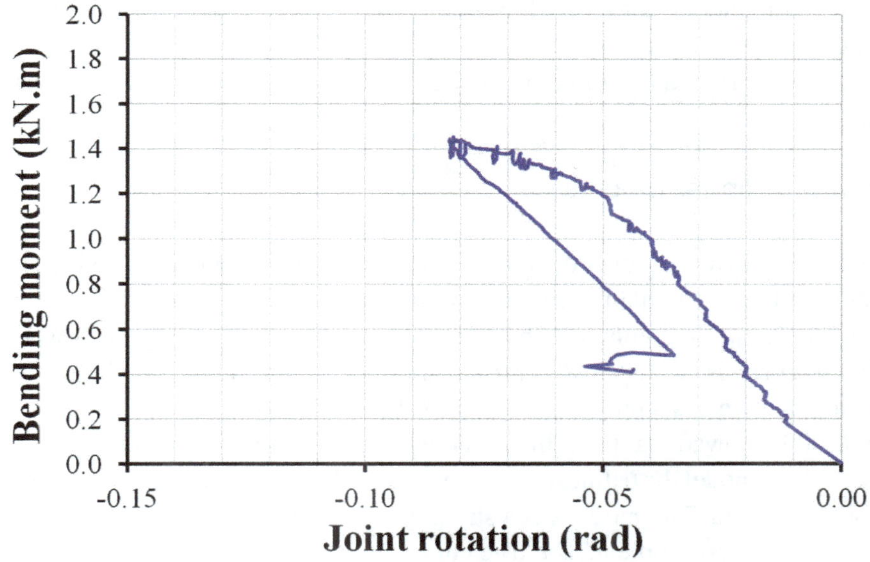

Figure 27. Schematic of baseplate test

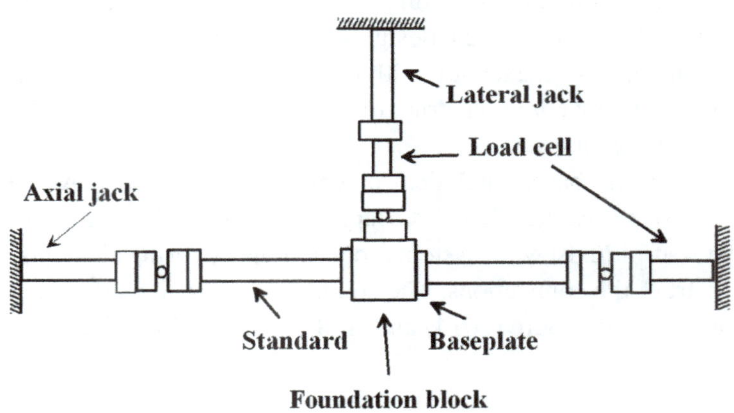

4.6.3.1 Linear Model

In this model, the connection behaviour is represented by a single straight line with a slope equal to the initial stiffness of the connection. The stiffness of the connection is determined by either conducting experiments or could be expressed in terms of beam stiffness (Lightfoot & Le Messurier, 1974). Thus in this model, the M-θ relation is given by the following relation:

$$M = k \cdot \theta \tag{20}$$

where k is the stiffness of the connection and remains constant throughout the analysis.

The model is very simple to use and can be easily incorporated in the analysis software. However, its validity is only within the serviceability limit state of the structure and may not give accurate results for large deflection analysis.

The simplest cases are rigid and simple connections, where the stiffness k is either very large or very low, respectively.

4.6.3.2 Bilinear/Multilinear Models

These models are an improvement over the previous model. The connection behaviour is approximated by a series of straight lines. The main limitation of this model is the sudden change in the stiffness of the connection at the transition points, which can lead to numerical instability. However, this method is most commonly used because of its close approximation to the true connection behaviour and it is also easy to incorporate in an analysis program (Godley & Beale, 2001).

Prabhakaran (2009) showed that the bilinear model defined in the pallet racking code BS EN 15512 (BSI, 2009a) was able to model the trilinear curve described above in Section 4.6.2 as well as a regression polynomial through the data. The model fits a straight line from the origin of the curve to a horizontal line at the maximum allowed moment. The true curve intersects the straight line at some intermediate point with curve having a higher gradient initially and a lower gradient after the intersection. The gradient of the line is determined by calculating the value which makes the two areas (the one above and the one below) equal so that work done by getting to the maximum moment is the same in both cases. The one proviso being that the rotation of the design straight line cannot be greater than 1.15 times the gradient of the regression curve's intersection with the maximum moment and the origin. The use of this straight line means that many common structural programs which do not allow multilinear curves can be used with reasonable accuracy.

A similar phenomenological-based analytical model was presented by André, Beale, & Baptista (2014), André et al. (2014) and André, Beale, & Baptista (2015a, 2015b) to simulate the behaviour of proprietary and tubular scaffold/falsework systems connections by spring elements. Figure 28 shows the model for the loading part of the connections. The complete hysteretic model, including unloading and reloading phases, is presented in André (2014) and André et al. (2014). Figure 29 shows a fit between

Figure 28. Multilinear model used to simulate the M vs. θ loading curves

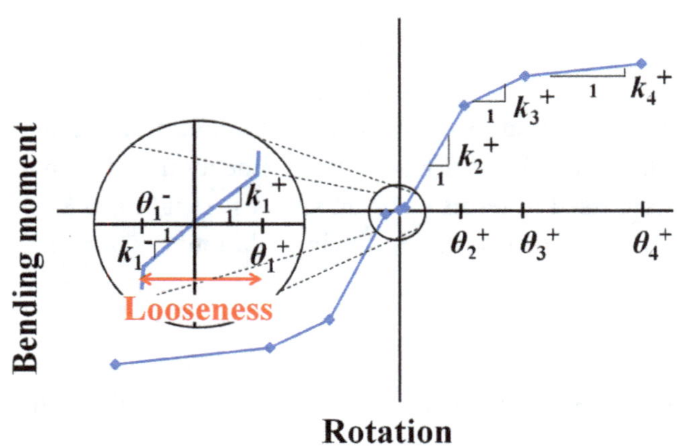

Figure 29. Comparison between the experimental behaviour and the one obtained using the proposed analytical model for the spigot joint under axial and lateral load

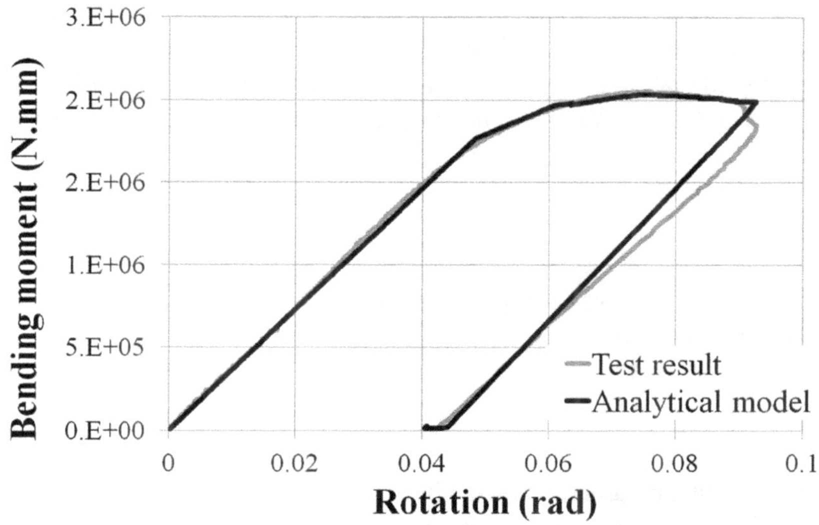

the experimentally observed behaviour and the one obtained from the analytical model for one type of connection. This model can be used to simulate the joint behaviour about all degrees of freedom, e.g. bending rotations and axial displacements.

Notwithstanding, alternative, more accurate but more complex, models to the multilinear model chosen to fit the diagrams may be used, see next Sections.

One reason in favour of using the simple multilinear model is because the parameters of the multilinear model are easily read, they directly represent characteristics of the behaviour of the joints (a one to one representation), and by using the multilinear model it is easy to analyse the influence of each characteristic of the joint, as the stiffness after looseness for example. On the contrary, if the more complex models were selected it would be more difficult to understand what the model parameters represent and how they influence each characteristic of the joint: stiffness evolution, maximum bending moment and ductility. Also, some of the more complex models do not respect the condition that the diagrams should start at zero bending moments for zero joint rotations.

4.6.3.3 Frye and Morris Polynomial Model

This model was originally developed by Frye & Morris (1975). The M-θ relationship is represented by a polynomial function expressed as:

$$\theta = C_1 \cdot (K \cdot M)^1 + C_2 \cdot (K \cdot M)^3 + C_3 \cdot (K \cdot M)^5 \tag{21}$$

where M is the moment, θ is the angular deformation, K is a standardisation parameter dependent upon the connection type and geometry, and C_1, C_2 and C_3 are curve fitting constants. These constants can be determined using the method of least squares. The model may represent the M-θ behaviour of the connection reasonably well. Alternately, the experimental data can be represented by a polynomial function using regression analysis. The main drawback of the method is that at some values of the connection moment, the expression could give negative values, which could lead to numerical instability. Hence care needs be taken while curve fitting the data.

The authors have found that for many common types of proprietary falsework that a simple regression polynomial curve can fit many experiments with a correlation coefficient better than 0.99 if a fifth order polynomial is used.

4.6.3.4 Power Model

Several power models have been developed for different types of connections. The simplest model is the two-parameter model expressed as:

$$\theta = a \cdot M^{b} \tag{22}$$

where a and b are curve fitting parameters and $a > 0$ and $b > 1$. This model cannot represent the behaviour of the connection accurately. Many improvements have been made to this model by introducing three parameters – the initial stiffness of the connection (k_i), the ultimate moment capacity (M_u) and a shape parameter (n). The three parameter model proposed by Kishi & Chen (1990) is of the form:

$$\theta = \frac{M}{k_i \cdot \left(1 - \left(\frac{M}{M_u}\right)^{n}\right)^{1/n}} \tag{23}$$

The main advantage of power model is that there is no sudden change in the stiffness of the connection and it always gives a positive value of the stiffness. However, this would rule out this model for modelling the unloading curve.

4.6.4 Metal Beam-To-Column Joints

Several experimental studies, see Table 6, have been carried out in the past to characterise the behaviour and resistance of these joints. Emphasis should be made to work of Beale and Godley in the UK, of Voelkel in Germany, of Chandrangsu et al in Australia and André et al again in the UK. Additional experimental studies have been published by Abdel-Jaber, Beale, Godley, & Abdel-Jaber (2009), Liu et al. (2010), Peng, Ho, Lin, & Chen (2015) and Peng, Wu, Chan, & Huang (2013). BS EN 12811-1 (BSI, 2003), BS EN 12812 (BSI, 2011b) include characteristics values of the resistance and stiffness for right-angle and swivel couplers conforming with BS EN 74-1 (BSI, 2005a) calibrated to be used in linear models of the joints.

Table 6. Published results of scaffolding/falsework beam-to-column joint tests

Reference	System type	Type of test	Model type	Stiffness (average values)	Resistance (average values)
Godley & Beale (1997) and Voelkel (1990)	Cuplok®	Joint bending tests (Cyclic) Frame tests (Cyclic)	Linear elastic spring	Strong axis: 78 kN.m/rad1)	2.9 kN.m
				Weak axis: 5.6 kN.m/rad	0.2 kN.m
		Joint axial tests (Monotonic)	Linear elastic spring	N/A	73 kN
Godley & Beale (2001)	Wedge type	Joint bending test	Linear elastic spring	Strong axis: 77 kN.m/rad (stiffness after looseness, clockwise rotation)	1.7 kN.m
				Strong axis: 27 kN.m/rad (stiffness after looseness, anti-clockwise rotation)	1.3 kN.m
Chandrangsu & Rasmussen (2011a)	Cuplok®	Joint bending tests (Monotonic)	Multilinear elastic spring	Strong axis Two ledgers: 77 kN.m/rad Three ledgers: 87 kN.m/rad Four ledgers: 102 kN.m/rad	3.5 kN.m
André et al. (2013)	Cuplok®	Joint bending tests (Cyclic)	Multilinear elastoplastic spring	See Table 4.7 to Table 4.10 and Figure 4.282),3)	
		Joint axial tests (Monotonic)	Multilinear elastoplastic spring	See Table 4.11, Table 4.12 and Figure 4.282),3)	

1) In tests performed by Voelkel and Beale, the initial stiffness associated with looseness was determined to be 10% of the stiffness value without or after looseness.

2) The values characterise the joint behaviour under quasi-static loading and adequate joint clamping by hammer blows. Joint bending behaviour under reloading, for dynamic actions and for inadequate joint clamping is provided in André (2014).

3) The values are valid for elements made of steel with nominal yield strength equal to or higher than 355 MPa.

Table 7. Results of the bending stiffness for the multilinear model of the cuplok joint (strong axis bending)

Joint configuration	k1		k2		k3		k4	
	Average (kN.m/rad)	COV	Average (kN.m/rad)	COV	Average (kN.m/rad)	COV	Average (kN.m/rad)	COV
Two ledgers (2L)	19.30	0.74	70.82	0.14	20.79	0.27	3.18	0.57
Three ledgers (3L)	10.56	0.65	83.44	0.24	12.76	0.44	2.52	0.44
Four ledgers (4L)	16.96	0.86	85.85	0.19	20.91	0.30	3.42	0.42

Table 8. Results of other parameters for the multilinear model of the cuplok joint (strong axis bending)

Joint configuration	Δθ01		Δθ02		Δθ03		Δθ04		kU		Mu	
	Average (rad)	COV	Average (rad)	COV	Average (rad)	COV	Average (rad)	COV	Average (kN.m/rad)	COV	Average (kN.m)	COV
All types	0.006	1.20	0.036	0.29	0.042	0.48	0.080	0.50	132.82	0.15	3.86	0.08

Table 9. Results of the bending stiffness for the multilinear model of the cuplok joint (weak axis bending)

Joint configuration1)	k1+		k1-		k2-		kU	
	Average (kN.m/rad)	COV	Average (kN.m/rad)	COV	Average (kN.m/rad)	COV	Average (kN.m/rad)	COV
All cases	10.88	0.59	8.19	0.43	2.29	1.43	19.00	0.32

1) Cuplok joints exhibit a asymmetric behaviour under bending loading (André, 2014). The values for the case when the joint rotation coincides with the torsion rotation applied to the free cup to lock the joint are shown with a – subscript, and for the opposite case with a + subscript.

Table 10. Results of other parameters for the multilinear model of the cuplok joint (weak axis bending)

Joint configuration1)	Δθ1+		Δθ2+		Δθ1-		Δθ2-		MU+		Mu-	
	Average (rad)	COV	Average (rad)	COV	Average (rad)	COV	Average (rad)	COV	Average (kN.m)	COV	Average (kN.m)	COV
All cases	0.02	0.60	0.18	0.93	-0.02	0.58	-0.18	0.74	0.1	–	-0.4	–

1) Cuplok joints exhibit a asymmetric behaviour under bending loading (André, 2014). The values for the case when the joint rotation coincides with the torsion rotation applied to the free cup to lock the joint are shown with a – subscript, and for the opposite case with a + subscript.

Table 11. Results of the tensile stiffness for the multilinear model (cuplok joint, tensile forces)

Joint configuration	k1		k2		k3		k4	
	Average (kN/mm)	COV	Average (kN/mm)	COV	Average (kN/mm)	COV	Average (kN/mm)	COV
All cases	3.57	0.91	33.37	0.35	7.76	0.35	2.77	0.63

Table 12. Results of other parameters for the multilinear model (cuplok joint, tensile forces)

Joint configuration	Δδ1		Δδ2		Δδ3		Δδ4		kU		Pu	
	Average (mm)	COV	Average (mm)	COV	Average (mm)	COV	Average (mm)	COV	Average (kN/ mm)	COV	Average (kN)	COV
All cases	0.23	1.47	1.39	0.43	3.20	0.66	2.33	1.26	127.85	0.31	70.71	0.16

4.6.5 Metal Column-To-Column Joints

4.6.5.1 General

When a temporary structure is to be constructed where the height of the structure exceeds the maximum length of standards then spigot connections are required. Traditionally, spigot connections were considered as pinned connections for simplicity of analysis and also to allow for the looseness at the connection. The prototype proprietary scaffold structure illustrated in Figure 30 was analysed in three ways, firstly as two 2-D structural frames using the principles described below in Section 4, secondly as a 3-D scaffold structure and thirdly as a 2-D scaffold frame but with the spigots modelled by contact elements. In the first and second models the spigot connections were ignored as is often the case in tubular scaffold structures. The results as presented in Figure 31. More details of the models can be found in Beale & Godley (1995). The interesting result of the analysis was that the different models all gave approximately the same maximum capacity of the frame but showed the importance of correct connection modelling to get an accurate prediction of the deflection profile.

4.6.5.2 Tubular Scaffolds

There are two types of joints in use for tubular scaffolds/falsework – external sleeves bolted to the outside of two tubes and parallel connections where two tubes lie beside each other and a connection bolts the two tubes together. In the case of parallel connections there is no moment capacity required in the European codes, only a slip capacity. To determine the rotation capacity of sleeved connections a test configuration similar to that used in baseplates is undertaken with a simple four-point bending test, where the transverse loads are applied directly midway between the centre of the sleeve and the supports (BSI, 2005a). No research has been reported into the effects of axial force on bending capacity. BS EN 12811-1 (BSI, 2003), BS EN 12812 (BSI, 2011b) include characteristics values of the resistance and stiffness for sleeve and parallel couplers conforming with BS EN 74-1 (BSI, 2005a) calibrated to be used in linear models of the joints.

4.6.5.3 Spigot Joints

Various models of spigot joints have been formulated. The simplest is the one in the model described above. In this case the spigot was modelled as a small tube inset into the larger tube and it was assumed that the spigot was pinned until the connection rotated so that the insert came in contact with the second tube whereupon the connection was then considered stiff. A more sophisticated model was constructed by Enright, Harris, & Hancock (2000), see Figure 32, which was based on the theoretical model given in the European code for steel props (BSI, 1999). In this model the spigot is considered to be rigid (300

Figure 30. Schematic of the Stuttgart prototype scaffold

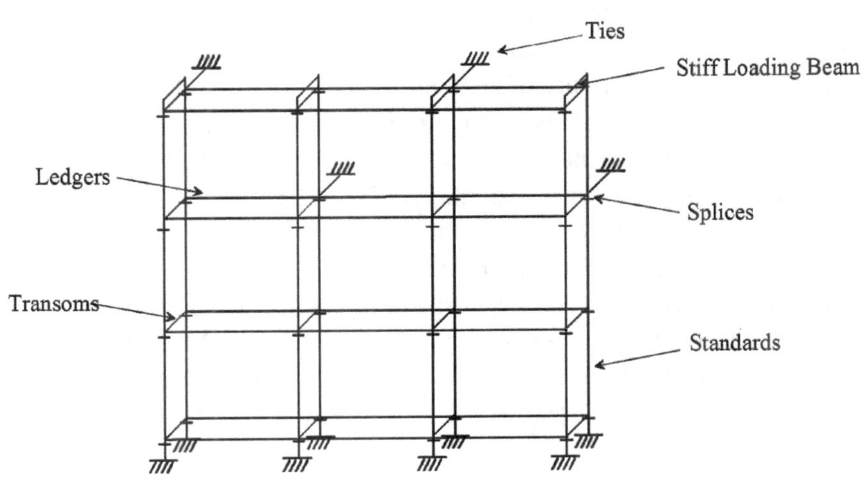

Figure 31. Comparison between different models for deflection of the prototype scaffold structure given in Figure 30

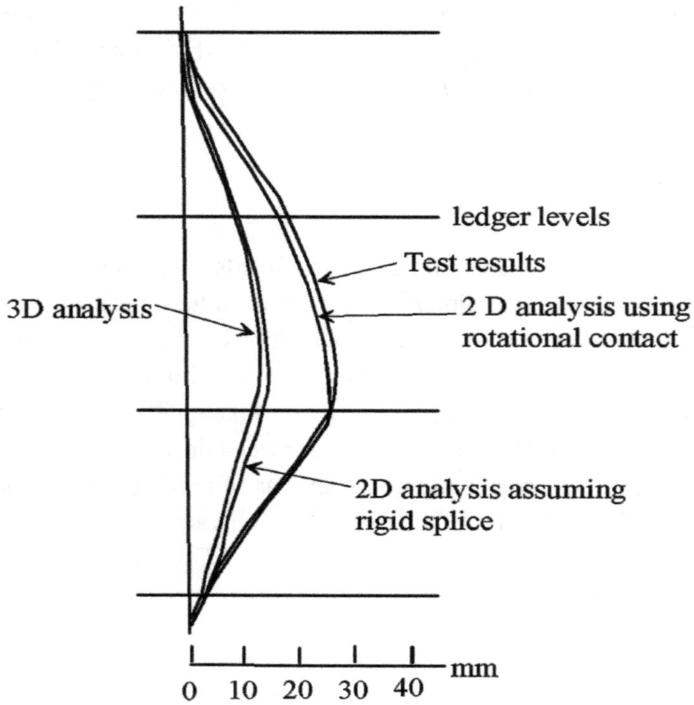

mm long) and is connected to the top and bottom standards by short elements with high rigidity (*E·I*) and high axial stiffness (*E·A*) which can only transfer lateral force into the spigot. The vertical load is transferred directly from the top standard to the bottom standard. The pinned connections between the standards and the spigot ensure that only lateral force and no vertical forces are transferred via the spigot but cause the standards to bend, the amount of bending depending upon the amount of out-of-plumb of the standards and the value of the load being transmitted from the top to the bottom standard. This model does not model the contact problem between the standard and spigot.

André's tests (André, 2014; André et al., 2013) showed that this model could not fully model the transmission of force effects of axial load on stiffness and resistance of the spigot joint and hence the Enright's model could return unsafe values of bending stiffness for low values of the ratio of moment to axial load and possibly conservative values for high ratios. He therefore constructed a statistical model based on the results of spigot joints bending tests which involved both new and used standards. As his tests only tested the cases of pure bending, and ratios of 20% and 50% lateral load to axial load he produced a table of average values of stiffness. Using the model presented in Figure 28, it was possible to simulate the experimental behaviour of the spigot. Table 13 and Table 14 contain the model parameters. Note that this model was only validated for spigot steel of the same material than the one the standards. The material had a nominal yield strength equal to or higher than 355 MPa. As the constitutive model depends on the ratio between the axial force and the bending moment, it is necessary to implement a special iteration process detailed in André (2014) and André et al. (2014).

Figure 32. Schematic of Enright's model

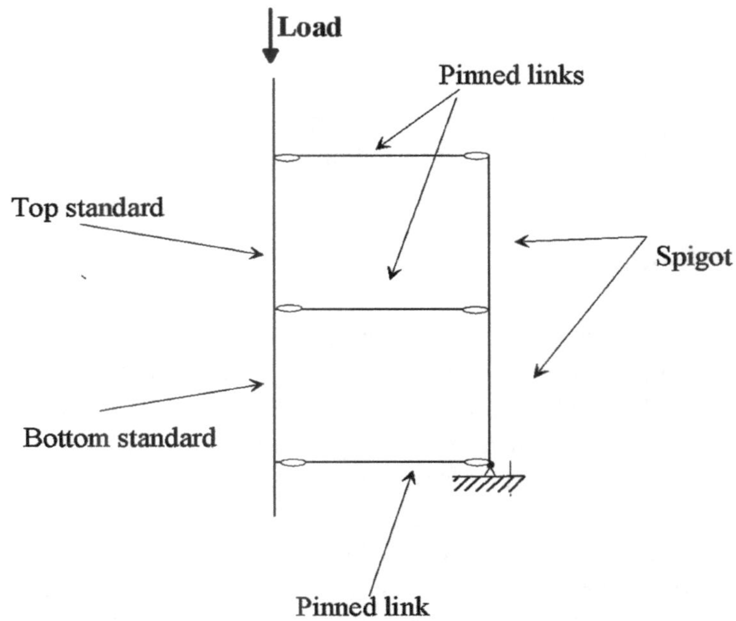

Unfortunately, neither Enright's nor André's tests cycled to determine looseness, θ_0, which was shown in Figure 14 to have significant effects of scaffold performance. This however, can be easily determined analytically by:

$$\theta_0 = \frac{D_{\text{int},Text} - D_{ext,Tint}}{L_o} = \theta_1 + \theta_2 = \left(\theta_{1,1} + \theta_{2,1}\right) + \theta_2 \tag{24}$$

where:

L_0 represents the overlap length of the spigot with the upper standard;
$D_{int,Text}$ represents the internal diameter of the upper standard;
$D_{ext,Tint}$ represents the external diameter of the spigot;
θ_1 and θ_2 are the rotations at the ends of the column, bottom and top, respectively.

If at the start there is no contact between the spigot and the upper standard, rotation θ_1 can be divided in two components: $\theta_{1,1}$ and $\theta_{2,1}$. The former makes the two elements get in touch at one initial point, while the latter in two points:

$$\theta_1 = \theta_{1,1} + \theta_{2,1} = \frac{D_{\text{int},Text} - D_{ext,Tint}}{2 \cdot L_1} + \frac{L_2}{L_1 + L_2} \cdot \left(\frac{D_{\text{int},Text} - D_{ext,Tint}}{L_o} - \frac{D_{\text{int},Text} - D_{ext,Tint}}{2 \cdot L_1} \right) \tag{25}$$

$$\theta_2 = \frac{L_1}{L_1 + L_2} \cdot \left(\frac{D_{\text{int},Text} - D_{ext,Tint}}{L_o} - \frac{D_{\text{int},Text} - D_{ext,Tint}}{2 \cdot L_1} \right) \tag{26}$$

where L_1 and L_2 represent the length of the lower and upper segments of the column, respectively.

1) N and M represent the axial force and the bending moment values at the spigot joint.

2) Assumed value.

4.6.6 Metal Brace Joints

A brace diagonal can be connected to a ledger element by a hook coupler or by a type of wedge coupler. In the case of the connections between a brace element and a standard element these consist in swivel couplers. BS EN 12811-1 (BSI, 2003), BS EN 12812 (BSI, 2011b) include characteristics values of the resistance and stiffness for swivel couplers conforming with BS EN 74-1 (BSI, 2005a) calibrated to be used in linear models of the joints. Voelkel (1990) analysed experimentally the behaviour of swivel joints under cyclic application of axial forces. From the tests results an average axial stiffness of the joint of 1360 kN/m and an average axial force resistance of 28 kN were obtained.

Results for an alternative test setup, using a 3-D frame assembly, have also been obtained but due to confidentiality cannot be made publicly available. BS EN 12811-3 (BSI, 2002) was used to analyse the results. The average values are:

Table 13. Results of the bending stiffness for the multilinear model of the spigot joint

N/M ratio (m-1) 1)	k1		k2		k3		k4	
	Average (kN.m/rad)	COV	Average (kN.m/rad)	COV	Average (kN.m/rad)	COV	Average (kN.m/rad)	COV
[0,20]	55.05	0.15	162.40	0.25	55.60	0.72	7.282)	–
]20,50]	102)	–	127.92	0.27	39.13	0.26	7.28	0.22
]50,+∞]	22)	–	27.92	0.26	9.50	0.59	2.09	1.11

1) N and M represent the axial force and the bending moment values at the spigot joint.

2) Assumed value.

Table 14. Results of other parameters for the multilinear model of the spigot joint

N/M ratio (m-1) 1)	Δθ1		Δθ2		Δθ3		Δθ4		kU		Mu	
	Average (rad)	COV	Average (rad)	COV	Average (rad)	COV	Average (rad)	COV	Average (kN.m/rad)	COV	Average (kN.m)	COV
[0,20]	0.022)	–	0.014	0.28	0.019	1.11	0.02)	–	73.932)	–	3.532)	–
]20,50]	0.022)	–	0.017	0.16	0.016	0.31	0.032	0.48	73.93	0.13	3.53	0.09
]50,+∞]	0.022)	–	0.054	0.31	0.044	0.81	0.151	1.48	24.32	0.66	1.79	0.24

- Service stiffness and load 1390 kN/m and 11.50 kN, respectively;
- Ultimate service stiffness and load 330 kN/m and 18.97 kN respectively.

Comparing the results of the two sets of tests, the service stiffness values are very similar whereas the maximum axial force resistance in the latter set of tests is substantially lower than the value obtained in the former set of tests. The reason for this could reside in the considerable looseness (over 25 mm) exhibited by the 3-D frame assembly when subjected to cyclic loads which could introduce localised over stresses in the joint components, thus reducing its resistance.

4.6.7 Metal Forkhead Joints

If the formwork beams are narrow enough so to allow the forkhead side plates to rotate without restraint and insufficient lateral confinement is enforced by means of the introduction of wood wedges between the beam and the forkhead side plates, the joint between the falsework and the formwork is indeed a pin connection. However, if, for small rotation values, there is interaction between the forkhead and the formwork beams then the joint is semi-rigid and can play an important role in stabilising the system against lateral loads. The model presented in André (2014) and André et al. (2014) can be used to simulate the nonlinear behaviour of the forkhead connection. Table 15 and Table 16 contain the model parameters.

The same comments can also be made when discussing the influence of the behaviour of the formwork system on the behaviour of the joint between the falsework and the formwork. If the formwork experiences severe stiffness degradation at early stages due to the interaction with the forkhead plate, then the joint, which could initially have a high stiffness value, will tend to a pinned joint.

4.6.8 Metal Baseplate Joints

As a rule, the foundations of temporary structures supporting vertical loads are shallow foundations, e.g. bases, footings or spread footings, or mats.

Table 15. Results of the bending stiffness for the multilinear model (forkhead joint, bending about axis 2)

N/M ratio (m-1) 1)	k1		k2		k3		k4	
	Average (kN.m/rad)	COV	Average (kN.m/rad)	COV	Average (kN.m/rad)	COV	Average (kN.m/rad)	COV
50%	0.0	–	29.33	0.19	11.30	0.46	6.68	0.08

1) N and M represent the axial force and the bending moment values at the forkhead joint.

Table 16. Results of other parameters for the multilinear model (forkhead joint, bending about axis 2)

N/M ratio (m-1) 1)	Δθ1		Δθ2		Δθ3		Δθ4		kU	
	Average (rad)	COV	Average (rad)	COV	Average (rad)	COV	Average (rad)	COV	Average (kN.m/rad)	COV
50%	0.022)	–	0.032	0.36	0.036	0.48	0.042	0.003	21.51	0.45

1) N and M represent the axial force and the bending moment values at the forkhead joint.
2) Assumed value.

Baseplates are often considered to be pinned connections at the base of each standard. Although this is a simplification it usually produces models which are conservative. In practice, however, the baseplate usually consists of a square flat plate welded onto the bottom of a piece of tube. A common fault in erecting falsework is to assume that the foundation on which the baseplate is erected is stiff and does not deflect. Frequently, baseplates are placed on timber spreader beams which lie on top of the foundation. If the foundation is very soft, such as a sandy soil, then it can displace leaving the baseplate suspended in air.

As the baseplate in reality is not a simply-supported connection as it has a flat plate welded onto the base then it can behave as a semi-rigid connection. Two models have been proposed and used in analysis.

The simplest model is to consider the baseplate as a rigid plate in contact with the foundation. It remains rigid until an overturning moment, M, is greater than the restraining force $N{\cdot}D / 2$ where N is the axial force in the standard and D the diameter of the baseplate. Moments above this moment cause the standard to rotate as if simply-supported. See Figure 33(a) where the rotation would occur about point A in the figure. A modification to this procedure is to assume that the foundation behaves as an elastic spring until loss of equilibrium occurs or plasticity sets in. BS EN 12811-1 suggests an elastic stiffness equal to $2{\times}10^7$N.mm/rad for scaffold baseplates.

An alternative approach suggested by André (2008) and André, Baptista, & Camotim (2009) was to use the procedure specified in BS EN 1065 for steel props. In this case the baseplate is considered to be supported on pinned connection distance $0.4{\cdot}D$ from the centre of the standard. See Figure 33(b). The standard is considered to be restrained from rotating by a spring attached to the ground with a rotational stiffness taken as $3{\times}10^7$N.mm/rad after an initial rotation of 1° has been exceeded. The difficulty with this model is that no experiments have been reported to validate it. Indeed, experiments conducted on baseplates for racking structures have shown that the rotational stiffness depends upon the axial force in the standard (Feng, Godley, & Beale, 1998) and on the stiffness of the foundation subgrade.

A more accurate approach is to simulate the contact between the surfaces, in both tangential and normal directions, see André et al. (2009) and Figure 34.

4.6.9 Bamboo Joints

The joints between scaffold bamboo elements are made by tying the poles together with nylon strips (formerly rattan strips) which according to the latest guidelines should have a strength of 0.5 kN per strip with a width of each strip of 5.5 to 6.0 mm and a thickness ranging from 0.85 to 1.0 mm. The

Figure 33. Baseplate joint models

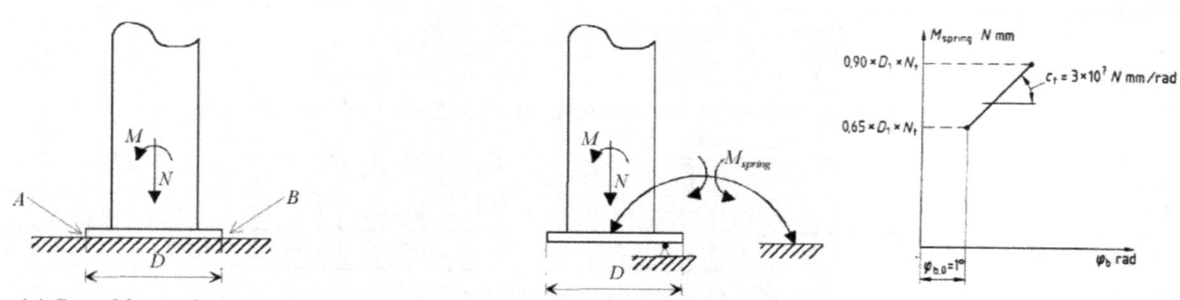

(a) Rigid baseplate *(b) Baseplate model used in BS EN 1065*

Figure 34. Contact modelling

design code states that all knots should be tightened with at least five rounds of nylon with the ends twisted and crossed to form a single twisted end (Hong Kong Labour Department, 2014). All joints can be considered to be pinned.

4.6.10 Anchor Joints

Anchor joints are necessary in most temporary structures systems: either to laterally restrain unstable scaffolds to rigid structures such as reinforced concrete façades or to hold down falsework baseplates to the ground.

There are many types of anchor available in the market: cast-in headed anchors and anchor channels as well as post-installed mechanical and chemical anchors and reinforcing bars. Some of these are proprietary solutions and specific analyses and design guides are available from the producers.

Recently, in Europe a set of rules for anchor connections were published as the European pre-standard CEN/TS 1992-4 (CEN, 2009). CEN/TS covers the design of post-installed fastenings (fasteners) and cast in situ fasteners (headed fasteners and anchor channels) in concrete components. The following types of fasteners are considered:

- Expansion fasteners, undercut fasteners, concrete screws, bonded fasteners, bonded expansion fasteners and bonded undercut fasteners
- Headed bolts as well as anchor channels with stiff connection of anchorage element and channel.

In the USA, the counterpart standard is the ACI 318 (ACI, 2014), Section 17, with design examples included in the ACI 355.3 (ACI, 2011).

The loads acting on the concrete component serving as anchorage ground can be static, cyclic (causing fatigue failure) or seismic.

In general, a linear elastic analysis may be used for establishing the loads on individual fasteners both at ultimate and serviceability limit states, provided element deformations are kept small. For a nonlinear analysis, the model presented in André (2014) and André et al. (2014) can be used to simulate the behaviour of anchor connections, see also Section 4.7.3. As there are multiple types of failure modes for anchor connections (e.g. steel tensile failure, pull-out/pull-through failure, concrete splitting failure), it is important to validate the model against experimental results so to obtain an accurate and safe representation of the behaviour and resistance of the anchor connection. An example of such study is presented in Ilick, Arora, & Dolejs (2015).

Further detailed descriptions of the load bearing behaviour and procedures for the calculation of anchor connections are presented in Eligehausen, Mallée, & Silva (2006) and Mallée, Fuchs, & Eligehausen (2013).

4.6.11 Joint Looseness

Looseness may exist in every type of joint in any degree of freedom (rotations and displacements). It is the expression of lack of fit between connecting elements that are caused inadvertently during fabrication, assembly and/or operation (including maintenance) activities or intentionally by existing gaps between elements. Looseness results in very small initial joint stiffnesses during loading and potential permanent deformations.

Models accurately simulating looseness may have convergence problems due to the non-uniqueness of a given internal force to the conjugate displacement. For beam-to-column joints, looseness was correctly modelled by Prabhakaran (2009) and Prabhakaran, Beale, & Godley (2011) where to ensure convergence the discontinuities in Moment-Rotation curves were smoothed out with simple curves and the looseness modelled by a linear segment with an artificial stiffness of 0.1 kN.m/rad.

An important result of Prabhakaran's research was the finding that the common structural model of replacing looseness by assuming an out-of-plumb geometry or an equivalent applied side-load resulted in different results for unbraced frames. In these cases the looseness must be applied in exactly the same way as occurs in practice. If a frame, however, is braced then the common assumptions of out-of-plumb or side-load are equally valid.

4.7 STRUCTURAL MODELLING OF TEMPORARY WORKS

4.7.1 Metal Props

The analysis and design of extendible props in Europe is governed by BS EN 1065 (BSI, 1999) for steel props and by BS EN 16031 (BSI, 2012) for aluminium props. The former standard was based on the German standard DIN 4424 (DIN, 1987). An exact solution of the differential equations governing the theoretical model is found in Chapter 8 of Feng's PhD thesis (Feng, 1994). Salvadori (2009) also derived a theoretical solution to the equations. The theoretical model, developed by Feng, is based on two intersecting columns as seen in Figure 35. There are two configurations, (a) with an exposed thread and (b) with an enclosed thread. Both however have the same model.

Figure 35. Prop types and computational model

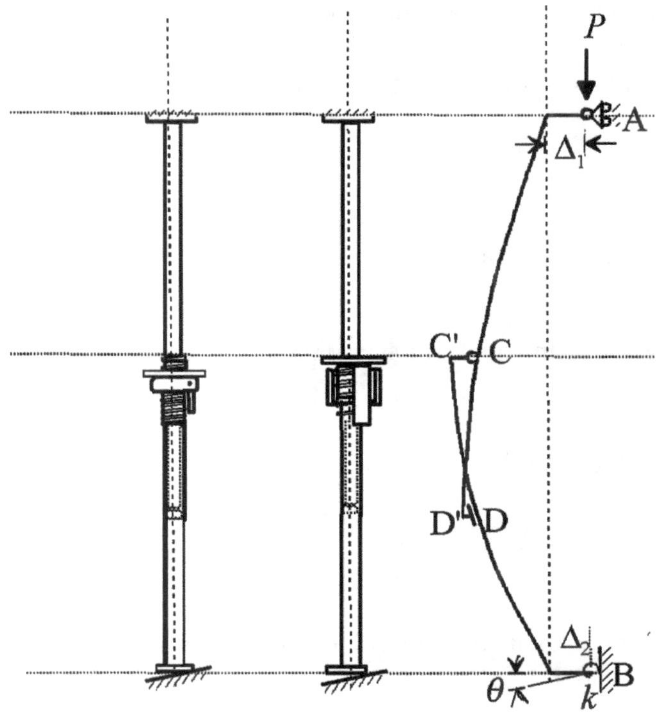

The baseplate at B is inclined at an angle θ to the horizontal and can be considered to be a rotational spring with stiffness k. The eccentricities occurring in the model are Δ_1 at A (with a value from the standard of 10 mm) and Δ_2 at the bottom, where Δ_2 is defined in Figure 36. $\Delta_{2,0} = 0.4{\cdot}D$ and $\Delta_{2,\text{crit}} = 0.5{\cdot}D+t$ where D is the diameter of the outer tube and t the width of the baseplate.

Figure 36. Eccentricity rotation-relation for the base of the prop

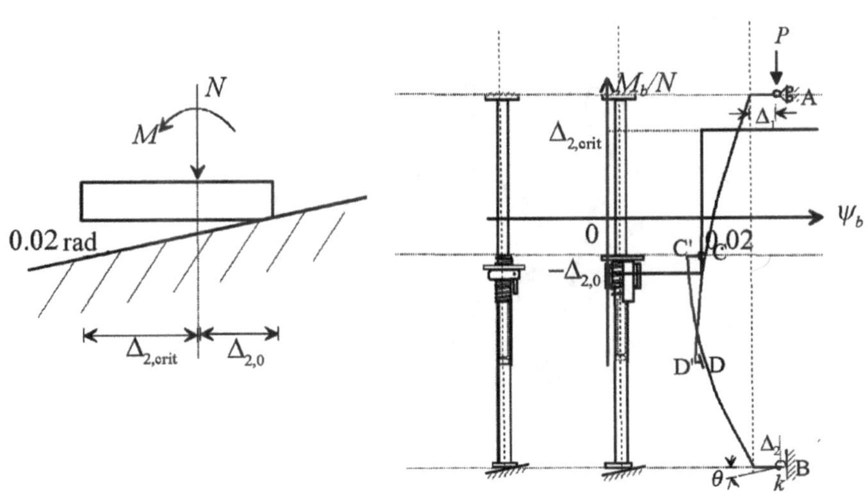

Assuming that the rotations are initially small and that the geometry of the prop can be treated by straight lines then from Figure 37 the following geometric relations apply:

$$\left(L_1 \cdot \sin \psi_1 + \Delta_1 - \Delta\right) = \left(L_2 - e\right) \cdot \sin \psi_2 + \Delta_2 \tag{27}$$

and

$$\left(L_2 \cdot \sin \psi_2 + \Delta_2 - \Delta\right) = \left(L_1 - e\right) \cdot \sin \psi_1 + \Delta_1 \tag{28}$$

Using the small angle approximations $\sin \psi_1 = \psi_1$ and $\sin \psi_2 = \psi_2$, and solving for ψ_1 and ψ_2 we get

$$\psi_1 = \frac{\Delta \cdot \left(2 \cdot L_2 - e\right) + e \cdot \left(\Delta_2 - \Delta_1\right)}{e \cdot L} \tag{29}$$

Figure 37. Detailed computational model showing inclinations and eccentricities

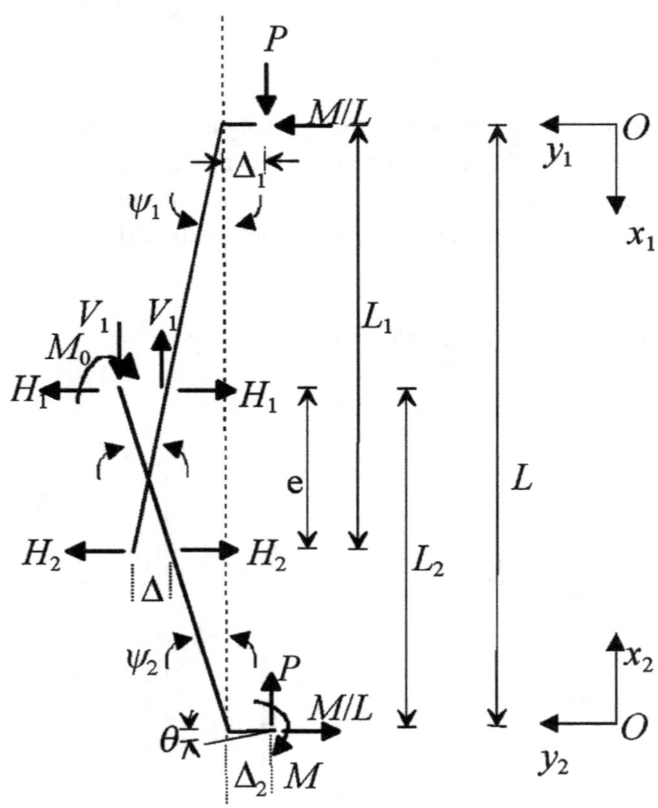

and

$$\psi_2 = \frac{\Delta \cdot \left(2 \cdot L_1 - e\right) + e \cdot \left(\Delta_1 - \Delta_2\right)}{e \cdot L} \tag{30}$$

The initial out-of-straightness of the prop is given by $y_0 = a_0 \cdot \sin\left(\pi \cdot x / L\right)$ with, from BS EN 1065 (BSI, 1999), $a_0 = L / 1000$.

The total initial deflections include out-of-straightness, eccentricities and angle of inclination at bottom support. Hence the total initial imperfection for the top half of the prop is given by:

$$y_{1,0} = a_0 \cdot \sin\left(\pi \cdot x / L\right) + \Delta_1 + \psi_1 \cdot x_1 \tag{31}$$

or

$$y_{1,0} = a_0 \cdot \sin\left(\pi \cdot x / L\right) + \Delta_1 + \frac{\Delta \cdot \left(2 \cdot L_2 - e\right) + e \cdot \left(\Delta_2 - \Delta_1\right)}{e \cdot L} \cdot x_1 \tag{32}$$

Similarly, the total initial deflection for the bottom half of the prop is given by:

$$y_{2,0} = a_0 \cdot \sin\left(\pi \cdot x / L\right) + \Delta_2 + \frac{\Delta \cdot \left(2 \cdot L_1 - e\right) + e \cdot \left(\Delta_1 - \Delta_2\right)}{e \cdot L} \cdot x_2 \tag{33}$$

The elements in Figure 35 are subjected to the action and reaction forces as shown which contribute to the bending equilibrium of the top half of the prop as:

$$E \cdot I \cdot \frac{\mathrm{d}^2 y_1}{\mathrm{d}x_1^{\,2}} = -P \cdot (y_1 + y_{1,0}) + \frac{M}{L} \cdot x_1, \ (0 \le x_1 \le L_1 - e) \tag{34}$$

and

$$E \cdot I_1 \cdot \frac{\mathrm{d}^2 y_{1,\mathrm{a}}}{\mathrm{d}x_1^{\,2}} = -P \cdot (y_{1,\mathrm{a}} + y_{1,0}) + \frac{M}{L} \cdot x_1 - H_1 \cdot (x_1 - L_1 + e) + \ldots$$
$$\ldots + V_1 \cdot (y_{1,\mathrm{a}} + y_{1,0} - y_{1,\mathrm{m}} - y_{0,\mathrm{m}}), \ (L_1 - e \le x_1 \le L_1) \tag{35}$$

where y_1 and $y_{1,a}$ are the deflections in the regions $0 \le x_1 \le L_1 - e$ and $L_1 - e \le x_1 \le L_1$ respectively, $y_{1,0}$ is the initial deflection and $y_{1,m}$ and $y_{0,m}$ are the values of y_1 and $y_{1,a}$ at the position $x_1 = L_1 - e$.

The corresponding equations for the bottom half of the prop are:

$$E \cdot I \cdot \frac{\mathrm{d}^2 y_2}{\mathrm{d}x_2{}^2} = -P \cdot (y_2 + y_{2,0}) - \frac{M}{L} \cdot x_2 + M, (0 \leq x_2 \leq L_2 - e) \tag{36}$$

and

$$E \cdot I_2 \cdot \frac{\mathrm{d}^2 y_{2,\mathrm{a}}}{\mathrm{d}x_2{}^2} = -P \cdot (y_{2,\mathrm{a}} + y_{2,0}) - \frac{M}{L} \cdot x_2 + M - H_2 \cdot (x_2 - L_2 + e), (L_2 - e \leq x_2 \leq L_2) \tag{37}$$

The general solutions of the four differential equations are:

$$y_1 = C_1 \cdot \cos(\alpha_1 \cdot x_1) + C_2 \cdot \sin(\alpha_1 \cdot x_1) + \Gamma_1 \cdot \sin\frac{\pi \cdot x}{L} + \frac{M}{P \cdot L} \cdot x_1 - \left(\Delta_1 + x_1 \cdot \psi_1\right), (0 \leq x_1 \leq L_1 - e) \tag{38}$$

$$y_{1,\mathrm{a}} = \frac{1}{E \cdot I_1} \cdot \left[\frac{M \cdot x_1{}^3}{6 \cdot L} - \frac{H_1 \cdot \left(x_1 - L_1 + e\right)^3}{6} - V_1 \cdot \left(y_{1,\mathrm{m}} + y_{1,0\mathrm{m}}\right) \cdot \frac{x_1{}^2}{2}\right] + \ldots$$
$$\ldots + C_5 \cdot x_1 + C_6, (L_1 - e \leq x_1 \leq L_1) \tag{39}$$

$$y_2 = C_3 \cdot \cos(\alpha_2 \cdot x_2) + C_4 \cdot \sin(\alpha_2 \cdot x_2) + \Gamma_2 \cdot \sin\frac{\pi \cdot x_2}{L} - \frac{M}{P \cdot L} \cdot x - \left(\Delta_2 + x_2 \cdot \psi_2\right) + \ldots$$
$$\ldots + \frac{M}{P}, \left(0 \leq x_2 \leq L_2 - e\right) \tag{40}$$

$$y_{2,\mathrm{a}} = C_7 \cdot \cos(\alpha_2 \cdot x_2) + C_8 \cdot \sin(\alpha_2 \cdot x_2) + \Gamma_2 \cdot \sin\frac{\pi \cdot x_2}{L} - \frac{1}{P} \cdot \left[\frac{M}{L} + H_2\right] \cdot x_2 + \ldots$$
$$\ldots. - \frac{1}{P} \cdot H_2 \cdot (e - L_2) + \frac{M}{P}, \left(L_2 - e \leq x_2 \leq L_2\right) \tag{41}$$

$$y_{2,\mathrm{a}} = C_7 \cdot \cos(\alpha_2 \cdot x_2) + C_8 \cdot \sin(\alpha_2 \cdot x_2) + \Gamma_2 \cdot \sin\frac{\pi \cdot x_2}{L} - \frac{1}{P} \cdot \left[\frac{M}{L} + H_2\right] \cdot x_2 + \ldots$$
$$\ldots - \frac{1}{P} \cdot H_2 \cdot (e - L_2) + \frac{M}{P}, \left(L_2 - e \leq x_2 \leq L_2\right) \tag{42}$$

where $\alpha_1 = \sqrt{\frac{P}{E \cdot I_1}}$; $\alpha_2 = \sqrt{\frac{P}{E \cdot I_2}}$; $\Gamma_1 = \frac{a_0}{\left(\frac{\pi^2 \cdot E \cdot I_1}{P \cdot L^2} - 1\right)}$; $\Gamma_2 = \frac{a_0}{\left(\frac{\pi^2 \cdot E \cdot I_2}{P \cdot L^2} - 1\right)}$ and

$C_1, C_2, \ldots, C_8$ are constants to be determined from the boundary and compatibility conditions.

The deflections must satisfy conditions at the top and the bottom of the prop (i.e. at A and B in Figure 35) and the compatibility conditions at C and D.

The top and bottom baseplates are not allowed to move horizontally which implies that $y_1\big|_{x_1} = 0$ and $y_2\big|_{x_2} = 0$.

The bottom column-baseplate system behaves semi-rigidly; according to BS EN 1065 (BSI, 1998) the rotational reaction moment, *M*, produced by the spring (see Figure 36) can be calculated from:

$$M_{\text{b,crit}} = N \cdot e_{\text{crit}} \leq M_{\text{pl}} \cdot \cos\left(\frac{\pi}{2} \cdot \frac{N}{N_{\text{pl}}}\right) \tag{43}$$

where M_{pl} is the moment resistance of the cross-section, *N* the actual normal force, N_{pl} the compression resistance of the cross-section and e_{crit} is 0.4·*D* the critical diameter of base of the prop. The following conditions apply:

(i) Before the gap is closed: $\frac{\mathrm{d}y_2}{\mathrm{d}x_2} < \theta_{\text{gap}}$, $M\big|_{x_2=0} = 0$, and $\Delta_2 = \Delta_{2,0}$

(ii) When the gap is closed: $\frac{\mathrm{d}y_2}{\mathrm{d}x_2} \geq \theta_{\text{gap}}$, $M\big|_{x_2=0} = k \cdot \left(\frac{\mathrm{d}y_2}{\mathrm{d}x_2} - \theta_{\text{gap}}\right)\Bigg|_{x_2=0}$ and $\Delta_2' = -\Delta_{2,\text{crit}}$

(iii) When $M \geq M_{\text{b,crit}}$, $M\big|_{x_2=0} = M_{\text{b,crit}}$

This system defines a trilinear moment-curvature relationship, i.e. a hinge until the rotational gap is closed (in BS EN 1065 (BSI, 1999) a gap of 0.02 radians) then a constant rotational stiffness, *k*, (taken as 40 kN.m/rad) followed by a plastic hinge moment defined by Eq..

The compatibility conditions at point D in Figure 35 are given by:

$$\begin{aligned} y_{1,\text{a}}\big|_{x_1=L_1} &= y_2\big|_{x_2=L_2-e} \\ y_2\big|_{x_2=L_2-e} &= y_{2,\text{a}}\big|_{x_2=L_2-e} \\ \frac{\mathrm{d}y_2}{\mathrm{d}x_2}\bigg|_{x_2=L_2-e} &= \frac{\mathrm{d}y_{2,\text{a}}}{\mathrm{d}x_2}\bigg|_{x_2=L_2-e} \end{aligned} \tag{44}$$

with the first two relations representing the deflection compatibility and the third the rotation compatibility.

Similarly, at hinge position C:

$$\begin{aligned} y_{2,\mathrm{a}}\Big|_{x_2=L_2} &= y_1\Big|_{x_1=L_1-e} \\ y_1\Big|_{x_1=L_1-e} &= y_{1,\mathrm{a}}\Big|_{x_1=L_1-e} \\ \frac{\mathrm{d}y_1}{\mathrm{d}x_1}\Bigg|_{x_1=L_1-e} &= \frac{\mathrm{d}y_{1,\mathrm{a}}}{\mathrm{d}x_1}\Bigg|_{x_1=L_1-e} \end{aligned} \tag{45}$$

Global static equilibrium gives:

$$V_1 = P \tag{46}$$

$$\frac{M}{P} + H_2 = H_1 \tag{47}$$

$$H_2 \cdot L_1 + V_1 \cdot \left(y_{1,\mathrm{m}} + y_{1,0\mathrm{m}}\right) = H_1 \cdot \left(L_1 - e\right) \tag{48}$$

From Eq. 30 at x_1=0:

$$C_1 = \Delta_1 \tag{49}$$

Similarly, from Eq. 31 at x_2=0:

$$C_3 + \frac{1}{P} \cdot M = \Delta_2 \tag{50}$$

When the gap at the baseplate is closed then differentiating Eq. 31 and using $\frac{\mathrm{d}y_2}{\mathrm{d}x_2} = \theta_{\mathrm{gap}}$ we get

$$\alpha_2 \cdot C_4 - \left(\frac{1}{P \cdot L} + \frac{1}{k}\right) \cdot M = \psi_2 - \Gamma_2 \cdot \frac{\pi}{L} + \theta_{\mathrm{gap}} \tag{51}$$

Substituting from Eqs. 31 and 32 into rotation constraint of Eq. 35 we get:

$$\begin{aligned} &C_3 \cdot \cos\left[\alpha_2 \cdot (L_2 - e)\right] + C_4 \cdot \sin\left[\alpha_2 \cdot (L_2 - e)\right] - L_1 \cdot C_5 - C_6 = \left(\frac{L_1^3}{6 \cdot E \cdot I_1 \cdot L} - \frac{L_1}{P \cdot L}\right) \cdot M + \ldots \\ &\ldots - \frac{e^3}{6 \cdot E \cdot I_1} \cdot H_1 + \Delta + (L_2 - e) \cdot \psi - \frac{V_1}{E \cdot I_1} \cdot \left(y_{1,\mathrm{m}} + y_{1,0\mathrm{m}}\right) \cdot \frac{L_1^2}{2} - \Gamma_2 \cdot \sin\frac{\pi \cdot (L_2 - e)}{L} \end{aligned} \tag{52}$$

Substituting from Eqs. 31 and 32 into the first part of 35 we get:

$$\left(C_3 - C_7\right) \cdot \cos\left[\alpha_2 \cdot (L_2 - e)\right] + \left(C_4 - C_8\right) \cdot \sin\left[\alpha_2 \cdot (L_2 - e)\right] = 0 \tag{53}$$

Similarly, using the second part of Eq. 35:

$$\alpha_2 \cdot \left(C_7 - C_3\right) \cdot \sin\left[\alpha_2 \cdot (L_2 - e)\right] + \alpha_2 \cdot \left(C_4 - C_8\right) \cdot \cos\left[\alpha_2 \cdot (L_2 - e)\right] + \frac{1}{P} \cdot H_2 = 0 \tag{54}$$

Using Eqs. 30, and 32 the third part of Eq. 35:

$$\begin{aligned} &C_1 \cdot \cos\left[\alpha_1 \cdot (L_1 - e)\right] + C_2 \cdot \sin\left[\alpha_1 \cdot (L_1 - e)\right] - C_7 \cdot \cos(\alpha_2 \cdot L_2) - C_8 \cdot \sin(\alpha_2 \cdot L_2) + \frac{e}{P} \cdot H_2 \\ &= \Gamma_2 \cdot \sin\frac{\pi \cdot L_2}{L} - \Delta_2 - L_2 \cdot \psi_2 - \Gamma_1 \cdot \sin\frac{\pi \cdot (L_1 - e)}{L} + \Delta_1 + \left(L_1 - e\right) \cdot \psi_1 \end{aligned} \tag{55}$$

Using Eqs. 30, and 31 the second part of Eq.35:

$$\begin{aligned} &C_1 \cdot \cos\left[\alpha_1 \cdot \left(L_1 - e\right)\right] + C_2 \cdot \sin\left[\alpha_1 \cdot \left(L_1 - e\right)\right] - \left(L_1 - e\right) \cdot C_5 - C_6 \\ &= \frac{1}{E \cdot I_1} \cdot \left[\frac{M \cdot \left(L_1 - e\right)^3}{6 \cdot L} - V_1 \cdot (y_{1,\mathrm{m}} + y_{1,0\mathrm{m}}) \cdot \frac{\left(L_1 - e\right)^2}{2}\right] - \Gamma_1 \cdot \sin\frac{\pi \cdot \left(L_1 - e\right)}{L} - \frac{M}{P \cdot L} \cdot \left(L_1 - e\right) + \ldots \\ &\ldots + \Delta_1 + \left(L_1 - e\right) \cdot \psi_1 \end{aligned} \tag{56}$$

Finally, Eqs 30., 31 and the third part of Eq. 35:

$$\begin{aligned} &-C_1 \cdot \alpha_1 \sin\left[\alpha_1 \cdot \left(L_1 - e\right)\right] + C_2 \cdot \alpha_1 \cos\left[\alpha_1 \left(L_1 - e\right)\right] - C_5 \\ &= \frac{1}{E \cdot I_1} \cdot \left[\frac{M \cdot \left(L_1 - e\right)^2}{2 \cdot L} - V_1 \cdot \left(y_{1,\mathrm{m}} + y_{1,0\mathrm{m}}\right) \cdot \left(L_1 - e\right)\right] - \Gamma_1 \cdot \frac{\pi}{L} \cdot \cos\frac{\pi \cdot \left(L_1 - e\right)}{L} - \frac{M}{P \cdot L} + \psi_1 \end{aligned} \tag{57}$$

Using the equations of static equilibrium shown in Figure 37 we get the bending moments and shear forces within the prop as:

$$\begin{aligned} &M_1 = P_1 \cdot \left(y_1 + y_{1,0}\right) - \frac{M}{L} \cdot x_1 && (0 \le x_1 \le L_1 - e) \\ &M_{1,\mathrm{a}} = H_1 \cdot \left(x_1 - L_1 + e\right) + V_1 \cdot \left(y_{1,\mathrm{m}} + y_{1,0\mathrm{m}}\right) - \frac{M}{L} \cdot x_1 && \left(L_1 - e \le x_1 \le L_1\right) \\ &M_2 = P \cdot \left(y_2 + y_{2,0}\right) + \frac{M}{L} \cdot x_2 && \left(0 \le x_2 \le L_2 - e\right) \\ &M_{2,\mathrm{a}} = P \cdot \left(y_{2,\mathrm{a}} + y_{2,0}\right) + \frac{M}{L} \cdot x_2 + H_2 \cdot \left(x_2 - L_2 + e\right) && \left(L_2 - e \le x_2 \le L_2\right) \end{aligned} \tag{58}$$

$$
\begin{aligned}
&V_1 = P\cdot\left(c\cdot q_1 + c\cdot q_5\right) - \frac{M}{L} && \left(0 \le x_1 \le L_1 - e\right)\\
&V_{1,a} = H_1 - \frac{M}{L}\cdot x_1 && \left(L_1 - e \le x_1 \le L_1\right)\\
&V_2 = P\cdot\left(c\cdot q_2 + c\cdot q_6\right) + \frac{M}{L} && \left(0 \le x_2 \le L_2 - e\right)\\
&V_{2,a} = P\cdot\left(c\cdot q_4 + c\cdot q_6\right) + \frac{M}{L} + H_2 && \left(L_2 - e \le x_2 \le L_2\right)
\end{aligned}
\tag{59}
$$

where:

$$
\begin{aligned}
&c\cdot q_1 = -C_1\cdot\alpha_1 \sin\left(\alpha_1\cdot x_1\right) + C_2\cdot\alpha_1\cdot\cos\left(\alpha_1\cdot x_1\right) + \Gamma_1\cdot\frac{\pi}{L}\cdot\cos\frac{\pi\cdot x_1}{L} + \frac{M}{P\cdot L} - \psi_1\\
&c\cdot q_2 = -C_3\cdot\alpha_2\cdot\sin\left(\alpha_2\cdot x_2\right) + C_4\cdot\alpha_2\cdot\cos\left(\alpha_2\cdot x_2\right) + \Gamma_2\cdot\frac{\pi}{L}\cdot\cos\frac{\pi\cdot x_2}{L} + \frac{M}{P\cdot L} - \psi_2\\
&c\cdot q_4 = -C_7\cdot\alpha_2\cdot\sin\left(\alpha_2\cdot x_2\right) + C_4\cdot\alpha_2\cdot\cos\left(\alpha_2\cdot x_2\right) + \Gamma_2\cdot\frac{\pi}{L}\cdot\cos\frac{\pi\cdot x_2}{L} - \frac{1}{P}\cdot\left(\frac{M}{L} + H_2\right) - \psi_2\\
&c\cdot q_5 = a_0\cdot\frac{\pi}{L}\cdot\cos\cdot\frac{\pi\cdot x_1}{L} + \psi_1\\
&c\cdot q_6 = a_0\cdot\frac{\pi}{L}\cdot\cos\cdot\frac{\pi\cdot x_2}{L} + \psi_2
\end{aligned}
\tag{60}
$$

In Europe, steel props are classified into Standard Type Props and Heavy Duty Props according to their safe working loads (P_N and P_G) respectively. They are calculated and constrained by:

$$P_j = \beta_j\cdot\frac{L_{max}}{L^2} \le P_{max} \tag{61}$$

with $\beta_N = 68.0$, $\beta_G = 102.0$ and $P_{max,N} = 51.0$ kN and $P_{max,G} = 59.5$ kN.

There are normally pin holes in the top prop to enable coarse adjustments to the length of the prop to be made before fine adjustments are made with the screw fitting. These reduce the moment of inertia of the prop, resulting in reduced moment resistance of cross-sections. Using Figure 38 these give the following reduced values according to (BSI, 1999):

$$A_{gr} = \pi\cdot\left(R^2 - r^2\right) \tag{62}$$

$$I_{gr} = \frac{\pi}{4}\cdot\left(R^4 - r^4\right) \tag{63}$$

$$A_n = 2\cdot\left(\varphi_R\cdot R^2 - \varphi_r\cdot r^2\right) - d_0\cdot\left(\sqrt{R^2 - \frac{d^2}{4}} - \sqrt{r^2 - \frac{d^2}{4}}\right) \tag{64}$$

Figure 38. Tube models

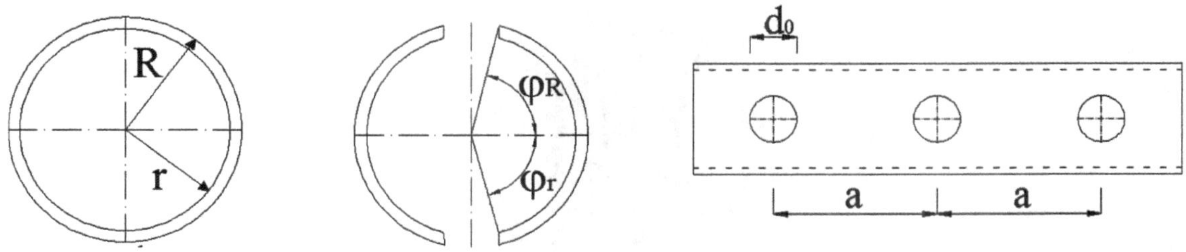

$$I_n = \frac{R^3}{2} \cdot \left[\varphi_R \cdot R - \frac{d_0}{6} \cdot \sin \varphi_R \cdot \left(3 + 2 \cdot \sin^2 \varphi_R \right) \right] - \frac{r^3}{2} \cdot \left[\varphi_r \cdot r - \frac{d_0}{6} \cdot \sin \varphi_r \cdot \left(3 + 2 \cdot \sin^2 \varphi_r \right) \right] \tag{65}$$

$$I_i = I_{gr} \cdot \frac{1}{1 + 2 \cdot \frac{d_0}{a} \cdot \left(\frac{I_{gr}}{I_n} - 1 \right)} \tag{66}$$

$$\varphi_R = \cos^{-1} \left(\frac{d}{2 \cdot R} \right) \tag{67}$$

$$\varphi_r = \cos^{-1} \left(\frac{d_0}{2 \cdot r} \right) \tag{68}$$

where A_{gr} and A_n are the gross and net cross-sectional areas, I_{gr} and I_n the corresponding moments of inertia and I_i the equivalent inertia moment of the cross-section.

When the results of these calculations were compared with a finite element analysis of the same prop in the fully extended case for a series of five props the mean difference was 2.1%.

Experimental tests can also be performed to analyse and design props. Figure 39 shows an example of a test as specified in BS EN 1065 (André, Baptista, & Camotim, 2007), and Figure 40 represents one of the results.

For aluminium props, the standard BS EN 16031 (BSI, 2012) introduces a structural model very similar to the one specified in BS EN 1065 for steel props and detailed above.

4.7.2 Metal Access Scaffolds

Access scaffolds are designed to enable work to be carried out on the side of buildings. They are commonly of two types – one row of standards called Putlog scaffolds and two rows of standards called independent tied scaffolds.

Figure 39. Telescopic prop under test

Figure 40. Behaviour under axial loading of an imperfect steel telescopic prop

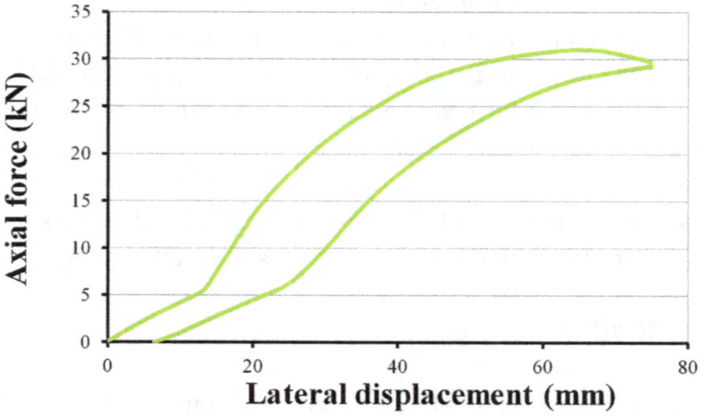

4.7.2.1 2-D Models

In the case of tubular scaffolds, see Figure 41, simple 2-D models have proved effective in the evaluation and design of these scaffolds (Beale & Godley, 2006).

In the case of the Putlog scaffold, buckling in the plane of the façade is restricted by the presence of the façade bracing so that the buckling length is equal to the storey height. Buckling normal to the façade is much influenced by the position of the ties. The standard which is not directly tied to the façade is the least restrained. Its horizontal restraints are provided at each level by the ledgers. The ledgers are normally put into bending as the buckling mode develops and act as horizontal linear springs. Because of the repetitive nature of the structure, for a putlog scaffold, the bucking may be represented by the two column model shown in Figure 42(a).

Figure 41. Typical access scaffolds

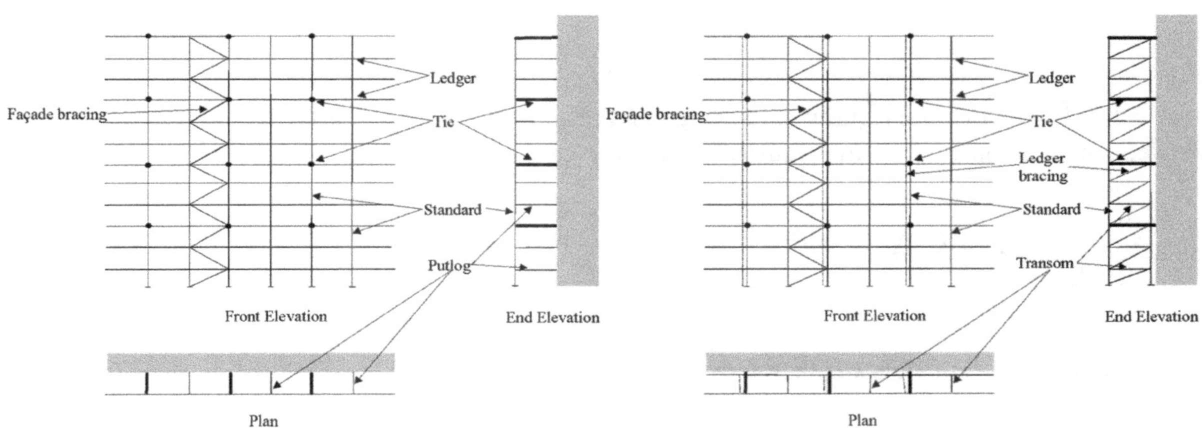

Figure 42. Double and single column models for access scaffolds

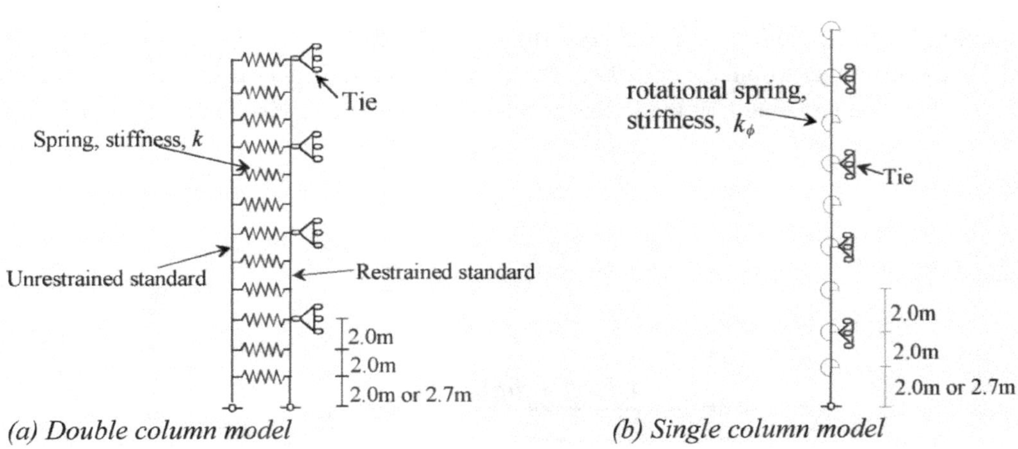

This model has one tied column and one free column. The tied column is restrained, for the example in Figure 42(a), at every third level and the free column is linked to it by linear springs with a stiffness *k*. If the springs are stiff enough, the free column and the tied column buckle together at a load corresponding to an effective length equal to the tie interval. In the case of the example in Figure 4.42(a), with stiff springs the effective length is 6 m and the ledgers do not flex but remain straight. If the springs are not stiff enough for this mode of buckling, the free column will buckle independently of the tied column at a load corresponding to an effective length greater than the tie interval.

Using a linear eigenvalue program it can be shown that that for a tie interval of 6 m, for example, log10 (spring stiffness) must exceed 0.34 if the two columns are to buckle together. The ledger is a continuous beam along the face of the scaffold at each lift. The lowest stiffness it can afford to the free standard occurs when alternate free columns buckle inwards and outwards and the ledger behaves as a simply supported beam as shown in Figure 43. From this figure the deflection, Δ, is given by:

$$\Delta = \frac{P \cdot \left(2 \cdot L_b\right)^3}{48 \cdot E \cdot I} \tag{69}$$

in which $E{\cdot}I$ is the flexural rigidity of the ledger, *P* the applied load and L_b is the bay width, the distance between the columns. Hence the required stiffness, *k*, can be found by:

$$P = \frac{48 \cdot E \cdot I}{\left(2 \cdot L_b\right)^3} \cdot \Delta = k \cdot \Delta \tag{70}$$

The value of the stiffness *k* is therefore:

$$k = \frac{48 \cdot E \cdot I}{\left(2 \cdot L_b\right)^3} \tag{71}$$

In the case of normal Putlog scaffolds, the stiffness afforded by the restraining ledgers is always sufficient for the columns to buckle together, except when the tie interval is 2 m. The highest value of stiffness that can be generated by the ledger is when it behaves as a fixed ended beam. In this case:

Figure 43. Derivation of stiffness for 'pinned' ledgers

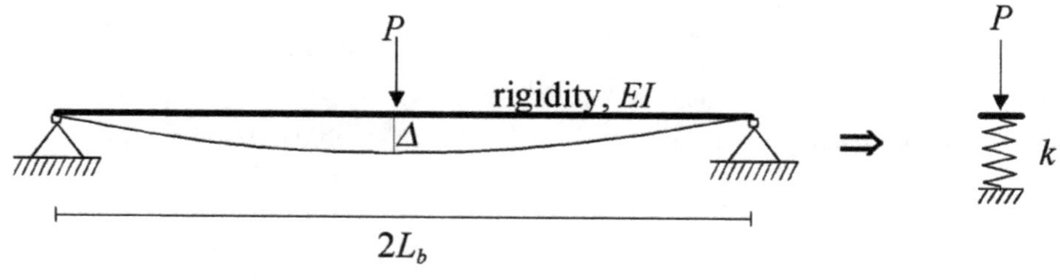

$$\Delta = \frac{P \cdot \left(2 \cdot L_b\right)^3}{192 \cdot E \cdot I} \tag{72}$$

which implies that:

$$P = \frac{192 \cdot E \cdot I}{\left(2 \cdot L_b\right)^3} \cdot \Delta = k \cdot \Delta \tag{73}$$

Hence in this case the stiffness is given by:

$$k = \frac{192 \cdot E \cdot I}{\left(2 \cdot L_b\right)^3} \tag{74}$$

Something close to this situation occurs if one free column buckles in isolation from the remainder of the columns, perhaps because it is carrying an additional axial load as part of a bracing system.

Calculations show that in nearly all cases the effective length can be taken as equal to the tie interval. This is because for many cases the stiffness provided by the ledger is sufficient to ensure that the free standard buckles with the tied standard. In those cases where there are ties at every level, bucking occurs in a mode where each free standard buckles alternately towards and away from the façade.

In the case of independent tied scaffolds, buckling normal to the façade takes place in a mode where alternate non-tied standards buckle in alternate directions. Consequently, for this case, the scaffold model in Figure 4.42(a) can be used, because the ledger bracing simulates ties normal to the façade at every lift, provided that there is at least one tie per ledger braced frame. In the direction parallel to the façade, buckling of the front face is governed by the façade bracing and the buckled length is equal to the lift interval. For the rear plane of standards, the buckling load is strongly influenced by the tie positions, and conservatively it could be assumed to be equal to the tie interval. However, the semi-rigid nature of the connection between the standard and the ledger reduces the buckling length. Figure 4.42(b), drawn for a scaffold tied at every other lift, presents a suitable single column model of the buckling of a rear standard. This model consists of a column with horizontal restraints at the tie levels and rotational restraints at all the intersections with the ledger. The rotational stiffness, k_φ, of this connection is calculated for ledgers in double curvature between standards as shown in Figure 44.

The rotation, θ, of the ledger is:

$$\theta = \frac{M}{2} \cdot \frac{L_b}{2} \cdot \frac{1}{3 \cdot E \cdot I} = \frac{M \cdot L_b}{12 \cdot E \cdot I} \tag{75}$$

where $E{\cdot}I$ is the flexural rigidity of the ledger, L_b is the bay width and M the restraining moment applied to the standard from the ledgers. The rotation, ϕ, of the coupler connecting the ledger and the standard of cruciform stiffness, k_c, between ledger and standard is given by:

$$\phi = \frac{M}{k_c} \tag{76}$$

Figure 44. Derivation of the rotation stiffness for a standard-ledger connection in double curvature

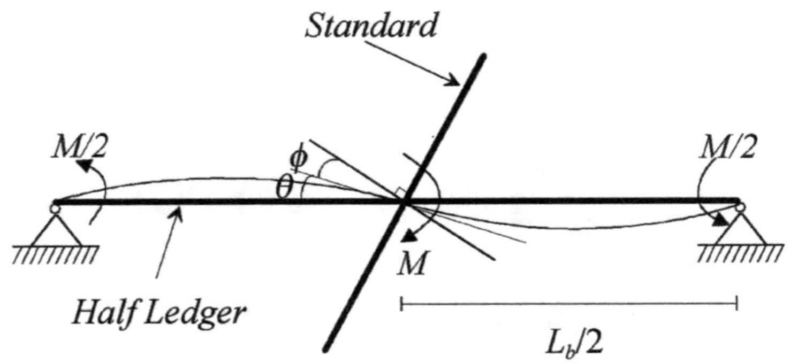

Hence the total rotation between ledger and standard is:

$$\theta + \phi = \frac{M \cdot L_{\text{b}}}{12 \cdot E \cdot I} + \frac{M}{k_{\text{c}}} = M \cdot \left(\frac{L_{\text{b}}}{12 \cdot E \cdot I} + \frac{1}{k_{\text{c}}} \right) \tag{77}$$

The equivalent rotational stiffness, k_{ϕ}, at the connection is therefore:

$$k_{\text{f}} = \frac{M}{\theta + \phi} = \frac{1}{\dfrac{L_{\text{b}}}{12 \cdot E \cdot I} + \dfrac{1}{k_{\text{c}}}} = \frac{k_{\text{c}}}{1 + \dfrac{k_{\text{c}} \cdot L_{\text{b}}}{12 \cdot E \cdot I}} \tag{78}$$

An analysis of this model shows that the buckling lengths are a little smaller than the vertical interval between ties.

4.7.2.1 3-D Models

The above 2-D models of access scaffolds were validated against 3-D nonlinear elastoplastic finite element models constructed using the program LUSAS. In these models, the standards and ledgers were modelled using four co-rotational beam elements per lift (the distance between successive horizontal levels) and per bay (the spacing of the standards parallel to the façade). Joint elements with three rotational and three translational degrees of freedom per element were used.

The eccentricity of 50 mm between the standards and ledgers in tube-and-fitting scaffolds was ignored as the research by Milojkovic (1999) and Milojkovic, Beale, & Godley (1996) showed that for scaffold assemblies the differences in results between including and ignoring eccentricity was marginal. Hence, in the verification analyses the standards and ledgers were co-planar. Note that this assumption has been successfully used by the authors in many occasions. Constraint equations were used to enforce the conditions of zero horizontal and vertical translations between ledger and standard.

The putlog connections from the standard to the façade or between front and rear faces in the ledgers of the tied scaffolds were modelled using bar elements with translational degrees of freedom at each end for the buckling analyses.

For nonlinear analyses, the transom connections were modelled by co-rotational beam elements with rotational releases at the ends. This change was introduced to allow for the possibility of transom buckling.

For both types of scaffold, the connection at the façade was restrained horizontally, parallel and vertically, and was free to move in a direction normal to the façade. In practice, there are small frictional resistances normal to the façade but these were ignored.

For the putlog scaffold, the ties were modelled as bar elements and pinned at each end. In the case of the independent scaffold, the ties were considered to be beams connecting the two faces parallel to the façade and going into the façade. At the façade the tie was pinned. Connections between each tie and the front and rear standards were modelled in the buckling analyses by using joint elements with constraint equations to remove translational degrees of freedom. For the nonlinear analyses, translational degrees of freedom were removed by giving the joints large axial stiffnesses.

Façade and internal bracing elements were modelled using beam elements. To allow for the reduction in axial stiffness due to bending in the diagonal elements, a reduced effective area was used. This is because a diagonal brace's eccentricity behaves in similar way to a pin-ended column loaded eccentrically. The formula for reduction was developed by Godley & Beale (Godley & Beale, 1997) and is given by Eq. 79.

$$\Delta = \frac{P \cdot L}{A \cdot E} + \frac{2 \cdot P}{k} = P \cdot \left(\frac{L}{A \cdot E} + \frac{2}{k} \right) = \frac{P}{A_{red} \cdot E} \tag{79}$$

where A is the original cross-section, P the axial force in the brace, L the length of the brace, k the axial stiffness of the brace determined from a frame test, E Young's Modulus of Elasticity and A_{red} the reduced area to be used. Hence:

$$A_{red} = \frac{L}{A} \cdot \frac{1}{\frac{1}{A \cdot E} + \frac{2}{k}} \tag{80}$$

Plan bracing was inserted into the model in every fourth lift, fourth bay as a horizontal diagonal bar element with a similar reduced area. The supports to the ground were pinned.

The loading from the scaffold boards, due to permanent and variable loads, was applied at nodal points. Wind loading was applied as a combination of point and distributed loads to all standards, see Chapter 6. The variable loads were applied to the top two levels only as vertical loads in agreement with the European standard BS EN 12811-1 (BSI, 2003). Imperfections were applied to the models.

Once a linear model was proved to be correct, a linear eigenvalue buckling analysis was carried out.

Depending upon the ratio of bay span to bay height and tying patterns, the scaffold either buckled normal to the façade or with a sway buckle of the rear row of standards parallel to the façade. There were no three dimensional combined modes. All buckles were two-dimensional. Convergence difficulties were encountered in obtaining buckling modes for load cases combining wind, imposed and permanent loads. These were due to the range of different stiffnesses in the structure – very low joint stiffnesses in combination with relatively high beam stiffnesses. However, for the buckling analyses, a good convergence was obtained for the loads applied at the top of the scaffold. See Figure 45. From this figure, it can be seen that buckling is confined to the lower elements for access scaffolds where the self-weight loading is predominant. Notice also that diagonal bracing prevents the front face from buckling in the independent tied scaffold.

Figure 45. Buckling modes

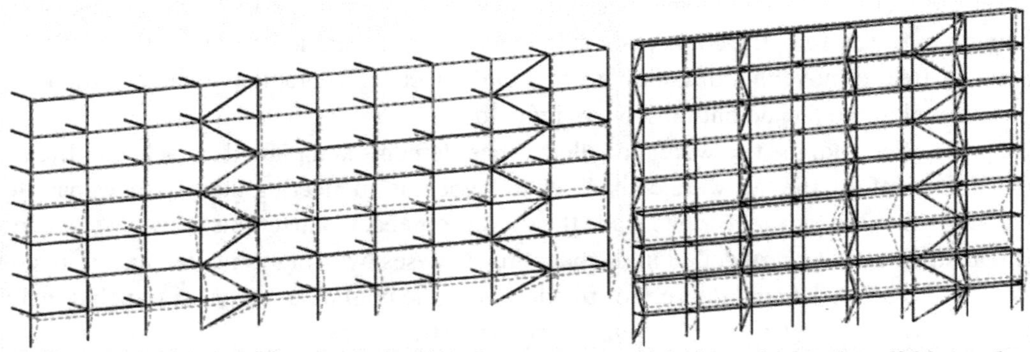

(a) Putlog scaffold, buckling normal to façade *(b) Independent tied scaffold rear face buckling*

The results of the finite element model were compared with the 2-D model results. They were a little higher in the 3-D finite element model because the height of the scaffold in the models was not always a multiple of the tie interval (the top lift was always tied in the finite element model as previous 3-D analyses had shown that the top lift of the scaffold would often fail under wind load normal to the façade if the top level was untied. In this case, standards in the top lift, when clad and subjected to the maximum wind, would simply fail as plastic hinges which would occur just above the highest tied lifts).

Imperfections must be included in analyses of falsework structures. Examples of common imperfections, namely out-of-plumb, sinusoidal member imperfection and imperfections associated with splices are given in Figure 46. Note that the common imperfection often applied to structures, that of the lowest elastic eigenmode, may not yield the correct mode of failure as initial deflections caused by loads such as the imposed load on the top of the structure or lateral wind loads may precipitate modes of failure below those of the eigenmode.

Figure 46. Common imperfections

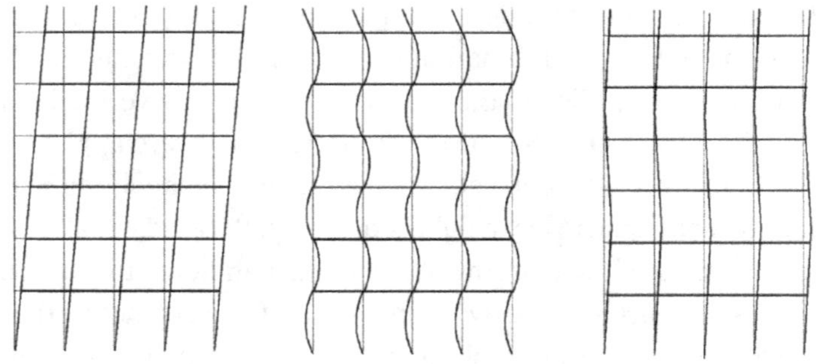

4.7.3 Metal Bridge Falsework

Numerical models of metal bridge falsework systems can be developed in various ways and adopting very different modelling options, finite element types, material models, etc. In the following an overview of the numerical models developed to study bridge falsework systems is presented. The information provided in this Section is applicable to shoring systems similar to bridge falsework systems.

4.7.3.1 Finite Element Types

4.7.3.1.1 Bridge Falsework Main Elements

The main elements of bridge falsework systems are standards (including jacks), ledgers and braces. All these elements can be modelled using first or second-order beam elements, usually with six degrees of freedom per node using the Euler-Bernoulli or the Timoshenko beam theory, suitable for finite strains and large rotations problems. The maximum mesh size should be determined by a mesh sensitivity analysis. For example, in André (2014) the mesh size did not exceed 50 mm. The sections of the different parts of the elements are included in the elements definitions.

The 3-D mesh allows to account for: (i) the exact relative positioning of the elements, for example brace joint eccentricities, (ii) nodal eccentricities, for example eccentricities at the interface between the bridge falsework and the formwork systems, and (iii) geometrical imperfections, either the ones specified in design codes or the ones measured in site surveys.

4.7.3.1.2 Formwork System

Different formwork systems are available. Taking as an example a formwork system consisting of plywood beams positioned in an orthogonal mesh on top of the bridge falsework system, and of plywood panels to which the construction loads were applied to: all the beam elements can be modelled using beam elements, and the plywood panels can be modelled using first or second order full or reduced integration shell elements with six degrees of freedom per node. For reduced integration elements care should be paid to avoid hourglass modes, whereas when using first order full integration elements, shear locking phenomenon must be avoided. Thin or thick shell theory can be used, suitable for finite strains and large rotations problems.

Since the centroids of the plywood beams and panel sections are not at the same height, multi-point constraints (MPC) between the nodes of the plywood beams and a surface defined on the plywood panels plan should be activated.

4.7.3.1.3 Joint Elements

The analytical model for connections presented in Section 4.6.3 can be used to simulate joint behaviour through nonlinear spring elements. Since the current spring elements available in FEA software packages often can only simulate simple models, it may be necessary to develop a universal spring user element. The formulation of such a finite element, consistent with arbitrarily large rotations, is overviewed in the following. It is presented in detail in André (2014) and André et al. (2014).

This spring user element is made of three nodes, labelled *node 1*, *node 2* and *node 3*, respectively, each with six degrees of freedom: three displacements and three rotations. The first two nodes are coincident and were used to control the constitutive behaviour of the user element. The third node of the user element is coincident with a node of a beam element attached to the user element. This third node is used to determine the initial directions of the *x*, *y* and *z* axis of the local coordinate system of the user element.

When arbitrary large rotations occur, relative rotations between *node 1* and *node 2* of the user element cannot be determined by simply subtracting the rotation of *node 1* by the rotation of *node 2*, because finite rotations are not additive.

Finite rotations are expressed by a finite rotation vector, $\boldsymbol{\phi}$, consist of a rotation magnitude, $\theta = ||\boldsymbol{\phi}||$, and a rotation axis or direction in space, P (Simulia, 2012). To characterise this finite rotation mathematically, the rotation vector is used to define an orthogonal transformation or rotation matrix by the Rodrigues formula (Crisfield, 1997):

$$\mathbf{R} = \mathbf{I} + \frac{\sin\left\|\mathrm{f}\right\|}{\left\|\mathrm{f}\right\|} \cdot \hat{\Phi} + \frac{1-\cos\left\|\mathrm{f}\right\|}{\left\|\mathrm{f}\right\|^{2}} \cdot \left(\hat{\Phi} \cdot \hat{\Phi}\right) \tag{81}$$

where $\mathbf{I}$ is the identity matrix, $\hat{\Phi}$ is the skew-symmetric matrix of f given by:

$$\hat{\Phi} = \begin{bmatrix} 0 & -\gamma & \beta \\ \gamma & 0 & -\alpha \\ -\beta & \alpha & 0 \end{bmatrix} \quad , \quad \mathrm{f}^{\mathrm{T}} = \begin{bmatrix} \alpha & \beta & \gamma \end{bmatrix} \tag{82}$$

To calculate the relative rotations, ϕ_i, between *node 1* and *node 2* of the user element, it is necessary that both rotation vectors are defined in the same basis. Let this basis be the local coordinate system of *node 1* of the user element. Further, let $\mathbf{R}^{12}$ be the rotation matrix that rotates the local coordinate system of *node 2* into the local coordinate system of *node 1*:

$$\mathbf{R}^{12} = \mathbf{R}^{1} \cdot \left(\mathbf{R}^{2}\right)^{\mathrm{T}} \tag{83}$$

Therefore, the rotation vector, $\boldsymbol{\phi}$, of the tensor $\mathbf{R}^{12}$ gives the relative rotations between *node 1* and *node 2* of the user element. Spurrier's algorithm is used to extract $\boldsymbol{\phi}$ out of $\mathbf{R}^{12}$ (Crisfield, 1997). The vector thus determined is expressed in the global coordinate system and must be transformed to the user element local coordinate system.

Relative displacements in the global coordinate system, δ_i, are obtained by:

$$\delta = \mathrm{r}^{2} - \mathrm{r}^{1} \tag{84}$$

where r^{i} is the position vector of node i in the current configuration, $i = 1,\ 2$.

The user element internal force vector expressed in the local coordinate system of the user element for a generic node is determined by:

$$\begin{bmatrix} \mathrm{f} \\ - \\ \mathrm{m} \end{bmatrix}^{L} = \left(\mathrm{K}\right)^{L} \begin{bmatrix} \delta \\ - \\ \end{bmatrix}^{L} \tag{85}$$

where f, m are the internal forces and moments of the user element;

$\left(\mathrm{K}\right)^{L}$ is the local stiffness matrix of the user element:

$$\left(\mathrm{K}\right)^{L} = \begin{bmatrix} K_{\delta x}^{L} & 0 & 0 & 0 & 0 & 0 \\ 0 & K_{\delta y}^{L} & 0 & 0 & 0 & 0 \\ 0 & 0 & K_{\delta z}^{L} & 0 & 0 & 0 \\ 0 & 0 & 0 & K_{\theta x}^{L} & 0 & 0 \\ 0 & 0 & 0 & 0 & K_{\theta y}^{L} & 0 \\ 0 & 0 & 0 & 0 & 0 & K_{\theta z}^{L} \end{bmatrix} \tag{86}$$

where K_i is the stiffness of the i degree of freedom of the node and is given by the user element constitutive model, which can be derived from the results obtained from the experimental tests presented in Section 4.6.2.

The user element stiffness matrix expressed in the global coordinate system is determined by:

$$\left(\mathrm{K}\right)^{G} = \begin{bmatrix} \dfrac{d\mathrm{f}^{G}}{d\mathrm{u}^{i}} & \dfrac{d\mathrm{f}^{G}}{d\theta^{i}} \\ \dfrac{d\mathrm{m}^{G}}{d\mathrm{u}^{i}} & \dfrac{d\mathrm{m}^{G}}{d\theta^{i}} \end{bmatrix} \tag{87}$$

where $\mathbf{du}^i$ is the displacements vector of node i of the user element and $d\boldsymbol{\theta}^i$ is the “linearised” rotation vector of node i of the user element (Simulia, 2012).

The derivatives for *node 1* of the user element are expressed by:

$$\frac{d\mathrm{f}^{1,G}}{d\mathrm{u}^{i}} = \sum_{j=x,y,z} \mathrm{e}_{j} \left[\sum_{k=x,y,z} \left(\frac{df_{k}^{1,L}}{d\delta_{k}^{L}} \frac{d\delta_{k}^{L}}{d\mathrm{u}^{i}} + \frac{df_{k}^{1,L}}{d\phi_{k}^{L}} \frac{d\phi_{k}^{L}}{d\mathrm{u}^{i}} \right) \right] + \sum_{j=x,y,z} f_{j}^{1,L} \frac{d\mathrm{e}_{j}}{d\mathrm{u}^{i}} \tag{88}$$

$$\frac{d\mathrm{f}^{1,G}}{d\theta^{i}} = \sum_{j=x,y,z} \mathrm{e}_{j} \left[\sum_{k=x,y,z} \left(\frac{df_{k}^{1,L}}{d\delta_{k}^{L}} \frac{d\delta_{k}^{L}}{d\theta^{i}} + \frac{df_{k}^{1,L}}{d\phi_{k}^{L}} \frac{d\phi_{k}^{L}}{d\theta^{i}} \right) \right] + \sum_{j=x,y,z} f_{j}^{1,L} \frac{d\mathrm{e}_{j}}{d\theta^{i}} \tag{89}$$

$$\frac{d\mathrm{m}^{1,G}}{d\mathrm{u}^{i}} = \sum_{j=x,y,z} \mathrm{e}_{j} \left[\sum_{k=x,y,z} \left(\frac{dm_{k}^{1,L}}{d\delta_{k}^{L}} \frac{d\delta_{k}^{L}}{d\mathrm{u}^{i}} + \frac{dm_{k}^{1,L}}{d\phi_{k}^{L}} \frac{d\phi_{k}^{L}}{d\mathrm{u}^{i}} \right) \right] + \sum_{j=x,y,z} m_{j}^{1,L} \frac{d\mathrm{e}_{j}}{d\mathrm{u}^{i}} \tag{90}$$

$$\frac{d\mathrm{m}^{1,G}}{d_{,}{}^{i}} = \sum_{j=x,y,z} \mathrm{e}_{j} \left[\sum_{k=x,y,z} \left(\frac{dm_{k}^{1,L}}{d\delta_{k}^{L}} \frac{d\delta_{k}^{L}}{d_{,}{}^{i}} + \frac{dm_{k}^{1,L}}{d\phi_{k}^{L}} \frac{d\phi_{k}^{L}}{d_{,}{}^{i}} \right) \right] + \sum_{j=x,y,z} m_{j}^{1,L} \frac{d\mathrm{e}_{j}}{d_{,}{}^{i}} \tag{91}$$

where $\mathbf{e}_j$ represents the vector *j* of the transformation matrix that transforms the tensors in the global coordinate system to the user element local coordinate system.

Additionally, traditional gap joints or more complex contact algorithms can be used to model the behaviour between surfaces that will experience opening-closing cycles during loading. Gap joints were placed at the following locations: (i) interfaces between the baseplate and the supporting ground (i.e. at baseplate joints), (ii) interfaces between the falsework system and the formwork system (i.e. at forkhead joints) and (iii) at the spigot joints.

As a simplification, it was considered that at the forkhead joints, no resistance was available to oppose separation along the vertical axis in the presence of tension forces. If the separation is higher than the height of the wood wedges used to lock the plywood beam to the forkhead, then all restraints are removed and the joint is free to move and rotate.

At the baseplate joint, the optional hypothesis that bolts were used to connect the baseplate to a suitable foundation element was included: anchor bolts for instance on concrete bedding. If these elements are present, then, only after they fail (in tension, bending or shear) is it possible for the joint to separate and move and rotate freely in the presence of tension forces.

At the spigot joint, the hypothesis that a pin was inserted to connect the spigot element to the upper standard tube was included. If these elements are present, then, only after they fail (in tension, bending or shear), there is no resistance available to oppose separation along the vertical axis in the presence of tension forces. If the separation is higher than the outstanding length of the spigot element, then all restraints are removed and the joint is free to move and rotate.

4.7.3.2 Verification of Numerical Models

In order to use the finite element models, it is good practice to verify them. Verification consists in the process of determining that a computational model accurately represents the underlying mathematical model and its solution. Usual checks typically consist in finite element mesh density analyses, see Figure 47 for an example, type of solvers analyses see Figure 48 for an example, and type of material models analyses.

4.7.3.3 Validation of Numerical Models

After being verified, the numerical models need to be validated. Validation is the process of determining the degree to which a model is an accurate representation of the real world from the perspective of the intended uses of the model. This is usually done by comparing the numerical behaviour with the behaviour measured in full-scale tests.

Figure 47. Mesh density sensitivity analysis: refined mesh model vs. reference mesh model

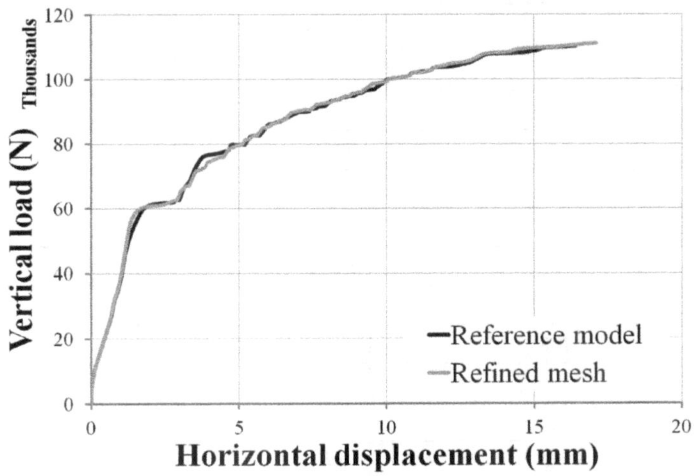

Figure 48. Implicit static solvers vs. Implicit dynamic solvers

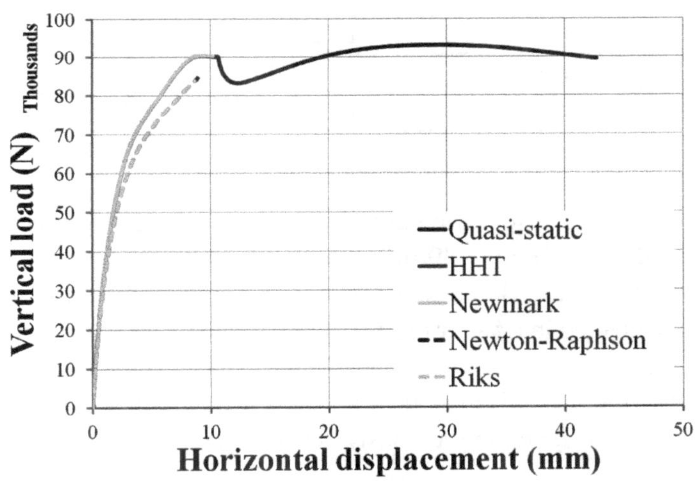

For example, André (2014) compared the numerical behaviour of bridge falsework models against 18 full-scale tests carried out at the University of Sydney in 2006, and published in Chandrangsu & Rasmussen (2011b). A summary of the test configurations which includes test number, lift height, number of lifts, top and bottom jack extension lengths, position of spigot joints, bracing arrangements, type of loading, and loading eccentricity is presented in Chandrangsu & Rasmussen (2011b).

Table 17 presents the results obtained with the numerical models and with the experimental tests. Results of tests A1, A7 and A17 are not presented because of testing problems. The maximum load obtained in the present study, and reported in Table 17, consists in the load value for which the first element failure occurred. After first failure the load does not increase significantly.

The statistical analysis of the ratio between the recorded maximum load and the numerically predicted value is presented in Table 4.18. It can be observed that the numerical models developed in this study can match the experimental resistance with a better precision and accuracy than the existing numerical

Table 17. Summary of results

Experimental tests results		Chandrangsu & Rasmussen (2011b) results		André (2014) results	
Test ID	Maximum load (kN)	Maximum load (kN)	Ratio (Test/Model)	Maximum load (kN)	Ratio (Test/Model)
A2	87	96	0.906	88	0.989
A3	91	91	0.995	90	1.006
A4	50	45	1.111	46	1.087
A5	60	60	1.000	56	1.071
A6	60	66	0.909	56	1.071
A8	130	138	0.942	135	0.963
A9	65	50	1.300	60	1.083
A10	70	64	1.094	72	0.972
A11	120	127	0.945	130	0.923
A12	120	129	0.928	133	0.900
A13	70	68	1.029	69	1.014
A14	160	160	1.000	156	1.026
A15	105	105	1.000	105	1.000
A16	100	100	1.000	104	0.962
A18	150	147	1.020	153	0.980

models. It is also important to analyse if the numerical models can return as accurate results in terms of the overall structural behaviour. Figure 49 illustrate the axial force *vs.* horizontal displacement of a selected node of the structure obtained in a selection of tests and by both numerical models developed in the present study and presented in Chandrangsu & Rasmussen (2011b). The complete comparison is available in André (2014).

It can be observed that the numerical models developed by André also predict better the overall behaviour the falsework systems than the previously developed numerical models.

Numerical Modelling Options

In this Section, the influence of various modelling options is presented and discussed. In particular, the influence of accounting explicitly the formwork, but also the influence of assuming continuous (rigid) or pinned connections to model the various types of joints is analysed.

Formwork Modelling

To compare the results of numerical models with and without formwork the loading distribution was made similar in the two models by applying in the latter model vertical loads on top of the falsework with a distribution proportional to the column influence area. Two models will be used in this study: Models A2 and A4, see André (2014) and Chandrangsu & Rasmussen (2011b) for details, both equal except the former has bracing elements.

Figure 49. Experimental and numerical tests results, tests A3, A6, A10 and A13

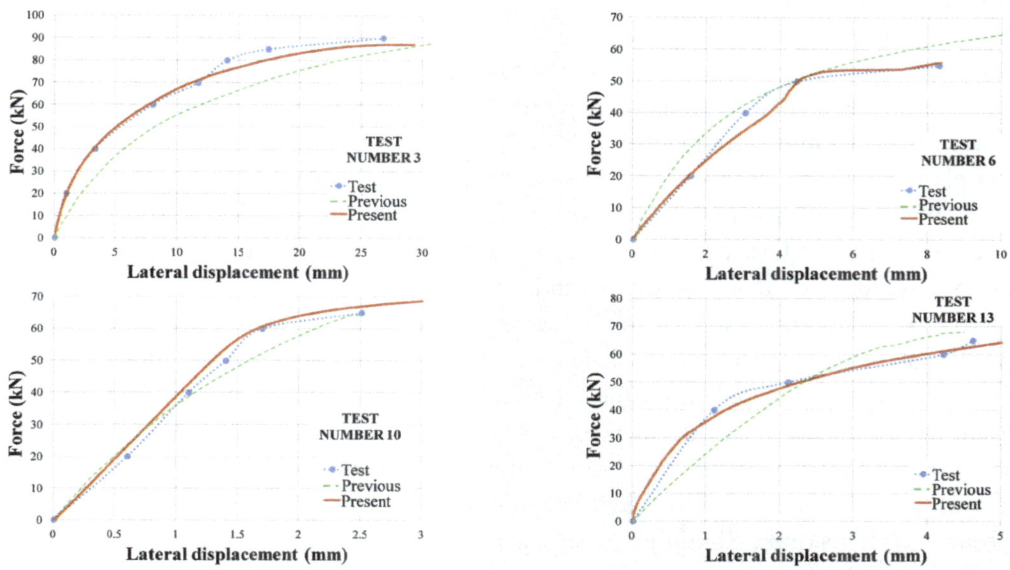

Table 18. Statistical analysis of the results ratio

Metric	Chandrangsu & Rasmussen (2011b) results	André (2014) results
Average ratio	1.012	1.003
Standard deviation of the ratio	0.100	0.057
COV of the ratio	0.098	0.057

The maximum resistance of the model without formwork, in terms of vertical loads, was 124.4 kN in the centre columns. Dividing this load with the influence of a centre column, 3.345 m^2, the equivalent maximum pressure was equal to 37.14 kN/m^2 which compared with the maximum pressure of 39.17 kN/m^2 obtained in the model with formwork. Performing the same analysis for a numerical model without any brace element, the maximum equivalent pressure obtained in the model with formwork was equal to 19.10 kN/m^2 which compared with the maximum pressure of 14.01 kN/m^2 obtained in the model with formwork.

This apparent paradox of obtaining a smaller resistance with the formwork included in the numerical model, is justified by the fact that the formwork is unrestrained. Therefore, in the unbraced model the falsework has no effective lateral restraint (other than the one provided by the ledgers) and follows the stiff formwork displacements, resulting in large rotations at the spigot joints which eventually fail. Thus, it can be concluded that in order to get an accurate estimate of falsework behaviour and resistance, and prevent obtaining unconservative resistance values, it is necessary to include the formwork system in the numerical model.

Joint Modelling

Various different modelling options can be used to simulate the joint behaviour: from the elastic model to the nonlinear elastoplastic model, and from the pinned joint to the rigid joint. It is therefore important to assess the influence that these modelling options have on the behaviour and resistance of bridge falsework predicted by numerical models.

Two models will be used in this study: Models A2 and A4, see André (2014) and Chandrangsu & Rasmussen (2011b) for details, both equal except the former has bracing elements. The analysis was performed considering a reference model, for each type (A2 and A4). After, each modelling option (e.g. continuous spigot joints) was considered individually while the remaining modelling options and parameters were kept the same as in the reference model.

From Figure 50 and Figure 51 it can be seen that for the tested models there is a significant influence on how the joints are modelled on the falsework behaviour and resistance. The most important joint seems to be the cuplok joint with increases in resistance of about 25% for Model A2 and 45% for Model A4 if it is modelled as continuous (infinite translational and rotational stiffness), and decreases in resistance of approximately 30% for Model A2 and 80% for Model A4 if it is modelled as a pinned joint (free rotations). This finding highlights once again the importance of correctly locking cuplok joints and not using damaged elements.

Another important joint is the spigot joint, with increases in resistance of about 2% for Model A2 and Model A4 if it is modelled as continuous (infinite translational and rotational stiffness), and decreases in resistance of approximately 50% for Model A2 and 45% for Model A4 if it is modelled as pinned joint (free rotations). From these results it is clearly seen the influence of using improper (damaged or shorter than normal) spigot elements in the behaviour and resistance of the falsework.

An interesting finding is that considering the brace elements as rigid for translation displacements (including axial displacements) result in a 10% overestimate of the falsework resistance. This highlights the importance of careful assessment of the behaviour of individual components and of rigorous structural analysis.

Producer documents recommend not placing brace elements more than 50 mm apart of the ledger-to-standard joints (SGB, 2006). Nevertheless, values much higher are often found in practice. Also, brace elements can be connected to a ledger element or to a standard element. Additionally to the bracing ec-

Figure 50. Results obtained for Model A2 (braced) considering different joint modelling options

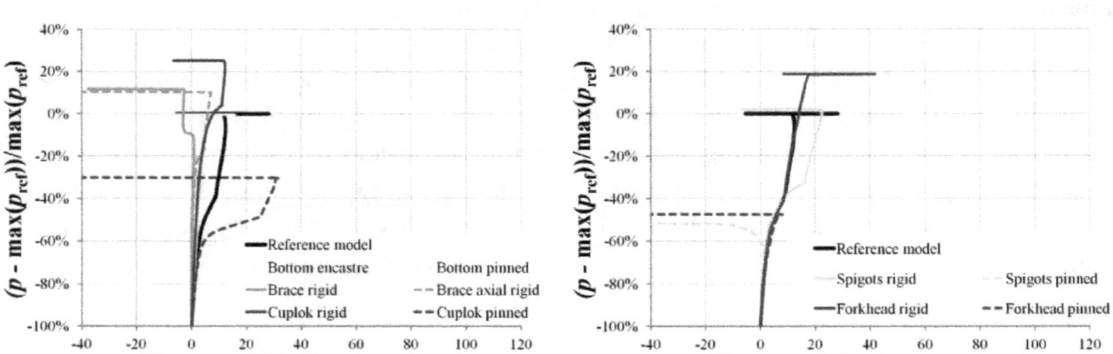

Figure 51. Results obtained for Model A4 (unbraced) considering different joint modelling options s

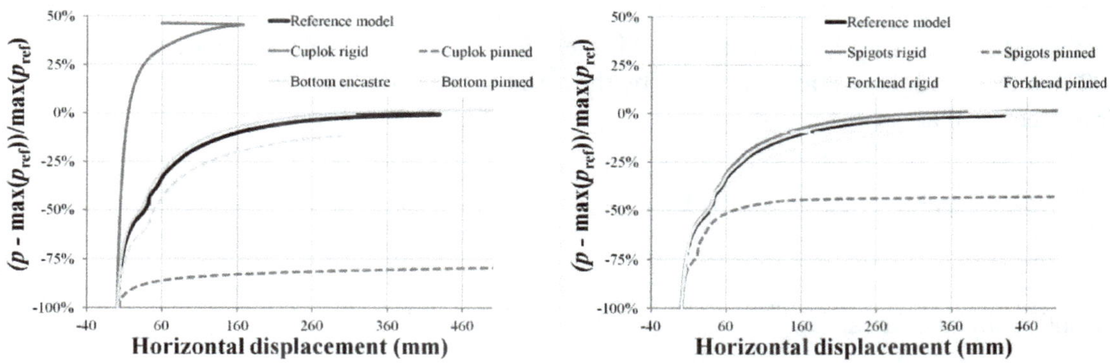

centricity, another variable was considered which is the positioning of internal braces when connected to ledgers, see Figure 52.

To analyse how these variables influence the system's resistance and robustness various models were developed and summarised in André (2014). The geometry, material and joint properties of the reference models are identical to the ones presented in the previous examples, where the brace elements are internally connected to the ledgers with an eccentricity equal to 60 mm.

The only action considered was the one due to the concrete weight placed on top of the formwork. This load was increased until collapse was attained.

From the results, see André (2014), it could be observed that resistance decreases with the increase of the brace eccentricity when brace elements are connected with ledger elements. Also, the inside positioning of the bracing was found to be beneficial when compared to the outside positioning. This happens because the bracing when the former positioning is adopted is more effective in providing lateral stiffness to the core columns (more stressed).

When bracing is connected to standard elements, resistance was also found to decrease with the increase of the brace eccentricity. In all cases, for 200 mm eccentricity, a 10% reduction in the resistance was obtained when compared with the one obtained for 60 mm eccentricity. For 400 mm, the reduction was higher than 30%.

Figure 52. Outside (Left) and inside (Right) positioning of the brace elements

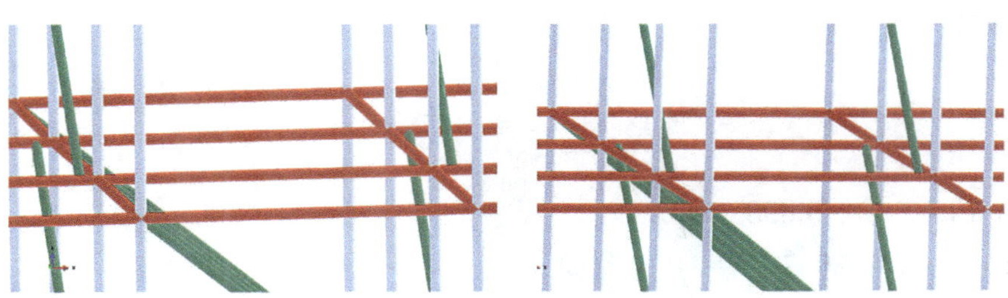

4.7.4 Bamboo Structures

As bamboo standards vary in diameter and thickness from the bottom of a standard to the top (Yu. et al., 2003) determined the section properties by the formulae provided below.

Cross-sectional area:

$$A_1 = \left[\frac{\pi}{4} \cdot \left(D_o^{\,2} - D_i^{\,2}\right)\right]_1 \tag{92}$$

Second moment of area:

$$I_1 = \left[\frac{\pi}{64} \cdot \left(D_o^{\,4} - D_i^{\,4}\right)\right]_1 \quad , \quad I_2 = \left[\frac{\pi}{64} \cdot \left(D_o^{\,4} - D_i^{\,4}\right)\right]_2 \tag{93}$$

Slenderness ratio:

$$\lambda_1 = \frac{L}{r_1} \text{ where } r_1 = \sqrt{\frac{I_1}{A_1}} \tag{94}$$

The subscripts 1 and 2 denote the upper (smaller) and lower (larger) cross-sections respectively. The subscripts o and i relate to the outer and inner diameters of the bamboo pole. The authors determined the elastic critical buckling strength of the column, f_{cr}, to be given by:

$$f_{cr} = \alpha \cdot \frac{\pi^2 \cdot E}{\lambda_1^{\,2}} \tag{95}$$

where the coefficient α was determined experimentally to be the minimum root of the cubic equation:

$$c_3 \cdot \alpha^3 + c_2 \cdot \alpha^2 + c_1 \cdot \alpha + c_0 = 0 \tag{96}$$

and the coefficients are:

$$\begin{aligned}
c_3 &= -0.2880 \\
c_2 &= 2.016 \cdot (2 + \rho) \\
c_1 &= -\left(14.11 + 14.11 \cdot \rho + 3.098 \cdot \rho^2\right) \\
c_0 &= 10.37 + 15.55 \cdot \rho + 7.047 \cdot \rho^2 + 0.932 \cdot \rho^3 \\
\rho &= \frac{I_2 - I_1}{I_1}
\end{aligned} \tag{97}$$

The compressive strength, $f_{c,d}$, of a bamboo scaffold is given by:

$$f_{c,d} = \frac{f_{c,k}}{\gamma_m} \tag{98}$$

where $f_{c,k}$ is the crushing strength of a section and γ_m is the material factor.

To determine the buckling strength, $f_{cc,d}$, of a bamboo column the authors used a modified Perry-Robertson formula:

$$f_{cc,d} = \frac{f_{cr} \cdot f_{c,d}}{\varphi + \sqrt{\varphi^2 - f_{cr} \cdot f_{c,d}}} \tag{99}$$

where $\varphi = \dfrac{f_{c,d} + (1+\eta) \cdot f_{cr}}{2}$ with the Perry factor $\eta = 0.001 \cdot a \cdot (\lambda_1 - \lambda_0)$, Robertson coefficient *a* = 15 for Mao Jue bamboo and *a* = 20 for Kao Jue bamboo.

The limiting slenderness ratio, $\lambda_0 = 0.2 \cdot \pi \sqrt{\dfrac{E}{f_{c,d}}}$.

Figure 53 shows a comparison of different buckling curves with experimental results, superimposed on non-dimensional plots. Note that for design, the values of the Robertson coefficient, *a*, specified above should be used and not the ones shown in Figure 53.

Figure 53. Column buckling curve for Kao Jue (Chung, Chan, & Yu, 2002)

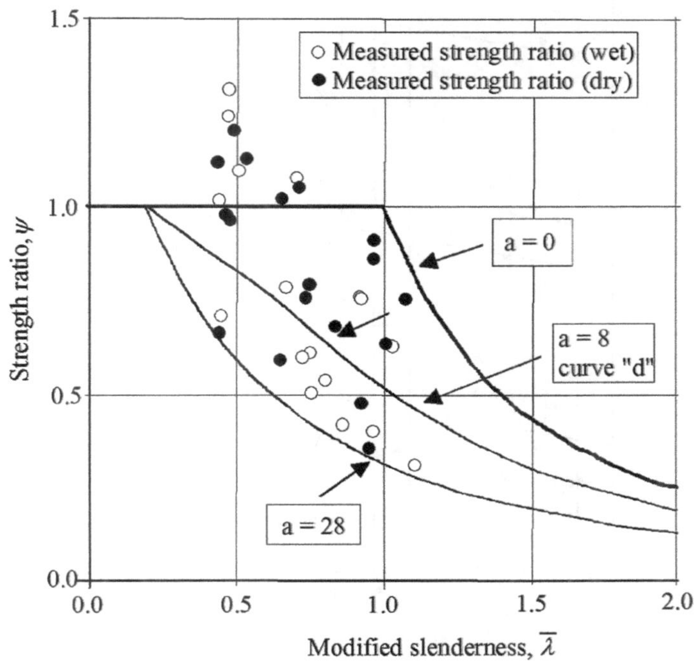

In Figure 53, the strength ratio, Ψ, is defined to be the ratio of the buckling load against the design compressive load. The modified slenderness ratio, $\bar{\lambda}$, is defined to be:

$$\bar{\lambda} = \sqrt{f_{c,d} / f_{cc,d}} \tag{100}$$

Common spacings for standards in access scaffolds are to have the standards set between 2.0 and 3.0 m apart. Because the Young's Modulus and ultimate load capacity of bamboo is significantly below that of metal standards it is common to use spacings of 1.2 to 1.8 m with horizontal ledgers at 1.8 to 2.1 m (Kao Jue bamboo) and standards at 1.5 to 2.4 m. with ledgers at 1.8 to 2.25 m (Mao Jue bamboo). In addition, one or two smaller diameter standards are placed between the main standards with two intermediate smaller diameter ledgers between the main ledgers (Chung, Chan, et al., 2002).

To ensure stability, ties are made to the façade in each main standard between 1.8 to 2.25 m in the vertical direction and 1.8 to 2.4 m horizontally. Diagonal braces are attached to provide additional stability. An isometric schematic of a typical bamboo scaffold is given in Figure 54.

Figure 54. Schematic of a typical bamboo access scaffold

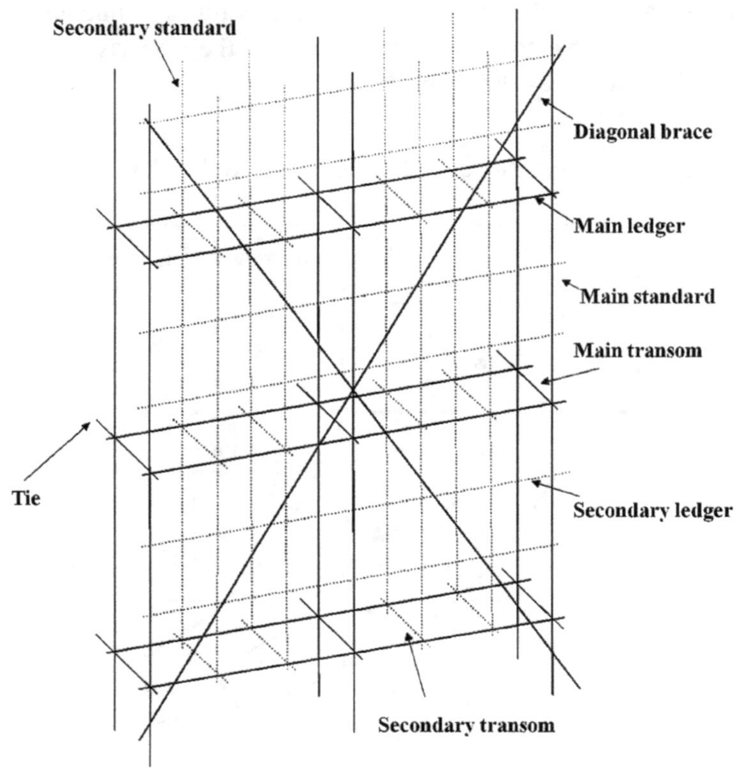

4.7.5 Bridge Construction Equipment

Regarding the design of BCEs, the use of numerical methods in the structural analysis is advisable, see Kindmann & Kraus (2012) and Rosignoli (2013) for complete guidance. As a minimum, a static second-order elastic analysis should be undertaken.

Initial imperfections should be considered in both system and element levels. In their definition, cooperation is recommended between BCE designers, suppliers of materials and specialised equipment, contractors and public agencies. It is important to define maximum acceptable values for the imperfections of elements from which rejection criteria could be established. The design values of the imperfections should then be set depending on the effectiveness and agreed level of the quality control, quality assurance, supervision, inspection and maintenance procedures to be implemented on site. However, both experimental and theoretical research is needed.

Dynamic effects due to impact loads or local failures can be determined by performing a dynamic analysis if no design guidance is found or if the available guidance leads to uneconomical structural solutions or demands application of unenforceable inspection and quality control procedures. An example of a dynamic analysis is presented in Rosignoli (2007).

The connections between the BCE and the permanent structure must also be correctly modelled. The assumptions made in the model must correlate with the conditions in reality. For instance, nominal pinned connections should be checked to assure that the available rotation is at least the required one. In some cases it might be necessary to include in the model the support structures, like concrete piers and temporary steel piers, in order to get accurate results owing to extreme differences in stiffness and material behaviour.

See Chapter 6 for additional guidance.

4.8 SOIL MODELLING

4.8.1 Soil types

Temporary structures can be supported in various types of ground: from natural rock and concrete pavements to soft and loose soils. Of these, soils are the most relevant in terms of modelling complexity and potential impact on the performance of temporary structures for standard design situations, and therefore this Section will focus essentially on them.

Soil is the result of mechanical and chemical weathering of rock (Bowles, 1997; Helwany, 2007). It is a three-phase composite material consisting of solid mineral particles, water and air. Its mechanical behaviour largely depends on the relative proportions of each material forming the soil. Solid particles are classically categorised based on the predominant particle size: gravel, sand, silt and clay (ordered in descending size of solid particle). If the particle size is less than 0.002 mm, the particle is categorised as clay; if they are larger or equal than 0.002 mm but smaller than 0.05 mm they are categorised as silt; if they are larger or equal than 0.05 mm but smaller than 2 mm they are categorised as sand and for sizes larger or equal than 2 mm as gravel.

Of course, natural soils are formed by a mixture of the abovementioned solid particles. The content of each particle in a soil is typically expressed by the texture (granulometry) of the soil, i.e. the size distribution or mass fractions of soil primary particles. Several methods are available to determine the soil texture, such as sieving or sedimentation methods.

Soils can also be divided into two main categories: cohesionless and cohesive. Cohesionless soils (also called frictional soils) have particles that do not adhere such as gravel, sand, and silt (stick) together even with the presence of water. On the other hand, cohesive soils (clays) are characterised by having particle size so small that generates binding forces between the particles.

A correct characterisation of the soils requires an adequate ground investigation (including testing). Guidance is available from AGS (2011), Bowles (1997), BSI (2007, 2009b, 2011a, 2015a, 2015b) and Dowrick (2009). Table 4.19 provides a preliminary characterisation of different types of soil based on Standard Penetration Test (SPT) results (N-value).

4.8.2 Soil Resistance and Stiffness

In this Section, it is only discussed the application of analysis methods to shallow foundations which are the most frequent solution in temporary works. Traditionally, soils have been modelled through very simple methods which analyse separately the bearing capacity of the soil and the soil deformations.

Focusing on the bearing capacity, the simplest way to model the ground is to consider it to be rigid. This enables the analysis to ignore ground behaviour and either (i) use spring elements to simulate the

Table 19. Soil characteristics, adapted from BSI (2011a)

Soil group	Density/compactness/strength		Bedding thickness (mm)	Composite soil types	Particle size (mm)	Soil type
	Term	Field test				
Coarse soils	Very loose (N-value: 0 to 4)	Borehole and SPT N-value	600 to 2000	Sandy (5 to 20%)	6	Gravel
	Loose (N-value: 4 to 10)		200 to 600		4	
	Medium (N-value: 10 to 30		60 to 200		2	
	Dense (N-value: 30 to 50		20 to 60	Very sandy (>20%)	0.6	Sand
	Very dense (N-value: > 50		6 to 20	Sand and gravel (50%)	0.2	
Fine soils	Un-compact		Alternating layers of different types of soils	Slightly sandy (<35%)	0.06	Silt
	Compact					Clay/silt
	Very soft (N-value: 0 to 20)					Clay
	Soft (N-value: 20 to 40)			Sandy (35 to 65%)	0.02	
	Firm (N-value: 40 to 75)				0.006	
	Stiff (N-value: 75 to 150)				0.002	
	Very stiff (N-value: 150 to 300)			Very sandy (>65%)		

rotational behaviour of the foundation element, restraining the translational degrees of freedom of the master element, or (ii) activating general contact algorithm (normal and tangential directions) between shell or solid finite elements used to simulate the foundation element and the rigid surface representing the soil. In this case, ground settlements are simulated as imposed displacements at the foundation nodes, most often ignoring the ground stiffness which results in a conservative design.

Concerning ground deformations, the soil can be simulated by a spring element with an equivalent linear elastic stiffness.

More rigorously, ground (typically soil) can be modelled using shell or solid finite elements and special constitutive relationships of the material behaviour (Helwany, 2007; Potts & Zdravkovic, 1999, 2001). As soil behaviour is highly nonlinear, linear elastic constitutive models are in general poor representations. Additionally, soil variability is much higher than structural materials like steel. As a result, anisotropic rather than isotropic material formulations should be used, meaning that material properties change not only with the magnitude of applied loading but also with its orientation.

Among the most popularly used constitutive models are the Tresca (for undrained conditions), the Mohr-Coulomb (for drained conditions, based on Coulomb failure criterion, see Eq. 101), and the Cam clay models (for both undrained and drained conditions) (Potts & Zdravkovic, 1999).

$$\tau = c + \sigma \cdot \tan\varphi \tag{101}$$

where τ and σ represent the shear and normal stresses on the failure plane, c represents the cohesion of the soil and φ represents the friction angle of the soil.

An in situ test method often used in temporary structures is to determine the vertical deformation and bearing resistance properties of soil and rock masses by loading the ground with a rigid plate that simulates the shallow foundation element and recording the load and the corresponding settlement throughout the test.

However, the results obtained using this test method may only be considered for design if the size of plate used is larger than every dimension of the shallow foundation element and if the superficial ground layer has a thickness of at least two times the width or diameter of the plate.

4.8.3 Soil Pressures

Two conditions will be discussed when soil loading is considered: active and passive pressure.

Active soil pressure is when the soil moves towards the retaining structure. The active soil coefficient, K_a, can be determined as follows:

$$K_a = \frac{1 - \sin\varphi}{1 + \sin\varphi} \tag{102}$$

Passive soil pressure is when the structure moves towards surrounding lateral soil. The passive soil coefficient, K_p, can be determined as follows:

$$K_p = \frac{1 + \sin\varphi}{1 - \sin\varphi} \tag{103}$$

Several theories have been developed over the years for soil pressure analysis. Rankine, Boussinesq, Coulomb, Terzaghi, and others have derived several theories that support how soil loads its support system. According to Rankine theory for active soil pressure, the pressure diagram that develops is given by:

$$p_a\left(h\right) = \gamma \cdot h \cdot K_a - 2 \cdot c \cdot \sqrt{K_a} \tag{104}$$

where:

p_a represents the active pressure at any given depth h;
γ represents the unit weight of the soil;
c represents the soil cohesion.

4.9 WIND ACTION MODELLING

4.9.1 Basis

One of the most important influences of temporary structures is that of wind action. Storms and their associated wind have often been blamed for causing some failures. It is therefore important to consider the effects of wind on such structures.

Wind loads on permanent structures have been described for many years (Baker, 2007; Simiu & Scanlan, 1996) with wind-tunnels used to enable predictions of the wind pressures to be made. However, when scaffolding structures are erected they are usually placed around the permanent structure and sheeted to provide access and support to permanent and temporary structures during different stages of construction in the UK and other parts of the world. Scaffolding, such as other temporary structures, are usually light in weight, easy to maintain, install, and dismantle.

Wind loads on temporary structures are usually taken from tables of loads for permanent structures such as BS 6399-2 (BSI, 1997). However, these loads do not take in to account the short lives of the structure and hence often overestimate the pressures. See Chapters 3 and 5 for a detailed discussion on this topic.

Many problems in Wind Engineering can be tackled by one of three approaches, or a combination of these: on-site measurements, reduced-scale wind tunnel measurements or numerical simulation based on CFD (called Computational Wind Engineering). Deciding which approach is most appropriate for a given problem is not always straightforward, as each approach has specific advantages and disadvantages. An important disadvantage of on-site measurements and wind tunnel measurements is that usually only point measurements are obtained. Techniques such as Particle Image Velocimetry (PIV) and Laser-Induced Fluorescence (LIF) in principle allow planar or even full 3-D data to be obtained, but the cost is considerably high and application for complicated geometries can be hampered by laser-light shielding by the obstructions constituting the model, e.g. in case of an urban model consisting of many buildings. Another disadvantage is the required adherence to similarity criteria in reduced-scale testing. This can be a problem for, e.g. multiphase flow problems and buoyant flows. Examples are the transport and deposition of sand, dust, rain, hail, and snow, and buoyancy-driven natural ventilation and pollutant dispersion studies.

Traditionally wind pressures on structures such as buildings and bridges have been determined using atmospheric wind tunnels. Most wind-tunnel models of buildings require that less than 3% of the tunnel is blocked by the building and hence scales of 1:30 or smaller are often used. Wind-tunnel tests on bare-pole and sheeted scaffold structures have not been often undertaken because of the scaling effect. For example, a scale of 1:50 requires the diameter of a model scaffold tube to be less than 1 mm and of sheeting to be 0.008 mm. At these scales the stiffness of the scaled scaffold tube cannot easily be made the same as that of the full-scale structure. Pressure taps on the scaled scaffold tubes and on netting/sheeting cannot easily be fitted. The aeroelastic nature of netting and sheeting requires wind-tunnel tests on clad scaffolds to be also aeroelastic because any question regarding the static or dynamic stability of the fabric can only be accurately answered by an aeroelastic wind-tunnel test. A rigid model test gives no information regarding the possibility of divergence or flutter, but can be used to predict fluctuating wind pressure due to buffeting.

Any model of wind pressure distribution must take account of the power law (Davenport, 1960):

$$V(z) = \left(\frac{z}{z_{\text{ref}}}\right)^{\beta} \cdot V_{\text{ref}} \tag{105}$$

where z_{ref} is a reference height above ground, and β is a function parameter which depends on the surface roughness at ground level. The wind speed is zero at ground level. Simiu & Scanlan (1996) give approximate values of β to be in the range 6.30 to 7.00 for open terrain, 3.60 to 4.50 in suburban areas, and 2.45 to 3.00 in inner city areas.

Atmospheric boundary layer wind tunnels must produce an equivalent roughness at the base which is often done by a series of blocks of different sizes. In addition, they must be large enough so that at top of the tunnel and at the sides the influence of the model on the air flow is minimal. For example, Figure 55 shows the method of obtaining correct surface roughness (Irtaza, 2009; Irtaza, Beale, & Godley, 2012). The wind tunnel in this experiment was 17.5 m long, 3.3 m wide and 2.2 m high. The particular experiment was to match a full-scale experiment conducted at Silsoe on a 6 m cubical structure (Richards & Hoxey, 2008). The roughness generated the same wind profile as was measured at Silsoe (see Figure 56).

Having obtained the correct velocity profile, a 1:30 scale model of the Silsoe cube was put into the middle of the wind-tunnel and pressure distributions across the surface obtained. These pressure distributions were compared with the Silsoe full-scale cube with good correlation. (see Figure 57 which also compares with the best correlation previously reported (Richards, Hoxey, Connell, & Lander, 2007)).

This procedure which has been discussed for the wind loads applied to a building can also be applied to models of sheeted scaffolds/falsework. However, the sheeting must be solid, for example acrylic (Irtaza, 2009; Irtaza et al., 2012; Wang, Tamura, & Yoshida, 2013, 2014). Most sheeting used to protect temporary structures from wind tends to be porous or netted as this allows some wind to go through and relieve the pressure distribution. Codes of practice give reduced pressures for netted scaffolds of up to 60% of the fully sheeted scaffold (BSI, 2003). Wind-tunnel tests have been reported on full-size porous façades by Gerhardt & Janser (1994), Irtaza (2009) and Richards & Robinson (1999) which give the reductions in pressure through the façade.

Figure 55. Surface roughness pattern. ©2009 H Irtaza. Used with permission

Figure 56. Comparison of velocity profiles of the wind-tunnel to the full-scale cube. ©2009 H Irtaza. Used with permission

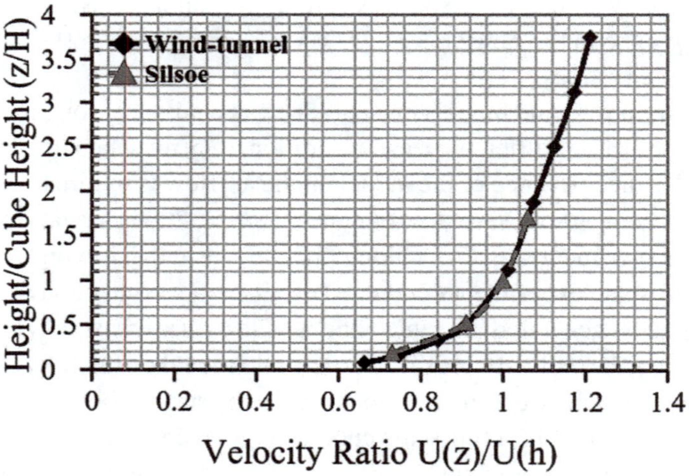

4.9.2 Introduction to CFD

Computational Fluid Dynamics (CFD) is a numerical approach to simulate or predict phenomena and quantities of a flow by solving the equations of motion of the fluid at a discrete set of points (Cao, 2013). It has wide applications in flow-related engineering fields including aeronautical, mechanical and civil/architectural fields, although the difficulties in applying it to particular problems in these fields are different (Blocken, 2011).

Figure 57. Comparison of mean pressures over the cube. ©2009 H Irtaza. Used with permission

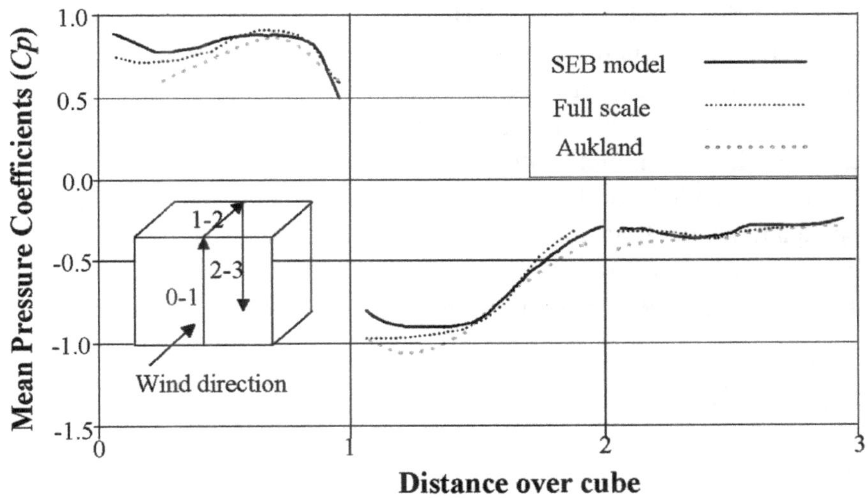

In theory, it is numerically possible to completely resolve all aspects of fluid dynamics problem including the rapid spatial and temporal variations of turbulence in the flow using a CFD technique known as Direct Numerical Simulation (DNS). This technique involves discretising the equations using the finite volume method at a mesh size below the smallest eddies in turbulent flow, the Kolmogorov length scale, and therefore resolving the flow down to the smallest spatial and temporal variations.

Unfortunately, the direct numerical simulation of practical turbulent fluid flows using the time dependent Navier-Stokes equations in their simplest forms is well beyond the capabilities of present day computing power. This is due to the fact that the amount of computer processing (CPU) time required is dependent on the degree of resolution of the small scale eddies. The smallest eddies in turbulent flow, the so-called "Kolmogorov microscales" are very small at about 0.1 to 1mm for natural wind (Versteeg & Malalasekera, 2007).

Therefore, the numerical discretisation of an entire wind engineering flow field with a complex geometry at high Reynolds numbers is at present well beyond the capabilities of even the most powerful supercomputers available. The only economically feasible way to solve this problem is to employ statistically averaged equations which govern the mean flow equations. Turbulence models are then required to achieve closure of the averaged equations and represent the action of turbulent stresses on the mean flow. Unfortunately, the mathematical models used in CFD are only able to perform as well as the physical assumptions and knowledge built into them will allow. In particular, the assumptions made regarding the modelling of the turbulent component of engineering flows have proved to be a major source of error in wind engineering simulations.

Presently, the most popular and widely used models use equations representing a single length and velocity scales are based on Reynolds averaging and the isotropic eddy viscosity concept (Versteeg & Malalasekera, 2007). Although many of these turbulence models have been used successfully in aeronautical applications, where fluid flow without separation may be a regular occurrence, the same is not true of wind engineering applications. Wind engineering flow fields are highly complex and are characterized by the presence of multiple recirculation zones embedded within a unidirectional flow. The addition of streamline curvature and favourable and adverse pressure gradients lead to flow fields possessing very different turbulence scales and structures. Consequently, such turbulence models have great difficulty in simulating wind engineering flow fields which are essentially transient and highly anisotropic. It is, therefore, apparent that one of the main obstacles in the use of CFD in wind engineering is that of turbulence modelling (Wright & Easom, 2003).

The numerical solution of any fluid flow problem requires the solution of the general equations of fluid motion, i.e. the Navier-Stokes and continuity equations. Fluid flow problems are described mathematically with a set of coupled non-linear partial differential equations with appropriate boundary conditions. These equations are derived from Newton's Second Law and describe the conservation of momentum in the flow (Acheson, 1990).

The general form of three dimensional incompressible instantaneous Navier-Stokes equations is:

$$\frac{\partial\left(\rho \cdot u_{i}\right)}{\partial t} = -\frac{\partial\left(\rho \cdot u_{i} \cdot u_{j}\right)}{\partial x_{j}} - \frac{\partial P}{\partial x_{j}} + \frac{\partial}{\partial x_{j}}\left[\mu \cdot \left(\frac{\partial u_{i}}{\partial x_{j}} + \frac{\partial u_{j}}{\partial x_{i}}\right)\right] + F \tag{106}$$

acceleration term convection term pressure gradient viscosity effects force and the continuity equation:

$$\frac{\partial \rho}{\partial t} + \frac{\partial \rho \cdot u_{i}}{\partial x_{i}} = 0 \tag{107}$$

All flows encountered in engineering practice become unstable above a certain Reynolds number (which gives a measure of the relative importance of inertia loads and viscous loads). At low Reynolds numbers, flows are laminar, i.e. flows are smooth. If the applied boundary conditions do not change with time the flow is steady. At values of the Reynolds numbers above a critical value, a complicated series of events takes place which eventually leads to radical a change of the flow character. A chaotic and random state of motion develops in which the velocity and pressure change continuously with time within substantial regions of flow. The motion becomes intrinsically unsteady even with constant imposed boundary conditions. This regime of flow is called turbulent.

The most accurate way of modelling fluid flow numerically is by DNS which involves discretising the equations at a mesh size below the Kolmogorov length scales and applying Eq. 107 along with suitable boundary conditions adjacent to the wall to the whole flow field. Unfortunately for practical wind engineering flows this is well beyond the capabilities of present day computers. Therefore, to reduce the amount of computational effort the effect of turbulence has to be modelled. Turbulence causes the appearance in the flow of eddies with a wide range of length and time scales that interact in a dynamically complex way. Given the importance of the avoidance or promotion of turbulence in engineering applications, it is no surprise that a substantial amount of research effort is dedicated to the development of numerical methods to capture the important effects due to turbulence. The methods can be grouped into the following three categories. (i) Turbulence models for Reynolds averaged Navier-Stokes (RANS) equations (ii) Large Eddy simulation (LES) (iii) Direct Numerical Simulation (DNS).

Of all the available turbulence models, the k-ε model is by far the most widely used and has been tested for a vast number of flow fields *where* k represents kinetic energy and ε the dissipation of the turbulent kinetic energy. It is favoured in industrial applications due to its relatively low computational costs and generally better numerical stability than more complex turbulence models such as the Reynolds Stress

Model. The model has proved a success in many applications, particularly in confined flows where the normal Reynolds stresses are relatively unimportant. Unfortunately, the opposite is true of wind engineering flow fields and the k-ε model performs poorly. Therefore, the assumption of a simple isotropic eddy viscosity term is insufficient to adequately describe the complexity of a highly anisotropic flow field and the results in the k-ε model's failure to accurately predict many turbulent flow fields, not least in wind engineering applications. One of the main problems with the standard k-ε model is the overproduction of kinetic energy in regions of stagnant flow (called the stagnation point anomaly). The renormalisation group (RNG) k-ε developed by Yakhot, Orszag, Thangam, Gatski, & Speziale (1992) is often used. There are some modified versions of the standard k -ε model that may provide improved predictions for some applications. The Reynolds stress model has a far greater universality than the models based on the eddy viscosity concept due to its more rigorous and detailed mathematical formulation. The inclusion of a great number of equations allows for a far greater description of the physics of turbulent flow.

The LES technique has the advantage of producing time dependent flow information of very high quality and accuracy even in complex flow fields such as those found in wind engineering. It has succeeded in reproducing the properties of a highly anisotropic flow field in wind engineering problems. The present difficulties in using the LES technique mainly revolve around the constraints on available computer processing time and storage capacity which effectively hold back its use and advancement. This technique, although being more economical than DNS, is still very resource intensive and as such is not yet used outside of the research community.

4.9.3 Application of Wind Engineering to Temporary Structures

Guidelines have been produced for the size of mesh required to get correct answers when modelling wind action on buildings or other structures by the COST C14 Consortium (Franke et al., 2004) and by the ERCOFTAC Special Interest Group (Casey & Wintergerste, 2000). They are summarised here. The region to be meshed in the vertical, lateral and flow directions is dependent upon the area being represented and the boundary conditions used. In particular, if the height of the building is H then the inlet, vertical and lateral boundaries should be at least $5 \cdot H$ away from the building and the outlet boundary at least $15 \cdot H$ behind the building. This is to allow for flow development, as the normal boundary conditions are for fully developed flow. If the height if the building is less than either the width or the length then the suggestion is that the obstruction to flow be less than 3% of the cross-section, similar to wind-tunnel obstruction ratios. Note that if other buildings are near to the one being analysed then the H should be the height of the tallest building and the recommendation is that if buildings behind the flow of height H_n then they be ignored if greater than 6-10H_n behind the building being analysed.

The boundary conditions recommended are that at the top and side boundaries symmetry conditions are applied which ensure parallel flow. Alternatively, outflow conditions may be applied which allow a velocity normal to the boundary to be applied. This latter condition must also not allow any inflow. At the final boundary behind the building the derivatives of all flow variables are forced to zero, implying fully developed flow at the outlet. At least $5H$ in front of the building an equilibrium boundary layer is prescribed. A logarithmic profile of the mean velocity corresponding to the surface terrain roughness is employed. Either wind-tunnel simulations or meteorological data is used to get the roughness length, z_0. Richards & Hoxey (1993) give the required formulae which apply to k-ε turbulence models. Richards & Norris (2011) updated the formulae so that RNG k-ε models were included. To ensure that the correct inflow boundary conditions are input it is recommended that the model should be run with no building and a check made that the outflow condition matches the input flow so that no dissipation has occurred.

Hexahedral elements are preferred over tetrahedral elements as they have smaller truncation errors. Adjacent to walls the elements should be perpendicular to the wall. To ensure accuracy of models grid convergence studies should be made. It is recommended that the ratio of cells from one model to the next should be at least 3 to 4.

Walls have significant effects on turbulent flows. The no-slip condition at the wall also affects the mean velocity distribution. Versteeg & Malalasekera (2007) comment that in close proximity to a wall that viscous damping reduces tangential velocity fluctuations, whilst normal fluctuations are reduced by kinematic blocking. In addition they state that in the outer part of the near wall region turbulence is increased by the production of turbulence kinetic energy due to large gradients in the velocity field. The Fluent User Manuel (ANSYS, 2016) suggests subdividing near wall regions into three areas, the viscous layer (innermost) where flow is almost laminar with molecular viscosity predominant, the fully turbulent layer (outermost) and an interim region where turbulence and molecular viscosity are equally important. To avoid having extremely fine grids adjacent to the wall the boundary conditions are often defined using wall functions. These functions assume that in the near wall region a constant shear stress exists and that the length scale of an eddy is proportional to its distance from the wall. The Fluent User Manual states that wall shear stress τ is related to the turbulent kinetic energy by:

$$\tau = \rho \cdot C_{\frac{1}{4}}^{1/2} \cdot k \tag{108}$$

where ρ is the air density, C_{μ} a constant in the turbulence equation and k the turbulent kinetic energy. By substituting for the turbulence constant and the turbulent kinetic energy (see Fluent, 2006) a scaled wall distance, y^{+} is given by:

$$y^{+} = \frac{y_{\mathrm{p}}}{\nu} \cdot \sqrt{\frac{\tau}{\rho}} \tag{109}$$

where y_{p} is the actual distance from the wall and ν the laminar kinematic velocity. A near wall flow is taken to be laminar if $y^{+} \leq 11.63$ and the wall stress is assumed to be viscous. On the other hand, if $y^{+} > 11.63$ the flow is turbulent. Versteeg & Malalasekera (2007) define the following wall function:

$$u^{+} = \begin{cases} y^{+}, & y^{+} < {y_0}^{+} \\ \frac{1}{k} \cdot \ln\left(E \cdot y^{+}\right), & y^{+} > {y_0}^{+} \end{cases} \tag{110}$$
$$k = {u_0}^{2} / \sqrt{C_{\frac{1}{4}}}, \quad \varepsilon = {u_0}^{3} / \left(\kappa \cdot y\right)$$

where ${y_0}^{+}$ is the cross-over between laminar sub-layer and the logarithmic region, κ is the von Karman constant and E the log-layer constant. The fluid is stationary at a solid surface and there no turbulent eddying motions there either.

A sample of mesh generation showing the increase in number of elements adjacent to walls is shown in Figure 58 and Figure 59 showing an overall mesh around a cubical building and the refinement near to the building (taken form Irtaza's PhD thesis – (Irtaza, 2009)).

Few examples of CFD analyses have been undertaken on temporary structures. The only reported ones on unsheeted scaffolds are those by Irtaza (2009) and Irtaza, Beale, & Godley (2007) where two dimensional analyses of bare pole scaffolds were undertaken to determine there was a shielding effect between columns in rows. Figure 4.60 and Figure 4.61 show the shielding effects of two bare pole scaffolds and two rows of scaffold poles surrounding a 6 m cubic building. It was not possible to include the horizontal ledgers and transoms as this would have exceeded maximum number of elements that the computer being used could handle. Unfortunately, shielding is only operative for two poles in a direct line.

Figure 58. Overall grid distribution and grid distribution in plan. ©2009 H Irtaza. Used with permission

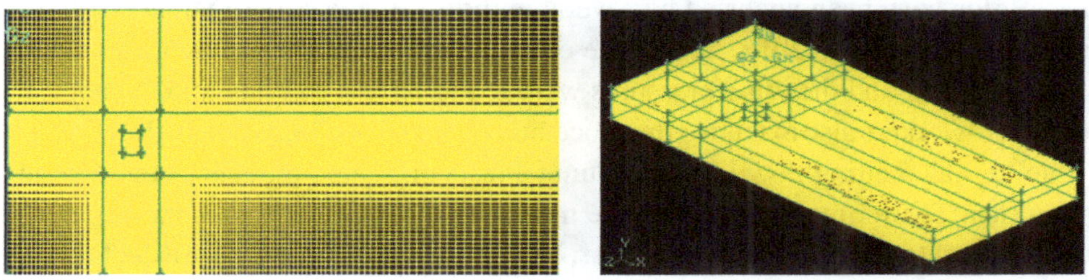

Figure 59. Detailed mesh in plan and elevation near to the building. ©2009 H Irtaza. Used with permission

Figure 60. X-velocity contour and turbulent kinetic energy around two scaffold tubes along the flow direction for unsteady RNG k-ε model. ©2009 H Irtaza. Used with permission

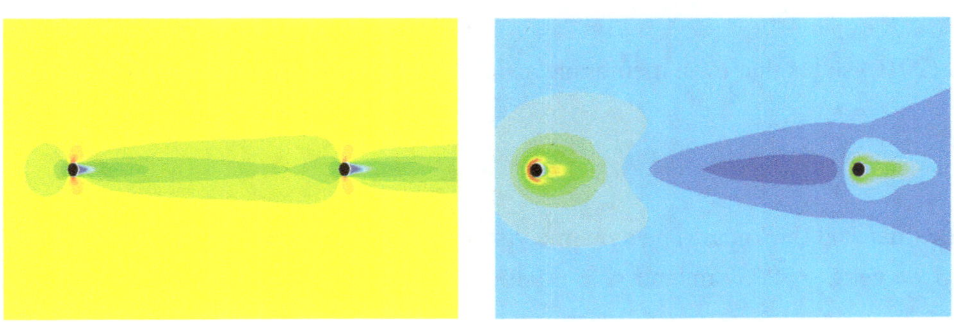

Figure 61. X-velocity contour and vector around the scaffold tubes surrounding a cubic building along the flow direction for unsteady RNG k-ε method. ©2009 H Irtaza. Used with permission

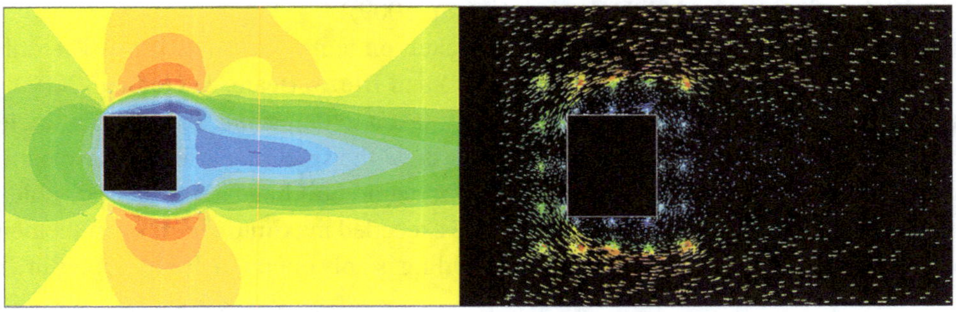

Sheeted scaffolds have been analysed by several authors (Huang, Li, & Xu, 2007; Irtaza, 2009; Irtaza, Beale, Godley, & Jameel, 2013, 2014; Yue et al., 2005). In most of these cases the sheeting was considered impermeable and the wind pressures were found on the sheeting and the adjacent building.

Irtaza and his co-workers also extended the procedure to netted scaffolds where the netting was porous. It is impossible to scale netting down for wind-tunnel tests on scaled models as the netting is only 0.4 mm to1.0 mm thick. Therefore, they can only be modelled using CFD. Porous media are considered to add a momentum term to the governing equations. For a homogeneous porous media this adds a viscous loss term and an inertial loss tem to give:

$$S_{i} = \left(\frac{\mu}{\alpha} \cdot v_{i} + C_{2} \cdot \frac{1}{2} \cdot \rho \cdot v_{mag} \cdot v_{i}\right) \tag{111}$$

where S_i is the momentum source term, μ the dynamic viscosity of air, α the permeability of the net, C_2 a constant, ρ the air density v_{mag} the absolute velocity and v_i the velocity in direction *i*. The term used in Fluent (ANSYS, 2016) for the pressure drop across a membrane or net is called "a porous jump". To obtain the constants used in Eq. wind-tunnel experiments on netting must be used where pressure drops across the net are measured for a given velocity of the fluid. The pressure drop equation is always a quadratic expression.

For example, when experiments on a net of thickness 0.42 mm were conducted at Oxford Brookes University (Irtaza, 2009) the following equation relating pressure drop Δp to velocity was:

$$\Delta p = 0.524 \cdot v^{2} + 2.429 \cdot v \tag{112}$$

ANSYS (2016) states that for a membrane a simplified form of the momentum equation is:

$$\Delta = -S_{i} \cdot n \tag{113}$$

where *n* is the material thickness. The dynamic viscosity of the air $\mu = 1.7894 \times 10^{-5}$. Hence from Eqs. 112 and 113 we get $C_2 = 2037\ \text{m}^{-1}$and $\alpha = 6.946 \times 10^{-9}\text{m}^{2}$.

To illustrate the use of the jump the model used in Figure 60 and Figure 61 was modified. Instead of having a fixed acrylic sheet a porous jump was applied. Figure 62 shows the pressure from the front to the back of the cube at mid height (0-1 is across the front, 1-2 is along the side and 2-3 is across the back). A permeability of 1.0×10^{-10} m^2 corresponds to a nearly impermeable sheet and a permeability of 1.0×10^{-6} m^2 corresponds to a very permeable sheet (almost non-existing). The two extremes matched the equivalent results from the cube with no sheeting around it to the clad scaffold with an impermeable sheet. There is a smooth transition between the different sets.

4.10 CONCLUSION

This Chapter presented an overview of structural analysis methods applied to temporary works. At the start a necessarily brief but broad introduction to the Finite Element Method (FEM) is presented, along with a description of the different types of structural analyses that can be performed, the assumptions of each type and the possible consequences in the results. In particular, the objectives of verification and validation of FEM models were presented. Also, the principles of first-order and second-order analyses were presented, the same for elastic and elastoplastic analyses.

Suitable material models for steel, aluminium, timber and bamboo were presented, ranging from the simple linear elastic model to complex three dimensional multiaxial flow rules. Next, a brief overview concerning actions modelling was provided.

Afterwards, several methods to characterise the behaviour and resistance of joints under quasi-static and cyclic regimes were detailed, including experimental and numerical techniques. Emphasis was given to scaffold and falsework joints, including column-to-column joints and anchor joints.

Figure 62. Pressure coefficients on the outer face of the net at different permeabilities. ©2009 H Irtaza. Used with permission

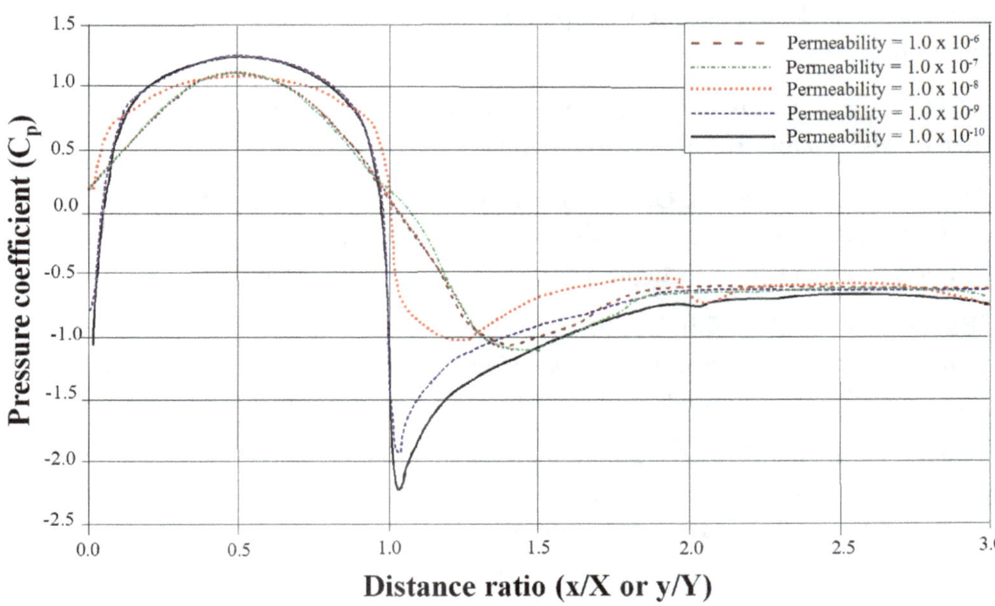

Detailed descriptions of different types of 2D and 3-D models have been developed and discussions on their relative accuracies and ease of use given. Applications were presented for scaffolds and falsework structures.

Soil modelling was overviewed, providing guidance on models for soil resistance and stiffness but also on soil pressures.

Finally, the use of Computational Fluid Dynamics (CFD) to simulate the wind action and its effects on temporary structures was detailed, and an application to scaffolding was presented.

REFERENCES

Abdel-Jaber, M., Beale, R., Godley, M., & Abdel-Jaber, M. (2009). Rotational strength and stiffness of tubular scaffold connectors.*Proceedings of the ICE - Structures and Buildings, 162*(6), 391–403. doi:10.1680/stbu.2009.162.6.391

Abdel-Jaber, M., Beale, R. G., & Godley, M. (2006). A theoretical and experimental investigation of pallet rack structures under sway. *Journal of Constructional Steel Research*, *62*(1–2), 68–80. doi:10.1016/j.jcsr.2005.04.008

Acheson, D. (1990). *Elementary fluid dynamics*. Oxford, UK: Oxford University Press.

ACI. (2011). ACI 355.3R–11 – Guide for design of anchorage to concrete: Examples using ACI 318 Appendix D. Washington, DC: American Concrete Institute (ACI).

ACI. (2014). ACI 318–14 – Building code requirements for structural concrete and commentary. Washington, DC: American Concrete Institute (ACI).

AGS. (2011). *UK specification for ground investigation* (2nd ed.). London, UK: ICE Publishing.

André, J. (2008). *Behaviour and quality of steel telescopic props. Analytical, numerical and experimental investigation* (M.Sc. Thesis). Instituto Superior Técnico, Technical University of Lisbon, Lisbon, Portugal. (in Portuguese)

André, J. (2014). *Determination of the main parameters affecting the performance of bridge falsework systems* (PhD Thesis). Oxford Brookes University, Oxford, UK.

André, J., Baptista, A., & Camotim, D. (2007). Experimental investigation of steel telescopic props (in Portuguese). In *Proceedings of the VI Portuguese National Conference of Steel and Composite Construction* (p. 12). Porto, Portugal: The Portuguese Steelwork Association (cmm).

André, J., Baptista, A., & Camotim, D. (2009). Numerical investigation of adjustable telescopic steel props. In *Proceedings of the 12th International Conference on Civil, Structural and Environmental Engineering Computing* (p. 25). Madeira, Portugal: Civil-Comp Press.

André, J., Beale, R., & Baptista, A. (2013). Experimental behaviour of bridge falsework joints. In *Proceedings of the IX Portuguese National Conference of Steel and Composite Construction* (p. 10). Guimarães, Portugal: The Portuguese Steelwork Association (cmm).

André, J., Beale, R., & Baptista, A. (2014). Numerical investigation of bridge falsework structures. In *Proceedings of the 5th Portuguese Structural Engineering Conference (JPEE 2014)* (p. 15). Lisbon, Portugal: Portuguese National Laboratory for Civil Engineering (LNEC).

André, J., Beale, R., & Baptista, A. (2015a). Numerical analysis of bridge falsework. In *Proceedings of the 8th International Conference on Advances in Steel Structures (ICASS 2015)* (p. 20). Lisbon, Portugal: Elsevier.

André, J., Beale, R., & Baptista, A. (2015b). Performance of bridge falsework structures under external hazards. In *Proceedings of the X Portuguese National Conference of Steel and Composite Construction* (p. 10). Coimbra, Portugal: The Portuguese Steelwork Association (cmm).

ANSYS. (2016). *ANSYS Fluent user's guide*. Pennsylvania, USA: ANSYS, Inc.

ASTM. (2014). ASTM D143–14 – Standard test methods for small clear specimens of timber. The American Society for Testing and Materials (ASTM).

Baker, C. (2007). Wind engineering: Past, present and future. *Journal of Wind Engineering and Industrial Aerodynamics*, *95*(9–11), 843–870. doi:10.1016/j.jweia.2007.01.011

Bathe, K. (2006). *Finite element procedures*. Klaus-Jurgen Bathe.

Beale, R., & Godley, M. (1995). The analysis of scaffold structures using LUSAS. *Proceedings of LUSAS*, *95*, 10–24.

Beale, R., & Godley, M. (2006). Numerical modelling of tube and fitting access scaffold systems. *Advanced Steel Construction*, *2*(3), 199–223.

Blocken, B. (2011). Computational wind engineering: Theory and applications. In C. Baniotopoulos, C. Borri, & T. Stathopoulos (Eds.), *Environmental wind engineering and design of wind energy structures* (pp. 55–93). Udine, Italy: Springer. doi:10.1007/978-3-7091-0953-3_3

Bowles, J. (1997). *Foundation analysis and design* (5th ed.). New York: McGraw-Hill.

BSI. (1997). BS 6399–2:1997 – Loading for buildings. Code of practice for wind loads. London, UK: British Standards Institution (BSI).

BSI. (1999). *Adjustable telescopic steel props – Product specifications, design and assessment by calculation and tests*. London, UK: British Standards Institution.

BSI. (2001).*Loose steel tubes for tube and coupler scaffolds – Technical delivery conditions*. London, UK: British Standards Institution.

BSI. (2002). BS EN 12811–3:2002 – Temporary works equipment – Part 3: Load testing. London, UK: British Standards Institution (BSI).

BSI. (2003). BS EN 12811–1:2003 – Temporary works equipment – Part 1: Scaffolds – Performance requirements and general design. London, UK: British Standards Institution (BSI).

BSI. (2004). BS EN 10025–2:2004 – Hot rolled products of structural steels - Part 2: Technical delivery conditions for non-alloy structural steels. London, UK: British Standards Institution (BSI).

BSI. (2005a). BS EN 74–1:2005 – Couplers, spigot pins and baseplates for use in falsework and scaffolds – Part 1: Couplers for tubes – Requirements and test procedures. London, UK: British Standards Institution (BSI).

BSI. (2005b). BS EN 1993–1–1:2005 – Design of steel structures – Part 1–1: General rules and rules for buildings. London, UK: British Standards Institution (BSI).

BSI. (2007). BS EN 1997–2:2007 – Geotechnical design – Part 2: Ground investigation and testing. London, UK: British Standards Institution (BSI).

BSI. (2008). UK National annex to BS EN 1993–1–1:2005. London, UK: British Standards Institution (BSI).

BSI. (2009a). *Steel static storage systems – Adjustable pallet racking systems – Principles for structural design*. London, UK: British Standards Institution.

BSI. (2009b). UK National annex to BS EN 1997–2:2007. London, UK: British Standards Institution (BSI).

BSI. (2011a). BS 5975:2008+A1:2011 – Code of practice for temporary works procedures and the permissible stress design of falsework (3rd ed.). London, UK: British Standards Institution (BSI).

BSI. (2011b). *Falsework. Performance requirements and general design*. London, UK: British Standards Institution.

BSI. (2012).*Adjustable telescopic aluminium props – Product specifications, design and assessment by calculation and tests*. London, UK: British Standards Institution.

BSI. (2013). BS EN 1999–1–1:2007+A2:2013 – Design of aluminium structures – Part 1–1: General structural rules. London, UK: British Standards Institution (BSI).

BSI. (2015a). BS 5930:2015 – Code of practice for ground investigations. London, UK: British Standards Institution (BSI).

BSI. (2015b). BS 8004:2015 – Code of practice for foundations. London, UK: British Standards Institution (BSI).

Cao, S. (2013). CFD applications in structural wind engineering. In Y. Tamura & A. Kareem (Eds.), *Advanced structural wind engineering* (pp. 301–324). Tokyo, Japan: Springer Japan. doi:10.1007/978-4-431-54337-4_11

Casey, M., & Wintergerste, T. (2000). *Best practices guidelines: ERCOFTAC special interest group on 'Quality and Trust in Industrial CFD*. Brussels, Belgium: ERCOFTAC.

CEN. (2009). CEN/TS 1992–4:2009 – Design of fastenings for use in concrete – Parts 1–5. Brussels, Belgium: European Committee for Standardization (CEN).

Chan, S., & Xian, X. (2004). *Engineering and mechanical properties of structural bamboo*. Hong Kong, China: Hong Kong Polytechnic University.

Chandrangsu, T., & Rasmussen, K. (2009). Full-scale tests and advanced structural analysis of formwork subassemblies. In *Proceedings of the 6th International Conference on Advances in Steel Structures* (pp. 1083–1090). Hong Kong, China: Hong Kong Institute of Steel Construction.

Chandrangsu, T., & Rasmussen, K. (2011a). Investigation of geometric imperfections and joint stiffness of support scaffold systems. *Journal of Constructional Steel Research*, *67*(4), 576–584. doi:10.1016/j.jcsr.2010.12.004

Chandrangsu, T., & Rasmussen, K. (2011b). Structural modelling of support scaffold systems. *Journal of Constructional Steel Research*, *67*(5), 866–875. doi:10.1016/j.jcsr.2010.12.007

Chen, W., & Han, D. (2007). *Plasticity for structural engineers* (2nd ed.). J. Ross Publishing.

Chung, K., Chan, S., & Yu, W. (2002). Design of bamboo scaffolds. In *Proceedings of the International Seminar 'Bamboo Scaffolds in Building Construction' - An Alternative and Supplement to Metal Scaffolds* (pp. 65–88). Hong Kong, China: Hong Kong Polytechnic University.

Chung, K., Yu, W., & Chan, S. (2002). Mechanical properties and engineering data of structural bamboo. In *Proceedings of the International Seminar 'Bamboo Scaffolds in Building Construction' - An Alternative and Supplement to Metal Scaffolds* (pp. 1–23). Hong Kong, China: Hong Kong Polytechnic University.

Crisfield, M. (1997). *Nonlinear finite element analysis of solids and structures* (Vol. 2). New York, USA: Wiley.

Davenport, A. (1960). Rationale for determining design wind velocities. *Journal of the Structural Division*, *86*(5), 39–68.

De Borst, R., Crisfield, M., Remmers, J., & Verhoosel, C. (2012). *Nonlinear finite element analysis of solids and structures* (2nd ed.). London, UK: Wiley. doi:10.1002/9781118375938

DIN. (1987).*Telescopic steel props – Safety requirements, testing*. Berlin, Germany: Deutsches Institut für Normung.

Dowrick, D. (2009). *Earthquake resistant design and risk reduction* (2nd ed.). New Jersey, USA: John Wiley & Sons. doi:10.1002/9780470747018

Eligehausen, R., Mallée, R., & Silva, J. (2006). *Anchorage in concrete construction*. Berlin, Germany: Ernst & Sohn.

Enright, J., Harris, R., & Hancock, G. (2000). Structural stability of braced scaffolding and formwork with spigot joints.*Proceedings of 15th International Specialty Conference in Cold-Formed Steel Design and Construction*, 357–376.

Faella, C., Piluso, V., & Rizzano, G. (1999). *Structural steel semirigid connections: Theory, design, and software*. Boca Raton, FL: CRC Press.

Feng, X. (1994). *The influence of semi-rigid connections on the behaviour of slender structures* (PhD Thesis). Oxford Brookes University, Oxford, UK.

Feng, X., Godley, M., & Beale, R. (1998). Rotational stiffnesses of semi-rigid baseplates.*Proceedings of 14th International Speciality Conference on Cold-Formed Steel Structures*, 323–334.

Franke, J., Hirsch, C., Jensen, A., Krüs, H., Schatzmann, M., & Westbury, P., ... Wright, N. (2004). Recommendations on the use of CFD in Wind Engineering. *Proceedings of the International Conference on Urban Wind Engineering and Building Aerodynamics: COST C14 – Impact of Wind and Storm on City life and Built Environment*, 11.

Frye, M., & Morris, G. (1975). Analysis of Flexibly connected steel frames. *Canadian Journal of Civil Engineering*, *2*(3), 280–291. doi:10.1139/l75-026

Gerhardt, H., & Janser, F. (1994). Wind loads on permeable façades. *Journal of Wind Engineering and Industrial Aerodynamics*, *53*(1-2), 37–48. doi:10.1016/0167-6105(94)90017-5

Gerrand, C. (1987). The equivalent orthotropic elastic properties of plywood. *Wood Science and Technology*, *21*(4), 335–348. doi:10.1007/BF00380201

Godley, M., & Beale, R. (1997). Sway stiffness of scaffold structures. *The Structural Engineer*, *75*(1), 4–12.

Godley, M., & Beale, R. (2001). Analysis of large proprietary access scaffold structures.*Proceedings of the ICE - Structures and Buildings, 146*(1), 31–39. doi:10.1680/stbu.2001.146.1.31

Godley, M., Beale, R., & Feng, X. (1998). Rotational stiffness of semi-rigid baseplates.*Proceedings of the 14th Specialty Conference on Cold-Formed Steel Structures*, 323–335.

Helwany, C. (2007). *Applied soil mechanics with ABAQUS aplications*. John Wiley & Sons. doi:10.1002/9780470168097

Hong Kong Labour Department. (2014). *Code of practice for bamboo scaffolding safety* (3rd ed.). Hong Kong, China: Occupational Safety and Health Branch.

Huang, S., Li, Q., & Xu, S. (2007). Numerical evaluation of wind effects on a tall steel building by CFD. *Journal of Constructional Steel Research*, *63*(5), 612–627. doi:10.1016/j.jcsr.2006.06.033

Ilick, J., Arora, V., & Dolejs, J. (2015). Validation of finite element updated models of the developed facade scaffold anchor.*IOP Conference Series: Materials Science and Engineering*, *96*, 9.

Irtaza, H. (2009). *Experimental and computational determination of wind loads on netted/sheeted scaffolds* (PhD Thesis). Oxford Brookes University, Oxford, UK.

Irtaza, H., Beale, R., & Godley, M. (2007). Determination of the effects of wind load on bare tube access scaffold structures using computational fluid dynamics.*Proceedings of the 5th International Conference on Advances in Steel Structures (ICASS 2007),*637–642.

Irtaza, H., Beale, R., & Godley, M. (2012). A wind-tunnel investigation into pressure distribution around sheet clad scaffolds. *Journal of Wind Engineering and Industrial Aerodynamics*, *103*, 86–95. doi:10.1016/j.jweia.2012.03.004

Irtaza, H., Beale, R., Godley, M., & Jameel, A. (2013). Comparison of wind pressure measurements on Silsoe experimental building from full-scale observation, wind-tunnel experiments and various CFD techniques. *International Journal of Engineering Science and Technology*, *5*(1), 28–41.

Irtaza, H., Beale, R., Godley, M., & Jameel, A. (2014). Pressure distribution around net clad scaffolds using computation fluid dynamics techniques. *IOSR Journal of Mechanical and Civil Engineering*, *4*, 54–62.

ISO. (2005). ISO 6507-1:2005 – Metallic materials – Vickers hardness test – Test method. Geneva, Switzerland: International Organization for Standardization (ISO).

ISO. (2009a). ISO 6892–1:2009 – Metallic materials – Tensile testing – Part 1: Method of test at room temperature. Geneva, Switzerland: International Organization for Standardization (ISO).

ISO. (2009b). ISO 8375:2009 Ed 2 – Timber structures. Glued laminated timber. Test methods for determination of physical and mechanical properties (2nd ed.). Geneva, Switzerland: International Organization for Standardization (ISO).

ISO. (2013). ISO 8492:2013 – Metallic materials – Tube – Flattening test. Geneva, Switzerland: International Organization for Standardization (ISO).

ISO. (2014). ISO 6506-1:2014 – Metallic materials – Brinell hardness test – Test method. Geneva, Switzerland: International Organization for Standardization (ISO).

Ivanyi, M., & Baniotopoulos, C. (2014). *Semi-rigid joints in structural steelwork*. Vienna, Austria: Springer Verlag Wien.

Kindmann, R., & Kraus, M. (2012). *Steel structures: Design using FEM*. Berlin, Germany: Ernst & Sohn Verlag. doi:10.1002/9783433600771

Kishi, N., & Chen, W. (1990). Moment-rotation relations of semirigid connections with angles. *Journal of Structural Engineering*, *116*(7), 1813–1834. doi:10.1061/(ASCE)0733-9445(1990)116:7(1813)

Lightfoot, E., & Le Messurier, A. (1974). Elastic analysis of frameworks with elastic connections. *Journal of the Structural Division*, *100*(ST6), 1297–1309.

Liu, H., Zhao, Q., Wang, X., Zhou, T., Wang, D., Liu, J., & Chen, Z. (2010). Experimental and analytical studies on the stability of structural steel tube and coupler scaffolds without X-bracing. *Engineering Structures*, *32*(4), 1003–1015. doi:10.1016/j.engstruct.2009.12.027

Mallée, R., Fuchs, W., & Eligehausen, R. (2013). *Design of fastenings for use in concrete - The CEN/TS 1992-4 provisions*. Weinheim, Germany: Wiley. doi:10.1002/9783433602737

Milojkovic, B. (1999). *Factors of safety for standard scaffold structures* (PhD Thesis). Oxford Brookes University, Oxford, UK.

Milojkovic, B., Beale, R., & Godley, M. (1996). Modelling scaffold connections. In *Proceedings of 4th ACME UK Annual Conference* (pp. 85–88). Glasgow, UK: Association of Computational Mechanics in Engineering (ACME).

Oberkampf, W., & Roy, C. (2010). *Verification and validation in scientific computing*. Cambridge, UK: Cambridge University Press. doi:10.1017/CBO9780511760396

Ota, M. (1950). Studies on the properties of bamboo stem. Part 4: Study on the form of specimen in the test of compressive strength. *Journal Japanese Forestry Society*, *32*, 65–69.

Ottosen, N., & Ristinmaa, M. (2005). *The mechanics of constitutive modeling*. Amsterdam, The Netherlands: Elsevier.

Peng, J., Ho, C., Lin, C., & Chen, W. (2015). Load-carrying capacity of single-row steel scaffolds with various setups. *Advanced Steel Construction*, *11*(2), 185–210.

Peng, J., Wu, C., Chan, S., & Huang, C. (2013). Experimental and numerical studies of practical system scaffolds. *Journal of Constructional Steel Research*, *91*, 64–75. doi:10.1016/j.jcsr.2013.07.028

Pieńko, M., & Błazik-Borowa, E. (2013). Numerical analysis of load-bearing capacity of modular scaffolding nodes. *Engineering Structures*, *48*, 1–9. doi:10.1016/j.engstruct.2012.08.028

Potts, D. M., & Zdravkovic, L. (1999). Finite element analysis in geotechnical engineering.: Vol. 1. *Theory*. London, UK: Thomas Telford Publishing.

Potts, D. M., & Zdravkovic, L. (2001). Finite element analysis in geotechnical engineering.: Vol. 2. *Application*. London, UK: Thomas Telford Publishing.

Prabhakaran, U. (2009). *Nonlinear analysis of scaffolds with semirigid connections* (PhD Thesis). Oxford Brookes University, Oxford, UK.

Prabhakaran, U., Beale, R., & Godley, M. (2011). Analysis of scaffolds with connections containing looseness. *Computers & Structures*, *89*(21–22), 1944–1955. doi:10.1016/j.compstruc.2011.03.016

Rasmussen, K. (2003). Full-range stress–strain curves for stainless steel alloys. *Journal of Constructional Steel Research*, *59*(1), 47–61. doi:10.1016/S0143-974X(02)00018-4

Richards, P., & Hoxey, R. (1993). Appropriate boundary conditions for computational wind engineering models using the k-ϵ turbulence model. *Journal of Wind Engineering and Industrial Aerodynamics*, *46–47*, 145–153. doi:10.1016/0167-6105(93)90124-7

Richards, P., & Hoxey, R. (2008). Wind loads on the roof of a 6 m cube. *Journal of Wind Engineering and Industrial Aerodynamics*, *96*(6–7), 984–993. doi:10.1016/j.jweia.2007.06.032

Richards, P., Hoxey, R., Connell, B., & Lander, D. (2007). Wind-tunnel modelling of the Silsoe Cube. *Journal of Wind Engineering and Industrial Aerodynamics*, *95*(9–11), 1384–1399. doi:10.1016/j.jweia.2007.02.005

Richards, P., & Norris, S. (2011). Appropriate boundary conditions for computational wind engineering models revisited. *Journal of Wind Engineering and Industrial Aerodynamics*, *99*(4), 257–266. doi:10.1016/j.jweia.2010.12.008

Richards, P., & Robinson, M. (1999). Wind loads on porous structures. *Journal of Wind Engineering and Industrial Aerodynamics*, *83*(1–3), 455–465. doi:10.1016/S0167-6105(99)00093-8

Rosignoli, M. (2007). Robustness and stability of launching gantries and movable shuttering systems - Lessons learned. *Structural Engineering International*, *17*(2), 133–140. doi:10.2749/101686607780680673

Rosignoli, M. (2013). *Bridge construction equipment*. London, UK: ICE Publishing. doi:10.1680/bce.58088

Salvadori, A. (2009). Ultimate strength of adjustable telescopic steel props according to standard EN 1065. *Journal of Constructional Steel Research, 65*(10–11), 1964–1970. doi:10.1016/j.jcsr.2009.06.006

Schardt, R. (1994). Generalized beam theory - An adequate method for coupled stability problems. *Thin-walled Structures, 19*(2), 161–180. doi:10.1016/0263-8231(94)90027-2

SGB. (2006). Cuplok: User's manual. SGB Group (now HARSCO).

Simiu, E., & Scanlan, R. (1996). *Wind effects on structures: Fundamentals and applications to design* (3rd ed.). New York: John Wiley and Sons.

Simões da Silva, L., Rebelo, C., Nethercot, D., Marques, L., Simões, R., & Vila Real, P. (2009). Statistical evaluation of the lateral–torsional buckling resistance of steel I-beams, Part 2: Variability of steel properties. *Journal of Constructional Steel Research, 65*(4), 832–849. doi:10.1016/j.jcsr.2008.07.017

Simulia. (2012). *Abaqus theory manual. Section 1.3.1: Rotation variables.* Simulia.

Szabó, B., & Babuška, I. (2011). *Introduction to finite element analysis: Formulation, verification and validation*. Chichester, UK: John Wiley & Sons, Ltd. doi:10.1002/9781119993834

Tamboli, A. (2009). *Handbook of steel connection. Design and details*. McGraw Hill Professional.

Versteeg, H., & Malalasekera, W. (2007). *An introduction to computational fluid dynamics: The finite volume method* (2nd ed.). New York: Pearson.

Vieira, R., Virtuoso, F., & Pereira, E. (2014). A higher order model for thin-walled structures with deformable cross-sections. *International Journal of Solids and Structures, 51*(3–4), 575–598. doi:10.1016/j.ijsolstr.2013.10.023

Voelkel, W. (1990). German assessment for the Cuplok scaffold system within the framework of the comparison of the procedures in different European countries - Part 1: Test report (No. T 2381/1). Berlin, Germany: Institut fur Bautechnik (IfB).

Wang, F., Tamura, Y., & Yoshida, A. (2013). Wind loads on clad scaffolding with different geometries and building opening ratios. *Journal of Wind Engineering and Industrial Aerodynamics, 120*, 37–50. doi:10.1016/j.jweia.2013.06.015

Wang, F., Tamura, Y., & Yoshida, A. (2014). Interference effects of a neighboring building on wind loads on scaffolding. *Journal of Wind Engineering and Industrial Aerodynamics, 125*, 1–12. doi:10.1016/j.jweia.2013.11.009

Wright, N., & Easom, G. (2003). Non-linear k–ε turbulence model results for flow over a building at full-scale. *Applied Mathematical Modelling, 27*(12), 1013–1033. doi:10.1016/S0307-904X(03)00123-9

Yakhot, V., Orszag, S., Thangam, S., Gatski, T., & Speziale, C. (1992). Development of turbulence models for shear flows by a double expansion technique. *Physics of Fluids A: Fluid Dynamics (1989-1993), 4*(7), 1510–1520.

Yu, W., & Chung, K. (2000). *Qualification tests on Mao Jue under compression and bending*. Hong Kong, China: Hong Kong Polytechnic University.

Yu, W., Chung, K., & Chan, S. (2003). Column buckling of structural bamboo. *Engineering Structures, 25*(6), 755–768. doi:10.1016/S0141-0296(02)00219-5

Yue, F., Yuan, Y., Li, G., Ye, K., Chen, Z., & Wang, Z. (2005). Wind load on integral-lift scaffolds for tall building construction. *Journal of Structural Engineering, 131*(5), 816–824. doi:10.1061/(ASCE)0733-9445(2005)131:5(816)

Yuen, T. (1994). *Bamboo scaffolding in Hong Kong. Part II* (Dissertation submitted for the partial fulfilment of the Bachelor Degree of Engineering). Hong Kong Polytechnic University, Hong Kong, China.

Zienkiewicz, O., Taylor, R., & Fox, D. (2014). *The finite element method for solid and structural mechanics* (7th ed.). Oxford, UK: Butterworth-Heinemann.

Zienkiewicz, O., Taylor, R., & Zhu, J. (2013). *The finite element method: Its basis and fundamentals* (7th ed.). Oxford, UK: Butterworth-Heinemann.

Chapter 5
Structural Safety

ABSTRACT

This chapter presents and discusses the principles, methods and the associated limitations that currently are seen as the state-of-the-art in structural safety. The basis for understanding the design philosophy of modern design codes is provided. Innovative concepts in safety, starting with definitions of risk, reliability, fragility and a new definition of structural robustness are presented. Uncertainties are discussed and a risk management framework for structural design is proposed. A probabilistic structural design philosophy is presented detailing a new methodology for analysing structural fragility and the robustness of structures against failure. An example is presented determining the robustness of a falsework structure against collapse. Strategies to enhance structural robustness and structural safety are given. An improved design methodology for temporary structures is presented and detailed, and an example is provided. Finally, the chapter discusses the use of reduction factors when determining design action values for the design of temporary structures.

5.1 INTRODUCTION

The design of engineering structures can essentially be defined as a continuous process of making difficult engineering decisions based on the available knowledge and under the severe constraints imposed by society and nature. Any structure can be analysed in an integrated system made of exposures, hazard events and consequences. However, no matter the existing time interval, budget size and analysis capacity it is not possible to determine precisely the behaviour of any structure due to uncertainties. The key element for structural safety is the impact of uncertainties in the available knowledge.

In the traditional approach, engineers resort to structural design codes to make decisions. These documents are developed specifically to address areas where significant past experience exists and where critical societal risks are not involved. Thereby, design codes are established for the purpose of providing a general, simple, safe and economically efficient basis for the design of ordinary structures under normal loading, operational and environmental conditions. Design codes not only greatly facilitate the daily work of structural engineers but also provide the vehicle to ensure a certain standardization within the structural engineering profession which in the end provides a uniformity of reliability of structural

DOI: 10.4018/978-1-5225-2199-0.ch005

Copyright © 2017, IGI Global. Copying or distributing in print or electronic forms without written permission of IGI Global is prohibited.

performance and enhances an efficient use of the resources of society for the benefit of the individual. In countries such as the USA, engineers are required by law to adhere to design codes, whereas in Europe, engineers are allowed to produce designs which deviate from the design code if they can demonstrate that the design fulfils the reliability levels specified in the design code. The latter may ease engineers in the process of producing innovative engineering solutions.

However, problems do exist. Current design codes are normally based on semi-probabilistic limit states design, such as the Limit State Design (LSD) methodology. However, design codes were calibrated to provide an adequate reliability only at the individual element level. Therefore, resistance safety checks are merely considered at a local level (e.g. a cross-section or an individual element) and the designer has insufficient control over the analysis and selection of preferred mode, or modes, of failure of the designed structure with respect to critical enabling/triggering hazard events. As a result, the global behaviour is not directly accounted for and the design efficiency and the global target reliability may not be achieved in practice.

As highlighted by Starossek (2006), the safety of the structure depends not only on the safety of all the elements against local failure but also of the system response to local failure. The implied assumption that the adequate resistance of the structure is guaranteed by the resistance of its elements is generally not valid, see Starossek & Wolff (2005). In addition, Ellingwood (2008) pointed out that

(...) no attempt was made to rationalise the calibrated reliabilities in explicit risk terms; thus, they are related to social expectations of performance only to the extent that reliability benchmarks obtained from member calibration to historical practice can be related to such expectations.

Current code design philosophies may also limit the tools at disposal of the designer to optimise the structure to specific performance objectives. To do so would require the use of different partial factors for each component type, size, structural arrangement, type of loading, type of usage, etc. (CIRIA, 1977), determined based on a risk informed decision-making process. As a consequence, the reliability target levels used during calibration of design codes are in practice the best estimates of the actual values. They are often called notional target reliabilities.

It can be concluded that the present basis for design does not assure optimal design in terms of resources allocation and risk acceptance. As a result, the traditional standards-based approach is becoming increasingly inadequate to handle the allocation of limited resources for structures design, operation, repair or improvement, in a climate of growing public scrutiny.

Furthermore, the prescriptive rules specified in present design codes if incorrectly applied or misunderstood can lead to unsafe design. Despite the latter being easily understood, it is sometimes forgotten due to the apparent unlimited safety, i.e. absence of risk, assured by the use of partial factors together with the fulfilment of a set of more or less standard requirements (CIRIA, 1977). When the engineer is confronted with an omission on the design code about a given problem, generally he/she has no option other than to resort to heuristic methods which are considered to represent good practice. The subtler aspects of this approach are based on intuition, and are often referred to as "engineering judgement". However, examples abound of new issues and new problems, such as the design of some temporary structures, where the experience of previous work does not provide an adequate guidance due to structural and/or economical specific characteristics. Furthermore, uncertainties can appear in the process of extrapolating past experience to existing problems due to differences in the current and past design methodologies.

It can also be concluded that existing design codes still have application limitations despite recent advances towards performance-based design and consequence-based design. This can be particularly relevant regarding temporary structures as the number of incidents and accidents that have occurred in the recent years (see Chapter 7) clearly highlight and demonstrate the need for a more comprehensive approach to design and safety evaluation.

Existing design codes certainly introduced improvements, but specific important aspects in assessing the safety and structural efficiency of temporary structures are still not adequately addressed. As mentioned previously, existing design codes were only calibrated with respect to structures where significant past experience exists. Sexsmith (1998) argues that

Calibration works where the variation of construction cost with safety is not particularly sensitive in the range of acceptable safety, and where there is a very large database of structures upon which to base the calibration. In the case of temporary structures such as bridge falsework, or parts of the permanent bridge that are subjected to temporary erection loads, calibration is unlikely to provide consistent results.

Furthermore, the safety and performance of temporary structures are more influenced than permanent structures by uncertainty sources which are difficult to fully cover in standards, related for instance to quality management errors (see Chapters 7 and 8). Therefore, the notional target reliabilities specified in existing temporary structures design codes can be insufficient if effective and rigorous quality control procedures are not developed and enforced in order to achieve adequate structural safety.

As a response to the abovementioned insufficiencies of existing design codes' philosophies, a risk informed design methodology may in certain cases be an advantageous alternative, see Faber (2009). In this approach, which will be explored in more detail in the following, the goal is to optimise the structure to achieve minimum risk, not minimum probability of failure which is the goal of reliability-based design.

In order to be able to minimise risk during design of temporary structures, recently developed structural robustness and structural fragility indices will be presented and the latter will form the basis of the proposed structural risk management framework. This new methodology is applicable, in principle, to all structures not only those concerning temporary structures.

This Chapter gives a complete insight to the risk management framework: from the principles and concepts that form the basis of risk assessment, to the general layout of the risk management framework and to the methods and procedures to be used to determine structural robustness, structural fragility and vulnerability. Guidance is included on how to address them from an economic cost-benefit point of view in order to achieve rational decisions in structural engineering.

As the currently available guidance relative to the overall methodology that constitutes the basis of structural risk assessment is limited and/or too specialised and therefore not easy to generalise and to apply by a design practitioner, the Chapter will provide sufficiently detailed information about this topic. Therefore, it will enable the reader to correctly interpret the safety format philosophy of the existing design codes, and furthermore it will contribute to the correct use of more advanced methods than the ones available in design codes to demonstrate structural adequacy.

On the basis of this Chapter it is expected that the reader will acquire knowledge on the following topics:

1. Definitions of risk, vulnerability, reliability, fragility and robustness.
2. Methods of modelling actions and resistance variables, including the associated uncertainties.

3. Safety format and basis of design of current design codes.
4. Structural risk assessment methods based on recently developed structural robustness and structural fragility indices.
5. Application of risk framework to temporary structures.

5.2 CONCEPTS AND DEFINITIONS

5.2.1 Basis

The structural performance of a structural system, no matter what definition is used, can be analysed considering the framework represented in Figure 1.

5.2.2 System Context

Every organisation functions within an environment which both influences the risks faced and provides a context within which risk has to be managed (HM Treasury, 2004).

The system context is defined by the external and internal variables that together govern the scope, behaviour and objectives of the system. External context can include the cultural, social, political, economic environment whether at international, national, regional or at local level. Internal context can include the organisational structure, the objectives and the strategies that are in place to achieve them and the capabilities understood in terms of available resources and knowledge (e.g. capital, time, people, processes, systems and technologies) (ISO, 2009b).

Figure 1. Framework for structural design.

5.2.3 Exposure and Hazard Scenarios

The exposure of the system is expressed as the number of different events that could act on the constituents of the system with potential consequences for the considered system. Each event may itself be a hazard scenario or may lead to one or more different hazard scenarios.

Hazard is understood as set of conditions with the potential of leading to undesirable consequences, i.e. a threat, danger or harm to the resistance and/or operation of part or the entire system, arising from a single event or from a combination of multiple events (ANCOLD, 2003; HSE, 2001; JCSS, 2001). In BS EN 1990 (BSI, 2005), hazard is defined as *"an unusual and severe event"*. Therefore, a hazard scenario is a critical situation at a particular time consisting of a set of events which can lead to unwanted consequences, if nothing stops it or reduces its consequences.

It is possible to distinguish between internal and external events. Internal events are those that stem from the structural system, whereas external events are those related to external actions.

Internal events can be related to all phases of the structure life: from design and construction to maintenance and decommissioning. In particular, the influence of human errors is very important. In the design phase, internal hazard scenarios can correspond to wrong design assumptions not matching the "real" structural behaviour, whereas in the construction and maintenance phases collapse can happen due to errors in assembly, planning, use of deficient or incorrect elements or components, etc.

External events can correspond to load types not accounted for in the design phase or loads intensity, duration, range or effects (forces, displacements, vibrations, etc.) larger than expected.

5.2.4 Damage, Failure Events and Consequences

Damage are associated with a given hazard scenario. Damage can be defined as an unfavourable change in the condition of a system that can affect the performance of the latter. Damage can be classified as direct and indirect damage. Direct damage are related to damage in those elements directly involved in the hazard scenario, which result in the first failures of elements, while indirect damage are the damage that result from the direct damage, due to the incapability of the system to sustain the latter without further damage.

Depending on the characteristics of the damage, failure events can take place when the performance of one, or more, of the structural system elements does not satisfy certain design objectives, regarding safety or serviceability for example.

Damage and failure events lead to consequences. Consequences can range from beneficial to adverse, and may be expressed in qualitatively or quantitatively terms to characterise loss of life, injury, economic loss, environmental harm, disruption of function or/and safety, etc. Both immediate consequences and those that arise after a certain time has elapsed, i.e. follow-up consequences, should be considered.

5.2.5 Load Paths and Failure Modes

Load paths can be defined as the integral of all elements of the system affected by internal and external action effects. They are described by element stresses, internal forces, reactions, etc., and can be identified by calculating or measuring those quantities from the point of application of the load to the boundaries of the system (Knoll & Vogel, 2009).

A failure mode describes how element failures can occur resulting in the total or partial collapse of the system. For a given hazard, for example overload or a construction flaw, a structure can exhibit very different failure modes depending on the critical elements and the primary load paths. The most common example is the weak beam/strong column concept, also known as the capacity design principle, adopted in most of the present seismic design codes.

5.2.6 Structural Fragility and Vulnerability

The structural fragility of a system is an expression of the system's structural performance, typically in terms of damage extension, for a given hazard event. Traditionally, fragility of a structure or element may be expressed by the conditional probability of failure for a given hazard event. Fragility is a system characteristic, independent of the probability of occurrence of the hazard event.

Vulnerability of a system describes the degree of susceptibility of the system to the occurrence of a specified level of loss (i.e. adverse consequence) following an initiating threat event (McGill & Ayyub, 2007). It provides a mapping between a given exposure event and a resulting consequence, typically economic losses or number of fatalities. Therefore, vulnerability links fragility to consequences.

The vulnerability of a system to a given degree of loss with respect to a specific initiating threat event requires all intermediate chains between cause and given consequence to fail. For the case of a single load path, if one chain does not fail the occurrence of the given degree of loss is prevented (McGill & Ayyub, 2007). If secondary load paths exist the risk control analysis must consider them.

In risk terms, the vulnerability of a structure to a hazard scenario is defined by the conditional probability of consequence c_i given hazard event e_j (McGill & Ayyub, 2007). Faber (2009) defines the vulnerability of a system as all possible direct consequences (consequences associated with direct damage) integrated (or summed up, depending whether the variables follow discrete or continuous functions) over all possible exposure events. In the present book, vulnerability is understood to be associated with all consequences, direct and indirect.

Fragility and vulnerability are both important components that need to be considered for a risk informed decision-making.

5.2.7 Structural Robustness

Structural robustness is defined in ISO 2394 (ISO, 2015) as the

ability of a structure to withstand adverse and unforeseen events (like fire, explosion, impact) or consequences of human errors without being damaged to an extent disproportionate to the original cause.

In this definition, structural robustness can be defined as a parameter indicating the sensitivity of the structure with respect to disproportionate collapse, for a limited set of hazard events.

Disproportionate collapse can be defined as a distinct disproportion between the triggering, spatially limited failure and the resulting widespread collapse (Starossek, 2009). In other words, the structural robustness of a system is a measure of the ability of a system to restrict the failures to those damaged elements directly involved with a given local hazard scenario.

Structural robustness was only defined and introduced as a design objective, even if just qualitatively, following the partial collapse that occurred at Ronan Point in the UK in 1968. As usual, following a peak of inflated focus the research attention given to robustness attenuated, until the Oklahoma City bomb attack on the Alfred P. Murrah Federal Building took place in 1995, and furthermore after the 2001 terrorist attacks to the Twin Towers in New York, turned the spotlights again towards robustness.

A general design method for structural robustness, or structural integrity, is not yet explicitly specified in the existing design codes. An exception to this general observation are the rules that most recent design codes provide regarding the structural analysis and structural requirements for accidental load cases, such as failure of a member (typically a column) due to an explosion or a vehicle (or ship) collision. In most design codes, however, robustness design is limited to the provision of general prescriptive design and detailing guidance to assure appropriate strength and ductility of the structural connections between members.

In the present Chapter, a new framework to determine structural robustness will be presented. As an initial step of this new framework, a broader definition of structural robustness is first introduced (André, Beale, & Baptista, 2015): Structural robustness is a measure of the predisposition of a structural system to loss of global equilibrium and global stability, as a result of a failure scenario, e.g. a failure of one or more elements of the structure, for a given hazard event.

It is the evaluation of the "what if" scenario, which is absent from present design codes and standards. This omission can lead to unsafe (damage intolerant) structures, since a non-robust structure can crumble in a progressive and disproportionate collapse fashion if submitted to a failure scenario under normal operation conditions, an accidental load case, with low probability of occurrence, an unexpected load case, or a load case unaccounted for.

Also, as a result of the new definition, structural robustness is now applicable to all design situations and not only those unforeseen, accidental, or concerning local failures all of which are difficult to define and to model.

5.2.8 Uncertainties in Structural Engineering: A Proposal of a Different Approach

Unlike the deterministic theoretical models considered by Newton and Laplace, the real behaviour of physical systems is uncertain. Chaos Theory, developed by scientists like Poincare, Liapunov, Lorenz and Mandelbrot clearly illustrates the importance of uncertainty.

This theory suggests that one system can behave very differently if the initial conditions vary slightly – the so-called "sensitive dependence on initial conditions". A tiny error or imperfection can have a tremendous influence on the behaviour of the entire system: the "butterfly effect". Therefore, in order to predict accurately the future states of a system it is not only necessary to know the laws of physics that govern the system but also the system's initial conditions. However, it is not always possible to know the initial conditions from observing the present behaviour of a physical system: irreversible systems are a good example. Also, initial conditions are never exactly the same and even small changes in the initial conditions of simple and ordered systems can produce unpredictable and complex behaviour; and the higher the number of random variables involved in a system, the higher is the chance it will become chaotic.

If the uncertainty associated with finite expression of real numbers is not considered, it is possible to postulate that the uncertainty from Chaos Theory derives from imperfect or insufficient information. Therefore, it can be reduced (and even "eliminated") if the necessary resources are made available, and eventually the system becomes "deterministic". The words eliminated and deterministic are written using quotation marks, because there are other types of uncertainties resulting from quantum mechanics which are impossible to be known. The basis of these types of uncertainties is the Heisenberg's Uncertainty Principle, which states that "one cannot assign exact simultaneous values to the position and momentum of a physical system. Rather, these quantities can only be determined with some characteristic 'uncertainties' that cannot become arbitrarily small simultaneously".

As a result of the above, it is not surprising that despite the recent advances in structural engineering, the presence of uncertainties during the design, construction and maintenance of structures such as temporary structures can greatly influence their expected performance.

Sources of uncertainty are present everywhere: from our perception of nature to human errors. In structural engineering this translates to uncertainty in the analysis methods and models used to assess the probabilities of occurrence of an event and of its consequences, and in the effectiveness of the control measures taken to manage the risk level.

Uncertainties are traditionally classified into two types of uncertainties: aleatory uncertainties and epistemic uncertainties. The former type corresponds to intrinsic variations in time and space of the properties of a given material, element geometry, and inherent variation of environmental loads, for example, that are not easily controlled and reduced. The latter type is related to knowledge-based insufficiencies (scientific and technological) and to human intervention (including human errors) which can be controlled and reduced. See Ang & Tang (2007) and Ayyub & McCuen (2011) for more details.

The classification of uncertainties in these two categories is not always obvious, straightforward and beneficial. Humans are not mere observers and should aim to improve their understanding of the world. Equivalent, but clearer definitions, are variability and uncertainty; the first being the result of random processes and the second being the result of approximations made to investigate the behaviour and properties of random processes.

However, the development of a more transparent model to analyse uncertainties is advantageous, such as the one illustrated in Figure 2.

Figure 2. Framework for uncertainty analysis

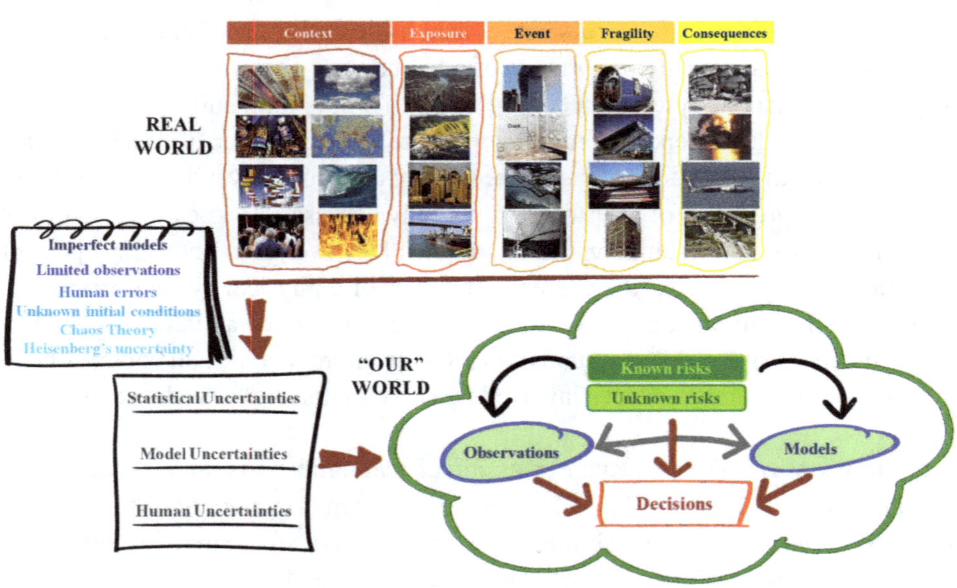

In this model, it is assumed that the level of uncertainties that mankind faces today stems mainly from our insufficient knowledge about natural systems, such as the atmosphere, materials, human mind, etc. Due to our lack of knowledge of the past, present and future, the models developed to interact with, or analyse, a given system are mere idealisations, often incomplete and inaccurate. This leads to (model) uncertainties. From Heisenberg's Uncertainty Principle and Chaos Theory, random variables exist and can only be expressed in terms of probabilistic models and statistical information, which may use objective and/or subjective information. This uncertain information will then be used in numerical or analytical models, or to test theoretical models in order to predict the behaviour (in time and in space) of the system, sub-system or components being analysed.

For example, model uncertainties can simply stem from incorrect measurand definition or from developing an inaccurate model, for example. Statistical uncertainties stem from various different sources. Probabilistic models are only approximations of the "true" model, since they are determined from samples and not from the entire population. Even the most accurate devices to measure lengths, temperatures or other random variables and random processes introduce uncertainties. All these uncertainties summed with the natural variability of the measurand add up to form statistical uncertainties. Human interaction with natural systems also introduces uncertainty: when applying a model, measuring a measurand, when analysing the data collected, or when performing other activities such as concrete casting of a bridge slab, erecting a bridge falsework structure, deciding between alternative solutions and implementing the selected measures, for instance. Human uncertainty is transversal since it is always present. The uncertainty of the output obtained from applying a given model is a combined uncertainty of model, statistical and human uncertainties.

Consequently, uncertainties can be classified in three basic categories: model uncertainties, statistical uncertainties and human interaction uncertainties. It is beneficial to keep these different sources separated in the analysis. The so-called aleatory uncertainty is here defined as the uncertainty to predict the future and describing the past; it represents the variability of natural phenomena and cannot be reduced, and is part of the statistical uncertainty. Epistemic uncertainty is present in all the three new uncertainty classifications.

Generally, it is very difficult to estimate most of the uncertainties mentioned above. In modern design codes some of them are only indirectly considered or are even ignored (e.g. human errors, phenomenological and decision uncertainties, etc.). This causes the associated target probabilities of failure used in design codes calibration to be notional estimates that does not reflect the actual failure rate of actual structures.

5.2.9 Probability in Structural Engineering

According to Faber (2009), probability theory forms the basis for the assessment of the likelihood of occurrence of uncertain events and thus constitutes a cornerstone in risk and decision analysis. Only when a consistent basis has been established for the analysis of the probability that events with possible adverse, or beneficial, consequences may occur it is possible to assess the risks associated with a given activity and thus to establish a rational basis for decision-making.

There are various possible interpretations of probability: classical, frequentist and subjective (Bedford & Cooke, 2001). In structural engineering, probability is best defined by a mathematical expression of the level of uncertainty (Bedford & Cooke, 2001). A more detailed meaning is given by McDonald, Bowles, Hartford, van der Meer, & Vrijling (2005) which define probability as

a measure of the degree of confidence in a prediction, as dictated by the evidence, concerning the nature of an uncertain quantity or the occurrence of an uncertain future event. It is an estimate of the likelihood of the magnitude of the uncertain quantity, or the likelihood of the occurrence of the uncertain future event. This measure has a value between zero (impossibility) and 1.0 (certainty).

All probabilities are conditional to the background information, including knowledge (Aven, 2004), and the use of the various types of probabilistic models is based on the degree of belief (confidence) that the analyst has on the available information. If the analyst is confident about the available data, then it may be argued that there is no need to propagate the uncertainty of the probabilistic distribution parameters, see Apeland, Aven, & Nilsen (2002) for details, which would only be considered if the analyst has vague information about the random variables.

5.2.10 Structural Reliability and Structural Safety

BS EN 1990 (BSI, 2005) defines reliability as

the ability of a structure or a structural member to fulfil the specified requirements, including the design working life, for which it has been designed.

As a measurand, reliability is commonly expressed as the complement of the probability of failure. Structural safety is defined in a report by the UK IStructE (2007) as referring

to the strength, stability and integrity of a structure to withstand the conditions that are likely to be encountered during its life-time. Structural safety is achieved through the proper procurement, design, construction, use and maintenance of the structure and the application of best practice.

5.2.11 Definition of Risk

Risk is defined in ISO 31000 (ISO, 2009b) as the effect of uncertainty on objectives, whether positive or negative. Risk depends on the system context and exposure to all relevant types of events, on the system behaviour, on the significance of consequences (beneficial and adverse) of a given event and on the uncertainties in the assessment of these variables over a certain period of time.

In general, considering an activity with only one event with potential consequences, risk is usually expressed as the probability that this event will occur multiplied with the consequences (beneficial or adverse) given the event occurs (Faber & Stewart, 2003). The activity can be the design, assembly, use, disassembly and maintenance of temporary structures, for example. However, in structural engineering, risk can also be expressed by the probability of structural failure (collapse) from all possible causes, usually in terms of the expected annual frequency (Melchers, 1999) or by the expected cost of consequences.

Risk is commonly expressed as an expected value. However, this practice can introduce distortions in the assessment, see Haimes (2009) and Savage (2009). It is highly recommended to include in the risk model the effect of uncertainties. The authors highly recommend including the effect of uncertainties in the risk model.

Risk can be classified as individual or societal risks, voluntary or involuntary risks; known risks, known unknowns and unknown unknowns (*force-majeure*) risks; recognised or unrecognised risks. See Blockley (1992) for an in-depth discussion of risk characteristics.

Individual risk is how individuals see the risk from a particular hazard affecting them and things they value in a given time period. It reflects the individual assessment regarding the expected benefits and the severity of the hazards, but also when (near future or long term?) and for how long is the individual in the proximity of the hazard sources.

Societal risk is the risk experienced in a given time period by a whole group of persons and is related to severe events that if it were to occur would cause widespread or large scale consequences and multiple fatalities. Studies, see Marsden, McDonald, Bowles, Davidson, & Nathan (2007), have shown that society is risk neutral, i.e. society considers that averting 1000 accidents with one fatality each is of nearly equal benefit to averting one accident with 1000 fatalities. However, society does not accept a large number of fatalities even if the risk per individual is small (Kumamoto, 2007).

Voluntary risks are self-willing risks and risks that the individual or society think they can control, while involuntary risks are imposed risks or unknown risks.

5.3 PROBABILISTIC STRUCTURAL DESIGN PHILOSOPHY

5.3.1 Basic Requirements of a Structural Design Philosophy

The basic requirements of the present state-of-the-art structural design philosophy are defined as (BSI, 2005): Structures and structural elements should be designed, executed and maintained in such a way that they are suited for their use during the design working life in an economic way. In particular, they should, with appropriate levels of risk, fulfil the following requirements:

- Remain fit for the use for which they are required;
- Withstand extreme and/or frequently repeated actions occurring during their construction and anticipated use;
- Not be damaged by accidental events like fire, explosions, impact or consequences of human errors, to an extent disproportionate to the triggering event.

Design working life is defined as the anticipated time period during which a structure, or parts thereof, need to fulfil its intended purpose without major repair being necessary (BSI, 2005).

There are plenty of aspects that should be considered when establishing a design philosophy that fulfils the above basic requirements. For example:

- Acceptable risk level and associated target reliabilities;
- Categorisation of classes of structural performance based on consequences of damage and the relative costs to increase safety;
- Basis for demonstrating safety in the design specifications;
- Development of models for action and resistance variables, including consideration of uncertainties;

- Development of methods of structural analysis, including consideration of uncertainties;
- Design working life of a structure and the associated durability requirements;
- Type of quality management during design, construction, maintenance and operation phases.

5.3.2 Design Philosophies

5.3.2.1 Semi-Probabilistic Design Method

5.3.2.1.1 Past Design Methods

One of the first methods for structural design to be developed was the method of permissible (or admissible) stresses that is based on the theory of linear elasticity. The basic design condition of this method can be written in the form:

$$\sigma_{max} < \sigma_{perm} = \frac{\sigma_{cr}}{k} \tag{1}$$

where σ_{max} and σ_{perm} represent the maximum imposed stress value and the permissible resistant stress value, respectively. The coefficient k (greater than 1.0) is the only explicit measure supposed to take into account all types of uncertainties associated with the design.

In this design method, no allowance was made to consider explicitly geometric nonlinearities, material plasticity and ductility, and most important uncertainties of individual basic variables.

An evolution was made with the development of the Allowable Strength Design (ASD) or Global Safety Factor method:

$$\frac{R}{\sum E_i} \geq FS \tag{2}$$

where E_i are effects due to the imposed load i, R is the resistance of the system against the imposed loads and FS is the global factor of safety.

However, as in the case of the permissible stress method, the main insufficiency of this method remains the lack of possibility of considering the uncertainties of particular basic variables. The probability of failure can be controlled by one explicit quantity only, i.e. the global safety factor FS. As in general only one value of the global safety factor was specified for each type of structural system, it was still not possible to take into account uncertainties of individual basic variables.

Therefore, both methods are in general conservative, but in some cases can lead to unconservative designs. For example, when a single controlling action dominates the design and its variability is not well represented by the assumed design action value implicitly considered in the design methods.

5.3.2.1.2 Limit State and Partial Factors Method

In this method, the fundamental condition to be verified in design calculations can be expressed by:

$$E_d\left(S, R, \theta, C, \gamma_E, \gamma_R, \gamma_I\right) < X_d\left(S, R, \theta, C, \gamma_E, \gamma_R, \gamma_I\right) \tag{3}$$

where:

E_d represents the design values of actions effects;
X_d represents the design values of structural performance;
S represents the actions variables;
R represents the resistance variables (material and geometrical variables for example);
θ represents the design uncertainties variables;
C represents a vector of serviceability constraints, e.g. acceptable mid-span deflection;
γ_E and γ_R represent partial factors of actions (or action effects) variables and resistance variables;
γ_I represents a coefficient to take into account reliability differentiation.

Various types of uncertainties are taken into account by establishing characteristic values for actions and resistance variables and determining adequate partial factors values. Thus, the bases are established for flexible design differentiation in the face of different design situations and requirements, but also for the harmonisation of safety for various types of structures made of different materials. However, the last statement can only be accomplished through the calibration of partial factors values, and/or characteristic values, against acceptable structural risk measures (e.g. probability of failure) for every single design project. As this procedure is unviable, the actual safety level ensured by this design philosophy should be understood as a notional value, not an accurate estimation of the actual value of the probability of failure of structures.

Adequate conservative models for resistance variables and actions variables are developed and used in the verification of relevant pre-established design criteria (defined by design performance requirements) expressed in the form of limit states for various design situations.

A limit state is generally understood as a state of the structure, or part of the structure, that no longer meets the relevant design performance requirements. Traditionally, the verification of design performance requirements by limit states is binary: it is verified or it is not verified. Each limit state is associated with a certain design criterion imposed on a structure.

Depending on the nature of the design criteria, different limit states can be defined. In general, a distinction is made between ultimate limit states and serviceability limit states. Both should be verified for all relevant design situations.

Ultimate limit states (ULS) are associated with collapse or other similar forms of structural failure and concern the safety of the structure (and possibly also of the surrounding structures) and/or the safety of people inside (and possibly outside the structure). Structures should be able to afford life safety to the occupants, in such a way that notwithstanding the occurrence of structural damage, the structure will not collapse and withholds its functionality at a level sufficient to support the response and recovery operations.

Serviceability limit states (SLS) correspond to service conditions of normal use. In particular, they concern the use of the structure or structural members, the comfort of the occupants and the appearance of the construction works. Examples are member deflections and vibrations, or cracks in reinforced concrete elements.

Although separated, there may be cases where both limit states are directly related, since a violation of a SLS can lead to a violation of an ULS.

Several ULS can be established, namely:

- Loss of equilibrium of the structure or any part of it, considered as a rigid body;
- Failure of the structure or part of it due to rupture, fatigue or excessive deformation;
- Instability of the structure or one of its parts.

Several SLS can be established, namely:

- Excessive deformation which can affect, for example, the appearance of the structure, comfort of users and function of the structure, and can cause damage to finishes and non-structural members;
- Excessive vibration which can, for example, cause discomfort to people and limit the function of the structure;
- Material damage that is likely adversely to affect the appearance, durability or function of the structure.

Concerning SLS, two sub-types may be further defined:

- Irreversible SLS, which are those limit states that remain permanently exceeded even when the actions which caused the infringement are removed (e.g. permanent local damage or permanent unacceptable deformations);
- Reversible SLS, which are those limit states that will cease to be exceeded when the actions which caused the infringement are removed (e.g. elastic deflections or elastic vibrations).

To account for the variability of actions, individually or concomitant, and the exposure and vulnerability of structures with respect to those actions throughout the design working life of a structure, separate design situations should be considered in the verification of limit states, representing the reasonably foreseeable hazard scenarios to occur in a certain time interval:

- Persistent situations. These refer to service conditions of normal use and are generally related to the design working life of the structure.
- Transient situations. These refer to temporary conditions of the structure, in terms of its use or its exposure (e.g. during construction or repair). This implies reference to a time period much shorter than the design working life.
- Accidental situations. These refer to exceptional conditions of the structure or of its exposure (e.g. due to earthquakes, fire, explosion, impact or local failure but also unspecified hazard events, including human errors).

Combinations of relevant actions effects must be defined for each design situation and applicable limit state. Each combination consists in adding the effects of permanent actions, leading and accompanying variable actions and possibly accidental actions. The latter are only relevant for accidental design situations.

Actions are defined by characteristic values. These constitute representative values of a given action, in general representing fractiles with a sufficiently small exceedance probability. Therefore, for uncorrelated variable actions, it is unrealistic to combine the characteristic values of each variable action; in addition, doing this would lead in many cases to highly uneconomical designs. As a result, characteristic, frequent and quasi-permanent values of variable actions are defined by multiplying the characteristic values of the accompanying (i.e. concomitant) variable actions with combination factors:

- The characteristic combination factor is calculated so that the probability of exceedance of the values of concomitant variable actions is approximately the same as the characteristic value of an individual variable action;
- The frequent combination factor is calculated so that the probability of exceedance of the values of concomitant variable actions is small for a given reference period of time;
- The quasi-permanent combination factor is calculated so that the probability of exceedance of the values of concomitant variable actions is large for a given reference period of time.

Traditionally, the first occurrence of an ULS or an irreversible SLS limit state is equivalent to failure. This is not the case for a reversible SLS. Therefore, distinct serviceability requirements can be formulated taking into account the differences between the occurrence of the first infringement and of failure. For example, the characteristic combination factor is used for irreversible SLS, the frequent combination factor is used for short-term reversible SLS, and finally the quasi-permanent combination factor is used for long-term reversible SLS.

Structural reliability is fulfilled by the verification of the design criteria (i.e. limit states) for all the relevant design situations. As a simple example, consider a structural element under a single action effect (E), e.g. a constant bending moment. The resistance (R) of the structural element is expressed by its bending moment resistance. Furthermore, consider that both action effect and element resistance variables are uncorrelated and follow a Normal probabilistic distribution, see Figure 3. It is possible to define a limit state function (G) given by:

$$G = R - E \tag{4}$$

Figure 3. Probability density functions of action effects and resistance

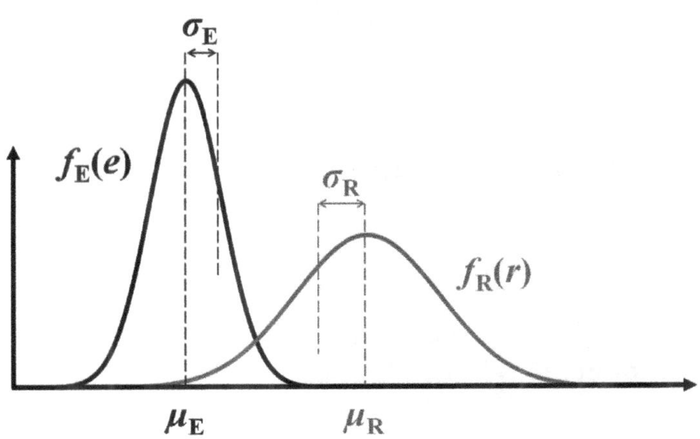

As G is also Normal distributed, its mean, μ_G, and standard deviation, σ_G, are given by:

$$\mu_G = \mu_R - \mu_E \tag{5}$$

$$\sigma_G^2 = \sigma_R^2 + \sigma_E^2 \tag{6}$$

The probability of failure, P_f, can then be expressed by following equation:

$$P_f = P(R - E < 0) = P(G < 0) = \Phi\left(\frac{0 - \mu_G}{\sigma_G}\right) \tag{7}$$

where $\Phi(\cdot)$ denotes the standard Normal cumulative distribution function.

A usual measure of reliability is the Hasofer-Lind reliability index, β, that is calculated based on the limit state function transformed into a standard normal space:

$$U_1 = \frac{R - \mu_R}{\sigma_R} \tag{8}$$

$$U_2 = \frac{E - \mu_E}{\sigma_E} \tag{9}$$

$$G = \mu_R - \mu_E + U_1 \cdot \sigma_R - U_2 \cdot \sigma_E \tag{10}$$

It can be shown that a geometrical interpretation of the reliability index is that it is equal to the minimum distance between the origin and the limit state function, where the design point lies (see Figure 4):

$$\beta = \frac{\mu_G}{\sigma_G} = \frac{\mu_R - \mu_E}{\sqrt{\sigma_R^2 + \sigma_E^2}} \tag{11}$$

and that the probability of failure is related with β by:

$$P_f = \Phi(-\beta) \tag{12}$$

The reliability index for a reference period of one year is related with the corresponding value for a reference period of n years by Eq. 13.

$$\Phi(\beta_1) = \left[\Phi(\beta_n)\right]^{\frac{1}{n}} \text{ and } \Phi(\beta_n) = \left[\Phi(\beta_1)\right]^{n} \tag{13}$$

Figure 4. Reliability index, adapted from Schneider (2006a)

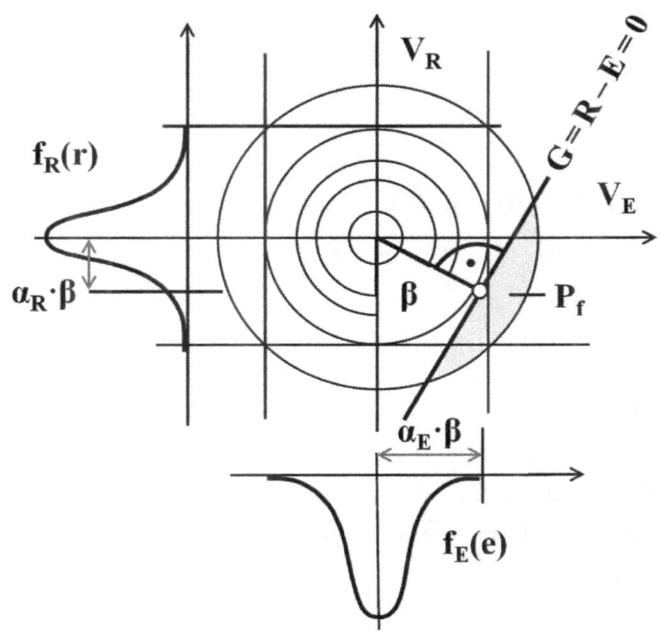

It has been assumed that the variables E and R are uncorrelated and both normally distributed. In general this is not the case, but only an approximation. However, through suitable methods every distribution may be transformed into the Normal distribution, see Melchers (1999) and Sudret (2007) for details.

Accepting the interpretation of the reliability index, the design point coordinates lying at the limit state surface are given by:

$$E_d = \mu_E + \alpha_E \cdot \beta \cdot \sigma_E \tag{14}$$

$$R_d = \mu_R - \alpha_R \cdot \beta \cdot \sigma_R \tag{15}$$

where α_E and α_R represent sensitivity coefficients of variables E and R, which give the proportion of the contribution of the variability of each individual variable to the variability of the limit state function:

$$\alpha_R = \frac{\sigma_R}{\sqrt{\sigma_R^2 + \sigma_E^2}} \tag{16}$$

$$\alpha_E = \frac{\sigma_E}{\sqrt{\sigma_R^2 + \sigma_E^2}} \tag{17}$$

As a simplification of the partial factors method, the values of the sensitivity coefficients are fixed constant values. For example, in BS EN 1990 (BSI, 2005) $\alpha_E = +0.7$ and $\alpha_R = +0.8$. The range of validity of this approximation is the existence of a single controlling variable and:

$$0.16 < \alpha_E \,/\, \alpha_R < 7.6 \tag{18}$$

The design values E_d and R_d of the variables E and R are thus defined as fractiles of each Normal distribution:

$$P\left(S > E_d\right) = \Phi\left(-\alpha_E \cdot \beta\right) \tag{19}$$

$$P\left(R < R_d\right) = \Phi\left(-\alpha_R \cdot \beta\right) \tag{20}$$

The partial factors γ_E (for actions effects or actions) and γ_R (for resistance) can be determined by (assuming that actions and resistance follow Normal distributions):

$$\gamma_E = \frac{E_d}{E_k} = \frac{1 + \alpha_E \cdot \beta \cdot V_E}{1 + 1.64 \cdot V_E} \tag{21}$$

$$\gamma_R = \frac{R_k}{R_d} = \frac{1 - k \cdot V_R}{1 - k_{d,R} \cdot V_R} \tag{22}$$

where:

E_d and E_k represent the design and characteristic values of the action effects, respectively;
R_d and R_k represent the design and characteristic values of the resistance variables, respectively;
V_E and V_R represent the values of the coefficient of variation of the actions effects and of the resistance variables, respectively, accounting for various types of uncertainties (see below);

k is a coverage factor used to obtain the fractile of the Normal distribution corresponding to the characteristic value of resistance. For example, according to BS EN 1990, the characteristic value of a resistance variable is expressed by the 5%-fractile of the resistance variable Normal distribution. Thus, a value of $k = 1.64$ should be used if a sufficiently large number (e.g. greater than 100) of experiments or numerical test results is available to estimate V_R. For smaller sample sizes, tables with values of k are provided in BS EN 1990;

$k_{d,R}$ is also a coverage factor used to obtain the fractile of the Normal distribution corresponding to the design value of resistance, expressed by $\alpha_R \cdot \beta$. Tables are provided in BS EN 1990 with values $k_{d,R}$ that take into account the size of the data available to estimate V_R. For example, if a sufficiently large number (e.g. greater than 100) of experiments or numerical test results is available, $k_{d,R} = 0.8 \cdot \beta$.

The partial factor applied to actions effects, γ_E, typically takes account of:

- The uncertainty in the characteristic values of actions;
- The uncertainty in the model of the action;
- The uncertainties in modelling the effects of the action.

The partial factor applied to resistance, γ_R, typically takes account of:

- The uncertainty in the characteristic values of materials;
- The uncertainties in the conversion of parameters derived from laboratory conditions to site conditions;
- The uncertainties of the design resistance model, plus effects of geometric deviations (if these are not explicitly modelled).

In order to calculate the characteristic values of actions and resistance variables, adequate probabilistic models should be developed. For most ultimate limit states, it is only of interest to consider rare events, such as the maximum values of the time-varying action. As a result, the characteristic values of actions are usually determined from the probabilistic distribution of the maximum values of the action within a given unit observation time τ, typically one year.

Characteristic values are representative values of a random variable which have a sufficiently small probability of not being exceeded (for action variables) or of being exceeded (for resistance variables) during a reference period t. The specification of the latter reference period should take into account the design working life of the structure and the duration of the design situation.

A commonly used fractile for the resistance variables is the 5%-fractile. For some resistance properties (e.g. stiffness related properties) and geometrical quantities, mean values may be used as characteristic values. Concerning the characteristic value of actions, the fractile varies with the nature of the actions. For example, for permanent actions they are represented by the mean value. For climatic actions (e.g. wind action) they are typically represented by the 98% fractile. For other actions, characteristic values may be defined as nominal values (e.g. for minimum notional lateral loads and for accidental actions). See Chapter 3.

In order to calculate values for partial factors, it is now only necessary to define the target reliability levels. These levels express acceptable risk values. Risk acceptance criteria take into account the following information:

- General context;
- Individual and societal preferences for investments into life safety;
- Number of persons at risk;
- Occupancy profile of the structure;
- Nature of the hazard and the type of knowledge of the hazard prior to its occurrence;
- Evacuation time for a given occupancy or occupancy profile;
- Usage and function of the structure;
- Exposure of the structure;
- Form and type of construction;
- External protection from hazards;

- Consequences of failure;
- Costs of measures to improve structural safety.

BS EN 1990 (BSI, 2005) establishes a reliability differentiation procedure by specifying three consequences classes, see Table 1.

Other classification schemes are provided in ISO 2394 (ISO, 2015), ASCE 7 (ASCE, 2010). Particular elements of the structure may be designated at the same, higher or lower Consequence Class than the one considered for the entire structure.

The difference between the design objectives of ULS and SLS, results in different values for the target reliability levels assumed in the verification of both types of limit states. In BS EN 1990 (BSI, 2005) the target reliabilities presented in Table 2 are recommended for a design working life of 50 years for ULS and SLS. In operational terms, the differences between ULS and SLS are typically translated in the use of actions and resistance partial factors equal to 1.0 for the latter limit states, whilst using the same representative values for the actions and resistance variables (i.e. the same characteristic values). Note that for some SLS it may be adequate to use representative values of resistance variables higher than the corresponding characteristic values, such as the concrete tensile strength value to be used for concrete cracking verification.

In BS EN 1990 (BSI, 2005), one strategy to achieve a reliability differentiation is by distinguishing classes of action partial factors to be used in design situations. The modified action partial factors are calculated by multiplying the normal values by an importance factor, γ_I, see Table 3.

Table 1. Consequence and reliability classification in accordance with BS EN 1990 for ULS

Consequences classes	Reliability classes	Reliability index, β		Examples of buildings and civil engineering works
		β_1 (1 year reference period)	β_{50} (50 years reference period)	
CC1 Low consequence for loss of human life, economical, social and environmental consequences	RC	4.2	3.3	Agricultural buildings, greenhouses
CC2 Medium consequence for loss of human life, economical, social and environmental consequences	RC2	4.7	3.8	Residential and office buildings, standard bridges
CC3 High consequence for loss of human life, economical, social and environmental consequences	RC3	5.2	4.3	Important bridges and public buildings

Table 2. Target reliabilities recommended in BS EN 1990 (50 years)

Limit State	Reliability index, β, for CC2	P_f
ULS	3.8	$7.2 \cdot 10^{-5}$
SLS	1.5	$6.8 \cdot 10^{-2}$

Table 3. Importance factors for reliability differentiation, according to Eurocodes

Design situations	Reliability class		
	RC1	RC2	RC3
Persistent	0.9	1.0	1.1
Seismic	0.8	1.0	1.2 to 1.4

Another strategy is to adopt stricter quality management procedures. To this regard, BS EN 1990 (BSI, 2005) specifies three design supervision levels and three inspection levels, related to the three consequences classes.

A final but important remark concerning the limit state and partial factors method, is to highlight that the requirements for reliability expressed in Table 1 and Table 2 are related to the structural verification of performance of the individual elements. It is usually assumed in design codes, that the latter generally leads to structures with a global reliability level similar to the one used for the individual elements, therefore acceptable. However, as referred in the introduction of this Chapter, this is not always the case. An example is a structural system built in series of brittle elements, such that failure of any individual element leads to the failure of the structure. Each element can be designed to have a very low probability of failure, however the failure probability of such a structure is related to the sum of the failure probabilities of the individual structural elements, which depending on the number of elements can add up to values much higher than anticipated.

5.3.2.1.3 Full-Probabilistic Method

Full probabilistic methods are based on the direct estimation of risk measures using advanced methods, see Sections 5.3.5 and 5.3.6.

The simplest case is to calculate the probability of failure, P_f, and compare it with the target (design) value $P_{f,d}$. More complex analyses methods are presented in Section 5.5.

5.3.3 Probabilistic Modelling in Structural Engineering

5.3.3.1 General

As mentioned previously, our perception and knowledge of the world is limited and subject to uncertainty. In order to predict accurately the behaviour of civil engineering infrastructures, engineers need to work with uncertainties in almost all engineering relevant variables. One way to accomplish this is through the use of the probabilistic theory.

Probability theory concerns with the description and modelling of random phenomena. Probability is a non-physical variable assigned to a random event to express the observed tendency or the degree of belief of its occurrence. The foundations of the modern probability theory were laid by Kolmogoroff (1933) which established the probability axioms:

- The probability value of a random event is a real number between 0.0 and 1.0;
- The sum of probabilities of all possible random events is equal to 1.0.

The formulation of probabilistic models may be based in existing data alone (*aka* frequentist approach), but most often data is not available to the extent where this is possible. In such cases it is also necessary to base the model building on physical arguments, experience and judgement. Ang & Tang (2007) and Haldar & Mahadevan (2000) present a comprehensive overview of probability methods applied to engineering. The challenging topic of how to derive a subjective probabilistic function by combining different expert opinions is discussed by Cooke (1991), Cooke & Goossens (2004) and Ouchi (2004).

5.3.3.2 Model Selection

A first aid to model selection is to construct probability plots (e.g. Q-Q plots). If more than one model is considered to be acceptable, further analysis can be developed by performing hypothesis testing on the selected distributions. Classical tests are the Kolmogorov-Smirnov test, the chi-square test and the Anderson-Darling test. A drawback of hypothesis testing is that it provides the analyst information regarding if there is significant statistical evidence to reject or not the null hypothesis (e.g. that the data follow a specified model), so it may be the case that more than one model cannot be rejected. These methods will not inform about which model is "true", but rather about the relative strength of each model given the information available (strong assumption!).

Other goodness of fit methods include selecting the model with the highest log-likelihood or using information theory methods (Ando, 2010; Claeskens & Hjort, 2008). However, these methods also do not tell if the models fit the data well, only which one is better considering the hypothesis of each method.

Frequentist and subjective probabilistic models are only approximations of the actual phenomenon being analysed. Therefore, there is statistical uncertainty. To include this uncertainty, Bayesian statistics methods take into account a set of good models and not only a single "true" model, e.g. Bayes factors methods or other Bayesian model selection methods (Congdon, 2006).

Point estimates of the model parameters can be obtained by the Method of Moments (MoM) or by the Maximum Likelihood Estimation (MLE) method for example. In Bayesian methods the analyst can assume that the model parameters have unknown distributions which can be updated with observed data.

However, as Box & Draper (1987) say: *"all models are wrong, but some are useful"*. The best model may be the wrong model because the data used to validate it, even if there is a large amount of data available, is often still insufficient or unrepresentative to derive with the required accuracy the tails of the frequentist probabilistic distribution. An example is model overfitting: if new data is included the model may no longer be able to return accurate results.

As a mean to circumvent the above difficulty, several standard probabilistic models have been proposed in the literature to model actions, actions effects, resistance variables, and time variant associated problems. Examples of standardised probabilistic models are given in the JCSS Probabilistic Model Code (JCSS, 2001). In any case, the analyst must use his knowledge and consider the nature of the physical or chemical problems at hand when deciding whether or not the selected distribution function is valid.

5.3.3.3 Bayes' Theorem

Bayes' Theorem, mentioned above several times, is given by Eq. 23 for continuous random variables:

$$f^{*}\left(\theta \mid x_1,\ldots,x_n\right)=\frac{f\left(x_1,\ldots,x_n \mid \theta\right)\cdot f\left(\theta\right)}{\int\left[f\left(x_1,\ldots,x_n \mid \theta\right)\cdot f\left(\theta\right)\right]d\theta} \tag{23}$$

where:

θ is a random variable;

x_j ($j = 1$ to n) are observations of θ;

$f(\theta)$ is called the prior probability distribution of θ which must be estimated either by expert judgement or by existing databases (which should refer to similar measurands as the one being analysed);
$f(x_1,...,x_n|\theta)$ is called the likelihood of x_j ($j = 1$ to n) observations given θ, also written as $L(x_1,...,x_n|\theta)$;
$f^*(\theta|x_1,...,x_n)$ is called the posterior probability distribution of θ given x_j ($j = 1$ to n) observations.

Bayes' Theorem has the advantages over classical probabilistic methods of being able to mix different sources of information, and thus for example providing a mean to update the analysis with new data, as well as of being able to incorporate in the analysis the influence of different types of uncertainties.

It is recommended that the existing dataset should be larger than the new dataset and that the quality of the new information be better than the existing information in order to justify the use of Bayesian Theory. Careful attention should be paid in the cases when there is no information about prior distribution and the posterior distribution is significantly affected by the prior distribution. If the prior distribution and its parameters are uncertain, a weighted fitting method can be used to include several distributions, in particular the approach of Bayes factors, see van Gelder (2000).

The evaluation of the posterior distribution can be difficult. In these cases, computational approaches, such as Markov Chain Monte Carlo (MCMC) can be used. Several open source programs are available that use Bayesian statistics such as WinBUGS or OpenBUGS (http://www.openbugs.net/).

Examples of probabilistic models in structural engineering

Several types of probabilistic models exist and have been proposed for various types of variables with interest for structural engineering. Table 4 provides a selection of conventional models.

Table 4. Conventional models of basic variables for time invariant reliability analyses (Cajot et al., 2003)

Class	Variables	Symbol	Distribution	Mean value μ_X	Standard deviation σ_X
Actions	Permanent (Self-weight of steel)	G	Normal	G_k	(0.01 to 0.03) μ_X
	Permanent (Self-weight of concrete)		Normal	G_k	(0.02 to 0.05) μ_X
	Imposed, 5 years	Q	Gumbel	0.2 Q_k	1.10 μ_X
	Imposed, 50 years	Q	Gumbel	0.6 Q_k	0.35 μ_X
	Wind, 1 year	W	Gumbel	0.6 W_k	0.26 $\mu_{X,1}$
	Wind, 50 years	W	Gumbel	1.1 W_k	0.15 $\mu_{X,50}$
Material strength of carbon steel	Yield strength	f_y	Lognormal	f_{yk}+2 σ_X	(0.07 to 0.11) μ_X
	Ultimate strength	f_u	Lognormal	1.5 μ_{fy}	0.05 μ_X
Geometry steel section	I profiles	A,W,I	Normal	0.99 X_{nom}	(0.01 to 0.04) μ_X
	L profiles, rods	A,W,I	Normal	1.02 X_{nom}	(0.01 to 0.02) μ_X
	Section thickness	t	Normal	X_{nom}	(0.01 to 0.02) μ_X
Model uncertainties	Load effect factor	$\theta_E = X_{model}/X_{real}$	Normal	1.00 to 1.20	0.05 to 0.10
	Resistance factor+	$\theta_R = X_{real}/X_{model}$	Normal	1.00 to 1.20	0.05 to 0.20

5.3.4 Uncertainty Propagation in Structural Engineering

Uncertainty propagation is a critical step in risk management. It is a signature of the quality of risk management and it has the power to possible control the decision-making process. Key elements to decision-making process involve knowing how different types and sources of uncertainty propagate in every step of the analysis, which of them contribute more, and what is their significance to the results. A very comprehensive report on different types of uncertainty and on various uncertainty propagation methods is presented in Hayes (2011).

Examples of existing methods that allow uncertainty propagation are Bayesian Theory (see Oakley & O'Hagan (2004)), Probability Bounds Analysis (PBA) (see Muhanna, Rao, & Mullen (2013), Rao, Mullen, & Muhanna (2011), Schweiger & Peschl (2005), Xiao, Huang, Wang, Pang, & He (2011) and Zhang, Mullen, & Muhanna (2010, 2012)) and second-order Monte Carlo simulations (also named probabilistic sensitivity analysis).

In the latter method, n runs of Monte Carlo simulations are carried out, each with different input probabilistic distribution model parameters, and for each run a second set of m Monte Carlo simulations are performed to cover the full range of each probabilistic distribution model. In the end, uncertainty from the input models is propagated to the distribution of the output results. In the context of finite element analyses to derive the relationship between input and output parameters it is possible to use Response Surface methods to determine an approximation of that relationship by running a much smaller number of calculations. The error of this approximation can be estimated by Bootstrap methods (Efron, 1979). To limit the number of n Monte Carlo runs it may be adequate to consider only distributions for the inputs that represent the average and two extreme quantiles of the possible distributions.

Probability Bounds Analysis (PBA) operates in a different way. Instead of approximating the output probability distribution, it provides bounds on that distribution. In a pure second-order Monte Carlo analysis, the types of probabilistic distributions of the input variables and the distributions parameters must be provided. In many cases the definition of this data is subject to a very large uncertainty. On the contrary, in probability bounds analysis the consequences deriving from the need to make strong subjective assumptions about the type of distribution functions or about the independence/dependence between input variables are limited since PBA uses conservative bounds to simulate the probabilistic distributions of the input variables. However, applications of PBA to finite element analyses are in its first steps and further work must be done in order to be a real alternative to other methods. In particular, progress is still needed in addressing the dependence between input variables to avoid obtaining overly conservative interval bounds for the outputs.

Methods also exist that try to reduce the uncertainty introduced when selecting the probabilistic model for the input parameters. For example, the arbitrary polynomial chaos expansions method (aPC), see Xie, Lu, Cóstola, & Hensen (2014), generalizes chaos expansion techniques towards arbitrary distributions with arbitrary probability measures. The aPC only demands the specification of the distribution moments.

5.3.5 Classical Structural Probabilistic Analysis

Although the origins of classical structural probabilistic analysis, i.e. reliability analysis, date back from early 1920s, the basis for its application as an accepted analysis method in structural engineering were mainly developed in the 1970s and 1980s decades, when several fundamental books were published (Ang & Tang, 1975; Benjamin & Cornell, 1970; Ferry Borges & Castanheta, 1968; Thoft-Christensen & Baker, 1982).

The classical reliability (time invariant) problem is defined by a structural system characterised by one resistance random variable, *R*, and subjected to only one load random variable, *S*. Both basic random variables, *R* and *S*, are considered independent and stationary random processes. Therefore, the probability of failure, P_f, of this system is given by:

$$P_f = P\left(G = R - S \leq 0\right) = \int_{-\infty}^{+\infty} \int_{-\infty}^{s \geq r} f_R\left(r\right) f_S\left(s\right) drds \tag{24}$$

However, it is not always possible to analytically solve the probability of failure integral and closed form solutions of the limit state function, *G*, often do not exist.

Therefore, it may be necessary to use simulation tools, such as the Monte Carlo method. The basic idea behind this method is to generate random simulations of the limit state function and observe the result. The disadvantage of this method is related with the high number of simulations, *N*, necessary to fulfil the accuracy requirements. In the crude Monte Carlo method, the confidence interval of the estimate of P_f is given by (Melchers, 1999):

$$P\left(\mu - k \cdot \sigma < P_f < \mu + k \cdot \sigma\right) = CL \tag{25}$$

where μ and σ are the estimates of the mean and standard deviation values of P_f and *CL* is the specified confidence level (95% for instance).

It was proved by Melchers (1999), that the value of σ decreases in proportion to $N^{-1/2}$. Additionally, the probability of a sampling a point at the failure region is in general very low, since $f_X(x)$ is the sampling function, so to achieve convergence it may be necessary to undertake more than 10^6 simulations.

As a result, several methods were developed to reduce the value of σ by other ways. Methods such as variance reductions techniques (Importance Sampling, Latin Hypercube Sampling or Adaptive techniques) try to reduce the value of σ by using additional information about the limit state function, restricting the sampling to be within the region of interest or adapting the sampling to the shape of the limit state function. The latter can also benefit from estimations of the shape of the limit state function; usually obtained by Response Surface methods and by other Design of Experiments methods (DoE). In certain cases, a description of the problem in the polar coordinate system is convenient; which led to the development of Directional Simulation methods.

Alternatively, to simulation methods, it is possible to simplify the limit state function by for example using the first-order or second-order Taylor series expansion about some point *x*. After, the first and second moments of the simplified equation of $G(X) = 0$ can be obtained and the reliability calculated. It is usual to perform a transformation of coordinates from the original space to a standardised space with zero mean and unit variance. If the distributions of the basic variables *X* are non-normal then they must be transformed into equivalent Normal distributions, using either the Rosenblatt transformation (when the joint probability function $f_X(x)$ is known) or the Nataf transformation (when only the marginal distributions and the correlation matrix are known) (Melchers, 1999). Examples of these methods are the First-Order Reliability Method (FORM) and the Second-Order Reliability Method (SORM) (Melchers, 1999). These simplified methods do not provide a measure of the prediction error and cannot guarantee that the critical design point is found.

It should be noted, that if the probability of failure is determined without accounting for all types of uncertainties, in particular the ones related with human intervention, the value of the probability of failure must be considered only as a nominal, or notional, value and not as an estimate of the actual failure frequency. As an indication, a CIRIA report indicates that the actual probability of failure is about one order (×10) higher than the probability of failure implicit in design codes (CIRIA, 1977). The use of a notional probability of failure for comparison of alternatives purposes must be done very carefully since the influence of ignored uncertainties on the total risk might vary considerably between alternatives. It might only be directly considered when the effect of ignored uncertainties is proportional to all considered alternative solutions.

5.3.6 Advanced Structural Probabilistic Analysis

In reality, several limit states exist in every structural system, for instance: bending moment resistance, shear resistance, fatigue resistance or equilibrium related limit states. In general, some of these limit states will not be independent from each other. Additionally, basic variables such as load values and resistance properties can vary over time, for instance: a structure can be subjected to various types of loads which can be applied at different times during the structure lifetime and their values can change over time; material's properties can also change over time by deterioration processes such as corrosion or damage by excessive usage.

Thus, the structural probabilistic analysis of complex structural systems, including geometrical and physical nonlinearities, load-path dependencies and the space-time variation of material and system properties, presents considerable challenges to classical reliability methods, such as FORM. However, the use of finite element methods coupled with advanced numerical algorithms can provide solutions to problems where classical methods fail to return accurate results.

Examples of these numerical algorithms are Response Surface methods (see Hastie, Tibshirani, & Friedman (2009) and Khun & Johnson (2013)), Polynomial Chaos Theory and Spectral Methods. Recently, Bayesian Probability Networks (BPN) have started to be used to assess the reliability, and risk, of complex and large systems which cannot be incorporated as a whole in the same analysis model. Application examples of BPN analysis are multiple hazard scenarios such as the ones encountered in the design of ships, tunnels or nuclear power plants. However, in problems with complex inter-dependent variables BPN's can underperform, see Hayes (2011).

Time dependent reliability problems are still an open area of research but some background can be found in Melchers (1999) and Sudret (2007). It is worth noticing that an action can change in time by simply varying its spatial distribution over time while its magnitude is kept constant. In these cases it is possible to model the action by different loads, each one considered as stationary variables, which are activated at different times in different positions. This can be useful when modelling construction loads, for example. However, in more complicated cases of basic variables with more intricate time-dependent functions, it is necessary to use appropriate models of time-dependent quantities and processes. For the probabilistic combination of actions either the Turkstra's rule or the Ferry Borges-Castanheta model may be used (Cajot et al., 2003; Ferry Borges & Castanheta, 1985; Melchers, 1999; Turkstra & Madsen, 1980).

5.4 RISK MANAGEMENT FRAMEWORK FOR STRUCTURAL DESIGN

5.4.1 Basis

Never before have humans lived longer and better. Housing and health care are available to millions of people and equipment, products, and food are quality tested and safer than ever before. Although there is an inherent risk in all human activities, there is an evident downward trend over the years. The common sense tells us that risk in the majority of our daily life activities is low and very well regulated.

As a result of this evolution, individuals and society are risk averse: they are willing to take advantage of advances in science and technology to reach certain objectives but only if the risks are small enough to be acceptable, or low and clearly controlled to be tolerable. However, not all the risks are known (or recognised) and the ones that are, are not always clearly explained and properly managed.

At the same time, individuals and society are much more risk reactive. The various media sources make information travel the world almost instantaneously and each severe accident is subject to public scrutiny and critics. Although, it is nowadays consensual that safety is not an absolute and infinite condition, but is instead a tolerated situation desirably balanced with low levels of residual risk (McDonald et al., 2005), society demands that proactive rather than reactive measures should be engaged so that risks with the potential to affect the welfare, safety and other interests of the community are kept under review and properly controlled.

In short, individuals and society at large expect risks to be properly managed; they are not willing to accept risk just based on economic factors and do not accept that risks have been hidden behind potential benefits.

It is thus necessary to integrate in the decision-making process the optimal allocation of available natural, economic and technological resources, balanced with the requirement to guarantee and to preserve a proper safety level. In structural engineering applications some gradual shifts are seen to meet these new societal expectations: great advances have been made in using probabilistic methods coupled with simulation analyses rather than pure deterministic ones, and reliability and risk analyses are increasingly gaining importance as decision support tools to ensure that structures' design, operation, maintenance and overall management are both economical and safe (Faber & Stewart, 2003). Therefore, it is not surprising that organizations which cannot demonstrate the rationale supporting its decision-making process, place themselves in a weak position should an adverse event occur.

Risk management is also an important tool of asset management. An asset can be defined as a physical system from which valuable services can be provided. According to a report by CIRIA (2009),

whole-life infrastructure asset management balances maintenance, repair, refurbishment, renewal, replacement, and upgrade activities to optimise the long-term value of an asset.

Asset management can be defined as

the systematic and co-ordinated activities and practices through which an organisation optimally and sustainably manages its assets and asset systems, their associated performance, risks and expenditures over their life cycles for the purpose of achieving its organisational strategic plan.

Every temporary structure is an important asset, either by itself or when it is used to temporarily support critical infrastructures since it may significantly influence the permanent structure whole-life value. It is thus logical that the structural design of temporary structures must be integrated in a broader scope of asset management to maximise the benefits and minimise the risks.

Risk management is the complete process of risk assessment and risk control that aims at achieving a balance between the need to protect against existing and future risks, and the benefits deriving from a given activity. In a framework of a specific context, risk management helps organizations to address risks and make efficient decisions to achieve the desired objectives with a limited and justifiable risk level. Risk management does not dictate decisions but rather contributes to a risk informed decision-making process.

Although one can never remove all uncertainties of a construction project – it is not technologically and economically possible to identify and eliminate all risks – systematic risk management improves the chances of a given project being completed on time and within budget, accomplishing the required quality, with proper provision for safety and environmental issues (CIRIA, 1996).

The process of risk management is complex and nonlinear, based on successive iterations until the correct balance between different inputs and generated outputs is judged to be found. Furthermore, risks cannot be addressed in isolation from each other. Risks are interlinked and the management of one risk may have positive and/or negative consequences on others. Clear communication, external consultation and constant review are recommended.

The optimal aim of risk management is to reduce as much as practicable the risk associated with a given activity or action. Depending on the legal context there are different ways of reaching this central objective. The UK's legal system is the Common Law legal system, where laws are written emphasising goals rather than detailing accepted actions to achieve those goals – the case of the rest of European countries where a Civil Law legal system is used, which details the required actions that must be taken. As a result, in the UK, Health and Safety at Work (HSW) regulations are based on risk management principles, namely the ALARP (As Low As Reasonably Practicable) principles (HSE, 2001).

In the UK legal framework, employers are responsible to ensure, that the risks to health, safety and welfare of their employees and of third persons, are managed to a level which can be justified as being acceptable or tolerable. This can be considered attained if the measures cost increase is not cost-effective or is disproportionate, respectively, in front of the expected risk reduction gains, e.g. the decrease of consequence costs of a hazard event (Bowles, 2003). The amount of risks which are judged to be tolerable is the "risk appetite" (HM Treasury, 2004). These residual risks must be accepted or insured against, monitored and controlled. Activities with an unacceptable or intolerable risk level should be terminated. Additionally, it is understood that one cannot use ALARP to justify not following good practice; ALARP should rather be used in cases where good practice is unclear, is only partially applicable or where higher levels of quality and safety are aimed.

Risk management should answer three fundamental questions (Kaplan & Garrick, 1981):

- What can go wrong?
- How likely is it to go wrong?
- What are the ensuing consequences?

ISO 31000 (ISO, 2009b) and ISO 13824 (ISO, 2009a) establish a risk management model of structural systems, see Figure 5. It is presented as having a hierarchical structure with various steps each one with different objectives, but in reality a global strategy must exist to make the process efficient and coherent.

Any activity, or project, is initiated to achieve certain objectives. The objectives should be SMART: Specific, Measurable, Achievable, Realistic and Time bounded (HM Treasury, 2003) and attributes (performance indicators of the objectives) should follow rules given in Goodwin & Wright (2010).

5.4.2 Risk Assessment

Risk assessment encompasses risk analysis and risk evaluation. Typically, risk assessment is an iterative process where risk is calculated by a structured, systematic examination of the likelihood of critical events and of the associated potential consequences on the planned objectives should these events occur. Another central characteristic of risk assessment is that it involves making trade-offs between risks to some individuals or groups and risks to the society; and between costs and benefits of different scenarios. It is also very important to appropriately identify, document and evaluate key types of uncertainty and then to consider them in an explicit and transparent way. At the end of risk assessment, the results must be properly documented and communicated.

Figure 5. Risk management framework, adapted from ISO (2009a, 2009b)

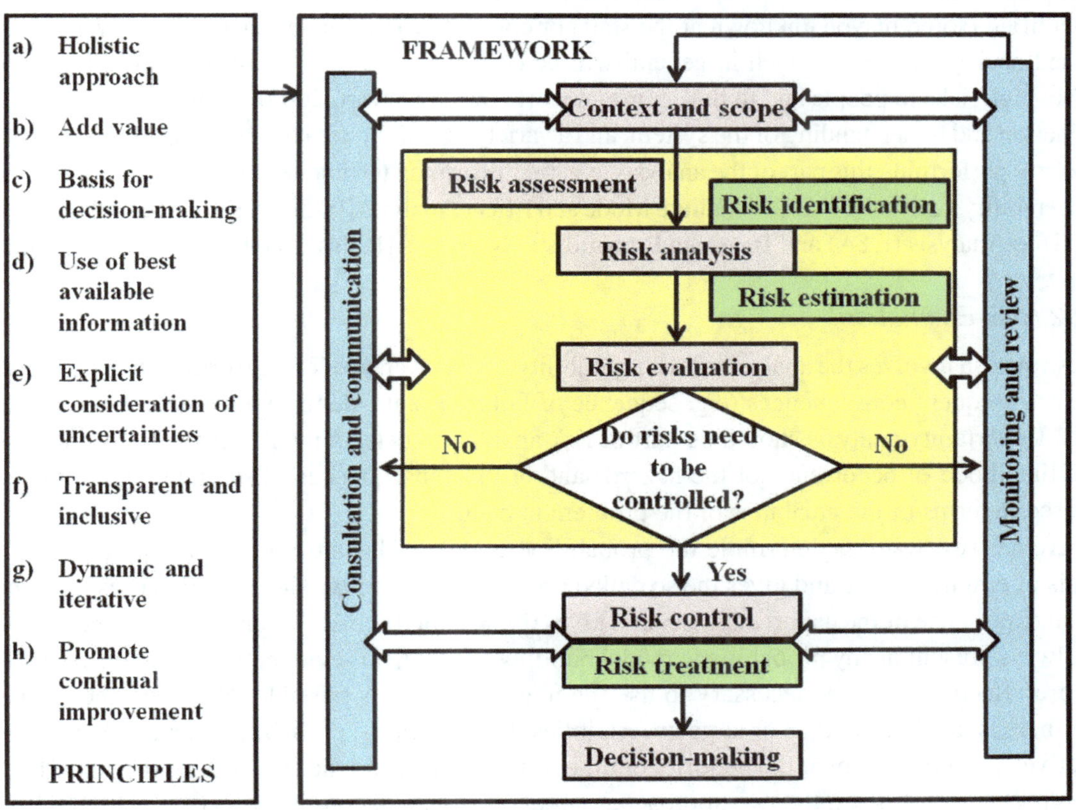

5.4.2.1 Risk Analysis

5.4.2.1.1 General

Generally, risk analysis is the start of risk management. Risk analysis usually begins with a careful description and examination of the system: context, activity objectives, performance requirements, methods of operation and development, structural components and their functions, design concepts, potential hazard events, possible failure modes and consequences over a certain period of time, e.g. corresponding to the design working life of the structure (McDonald et al., 2005). Several practical problems can arise at this phase, some of which are described in Blockley (1992).

5.4.2.1.2 Risk Identification

Risk identification is the compilation, review and use of the available information concerning relevant hazard scenarios, with appropriate consideration of the uncertainties involved, for characterisation of what is known and what is uncertain about the present and future performance of the structure. It generally involves a systematic approach for describing the system context, for identifying and describing the relevant hazard scenarios: what can happen, how, why and who will be involved? At this phase, links are established between hazards, consequences and causes, and their sensitivity to each individual contribution is evaluated.

Civil engineering contains many potential risks, related for instance with: design, construction, testing, maintenance, third party activities, environment, health and safety, finance, legal contracts, management and political organisation (Artamonov et al., 2008; Bunni, 2003). If all the relevant hazard scenarios are not identified (some maybe unknown at the start) or correctly characterised then risk analysis will result in biased decision-making, which in general will be cost inefficient and ultimately could lead to unacceptably high risks to people and to the environment (Faber & Stewart, 2003). Thus, it requires a detailed examination and understanding of the system, and a variety of techniques have been developed to assist the engineer in performing this part of the analysis, e.g. brainstorming techniques, morphologic boxes, Hazard and Operability analysis [HAZOP], Failure Mode and Effect Analysis [FMEA], Fault Tree Analysis (FTA), Event Tree Analysis (ETA) and Bayesian Probabilistic Networks [BPNs], see ISO 31010 (ISO, 2009c).

5.4.2.2 Risk Estimation

Risk estimation involves the analysis of the probability of occurrence of certain critical hazard events and of their subsequent consequences (e.g. sequence of failure events, damage to functionality, health and safety). Uncertainty analysis should be part of risk analysis to determine the influence of uncertainties on the likelihood of occurrence of the hazards and of the consequences. Risks must be estimated and expressed in terms of the attributes of the problem in hand.

There are two ways to determine the probability models of hazard scenarios. One uses statistical analysis of empirical data and gives the so called objective probability. The other one uses intuition and relevant experience of the expert engineer and gives the so called subjective probability. Since subjectivity is always present in any probabilistic model building process, it is evident that objective probabilities are more effective and it is necessary to use them every time it is possible. However, there are cases where there is insufficient data or large uncertainties. In these cases, probabilities can only be estimated subjectively, quantitatively or the majority of times qualitatively, and the engineer's experience becomes very important. In general, the two approaches are often used in a complementary way.

As a first approach, the probability of occurrence of given hazard scenario and the significance of its expected consequences can be estimated qualitatively, by one of the various available methods, such as FMEA. Doing this for all hazard scenarios one can build a risk matrix, see Figure 6 – "5×5" matrices are often used, with consequences on a scale of "insignificant / minor/ moderate / major / catastrophic" and likelihood on a scale of "rare / unlikely / possible / likely / almost certain" (HM Treasury, 2004).

Alternatively, or as a subsequent second step, the probability of occurrence of hazard scenarios can be estimated quantitatively by probability analysis. Several methods exist to achieve this goal: First Order Reliability Method (FORM), Second Order Reliability Method (SORM), Monte Carlo methods (MC) or Stochastic Finite Element Methods (SFEM). Probability analysis can also be performed using ETA, FTA or BPN to obtain estimates of the system's reliability. A detailed discussion about these methods is given in Bedford & Cooke (2001), Det Norske Veritas (2002) and Hartford & Baecher (2004). With this information, the most important (critical) risks can be identified and risks concerning the different attributes can be ranked, possibly with the introduction of weights to allow for multi-criteria analysis.

5.4.2.2.1 Risk Evaluation

Next follows risk evaluation. It is the process of examining and judging the significance of risk (McDonald et al., 2005). First, the risk acceptance criteria are established and the acceptable and the unacceptable risk levels are defined. The UK Health and Safety Executive (HSE, 2001) present three risk criteria, explained in greater detail in Bowles (2007):

- "An equity-based criterion, which starts with the premise that all individuals have unconditional rights to certain levels of protection (...);
- A utility-based criterion which applies to the comparison between the incremental benefits of the measures to prevent the risk of injury or detriment, and the cost of the measures (...);
- A technology-based criterion which essentially reflects the idea that a satisfactory level of risk prevention is attained when 'state of the art' control measures (technological, managerial, organisational) are employed to control risks whatever the circumstances".

Figure 6. Example of a risk matrix

PROBABILITY \ CONSEQUENCES	Insignificant	Minor	Moderate	Major	Catastrophic
Almost certain	M	H	E	E	E
Likely	M	M	H	E	E
Possible	L	M	H	H	E
Unlikely	L	L	M	H	E
Rare	L	L	M	H	H

L Low Risk M Moderate Risk H High Risk E Extreme Risk

A fourth criterion, a sustainability criterion, should be also considered. This criterion involves the consideration of problems such as intergenerational equity and allocation of resources in the long-term, for example to maximise the design working life of civil engineering infrastructures at a minimum cost (durability and debt problems) and green engineering (climate change problems), see Nishijima (2009) for examples.

Acceptable and unacceptable (i.e. limit of tolerability) risk levels must be defined by taking into account the context: nature of risk, type of stakeholders, amount of available resources and magnitude and distribution of consequences to individuals and society. In order to help establishing acceptable and unacceptable risk limits, the authors recommend assessing the individual and societal perception and aversion to risks (distinguishing between voluntary or imposed, known or unknown risk scenarios) and afterwards determine the risk limits using the Value for Preventing a Fatality (VPF), the Life Quality Index (LQI) or other suitable methods.

Risks are classified as acceptable, as unacceptable or as being in the tolerability range by comparing the estimated risks with the risk criteria. It should not be forgotten that it is the total risk that matters in the end. The risk associated with a single hazard scenario can be acceptable but when the risks associated to all identified hazards are summed up the total risk can be higher than the acceptable level.

Furthermore, a list with a range of alternative measures for managing the risks which are higher than the acceptable risk level is developed. Figure 7 illustrates a possible framework (the "safety cube") to define the breadth of measures, their application opportunity and the tools involved in their implementation. Measures can also be defined as active or passive, preventive (proactive) or protective (reactive), see Todinov (2007) for more details.

5.4.3 Risk Control

The last phase of risk management is *risk control*. Here all the information made available in the previous phases is gathered and reviewed. If the estimated risk is greater than the acceptable risk level, and because citing Fischhoff, Lichtenstein, Slovic, Derby, & Keeney (1981)*"One accepts options, not risks"*, the risk level must be modified by suitable proportional measures. This process is referred as *risk treatment* and can involve different approaches – see HSE (2001) for more details –, which are summarised below:

- **Risk Mitigation**: In essence, risk mitigation is implemented by reducing the probability of the occurrence of the hazard scenario to nominally zero, by for instance restricting the use of the structure (i.e. changing the exposure).
- **Risk Reduction**: This may be implemented by reducing the probability of the occurrence of the hazard scenario and/or of its consequences. In practice risk reduction can be performed by decreasing the fragility and vulnerability either by changing the exposure of the system, by increasing the reliability and the robustness of the system, and by non-structural measures such as: monitoring, surveillance, and periodic inspections, and planning post-failure measures. Considering the risk of collapse of slender steel frame structures, such as bridge falsework systems, due to instability and second-order effects, this might be reduced by bracing critical elements.
- **Risk Transfer**: This may be performed by insurance or other financial arrangements where a third party, normally an insurer, takes over the risk. Therefore, risk transfer is normally associated with a premium cost.

Figure 7. Safety cube, adapted from(Schneider (2006b)

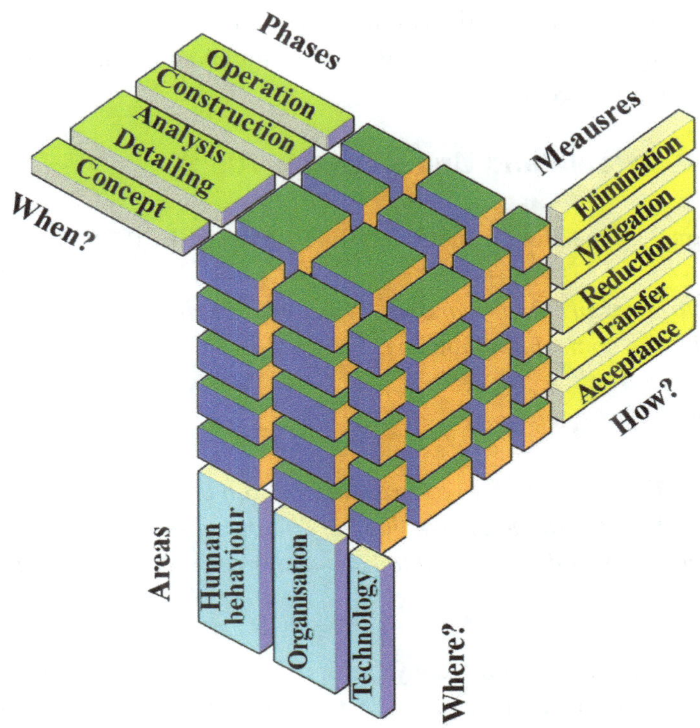

- **Risk Tolerance and Risk Acceptance**: Risk can be tolerated if it is ALARP. Risk acceptance may be an option in the case of activities with a low risk profile, close to the acceptable risk level, and for which it can be demonstrated that pursuing with any of the other options would lead to unacceptable economic losses.
- **Risk Elimination**: Decision not to start or continue the activity (decommissioning).

Risk control incorporates the selection of the measures most suitable to manage risks as well as the definition of the performance objectives and requirements of the implementation methods, but also the definition of the monitoring, evaluation criteria and review methods of the selected measures (accounting for possible updates when relevant information becomes available).

It must be kept in mind that the greater risk reduction achieved by a single measure the more critical this measure becomes. Therefore, these measures should be very reliable, possibly involving the adoption of safeguard measures and the implementation of a comprehensive and continuous monitoring and review system is recommended. Additionally, it should not be forgotten that in certain circumstances reducing a risk source, by increasing the reliability of a critical element, for example, could create other hazard scenarios which were not accounted for in the initial risk analysis. This is one of the reasons why risk management is an iterative process.

For each one of the selected risk treatment measures, residual risks are estimated and resource allocation is optimised. After consultation with the interested stakeholders, a decision needs now to be made about whether these reminiscent risks are unacceptable, intolerable, ALARP or acceptable; the concept of "risk appetite".

Risk management should be reviewed on a regular basis throughout the duration of the project: typically at the project development stage, at the contract procurement stage, at the design stage (including a regular review of the temporary structures design, for example) and at the construction, operation and maintenance stages (Artamonov et al., 2008).

5.4.4 Acceptable, Tolerable and Unacceptable Risks

Acceptable risk is defined by the Health and Safety Executive (HSE, 1995) as

a risk, which for the purposes of life or work, everyone who might be impacted is prepared to accept assuming no changes in risk control mechanisms.

In turn, tolerable risk

refers to a willingness to live with a risk so as to secure certain benefits and in the confidence that it is being properly controlled. It is a range of risk that we do not regard as negligible or as something we might ignore, but rather as something we need to keep under review and reduce it still further if and as we can.

The acceptability of risk is affected by many factors such as the nature of the hazard, the exposure level to the risk (voluntary or involuntary risk, short or long periods), the importance of the possible benefits and the scale of the associated consequences (individual and societal risks: who is affected?), the state of knowledge about the risk (known and unknown risks), the individual and societal awareness, degree of control and fear about the hazard, individual and societal moral and ethical values (Das, 1997; Sommer et al., 1999).

It is therefore very difficult to assign values to individual acceptable risks or to individual unacceptable risks. However, guidance can be found in specialised textbooks concerning dam, nuclear or bridge engineering, or in regulatory reports produced by public institutions (no published guidance is available concerning temporary structures). Based on these documents, the value commonly assigned to individual acceptable risk (i.e. the broadly acceptable risk limit) ranges from 1 in 10^6 to 10 in 10^6 deaths per year, and to individual unacceptable risk (i.e. the limit of tolerability) ranges from 100 in 10^6 to 1000 in 10^6 deaths per year, see ANCOLD (2003), Das (1997), HSE (1992, 2001) and UK DfT (1999).

It is also a very difficult task to define acceptable risks for society. A commonly used approach is the definition of *F-N* curves, being *F* the annual probability of exceedance of *N* or more fatalities. Vrijling, van Gelder, Goossens, Voortman, & Mandey (2004) present a methodology to evaluate societal risks based on *F-N* curves. However, *F-N* curves have some inconsistencies when different risk scenarios are combined, see Bedford & Cooke (2001).

The maximum allowable annual probability of structural failure, P_f, depends on the conditional probability of a person being killed, given the failure of the structure, $P(d \mid f)$, and can be obtained by (ISO, 1998):

$$P_f \leq \frac{10^{-6}}{P(d \mid f)} \tag{26}$$

In an excellent report published by CIRIA (1977), a method to determine a rational target of probability of failure of civil engineering structures is expressed by:

$$P_f = \frac{10^{-4}}{n_r} \cdot K_s \cdot n_d \qquad (27)$$

where P_f is the acceptable probability of failure due to any cause during the design working life (n_d years), n_r is the number of people at risk in the event of failure and K_s is equal to:

- Places of public assembly, dams: $K_s = 0.005$;
- Domestic, office or trade and industry: $K_s = 0.05$;
- Bridges: $K_s = 0.5$;
- Towers, masts offshore structures: $K_s = 5$.

For temporary structures used for building construction, $K_s = 0.05$ can be used, whereas for temporary structures used for bridge construction $K_s = 0.5$ can be used.

Another method presented in McDonald et al. (2005) is expressed by the following formula:

$$P_f = \frac{\eta_i}{P_{d|fi}} \cdot 10^{-4} \qquad (28)$$

where P_f is the acceptable probability of failure due to any cause during the design working life, $P_{d|fi}$ denotes the probability of being killed in the event of an accident and η_i is a policy factor which varies with the degree of voluntariness with which an activity *i* is undertaken and with the manner the benefit is perceived. It ranges from 100 in the case of complete freedom of choice, to 0.01 in case of an imposed risk without any perceived direct benefits. Vrijling, van Hengel, & Houben (1998) proposes the values for η_i given in Table 5.

The concepts of acceptable, tolerable, intolerable and unacceptable risks (written in ascending risk order) are used to assess the trade-offs between the importance of expected benefits and the significance of the expected adverse consequences, not forgetting the resources involved.

As the UK HSE emphasises "tolerable does not mean acceptable" (HSE, 2001). Acceptable risk is typically associated with residual risk to life, property or other fundamental values, either because the probabilities of occurrence of the hazards are so small or whose consequences are so slight, whereas tolerable risk is associated with greater risk levels which can be tolerated if certain conditions are met in order to achieve a given set of relevant benefits. In the latter case, the focus is set more on the analysis of the consequences rather than on the computation of the likelihood of the hazards.

When the risks are acceptable, structural reliability can be optimised solely based on economical constraints. However, when the risks are higher than the broadly acceptable risk limit, societal concerns come into play. For the risk to be tolerable the amount of resources used to reach certain desired benefits must guarantee that the level of risk to life and property is not unacceptable and moreover is reduced as reasonably as practicable, what is usually called the "ALARP" principle, see Section 5.4.5. Otherwise risk is classified as intolerable or unacceptable.

Table 5. Values for policy factor η_i (Vrijling et al., 1998).

η_i	Voluntariness	Benefit	Example
100	voluntary	direct benefit	mountaineering
10	voluntary	direct benefit	motor biking
1.0	neutral	direct benefit	car driving
0.1	involuntary	some benefit	factory
0.01	involuntary	no benefit	LPG-station

5.4.5 Risk Informed Decision-Making Framework

5.4.5.1 Basis

Decision-making is the process of committing resources available today to reach results in the future. Therefore, decision-making involves uncertainties; and risk management is a way of analysing and judging these uncertainties. In order to achieve a rational, efficient and transparent decision-making under uncertainty that maximises the benefits and minimises the losses, several decision support tools can be used. These include Cost-Benefit Analysis (CBA), Cost-Effectiveness Analysis (CEA), Utility and Prospect Theories and Life Quality Index (LQI). Examples of these models can be found in Bedford & Cooke (2001) and McDonald et al. (2005) and in Nathwani, Lind, & Pandey (1997). Additionally, as mentioned earlier, in the UK the ALARP principles need to be taken into account in risk management. It is important to emphasise again that these decision support tools are just that, tools, they do not force a decision. An aid to set up the decision-making criteria is given in UKOOA (1999), see Figure 8.

Figure 8. Decision-making aid, adapted from UKOOA (1999)

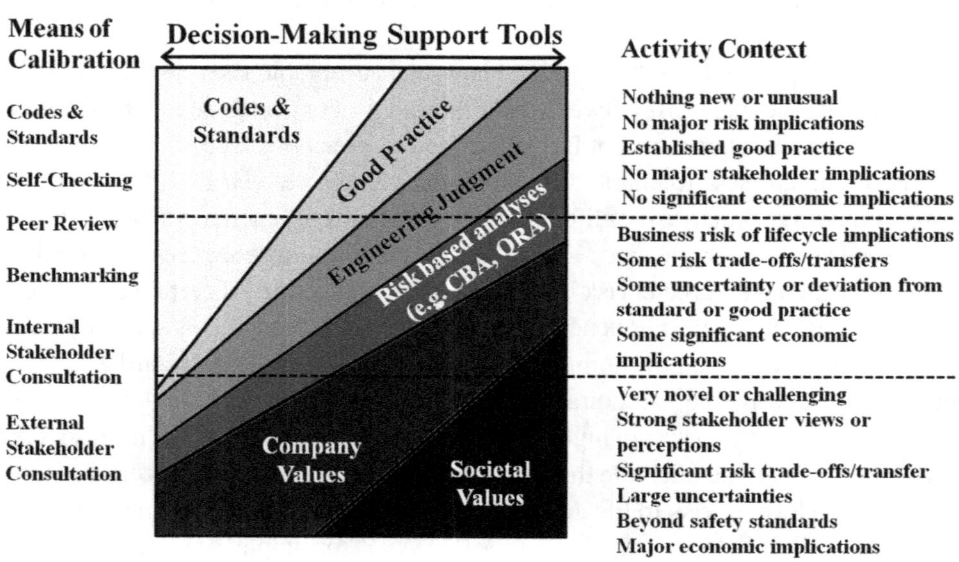

Decision-making should allocate resources efficiently within the context of a given activity, but also in the broader context of society priorities and preferences: how much benefit does it buy, and could the same resource, if directed elsewhere, result in better gain for society as a whole? This is a challenging question with a very difficult answer (if it is indeed possible to give a definitive answer). A way society found to efficiently distribute resources was the development of a general regulatory and normative environment that manages most of human related activities. In this context, acceptable minimum standards of operation are specified. It is, therefore, a requirement that activities, such as bridge and building construction, follow specific rules and codes of practice to ensure that the benefits for society are greater or equal to the total losses that society may bear.

However, more often than not, risk governance is complicated, see Berry, Lindsay, Salomon, & Veale (2006). Despite the abovementioned society risk neutrality, often public decision-makers assign priorities to activities with a potential for large loss of life in a single event and to activities with the potential for saving a large number of lives. However, it may be argued that this form of decision-making lacks fundamentals. The growing use of CBA, CEA and other methods may be seen as a step forward in this regard. Nevertheless, because decisions are dictated by multiple stakeholders' preferences defined in a specific context it is clear that decisions cannot be determined by solving a more or less complex equation. It is clear that decision-making must move toward a broader consultation with society in order to achieve optimal allocation of resources with ample consensus across society. The lack of active participation of society in the decision-making process is the main reason why some projects did not return the expected benefits and society confidence in the engineering community abilities may be degraded.

A commonly used approach for decision appraisal where different alternatives are compared is to define a baseline scenario. The basic approach is to consider a future scenario without incurring additional costs, the "business as usual" forecast also known as "do-nothing" scenario. If it is considered that the activity must comply with regulatory requirements at all times, then certain future costs such as maintenance costs or upgrade costs should be included: this scenario is often termed "do-minimum".

The potential benefits that come from risk informed decision-making are multiple and wide-ranging (Goodwin & Wright, 2010). For example, the analysis can provide guidance on what new information is worth to gather before a decision is made. For example: is it worth performing more advanced reliability analysis, further testing or measurements? If the cost of obtaining additional information is more than the expected benefits (monetary savings, safety, time or other tangible or intangible objectives) which arise from this additional information, then it may not be worth obtaining it. The process of determining whether it is worth obtaining new information is referred to as preposterior analysis or as *value of information* analysis (Goodwin & Wright, 2010; Konakli & Faber, 2014).

5.4.5.2 Cost-Benefit Analysis

According to Jones-Lee et al. (2008) cost-benefit analysis (CBA) is the welfare economic model that currently provides the normative basis for much of UK public policy. CBA *"seeks to quantify in monetary terms as many of the costs and benefits of a proposal as feasible, including items for which the market does not provide a satisfactory measure of economic value"* (HM Treasury, 2003). In a CBA, a proposal should only be implemented if all of its benefits are equal to or greater than all of its costs (Eales et al., 2003). Therefore, among a set of competing alternatives, the preferred option should be the alternative with the highest positive risk adjusted Net Present Value (NPV), including the effect of uncertainties.

In a CBA, it is necessary to assign a value to costs and benefits. This can be done on the basis of individuals' preferences, namely by using the concepts of willingness to pay (WTP) to obtain a benefit or willingness to accept (WTA) the loss of that benefit. Values of WTP and WTA are usually obtained by examining people's attitudes towards risk, either by observing people's revealed preferences or by testing their stated preferences, respectively. The latter approach is by and large the preferred in Europe (Spackman et al., 2011). As these values are generally given in market prices, costs must contain indirect taxes (Spackman & Holder, 2007).

As Bedford & Cooke (2001) note, there is a significant difference between WTP and WTA. The choice between these two concepts, depends on the weights (preferences) given to innovation and to risk aversion. If the former is preferred then WTP might be used, whereas if the latter is favoured WTA might be used instead.

It should be mentioned that CBA by using WTP, or WTA, as relevant measures of strength of preference, incorporates elements of Utility Theory and of Prospect Theory (see next Section). In fact, the

individual's maximum willingness to pay for a good or service is a clear reflection of what that good or service is worth to the individual relative to other potential objects of expenditure, taking account of the individual's ability to pay – which is, of course, ultimately a reflection of society's overall resource constraint. Obtaining data concerning individuals' maximum willingness to pay for safety is therefore a natural way in which to feed information concerning individual preferences – and, more particularly, strength of preference – into the allocative decision making process (Jones-Lee et al., 2008).

Critics of CBA refer that valuing monetarily human, environmental and cultural matters raises ethical issues. Additionally, the methods used in CBA to express losses or benefits to these matters are subject to high uncertainties. There are plenty of past examples where CBA has not been used properly, but it also true that there are many past examples where CBA suggestions have been refused based on environmental and cultural matters but for which today there is a general consensus that the opportunity costs of not having started the suggested activities are very large and are not compensated by the resulting benefits.

In CBA, the concepts of Pareto efficiency and/or Kaldor-Hicks efficiency could be used. An outcome of a given measure is considered Pareto efficient if at least one individual is made better off with no individual being made worse off. A less strict principle is the Kaldor-Hicks efficiency in which an outcome is more efficient if those that are made better off could in principle compensate those that are made worse off (Bellinger, 2007).

The analysis of the optimum level of risk can be formulated as an economic optimisation problem. The investment costs (construction costs, insurance costs, debt payment costs, etc.) are compared with the expected costs of damage, including, maintenance costs, repair/retrofit costs, reconstruction costs, penalties and compensation costs, user costs, etc. Other costs such as operation costs (including inspection costs and costs of decommissioning activities) should be considered. The function that relates consequences to costs is termed the cost function. An example of a possible cost function is presented in André (2014). Finally, the benefits from the activity may also be included in the economic optimisation equation:

$$\arg\ \max\ f(T) = \mathrm{B}(T) - {}^{\circ}(T) \tag{29}$$

where B and κ are the benefits and costs functions, respectively. $_f$ is the

Several constraints will need to be included such as those specified in the previous Section. Solving this optimisation problem is complex since it requires running a time variant problem or multiple time invariant problems. A simple comparison of the average values of the benefits and costs may not be sufficient and considerable intolerable risks may eventually end up being tolerated where otherwise an analysis taking full account of the probabilistic distributions could produce very different outcomes. Uncertainty propagation should also be considered.

A simplification is to consider the benefits as constant in all alternatives. The validity of this hypothesis must be checked before being used. When it holds acceptable, the decision problem is made easier (see Section 5.6 for details and an application example).

5.4.5.3 Other Methods

In contrast to CBA, in Cost-Effectiveness Analysis (CEA), the benefits do not have to be expressed in monetary terms. Limitations of CEA are given in House of Lords (2006) and Spackman et al. (2011).

When not all decision variables are expressed in the same units, such as in CEA, a useful tool to compare several efficient choices in a multi-criteria decision-making framework is to develop Pareto sets or to use Utility Theory. Utility (von Neumann & Morgenstern, 1944) is a measure of stakeholder satisfaction. A utility function should incorporate all relevant decision criteria, including the various constraints (rationality requirements), express the hierarchy of objectives and preference ordering, and finally the trade-offs between different criteria and uncertainties. For example: for an investor the utility function could have only two variables, the expected return of the portfolio and the associated risk. Thus, the decision-making problem is to maximise the expected return of the portfolio and minimise the corresponding risk. Utility Theory is appealing but it may be difficult to determine consistent utility functions and also utility theory has its own limitations as clearly evidenced by the Allais's paradox (Goodwin & Wright, 2010), see also Kahneman & Tversky (1979). As a final remark, if the utility function is linear, meaning a risk neutral attitude as for example society attitude towards fatalities, then Utility Theory will return the same results obtained by a cost-benefit analysis.

As an answer to the limitations of Utility Theory, the Cumulative Prospect Theory (CPT) was developed, see Tversky & Kahneman (1992). This theory suggests that people make decisions based on a reference context and value gains and losses from this reference point rather than absolute wealth considerations. Also, it is postulated that people are risk seeking towards high probability losses and towards low probability gains but risk averse towards high probability gains and towards low probability losses. Nevertheless, several questions have also been raised regarding the adequacy of CPT to decision-making problems, see Birnbaum (2008), Goda & Hong (2008) and Nwogugu (2006).

Classical decision support tools are the maximin, minimax criteria, however they have a number of limitations see Goodwin & Wright (2010) and Levy (2006). Other methods include the first and second-degree stochastic dominance to compare cdfs of different alternatives (Goodwin & Wright, 2010; Levy, 2006).

Another tool was presented by Schneider (2006b), where alternative risk measures are ordered by the so-called "rescue cost" (RC_M) which is given by the ratio between the safety costs of the measure and the variation of risk achieved determined in relation to a reference state. The smaller the value of RC_M the more efficient is the measure.

As a final example, Nathwani et al. (1997) developed the LQI method. They considered that the maximisation of healthful life for all is the proper basis for managing risk in the public interest, and this criterion is considered *"achieved when the net contribution to the total saving of life from the wealth produced is balanced against the loss of life from the risk of operation"*. The LQI method is expressed by:

$$LQI = g^{w} \cdot E^{(1-w)} \tag{30}$$

where *g* represents the personal income (GDP per capita), *e* is the national life expectancy and *w* is the national average working time. See Nathwani et al. (1997) for details.

Developments of the LQI to decision-making are given in Ditlevsen & Friis-Hansen (2005), Rackwitz (2004) and Vrijling et al. (2004). Limitations of LQI method, similar to CEA, are presented in House of Lords (2006), Jones-Lee et al. (2008), Spackman (2009) and Spackman et al. (2011). Namely, it is argued that *"the method is too simplistic to be a competitor to the methods now established in the UK and elsewhere for the valuation of fatality risks"* (Spackman, 2009). For example, the LQI method is based on historic economic data, but society expectations and preferences regarding the future can be quite different from past situations. Other doubts relate to philosophical issues of conditioning human preferences mainly to economic data which society cannot control completely and therefore may not represent true human preferences.

5.4.5.4 The ALARP Principle

The ALARP principle, see Figure 9, reject the simplistic and non-dynamic idea that there can only be two possibilities in the end of risk assessment: the risk is either acceptable or unacceptable. It enforces the consideration of an intermediate region in which risks could be tolerated in order to gain benefits (Rimington, McQuaid, & Trbojevic, 2003).

Figure 9. ALARP principle, adapted from ANCOLD (2003)

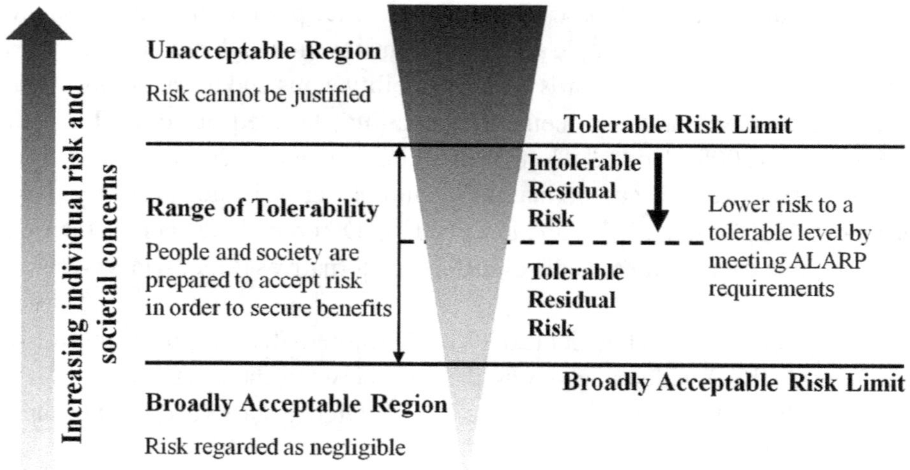

In general, the application of the ALARP principle involves three essential requirements. The first relates to the cost effectiveness of a solution and can be determined by a Cost-Benefit Analysis (CBA), comparing for example the Cost of Preventing a Fatality (CPF) with the accepted Value of Preventing a Fatality (VPF). The second corresponds to the assessment of the disproportionality of a solution and can be evaluated by comparing the CPF with the VPF multiplied by a proportion factor. The third requirement is related with the quality of the analysis and the competence of the analysts, the level of uncertainty attached to the options, the effectiveness of the risk treatment measures and also the time feasibility, i.e. the time available and the time necessary to implement alternative options.

As seen in a previous Section, these three requirements are necessary but not necessarily sufficient to make a risk informed decision, for instance to consider a risk level tolerable. During the decision-making process they will be important, but also the significance of the benefits *vs.* the significance of the adverse consequences, the consideration of state-of-the-art technology and of existing good practice, as well as sustainability, political, societal, equity, moral, ethical and other intangible matters will be considered. See HSE (2001) for an in-depth analysis.

For less risky activities, i.e. corresponding to a risk level near the acceptable risk level, one may not reduce the risk further if it is demonstrated that it is not cost effective. Nonetheless, the implementation of monitoring and control measures is enforced, especially when the nature, scale and the likelihood of the hazards are extremely uncertain. For activities with a higher risk level, just below the limit of tolerability, as the consequences are so severe, or so uncertain, the precautionary principle stipulates the need to reduce the risk level, even if it is by a very small amount. If not the risk should be classified has intolerable unless it can be demonstrated that:

- The costs of the additional risk reduction solutions are in gross disproportion with the amount of risk averted;
- The expected benefits to society are of such fundamental importance so to justify the increase of risk exposure and the associated expected adverse consequences;
- The risks are distributed equitably;
- Relevant good practice is followed, or
- It must be demonstrated that ALARP principle is not applicable or it is overly conservative.

In short, the higher the risk the more biased is the decision-making methodology towards health and safety, and more stringent measures to reduce the risks are required.

The value of CPF is determined by dividing the total final cost with the total number of fatalities prevented. The cost of a solution is disproportionate to its benefits if the following criterion is not met:

$$\frac{CPF}{VPF} \geq F \tag{31}$$

where F is a proportion factor which increases proportionally with the increase of the risk level, to provide a higher margin of safety, and is in the range of two (one) to ten (HSE, 2009b) for typical problems where ALARP principles must be used.

ANCOLD (2003) presents tentative monetary values for which the justification of the ALARP principle varies from strong to poor. It is the opinion of some economists, see Viscusi & Gayer (2002), that demanding safety improvements corresponding to CPF values higher than a certain amount will result in a net harm to society by drawing resources away from more cost-effective improvements to health and safety.

VPF is often misunderstood to mean that a value is being placed on a life. This is not the case. It is simply another way of saying what people are prepared to pay to secure a certain averaged risk reduction. For example, a VPF of £1 million corresponds to a reduction in risk of one in 100 000 being worth about £10 to an average individual (HSE, 2001).

Several methods have been presented in the literature to determine the value of VPF. The values recommended by the various experts are widely different: from less than 1 million € to over 10 million €, see Chilton et al. (1998), Cropper & Sussman (1990), Le Guen (2008), U.S. DOT (2009), Viscusi (1993) and Viscusi & Aldy (2003) and Table 6. In the UK, the Department for Transport (DfT) publishes the value of VPF. The latest value is equal to £1 585 510, at 2009 prices. Figure 10 shows the evolution of the value of VPF in the UK over the years. According to UK DfT (2011), future values of VPF can be obtained by multiplying the present values by a factor equal to:

$$1+\frac{\text{\% increase in nominal GDP per capita}}{100} \tag{32}$$

A question can also be raised concerning the treatment of injuries (major and minor). Documents from UK public institutions such as HSE, DETR, DfT, Highways Agency and the Railways Inspectorate indicate the following weights: VPF $\approx$ 10 major Injuries $\approx$ 200 minor Injuries, see Table 7. In comparison, Viscusi & Aldy (2003) showed that the majority of the existing research considered one injury in the range of \$20,000–\$70,000. Other interesting data is reported by Steven (2010). In this study approximately two million accidents in the USA were analysed. This research showed that for every major injury there can be as many as 10 causing minor injury, 30 causing property damage and 600 near misses that resulted in neither injury nor damage.

5.5 FRAMEWORK FOR FRAGILITY AND VULNERABILITY ANALYSIS

5.5.1 Structural Robustness Analysis

Structural robustness is a measure of the predisposition of a structural system to progressive and disproportionate collapse. It is an essential tool to design damage tolerant structures because citing Todinov (2007) "maximising the reliability of a system does not necessarily guarantee smaller losses from failures".

Using the newly proposed definition, see Section 5.2.7, structural robustness can be determined for any given deterministic combination of loads, or loads with a given conditional probability, which cause a failure in the structural system, irrespective of the system context and exposure. If the uncertainty of the resistance properties of the structure, and consequently of the system's response, is accounted for, then robustness is not defined by a single value but by a probabilistic distribution of possible values.

Using the new definition, it is clear that structural robustness and reliability are two different concepts, although related.

Table 6. Comparison of international values of VPF divided by GDP per head (Spackman et al., 2011)

Countries	VPF (PPP[1)] adjusted) (US$ at 2008 prices)	GDP[2)] per head (PPP[1)] adjusted) (US$ at 2008 prices)	VPF/GDP[2)] per head	(VPF/GDP[2)] per head)/ UK value
Austria	3.17	36,617	87	1.20
Belgium[3)]	6.31	33,997	186	2.61
Canada	3.95	39,950	99	1.39
Denmark	1.40	36,362	38	0.54
France	1.26	29,936	42	0.59
Germany	1.36	31,310	43	0.61
Netherlands	2.84	34,760	82	1.15
New Zealand	2.11	26,651	79	1.11
Norway	3.62	57,524	63	0.88
Singapore	1.26	33,767	37	0.52
Sweden	2.41	36,618	66	0.92
UK	2.59	36,362	71	1.00
USA	5.80	47,186	123	1.73

1) PPP indicates purchasing power parity.
2) GDP indicates gross domestic product.
3) The value for Belgium is not an official value

Figure 10. UK's official values of preventing a road fatality, major injury and minor injury: 1987-2009, £ at 2011 prices (Spackman et al., 2011)

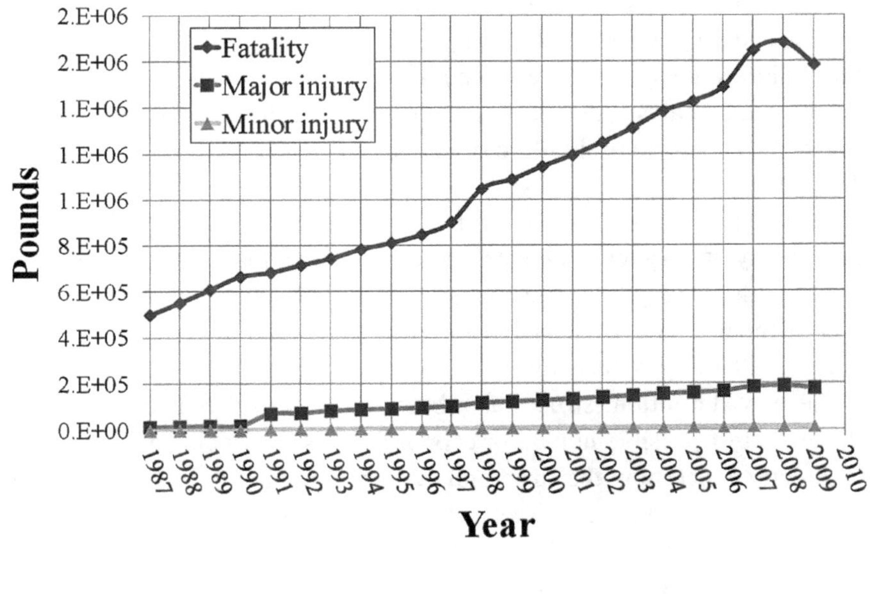

Table 7. Injury classification, weights and values (UK DfT, 2011)

Injury	Description	Weight	Average value, £ at 2009 prices
Fatality	Fatality within one year of the causal accident	1.0	1 585 510
Major injury	An injury as defined in schedule 1 of RIDDOR 1995, or where the injury resulted in hospital attendance for more than 24 hours	0.1	158 551
Reportable minor injury	For workforce, any injury resulting in more than 3 days off work, which is not a major injury. For passengers and members of the public, any injury that leads to a person being taken from the site of the accident to hospital for treatment, which is not a major injury	0.005	7 928
Non-reportable minor injury	Any other physical injury that is not a fatality, major or reportable minor injury	0.001	1 586
Class 1 shock/trauma injury	Shock/trauma injuries due to witnessing all fatal incidents, attempted suicides, passengers struck by trains, train accidents	0.005	7 928
Class 2 shock/trauma injury	Shock/trauma injuries due to physical and verbal assaults, witnessing non-fatal incidents of near misses, assaults, trespasser and workers struck by train, and all other miscellaneous events	0.001	1 586

Reliability of a system is associated with the probability of structural failure, which depends on not only the definition of structural failure (local or global level), the resistance properties of the system but also on the likelihood of occurrence of the hazard scenario.

In general, reliability analysis consists in a comparison between a value (p) of an applied action, P, defined within a particular hazard scenario, H, with a given probabilistic distribution, and the ultimate resistance of the structural system expressed as the value (p_{max}) of the action P when structure failure state is attained. The probability of structural failure can be determined based on the differences between corresponding p values and p_{max} values. What happens between the first structural failure state and the complete structural failure state, and also between p and p_{max}, is not assessed.

In a structural robustness analysis, the focus is not in assessing the probability of structural failure but in what happens after the first structural failure state, measuring the predisposition for failure (damage) propagation within the system.

However related these two concepts are different. A system can have a very high reliability but if it is governed by the reliability of very few elements, there is always a load scenario for which the system's structural robustness might be very low. The opposite is also true, a system can have a very large structural robustness but if the ultimate resistance of the system is small, there might be a hazard scenario for which the system's reliability is very low.

According to the new definition, structural robustness is considered to be a structural attribute (property), not dependent of possible human or economical risks associated with a failure or collapse (other than those directly concerning the structural system, e.g. damage to structural elements). Structural robustness is a measure of the total structural damage and not of total consequences, contrary to what is suggested by Baker, Schubert, & Faber (2008), since the latter concept has a broader scope as it encompasses structural consequences (structural damage) but also life safety, social and economic consequences, for example.

As robustness is understood to be an attribute (property) of a structural system, it cannot be controlled by external measures such as: the reduction of the structure exposure to hazard events or the introduction of external elements to minimise the effects of those hazard scenarios on the structure. However, these types of measures would increase the reliability of the structure and decrease the risk of disproportionate collapse.

In conformity with the new definition, it is possible to have a system with a very high resistance but with a very low structural robustness, and vice-versa. Increasing the resistance of all the elements of a structure, for instance by choosing materials with a higher tensile strength, although a sufficient condition to increase the system's resistance (and reliability for the same hazard exposure), might not be a sufficient condition for increasing the system's structural robustness, see André et al. (2015).

Additionally, increasing material strength, or member resistance, is not always the most cost-effective approach to increase structural robustness. In some structures, it may even be counterproductive. For instance, structural robustness may decrease by increasing the resistance of joints connecting different parts of a structure since the collapse might propagate to other initially undamaged (or even unloaded) areas. Finally, resistance is not a suitable property to measure structural robustness since it must always have to be expressed in terms of the local behaviour of the structure, which might differ greatly within the structure and between different structures.

The main advantages of the proposed new definition of structural robustness in relation to the existing definitions are:

- Structural robustness, structural resistance, reliability and risk (or vulnerability) can now be considered to be four different concepts. The existing structural robustness definitions mixed these four concepts which made the analysis, interpretation and evaluation of the former variables difficult tasks. Furthermore, by coupling in the same definition of structural robustness up to four different concepts the benefits of determining robustness was not clear. The present definition makes structural robustness a property than can be measured independently of the system's resistance, reliability and risk. Structural robustness can for the first time be considered an independent requirement for the structural performance of civil engineering infrastructures. Together with the structural resistance, reliability and vulnerability they become powerful tools that can, and should, be used in the risk management of civil engineering infrastructures.
- The second advantage of the new definition, is that for the first time, progressive and disproportionate collapse analysis is clearly defined as a requirement not only for unforeseen and accidental situations affecting localised areas of a given structure, but also for normal service conditions covering for instance design cases where the permanent load is the dominant action.

Structural robustness is a function of resistance variables, R, of the structural system (Knoll & Vogel, 2009) and also a function of the hazard scenario, H: loads, imposed displacements, etc.

5.5.1.1 Traditional Methods for the Analysis of Structural Robustness

Robustness has been present directly or indirectly in several structural codes throughout the last thirty years. However, to date there is not one document that specifies a general-purpose design method for determining robustness in a consistent and effective manner.

Starting from the first structural codes adopting limit state design theory, a structural insensitivity requirement was incorporated to avoid progressive collapse scenarios, i.e. the structure would not collapse if subjected to a limited damage (Comité Euro-International du Béton, 1993). Thus robustness was treated qualitatively and indirectly by specifying standard prescriptive detailing rules for members; linked with an undesirable failure mode. No rules for design and verification were specified.

More recently, BS EN 1990 (BSI, 2005) establishes "*robustness (structural integrity)*" as a way to achieve required levels of reliability relating to structural resistance and serviceability. BS EN 1991-1-7 (BSI, 2014) defines robustness as

the ability of a structure to withstand events like fire, explosions, impact or the consequences of human error, without being damaged to an extent disproportionate to the original cause.

Depending on the consequence class of the structure, see BS EN 1990 (BSI, 2005), different types of design rules must be followed. For consequence class CC3 (the most stringent class) a complete risk assessment is required. Typical structures that fall in this class are grandstands, public buildings where consequences of failure are very high. Specific guidance for buildings subject to accidental actions is given in Annex A of BS EN 1991-1-7 (BSI, 2014). An additional review of robustness related rules present in the structural Eurocodes is given by Narasimhan & Faber (2009).

In the USA, the following national codes for the design of buildings were prepared, defining requirements, rules and verification procedures to achieve collapse-resistant buildings in the event of abnormal loading: the NIST "*Best Practices for Reducing the Potential for Progressive Collapse in Buildings*" report (Ellingwood et al., 2007) and the US Defence Department United Facilities Criteria (USDOD, 2010).

The first document defines progressive collapse as

the spread of local damage, from an initiating event, from element to element resulting, eventually, in the collapse of an entire structure or a disproportionately large part of it; also known as disproportionate collapse.

Therefore, structural robustness analysis is yet not fully implemented in the existing structural codes, except for the scenario of an accidental action.

5.5.1.2 Advanced Methods for Analysis of Structural Robustness

There are however, methods available to analyse structural robustness, some of them developed quite recently. All of these methods are based on structural robustness definitions different than the one introduced in the present paper, and all of them suggest different approaches to measure structural robustness.

As observed by Starossek & Haberland (2008) existing methodologies can be based on structural behaviour or be based on structural attributes or assume an initial local damage or be based on the identification of a collapse sequence. Finally, it is also possible to distinguish between deterministic approaches and probabilistic approaches.

A common trend in structural robustness evaluation is to define a structural robustness measure, often in the form of a structural robustness index. The first question that needs an answering is why is it useful to measure quantitatively the structural robustness of a system? One could simply compare resistances or load *vs.* displacement curves associated with different failure scenarios of the same structure. The answer to this question is found in the definition of structural robustness presented previously. To assess structural robustness, i.e. its global stability reserve, complex nonlinear analyses must be performed. It is thus a requisite that the maximum information should be extracted in order to obtain return of knowledge for the additional computational and analysis effort. Therefore, the results obtained should be informative and easy to use, e.g. should allow a straightforward comparison between different structures. All of the latter favour the development and use of a simple measure of structural robustness.

Smith (2006) presented an analytical procedure coupled with finite element simulations to analyse the progressive collapse of building structures based on the parallelism of progressive collapse with the theory of unstable fast fracture in fracture mechanics. The idea behind the method is:

if the energy released by loss of a damaged member is greater than the energy absorbed by the destroyed member and other damaged members, then progressive collapse will occur (Smith, 2006).

Smith determined the energy required to destroy sufficient structural members to develop an unstable mechanism (which he named damage energy) and using a minimisation process, consisting basically in continuously deleting damaged elements from the finite element mesh, coupled with a sorting procedure, he identified the sequence of damage events that required the least amount of damage energy. Smith then used this minimum damage energy as a measure of the structural system robustness. However, this measure has some limitations, see André et al. (2015).

In terms of probabilistic-based approaches, two different methodologies exist: one focusing in the probability of failure and the other on the risk of the structure.

There have also been developed measures focusing in the probability of failure, notably the ones proposed by Frangopol & Curley (1987) and Fu & Frangopol (1990):

$$RI = \frac{P_f\left(\text{damaged}\right) - P_f\left(\text{intact}\right)}{P_f\left(\text{intact}\right)} \tag{33}$$

$$\beta_R = \frac{\beta_{\text{intact}}}{\beta_{\text{intact}} - \beta_{\text{damaged}}} \tag{34}$$

where P_f represents the probability of failure and β represents the reliability index.

Finally, in 2008, a new structural robustness index was presented based on a complete risk analysis where the risks (or consequences) are divided into direct and indirect risks (R_{Dir} and R_{Ind}, respectively) and the measure is given by the ratio between the direct risks with the total risk (sum of the direct and indirect risks) (Baker et al., 2008):

$$I_{Rob} = \frac{R_{Dir}}{R_{Dir} + R_{Ind}} \tag{35}$$

It may not be immediately transparent to the practitioner of the usefulness of this index in the context of risk management. In risk management, it is in general given more importance to total risks and not to the relative balance between direct and indirect risks. A possibility may exist of overlapping rules, analyses and interpretations for risk and robustness.

In addition, the use of the index may weaken the value of robustness as a "*structural concept*" since it makes robustness dependent of variables external to the structure. In theory, less emphasis could be given to properly design the structure and more to limit the consequences of failure. For example, it is possible to reach a high value of the robustness index by only having a small value of indirect risks when compared with the value of the direct risks, independent on the absolute value of the latter. In addition, it is possible to achieve a higher value of robustness if it is decided to increase the direct damage while keeping the same indirect damage, which seems to favour a less efficient structural design.

5.5.2 A New Measure of Structural Robustness

5.5.2.1 Formulation

From the analysis of the existing structural robustness measures, it can be concluded that the reviewed structural robustness evaluation strategies are not consistent with the new structural robustness definition presented in this book. In some indices, mainly in the newly developed risk-based index, see Eq. 35, it is considered that robustness of a structural system depends not only on the structural characteristics of the structure but also on the variation of the loads and the exposure of the structure (probability of occurrence of the loads). In the risk-based index, robustness is also linked with the consequences (life safety, economic, social, etc.) of the collapse. Therefore, an alternative structural robustness measure is proposed.

The basis for the development of the new structural robustness index is the analysis of the structural behaviour in terms of energy balance. There are plenty of advantages of energy-based measures over resistance-based or reliability-based robustness measures. Energy-based measures concern the global behaviour of the structure, which removes the need for subjective selection of the parameter which structural robustness depends on with all the possible loss of objectivity, expressiveness and generality that comes with it.

The general expression of the structural robustness index, $\mathrm{I_R}$, is given by (André et al., 2015):

$$\mathrm{I_R}\left(\mathrm{A}_L \mid H\right) = \frac{D_{uc} - D_{\text{1st failure}}}{D_c - D_{\text{1st failure}}} \text{ with } \begin{cases} 0 \leq \mathrm{I_R} \leq 1 \\ D_c - D_{\text{1st failure}} = 0 \Rightarrow \mathrm{I_R} = 1 \end{cases} \tag{36}$$

where:

A_L represents the leading action;

$H = \{h_1, h_2, \dots, h, \dots, h_n\}$ is a set of hazard scenarios. For example a set of different actions with determined values applied in a given sequence;

$D_{\text{1st failure}}$ represents the damage energy of the structure when the "first failure" state takes place for the hazard scenario considered;

D_{uc} represents the damage energy corresponding to the state where collapse is unavoidable, the "unavoidable collapse" state, for the hazard scenario considered;

D_c represents the damage energy corresponding to the collapse state for the hazard scenario considered.

The selected criteria for monitoring the damage and the collapse of a structure is the system's damage energy (D) evolution, because it gives a good estimate of the capability of the system to redistribute forces by alternative load paths and resistance mechanisms.

The damage energy is given by the sum of the plastic strain energy (non-decreasing function) with the internal energy released in each failure (stepped function). Therefore, the damage energy is a non-decreasing function. Failure is any state where there is a release of internal energy: it can be the formation of a crack, failure of a joint, failure of a cross-section, for example.

A value of the structural robustness index equal to 1.0 means that the structure is completely optimised in terms of structural robustness, for the hazard scenario considered. In the contrary, a value of the structural robustness index equal to zero may indicate that the structure completely lacks optimisation in terms of structural robustness, for the hazard scenario considered.

It is assumed that the damage energy is zero if there are no plastic strains within the structure and if no failures have occurred. Therefore, a system whose collapse is solely triggered by elastic instabilities lacking post-buckling resistance or by loss of overall stability of foundations, without prior failures, has a robustness index equal to zero, because no system damage energy is needed to attain the "unavoidable collapse" state. On the other hand, a system where all elements are brittle, i.e. which cannot deform plastically, and in which the collapse trigger mechanism involves the failure of all elements, has a robustness index equal to 1.0, because the entire damage energy available in the system has been used to attain the "unavoidable collapse" state.

In order to calculate the structural robustness index a three-step procedure must be followed (André et al., 2015):

- **First Step:** Define the *nominal* loading conditions (hazard scenario), i.e. the sequence of action (loads) application is rationally chosen and the initial values of the various actions, material properties, system imperfections, etc. are generated, corresponding to values obtained from the probability density functions (actions are modelled with uniform probability density distributions and resistance variables can also be modelled with uniform probability density distributions but preferably with more informative distributions).
- **Second Step:** While holding everything constant ("*ceteris paribus*"), a leading action that can cause the structure to collapse (if it has not already occurred during the first step) is selected and increased until the "unavoidable collapse" state is attained. Alternatively, several actions can be defined as leading actions and increased simultaneously if it is considered appropriate (if actions are correlated for example). The aim should be to obtain a realistic safety assessment of the structure and therefore of the most likely damage propagation within the structure. The value of D_{uc} is determined.
- **Third Step:** The structural robustness index is determined from Eq. 36 based on the adopted limit state which defines the first failure state (the value of $D_{1\text{st failure}}$ is determined).

The value of D_c can be determined in any of the three steps, depending on the method used.

This index is also flexible since the inputs can change, for example: the "first failure" state can be replaced by another criterion, possibly related to a particular element failure or simply the first material yield strain, and the "unavoidable collapse" and "collapse" states can also be changed to represent a maximum limit of acceptable damage, D_{max}, for instance, with $D_{max} \leq D_{uc} \leq D_c$.

The value of D_{uc} can be estimated from the theory formulated by Dusenberry & Hamburger (2006).

Applying this theory to a framed structural system, see Figure 11 for example, such as a building or a bridge falsework, the collapse of a lower level of elements due to the failure of an upper level of elements can only be arrested if and if only (Bažant & Verdure, 2007):

$$W\left(t\right) < E_S\left(t\right) \tag{37}$$

where:

$W(t)$ represents the value of the work done at time t by external actions on the lower level elements, including the potential energy associated with the kinetic energy of the moving mass of the upper level elements.

$E_S(t)$ represents the value of the internal energy of the lower level non-failed elements at time t.

For such a system, the progressive collapse criterion given by Eq. 38 can be expressed by (André et al., 2015):

$$W\left(t\right) < E_{S,\max} \tag{38}$$

where $E_{S,max}$, represents the maximum internal energy dissipated by all the non-failed elements of the lower level.

In each level of a typical framed structural system, the energy can potentially be dissipated by the floor elements (slabs and beams), the columns and the joints between these structural elements. The contribution of non-structural elements can also be considered. This characteristic structural layout leads

Figure 11. Example of a framed structural system (André et al., 2015)

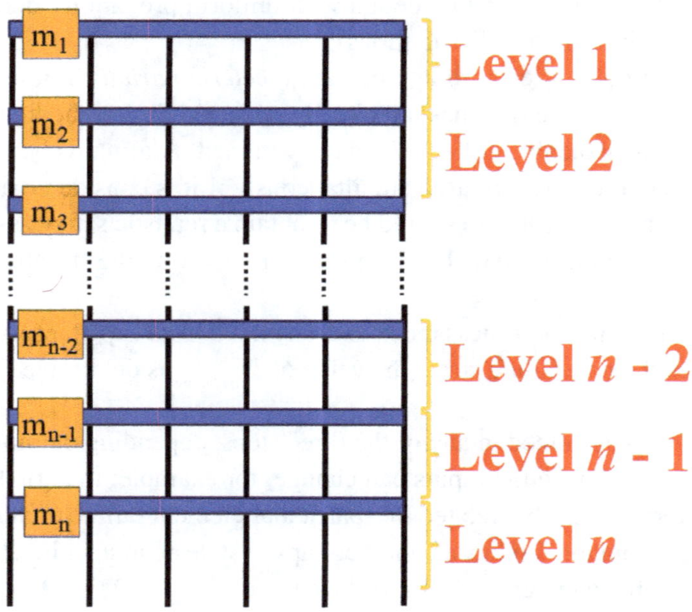

to two different mechanisms that control the value of $E_{S,max}$. One concerns the column elements of the level – their collapse leads to the level collapse. The other relates to the elements of the level floor – if all beam-to-column joints fail it is very likely that the entire floor collapses, for example.

In general, it is necessary to consider in the analysis all of the above-mentioned mechanisms. With respect to the energy dissipation by the column elements, for a steel frame and assuming a three plastic hinge dissipative mechanism for the columns of each level, $E_{S,max}$ of level j can be estimated by (Bažant, Le, Greening, & Benson, 2008):

$$E^{j}_{S,\max} = \sum_{i=1}^{N^{j}_{columns}} \left[2 \cdot \int_{0}^{\theta_{e,pl}} M^{j}_{e,i}\left(\theta\right) d\theta + \int_{0}^{\theta_{m,pl}} M^{j}_{m,i}\left(\theta\right) d\theta \right] \tag{39}$$

where:

θ represent the rotation at the column i extremities and middle of column i, respectively;

$\theta_{e,pl}$ and $\theta_{m,pl}$ represent the maximum rotation capacity at the column i extremities and middle of column i, respectively, with $\theta_m = 2\,\theta_e$;

$M_{e,i}$ and $M_{m,i}$ represent the bending moments at the extremities and middle of column i;

$N^{j}_{columns}$ represents the number of columns of level j that have not failed.

It is assumed that failures within each element will only occur due to excessive bending rotations. However, failure modes of joints can be of any kind and the failure of each joint has an influence on the value of $E_{S,\max}$ of level j. This plastic hinge model neglects any contribution of the axial and shear deformation energy capacity to the internal energy and therefore may underestimate the actual dissipated energy (Korol & Sivakumaran, 2014).

Methods used to improve the accuracy of the $E_{S,max}$ value are provided in André et al. (2015). The maximum rotations at the plastic hinges, $\theta_{m,pl}$ and $\theta_{e,pl}$ in Eq., can be determined by the following procedure:

Under the Bernoulli hypothesis, the extensions, e, and deformations, ε, at any cross-section of a linear element are given by (André et al., 2015):

$$\text{Extension: } e\left(z\right) = e_N - \theta \cdot z \quad , \quad e_N = e\left(z = 0\right) \tag{40}$$

$$\text{Deformation: } \varepsilon\left(z\right) = \varepsilon_N - \chi \cdot z \quad , \quad \varepsilon_N = \varepsilon\left(z = 0\right) \tag{41}$$

where:

z represents the coordinate along an axis perpendicular to the longitudinal axis of the element with origin at the cross-section geometric centre;

ε_N and e_N represent the deformation and extension due to the axial force, respectively;

χ and θ represent the curvature and rotation at a given cross-section, respectively.

Functions of ε over the plastic hinge length are provided in André et al. (2015). Knowing that the extensions are a function of the deformations:

$$e(z,x) = \int_l \varepsilon(z,x)\,dx \tag{42}$$

Introducing Eqs. 41-42 in Eq. 40 gives:

$$\theta_{m,pl} = \left[\frac{(\varepsilon_u + \varepsilon_y)}{2} \cdot l_{pl} - e_N\right] \Big/ \left(\frac{d_{ext}}{2}\right) \tag{43}$$

Having determined an estimate of $E_{S,max}$, D_{uc} can be estimated by the following procedure:

Energy demand at level h, $E_D^h \equiv W^h$:

$$E_D^h(t) = K_D^h(t) + E_S^h(t) \tag{44}$$

where K_D^h represents the kinetic energy at level h transferred by the levels above level h .

Considering levels as rigid blocks, K_D^j is determined by:

$$\text{If: } E_D^{j-1}(t) < E_{S,\max}^{j-1} \Rightarrow K_D^j(t) = \sum_{i=1}^{j-1} K^i(t) \;\; \text{Else: } K_D^j(t) \equiv \sum_{i=1}^{j-1} m_i \cdot g \cdot h^{ij} \tag{45}$$

where:

m_i represents the moving mass of level i;
g represents the gravitational acceleration;
h^{ij} represents the vertical distance between levels i and j.

Finally, an estimate of D_{uc} is obtained by:

$$E_D^{n_{level}}\left(t^{n_{level}}\right) = E_{S,\max}^{n_{level}} \Rightarrow \begin{cases} t_{uc} = t^{n_{level}} \\ D_{uc} = D(t_{uc}) \end{cases}, \;\; n_{level} \text{ represents the bottom level index} \tag{46}$$

In order to further improve the accuracy of the procedure, to capture local effects due to gravity action, the values of the leading action should only be increased when a static equilibrium between the applied loads and the internal forces has been reached. For time t between instants t_i and $t_i + \Delta t$ when the same action value is applied, only the damage energy dissipated at the first time instant, t_i, should be considered for the value of the damage energy terms, D, that appears in Eq. 36. As a result, the damage energy that is dissipated between $t_i < t \leq t_i + \Delta t$ is not accounted for since it corresponds to unstable unequilibrated behaviour of the structural system.

A suitable failure search and detection algorithm should also be developed and included inside the numerical protocol in order to update the values of E_D and E_S of each level, by for example considering the effect of the failed beam-to-column connections in the columns' energy demand and in the column's length, but also that the failed column elements (attained either by joint failure or plastic failure) do not enter in the calculation of E_S.

If the analysis additionally considers the dissipative energy capacity of the floor elements, a similar model as the one described above can be used for the beam elements, whereas for the slab a specific model has to be developed for each type of structural solution considered.

In all cases, it is assumed that the upper level of elements will fall onto a lower level of elements. If not, the upper level of elements will free fall until reaching the ground. In this case, the kinetic energy of the upper level of elements should not be considered when evaluating the resistance capacity of the lower level of elements. In addition, care should be paid in choosing the leading action, since the entire mass where the selected leading action is applied may be free falling and thus might not cause the complete collapse of the system. In addition, the damage energy dissipated by a structural component (element or joint) after its total collapse (during free fall or impact with other elements of the system) should not be considered for the damage energy terms, *D*, that appears in Eq. 36. However, if relevant, this energy may be used to reduce the value of *W*, but only if properly considered in Eqs. 37, 38 and 44. An example of such a case is presented in Bažant et al. (2008).

The variable D_c included in the denominator of Eq. 36 is also difficult to determine. In André et al. (2015) methods to determine D_c are presented.

Structural robustness calculated by Eq. 36 may be considered to have a potential limitation. Recall that structural robustness was defined in this work as a measure of the predisposition for structural damage propagation within the system and structural damage is calculated using damage energy and not the number of damaged elements. Eq. 36 expresses the structural consequences to the system subject to a given hazard scenario as a function of the total damage energy available in the system, irrespective of the total number of elements in the system and how the available damage energy is distributed between the elements of the system. Thus, the collapse disproportionality is assessed by evaluating the relationship between the structural damage needed to trigger global collapse and the maximum structural damage possible measured in terms of damage energy and not in terms of the number of damaged elements. Applying Eq. 36 directly to a system where the entire available damage energy of the system is almost concentrated on a single massive element, a value of I_R close to 1.0 might be obtained if the stability of this controlling element does not depend significantly on the resistance of the remaining system elements, since in this case $D_c \approx D_{uc}$, regardless of whether the failure of the remaining elements might have or might not have occurred during damage propagation given that their contribution to the available damage energy of the system is irrelevant. As a result, by Eq. 36 such a system may be robust even if only one massive element triggers the collapse of the entire system.

In systems such as the above, characterised by having structural components (e.g. entire parts, elements or simply critical joints connecting different parts) with very dissimilar available damage energies, i.e. some with very high values and the remaining with very low values, it may be more sensible to assess structural robustness by a segmented calculation using the same analysis model. Accordingly, robustness of the very weak structural components should be calculated separately from the robustness of the very strong structural components, and vice-versa.

Structural robustness as defined in this work is an attribute (property) of a structural system, independent on the consequences of collapse beyond those directly concerned with the structural system. Therefore, costs of consequences are only given by costs of materials and costs of repairs and in the assumption that the latter are proportional to the available damage energy of each element, Eq. 36 is a correct representation of the system's structural robustness. Of course, that other types of consequences need to be accounted for when assessing the vulnerability of the system, see Section 5.5.4.

It is very important to be able to establish such a clear distinction between important concepts as system structural robustness and system vulnerability, and this is one of the merits of the structural robustness definition and structural robustness calculation method (Eq. 36) presented in this work. For example, a structural system with a high structural robustness value and concurrently a high vulnerability value (for a hazard scenario not as severe as the one associated with the "unavoidable collapse" state), means that the damage energy capacity is not well distributed between critical elements (i.e. those whose failure lead to high follow-up consequences), or in another words that the structural design is not effective although it is efficient. Additionally, a structural system with a low structural robustness value and concurrently a high vulnerability value for a hazard scenario not as severe as the one associated with the "unavoidable collapse" state, means that the structural design is both not effective and not efficient.

However, Eq. 36 could be modified to account for a somewhat more direct representation of the vulnerability of the system, by including the total number of elements in the system and the relationship between the damage energy imposed on each individual element and the available damage energy of the corresponding element. These alternative variables account for the propagation of element failures in the system relative to the overall structural configuration of the system. Thus, they can be linked to consequences beyond the ones of the structural system, such as life safety, social and economic consequences. The alternative expression to calculate structural robustness is presented by Eq. 47.

$$\mathrm{I_R}\left(\mathrm{A}_L \mid H\right) = \frac{1}{n_{total}} \cdot \sum_{i=1}^{n_{total}} \mathrm{I_R^i} = \frac{1}{n_{total}} \cdot \sum_{i=1}^{n_{total}} \left(\frac{D_{uc}^i - D_{\text{1st failure}}^i}{D_c^i - D_{\text{1st failure}}^i} \right) \text{ with } \begin{cases} 0 \leq \mathrm{I_R}, \mathrm{I_R^i} \leq 1 \\ D_c^i - D_{\text{1st failure}}^i = 0 \Rightarrow \mathrm{I_R^i} = 1 \end{cases} \quad (47)$$

where:

i is the element i of the system;
n_{total} represents the total number of elements present in the system.

By Eq. 47 a system where the entire available damage energy of the system is almost concentrated on a single element to be robust ($\mathrm{I_R} \approx 1$) requires that the great majority of the elements of the system contribute to attain the "unavoidable collapse" state, and no longer only the massive single element as Eq. 36 necessitates. If $D^1_c \approx D^2_c \approx .... \approx D^i_c \approx (D_c / n_{total})$ then structural robustness calculated by Eq. 36 and by Eq. 47 are very similar values.

However, Eq. 47 and indeed every other expression that includes additional terms to the calculation of structural robustness other than those related to damage energy, has limitations. Whereas the structural robustness values obtained by Eq. 36 have always a single interpretation linked with the relationship between the damage energy needed to attain the collapse and the total available damage energy, the values obtained by Eq. 47 can in extreme cases have a double interpretation. For example, in systems with one massive element and a large number of weak elements, a robustness value equal to 1.0 can result from failure modes involving only the weak elements or from failure modes involving also the massive controlling element. The former result is not appropriate since the failure of the most important element of the system, that controls the global collapse of the system, might not be robust. This occurs because the individual contribution of each element to the structural robustness in Eq. 47 is not as relevant as in the case of Eq. 36. Of course, the structural segmentation method formulated earlier can also be used to obtain the correct value of structural robustness through Eq. 47. This solution is mandatory for the extreme cases such as the one abovementioned when using Eq. 47.

In addition, the calculation of structural robustness by Eq. 47 is more laborious than by Eq. 36. Therefore, Eq. 36 is preferred over Eq. 47 to calculate structural robustness because it is directly in agreement with the robustness definition presented in this work. It offers results that are more transparent to interpret and less elaborate to calculate. However, Eq. 47 applied with caution is also considered to be a valid means to determine structural robustness.

From the results of advanced finite element analysis programs it is also possible to follow the damage (failure) path throughout the system as the loading increases, for instance by using flag variables in the numerical model which are activated if a given damage criterion is met. This information can be used to modify the value of the robustness index by giving more emphasis to the existence of damage in selected critical areas or critical elements of the system, see André et al. (2015). However, this type of differentiation should preferably be done only in the vulnerability analysis where the direct and indirect costs associated with structural damage are calculated to avoid introducing risk related parameters or subjectivity into the determination of structural robustness.

The robustness index presented does not exhibit the same limitations as the existing indices. Thus, it is thought that the new structural robustness index fulfils all conditions listed by Starossek & Haberland (2008) and can be used as a measure of the robustness of a structural system.

5.5.2.2 Application Example

Bridge falsework systems, see Figure 12, are typically low robust structures since they are an assemblage of similar and slender steel linear elements prone to instability phenomena, joined by weak loose connections, where the critical design load occurs often at its maximum value, uniformly distributed over the entire, or a significant part, of the structure. Additionally, since the elements of these systems are linear elements they do not possess alternate resistance models like the slabs of bridge decks, and thus complete failures of elements are more easily reached in bridge falsework systems than complete failures of bridge elements.

Additionally, factors such as lack of competence in design, absence of rigorous quality control supervision, reuse of damaged elements, etc. have large impact on the structural behaviour of bridge falsework systems, which contribute to the existence of high levels of uncertainty associated with these temporary structures, see André, Beale, & Baptista (2012, 2013b) and Beale (2014).

Therefore, it is extremely important to evaluate the robustness of these temporary structures since their margin of safety given an irreversible event (e.g. a failure) is usually much lower than the one achieved in permanent structures, the failure of one element may lead to the progressive and disproportionate collapse of the entire, or the majority, of the structure, and their exposure to critical hazard scenarios is also much larger than the one of permanent structures.

Figure 12. Example of bridge falsework Cuplok® systems

As demonstrative examples, the Model A2 tested in the University of Sydney, see Chandrangsu & Rasmussen (2011), will be considered. The cross-section geometrical characteristics as well as the material properties of the various elements which are part of the falsework system are given in Chandrangsu & Rasmussen (2011). The properties of the finite element model used are detailed in André (2014) and André, Beale, & Baptista (2014a). The formwork was explicitly modelled, with an equivalent thickness equal to 100 mm, and the joint characteristics considered were taken as the average values of the joint tests results reported in André, Beale, & Baptista (2013a). The value of the top and bottom jacks' extension lengths was equal to 600 mm. Figure 13 illustrates the numerical representation of Model A2.

The only action considered in this example, besides the materials' self-weight, was the weight of the concrete slab. This action was selected as the leading action, applied uniformly to the formwork elements and increased monotonically over a step period of 100 s (using an implicit dynamic quasi-static analysis).

Figure 13. Overview of Model A2

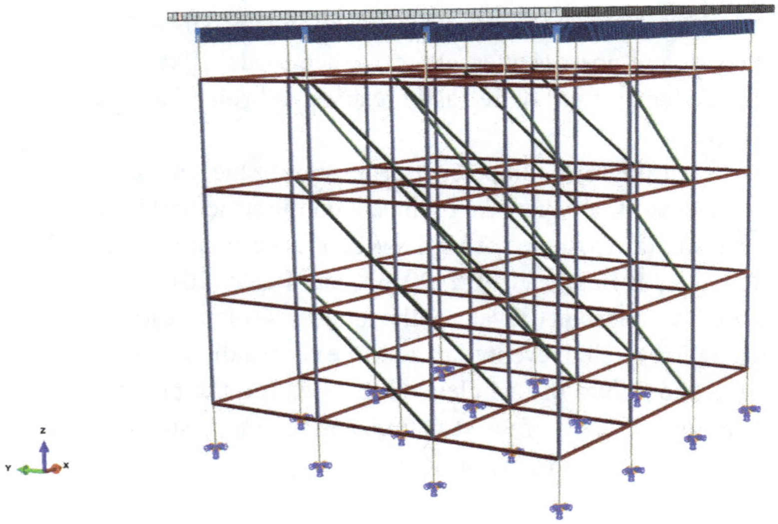

The procedure developed to detect the "unavoidable collapse" state is presented in André et al. (2015). The structural robustness index of Model A2 is calculated by Eq.. From the analysis results, the value of D_{uc} is equal to 9.15×10^6 mJ. The value of D_c is equal to 1.12×10^8 mJ. The latter value was determined using Eq. 48. which was considered to represent a reasonable estimate of the true value (André et al., 2015).

$$D_c = \sum_{i=1}^{N_{levels}} E_{S,\max}^{i} \tag{48}$$

Consequently, a robustness index equal to 0.082 is obtained (considering $D_{1st\ failure} = 0$). The maximum resistance to concrete pressures applied to the formwork of Model A2 is equal to 0.03909 N/mm^2.

It can be observed that Model A2 has a small robustness index value, meaning that the systems have a small structural robustness against uniformly applied actions to the formwork. This is justified because the critical elements to the collapse resistance of this model were the forkhead plates and top jacks. The jacks have an energy deformation capacity lower than the standard elements since their cross-section dimensions are smaller. Therefore, damage was concentrated in few elements that have a lower energy deformation capacity than the rest of the elements of the structure, see Figure 14.

Structural robustness is a function of the hazard scenario, *H*, in particular of the actions values, A, which have an impact on the initial damage mechanism (e.g., an explosion of magnitude *a*) and damage propagation, and of the resistance variables of the structural system, R. Resistance variables are random variables and structural robustness directly depends on the resistance variables of the system. Therefore, structural robustness is a random variable, function of resistance variables and action variables.

Determining analytical expressions for the functions relating actions with resistance is quite difficult. Therefore, simulation schemes, like Monte Carlo or other, are a viable alternative solution to determine the structural robustness probability density function (pdf), see Figure 15 for example.

Figure 14. Deformed shape and plastic strain distribution at "unavoidable collapse" state (High plastic strain are shown in dark gray)

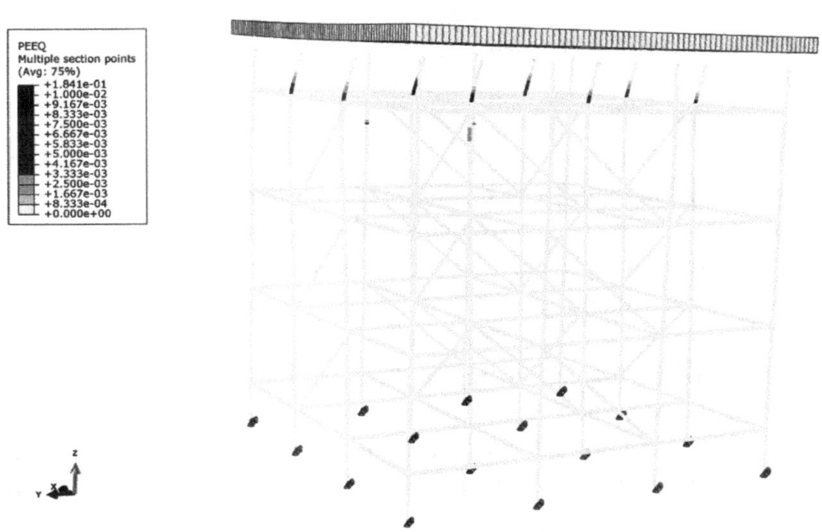

Figure 15. Illustrative example of robustness index probability density function (pdf) (André et al., 2015).

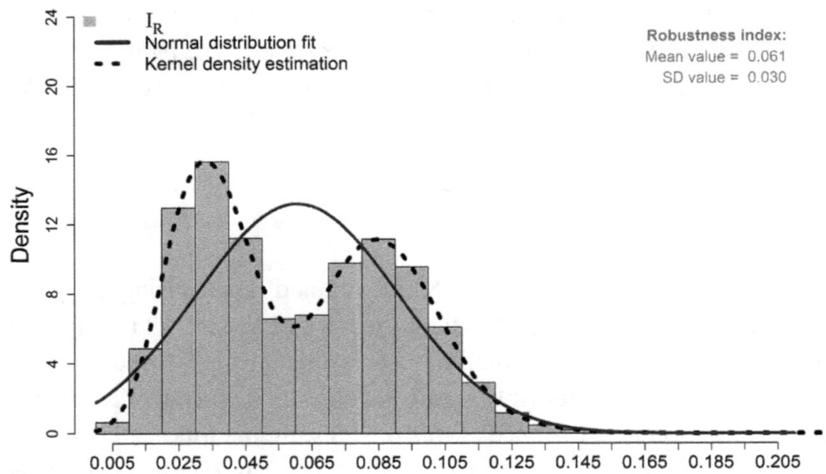

5.5.3 Structural Fragility Analysis

Structural robustness is a measure of the predisposition of a structural system to progressive and disproportionate collapse. Therefore, it is not the best parameter to evaluate when the objective is to assess the system's resistance against the applied actions. The development of such a measure is of great benefit, and even more, if this measure could relate directly to the damage extension within the system for a given action combination. A structural fragility index, F_R, which is capable of addressing adequately these objectives, was developed. The general expression is given by (André et al., 2015):

$$F_R\left(A_R, A_L \mid H\right) = \frac{D_p - D_{\text{1st failure}}}{D_c - D_{\text{1st failure}}} \text{ with } \begin{cases} 0 \leq F_R \leq 1 \\ D_c - D_{\text{1st failure}} = 0 \Rightarrow F_R = 1 \\ A_L \geq A_{L,uc} \Rightarrow F_R = 1 \end{cases} \tag{49}$$

where:

A_R represents the reference action;
A_L represents the leading action, which can be different from the reference action. $A_{L,uc}$ represents the value associated with D_{uc}, i.e. the "unavoidable collapse" state;
$H = \{h_1, h_2, \ldots, h, \ldots, h_n\}$ is a set of hazard scenarios. For example a set of different actions with determined values applied in a given sequence;

D_p represents the value of the damage energy of the structure when the new static equilibrium state is reached for value p of the reference action within the considered hazard scenario. Even though in the remainder of the Section the latter definition is used, for certain analyses cases, for example when seismic actions are involved (or other non-monotonic actions), to attain an improved interpretation, D_p can alternatively represent the damage energy associated with: a given analysis time value, a displacement (absolute or relative) value, a force value, or a value of other variable considered appropriate. The value of D_p can also be given by the damage energy at the end of the application of the reference action;

$D_{1st\ failure}$ represents the damage energy of the structure when the "first failure" state takes place for the hazard scenario considered;
D_{uc} represents the damage energy corresponding to the state where collapse is unavoidable for the hazard scenario considered, i.e. the "unavoidable collapse" state;
D_c represents the damage energy corresponding to the collapse state for the hazard scenario considered.

In order to calculate the structural fragility index a three-step procedure must be followed, similar to the one detailed above for the structural robustness index. The main differences are:

- **First Step:** Equal to structural robustness index. Additionally, the reference action must be also chosen and the value of D_p determined. The system performance should preferably be sensitive to the selected reference action values.
- **Second Step:** Equal to structural robustness index.
- **Third Step:** The fragility index is determined from Eq. 49 based on the adopted limit state, which defines the first failure state (the value of $D_{1st\ failure}$ is determined).

For damage energy values higher than D_{uc}, the value of the fragility index is equal to 1.0; for values close to $D_{1st\ failure}$, the value of the fragility index is in general very small.

This index is also flexible since the inputs can change, for example: the "first failure" state can be replaced by another criterion, possibly related to a particular element failure or simply the first material yield strain, and the "unavoidable collapse" and "collapse" states can also be changed to represent a maximum limit of acceptable damage, D_{max}, for instance, with $D_{max} \leq D_{uc} \leq D_c$.

This flexibility is important. For example, in structural systems where there is a large discrepancy between the available damage energy of great part of the elements (e.g. they are very brittle and weak and thus have a very low damage energy) and the remaining few (e.g. they are very resistant and ductile and thus have a very high damage energy), a hazard scenario may occur where only the majority of the weak elements fail. Since the sum of their available damage energy is only a fraction of the sum of the available damage energy of the strong elements, the fragility index will still be close to zero despite the bulk of the elements have failed. In these cases, the preferred solution is to apply the structural segmentation referred to in the previous Section. As an approximate alternative, it is possible to define a maximum limit of acceptable damage, D_{max}, namely the sum of the available damage energy of the weak elements, and assign it to D_{uc}. In this way, for action values higher than A_{max} (associated with the attainment of D_{max}) structural fragility index is equal to 1.0.

Another extreme case is where there are very few controlling elements. In these cases, there is no need to adapt the parameters of the fragility index since D_{uc} and D_c are almost only defined by the damage energies of these controlling elements.

The structural fragility index also shares many of other features of the structural robustness index. An additional remark should be made about the analysis of structural fragility using the proposed index. It is possible that the same increment of the action value (A: load, displacement, rotation, temperature, etc.) causes different increments in the fragility indices for structures *A* and *B*, despite having the same yield and ultimate energy values. This translates to structural fragility (damage accumulation) sensitivity to action values, which may be important when performing risk analysis.

Structural fragility is also a random variable, function of resistance variables and action variables. The probabilistic description of fragility follows closely the one described for structural robustness. From the cumulative distribution function (cdf) of fragility a graphic representation of fragility curves can be obtained, which could express simply fragility as a function of action values, or the probability of non-exceedance of fragility values as a function of the actions values, see Figure 16, for example.

5.5.4 Vulnerability Analysis

Vulnerability, in terms of costs of consequences, is related with structural fragility by a cost function κ(C), which translates levels of structural damage to costs of consequences. An example of a cost function is detailed in André (2014).

5.5.5 Risk Measures

Risk is generally expressed in terms of the probability of structural collapse times the cost of the consequences given the collapse. Additionally, in the classical approach, risk can also be expressed by a probability of failure. However, these definitions are quite limited since they do not account for the various damage states that might occur (damage is a continuous function) but that do not directly imply the global collapse of the structure. Therefore, valuable information is lost that could be used during the risk informed decision-making process. For instance, two structural systems *A* and *B* can have the same probability of failure but the damage evolution in *A* can be quite different than in *B*.

In the suggested framework, if actions and resistance variables are simulated by their real probability distributions and not by uniform probability distributions, structural fragility becomes an expression of

Figure 16. Example of a representation of fragility curves (André et al., 2015).

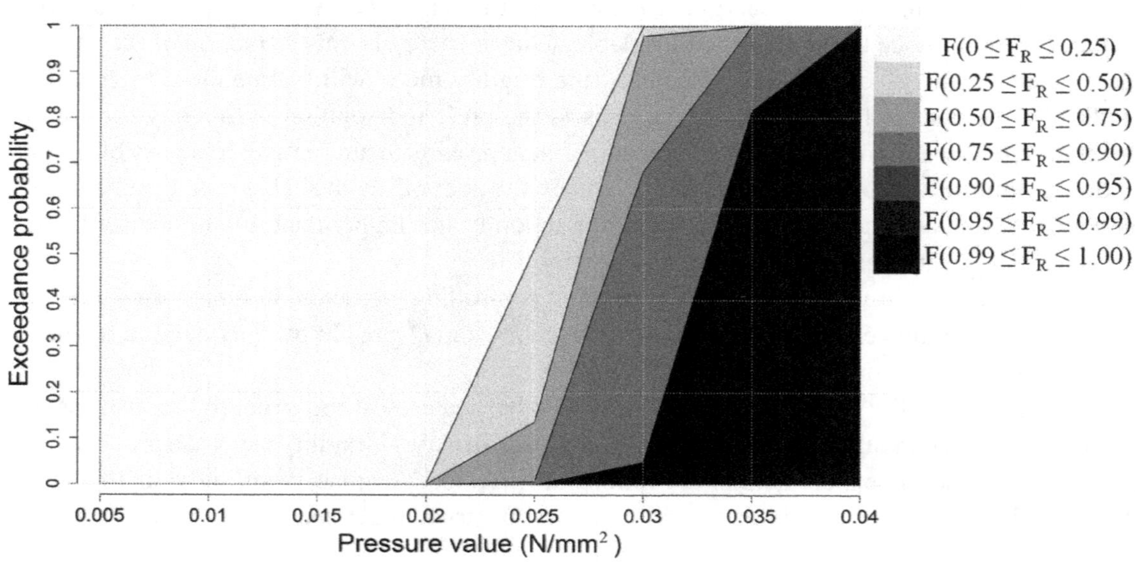

the damage extension (D) of the structural system, a measure of the system's structural risk and damage tolerance, and vulnerability becomes a measure of risk that can be used in a Cost-Benefit analysis (CBA). The newly developed structural robustness index can be used as a design option to reduce the structural risk and the newly developed structural fragility index is an analysis tool that should be used to assess the structural risk.

With this approach, it is possible to analyse how risk changes with structural robustness or with other risk control measures thus contributing to a better decision-making process. Using this framework, it is possible to perform comparative analysis between alternative scenarios. For example, Figure 17 illustrates the results of a risk analysis of a bridge falsework of similar layout as the one presented in Figure 13, expressed in terms of the cdf of the relative net value between two scenarios: a baseline scenario where no specific quality management requirements are defined and an alternative scenario where quality management requirements were enforced to control the variability and magnitude of variables that control the structural behaviour of the system under consideration, namely initial geometrical imperfections, ledger-to-standard joint looseness, stiffness and deformation capacity, see Section 5.6.2 and André (2014) and André, Beale, & Baptista (2014b) for details. Using the suggested risk assessment methodology, it was clearly demonstrated that if the cost of the permanent structure significantly exceeds (about one order higher, i.e. ×10) the cost of the temporary structure, the extent of improvements in terms of structural and economical risks completely justifies the extra costs incurred by adopting better quality management procedures.

The potential benefits of using the suggested fragility index over the traditional risk measures, i.e. probability of failure × total cost, can be readily observed. In the traditional risk framework, only one damage state is usually analysed, typically structural collapse. This corresponds to a single value of cost of consequences. With the new proposed methodology, several damage states are already included in the fragility index calculation and therefore it is possible to obtain with no added effort additional and important information for a wide range of probable damage states that if not accounted for in the decision-making process could lead to inefficient solutions.

Figure 17. Example of a representation of cdf of relative Net Value (André, Beale, & Baptista, 2017)

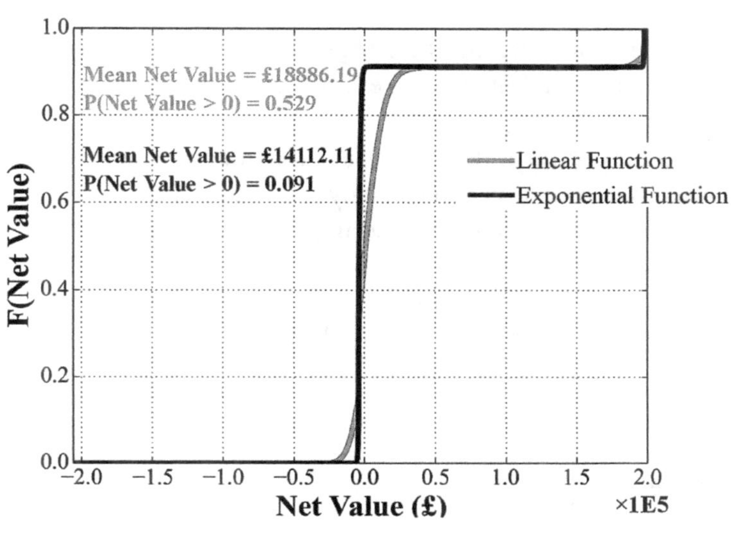

Of course, more intricate, complex cost functions can be used depending on the problem. For instance, flag variables can be included in the numerical model so to indicate that a given criterion (structural, e.g. nature of damage, or other such as type of operation, e.g. number of persons in the affected area) in a certain critical location of the system has been met. Different cost functions can be attributed to each criteria and location, and the overall cost is determined by the sum of all these particular functions. In the limit, a different cost function can be used for each element.

Multiple failure criteria can be used simultaneously, and failure is attained when the first criterion is met. As in general, there is not a univocal (single) relationship (function) between the consequence costs and the damage intensity (fragility) for every failure criterion, a slight change must be considered in deriving the probabilistic models for vulnerability. Instead of determining the probabilistic models for vulnerability based on the probabilistic model for fragility, the vulnerability must be determined for each combination of input values, for which the function between the consequence costs and the fragility is known. Having a large sample of vulnerability values, obtained from a surrogate model for example, it is possible to estimate the probabilistic model for vulnerability.

In addition, with the new definition of fragility, different possible definitions for failure can be used. If the objective is to analyse a structural system until a damage state other than the complete collapse, for instance to control the rotation of a particular joint, then, as was already mentioned, it is just necessary to assign a fragility index equal to 1.0 to that target damage state.

Furthermore, in the existing probability of failure based design methods it is not straightforward to analyse the sensitivity of the system's probability of failure to a change in the input variables, as well as to perform uncertainty propagation analysis. Consider a structural system to which an acceptable probability of failure was determined using certain input probabilistic models and model parameters. What would happen to the system's probability of failure if these initial hypotheses change? Also, uncertainty may be unevenly distributed across all possible damage states.

In the majority of cases it is not possible to know with appropriate confidence the types of probabilistic models of the input variables and of the distributions parameters, or the degree of dependence/independence between input variables. Knowing that many engineering problems are governed by the extreme values of the input variables, the analyst choices play a crucial role in the follow-up assessment of the results and in the decision-making process. Therefore, uncertainty propagation needs to be considered in the analysis.

The new fragility index gives a direct insight to the consequences of changing the initial hypotheses and if coupled with simulation methods it can also easily incorporate directly the influences of different uncertainties sources.

It is important to emphasise that, in principle, risk in structural engineering can be controlled without structural robustness. This can be readily seen by analysing how risk of consequence X associated with the collapse of the structure is determined considering just a single hazard event (consequence X is in general expressed in terms of costs):

$$RISK\left(X\right) = X \cdot \mathrm{P}\left(X \mid CL\right) \cdot \mathrm{P}\left(CL \middle| F \cap H\right) \cdot \mathrm{P}\left(F \middle| H\right) \cdot \mathrm{P}\left(H\right) \tag{50}$$

where:

H represents the hazard event;

P(*H*) represents the probability of occurrence of the hazard event;
P(*F*|*H*) represents the conditional probability of occurrence of a local failure given *H*;
P(*CL*|*F*∩*H*) represents the conditional probability of occurrence of a collapse (CL) given *H* and *F*;
P(*X*|*CL*) represents the conditional probability of occurrence of consequence *X* given *CL*.

Note that an adequate level of consequence *X*, e.g. maximum admissible level, is often predefined for each specific project and hazard event (e.g. using methods presented in Section 5.4.5). Therefore, risk of consequence *X* is usually evaluated by analysing the probability of occurring consequence *X*. In these cases, assuming $E[RISK(X)] \leq 10^{-6}$ and $E[P(H)] = 10^{-5}$, for example, then the left part of Eq. [51] must be equal or lower than 10^{-1}:

$$E\left[P\left(X \mid CL\right) \cdot P\left(CL \middle| F \cap H\right) \cdot P\left(F \middle| H\right)\right] \leq \frac{10^{-6}}{E\left[P\left(H\right)\right]} \tag{51}$$

Risk can be controlled by the following design strategies:

- At the source, i.e. the hazard event, by diminishing its probability of occurrence, P(*H*), by eliminating the hazard source or by reducing the hazard source, e.g. by better control of the application of concrete casting loads, reducing the dynamic load effects and their variability, or by specifying maximum working wind velocities for the assembly and operation phases and implementing early-warning systems based on monitoring and surveillance.
  This strategy does not increase the intrinsic resistance of a structure to damage with disproportionate consequences, it prevents or limits the likelihood of occurrence of an adverse hazard event.
- By diminishing the severity of the hazard, P(*F*|*H*): externally by reducing the magnitude of the actions effects, adopting shielding barriers outside the structure for example, or internally (structurally) by increasing the structure's resistance (e.g. by using the concept of *key elements* which imply using higher values of load partial factors for all design situations) and/or reducing the resistance variability of each element of the system using quality management tools (especially in the lower-tail region of the probabilistic distributions). It is also possible to use passive isolation techniques such as base isolation of the structure.
- By managing the consequences of the hazard applying protective (reactive) measures: (i) structurally by increasing the resistance (reliability) and/or the robustness of the structure, i.e. by modifying P(*CL*|*F*∩*H*), or (ii) by changing the context (e.g. by moving valuable goods, people to safer areas or by installing alarm systems and defining efficient exit routes), i.e. by modifying P(*X*|*CL*).

The first two design strategies consist in preventive measures (proactive, i.e. that reduce the likelihood of failure) and will increase the system's structural reliability. The other strategy consist in protective measures (reactive, i.e. that reducing the consequences from failure). Reactive measures are often preferred in cases where the likelihood of damage is significant and little or no control exists over the occurrence of the hazard event or over the hazard event direct consequences. Risk can be reduced and controlled to an acceptable (or tolerable) level by implementing one or a suitable combination of the above design strategies.

The reliability of the preventive methods used (e.g. early-warning systems, quality management, barriers, key elements) should be commensurate with the consequences to the structure in case the intended performance of the methods is not verified. For example, it is recommended that the reliability of the barriers and key elements should be equal to or greater than the reliability associated with the Consequence Class immediately above the Consequence Class of the structure, and maintained during the service life of the structure unless adequate alternative strategies are enforced in replacement.

Practical experience shows that implementing a quality system including organisation measures and controls at the stages of design, execution, use and maintenance of the structure is one of the most significant tools to reduce errors and negligence. In general, quality management methods are used in association with other methods in order to improve structural safety. See Chapter 8 for a list of examples of measures that can be taken for quality management.

It is also necessary to recognise that material properties, geometrical characteristics and actions values vary with time. This fact implies that the behaviour, resistance, reliability, robustness and risk of a structural system changes with time. Therefore, it is important that risk management includes prediction of the risk measures over time: a time variant problem. Here, it is beneficial to refine and to update the models with information, new and more accurate, acquired over time.

5.5.6 Design Strategies to Enhance Structural Robustness

5.5.6.1 General

In this Section, some selected examples of design strategies to enhance robustness are presented. Robustness is especially important in structures where it is economically unfeasible to adopt measures to reduce the probability of occurrence of the critical events, or to minimise the structural damage by adopting a higher target reliability level for the critical elements. In such cases damage can be limited instead, by activating secondary load paths and structure redundancy, or by using knock-out elements for example. Additionally, it should always be acknowledged that absolute safety against local failure cannot be achieved, and thus in the face of unknown future actions, strategies such as increasing resistance of key elements can underperform as the expected safety may not be as high as hoped for.

It should also be highlighted again that, in principle, risk in structural engineering can be controlled without structural robustness, see previous Section for various possible design strategies by which risk can be controlled.

5.5.6.2 Resistance

One way to enhance the robustness of an element or structure is to selectively increase the resistance of some elements, either by increasing the elements strength, or the elements stiffness. The former can be achieved by choosing materials with higher mechanical properties (strength and deformation capacity), while the latter can be attained by adding local or global reinforcements.

In the absence of elastic instability, mechanical properties control the structural resistance of elements and the failure mode of the structural system.

Stiffness is an important property since it provides structural stability. In comparison with a more flexible structure, the structure with a higher stiffness could exhibit a more direct load path if the structural form is properly chosen. Also, second-order effects would be smaller meaning that elements could be subjected to less stresses. Another aspect is the stiffness distribution within the structure. It is well known that structures with abrupt changes in stiffness and irregular structures in general are more vulnerable to hazards.

A usual misconception is to assume that increasing the mechanical properties of all elements will always lead to an increase of the system's robustness. As shown previously this may not be the case. This is also applicable to measures targeting increasing the strength (reliability) of components upon which the structure's stability depends (so called *key elements*). More efficient alternative options are available. Examples are the reinforcement of brittle elements in critical load paths, and the strengthening of the beam-column joints.

Classification of key elements should not be restricted to those elements directly affected by the action, but it should also be possible to classify secondary elements as key elements as this may be more efficient strategy to achieve robustness, by limiting the follow-up damage to an acceptable level, than by reinforcing only the directly affected elements. The advantage of the alternative strategy is that it is likely that the reinforcement of secondary elements may also be beneficial to other hazard scenarios, whereas the traditional key elements design is a more hazard specific driven approach.

5.5.6.3 Structural Integrity

This strategy primarily concerns elements continuity. This can be assured by specifying appropriate levels of tying strength between structural elements to avoid physical separation between elements of the structure. Additionally, integrity relates to soil stability, to avoid collapse due to insufficient resistance of the foundation. Care should be taken to ensure that structural integrity is not dependent on only one (or a few) element(s).

5.5.6.4 Redundancy

Redundancy means multiple load paths. If an element fails the stresses are transferred to neighbouring elements, the goal being to maintain the overall stability of the structure. However, in order for this strategy to be effective the elements and connections which will be overloaded must possess appropriate strength reserves and ductility. If not, for instance if the elements are brittle then the failure of one element can lead to the progressive collapse of adjacent elements in a domino fashion. Another important thing to bear in consideration is the distribution of redundancy within the structure: critical regions may lack redundancy whereas non-critical regions can be over-redundant.

5.5.6.5 Second Line of Defence

Another strategy is to introduce elements with *second line of defences* (Knoll & Vogel, 2009) on the system, i.e. elements with secondary load paths. For example, a slab can resist to vertical loads by bending, but when a central column is removed it behaves as a membrane (catenary action). One should be aware that to activate secondary load paths, the elements, including supporting elements, and their connections must undergo significant deformations, which mean that they need to be ductile enough.

5.5.6.6 Ductility

Ductility can be defined as the capacity of one material to continue to resist after yielding by absorbing energy and thus allowing energy to be dissipated in a stable fashion and stresses to be redistributed without significant deterioration of the structure's performance. Material ductility can be achieved by material strain-hardening and/or by material deformation capacity.

Ductility have a relevant role when designing for robustness, since ductility allows the structure (element) to absorb and dissipate energy in a controlled way, avoiding brittle failures and giving a early warning to users of the structure about the distress of the structure. Additionally, strain-hardening represents a resistance reserve after yielding until the fracture of the material. Ductility is a key material property to enable structural redundancy and second line of defence resistance models.

Attention should be paid to strain rate sensitivity, cyclic behaviour, strength and stiffness degradation, low cycle fatigue of materials and joints.

5.5.6.7 Capacity Design Principles

The principle of capacity design is to avoid energy dissipation mainly by brittle elements. Material ductility, well defined redundancy and wise choice of energy dissipating regions are key aspects to reach this goal. For example: it may be preferable to avoid concentrating maximum stresses on the connection zones where these are made by several components with brittle behaviour.

5.5.6.8 Segmentation

Robustness can also be achieved by limiting the damage to restricted areas. In some structures structural continuity can produce opposite results from the ones expected and contribute to a disproportionate collapse of the system. An example of this behaviour is a structure which was not designed to resist the additional forces redistributed after an element failure.

Therefore, a possible solution could be to limit the extent of the tolerable collapse progression. To achieve this goal, continuity of internal forces between elements is eliminated or minimised, by reducing the possible load paths i.e. reducing the redundancy of parts of the structure, or by disrupting or reducing the elements continuity (Starossek & Wolff, 2005). The structure would be made of a series of low robust parts, or with a limited number of low robust parts being the others high robust parts.

In this structural concept, a particular area of the structure would collapse in case of a failure event without damaging the nearby structure. However, the remaining structure must remain in place and operational, possibly under higher loads e.g. it could be subjected to impact loads. If not, a domino like progressive collapse of the entire structure could take place.

The type and location of structural discontinuities should be appropriately chosen and suitably designed, detailed and executed in order to ensure that the intended hierarchy of resistance of the various structural components and configuration of failure modes necessary to implement the structural segmentation strategy are verified.

5.5.6.9 Quality Management

Practical experience shows that a quality system including organisation measures and control at the stages of design, execution, use and maintenance is the most significant tool to achieve an appropriate level of structural reliability. See Chapter 8 for a list of examples of measures that can be taken for quality management.

5.6 APPLICATION TO TEMPORARY STRUCTURES

5.6.1 Design Philosophy

5.6.1.1 Basis

Following the classification scheme presented in BS EN 1990 (BSI, 2005) for reliability differentiation, in this Section an adaptation of those principles to temporary structures is given. The proposal is also based on information presented in BS 5975 (BSI, 2011) and in published reports from the UK's Temporary Works Forum (TWf), in particular (TWf, 2014).

For each relevant design failure mode scenario, a single Design Approach, selected from the ones given in Table 8, should be applied to ensure an adequate level of safety of a temporary structure. Design failure mode scenarios are scenarios of evolution of damage and failures of structural elements associated with consequences that are in the same order of magnitude as a given Consequences Class.

The choice of the Design Approach for each relevant design failure mode scenario should be based on the significance of the resulting expected consequences. The latter may be associated with a Consequence Class and the coupled global structural reliability levels (e.g. reliability of the structure against unacceptable collapse), as indicated in Table 9, and adequate levels of total damage. Table 9 also presents a non-exhaustive categorisation of temporary structures in the different Consequences Class. When no guidance is available or where appropriate, the categorisation of temporary structures in the different Consequence Classes can be agreed with the client and/or the regulatory authority.

Application of Design Approach 2 and Design Approach 3

In Design Approach 2 and Design Approach 3, an adequate level of global structural safety (i.e. reliability of the structure against unacceptable collapse, see Table 9, and acceptable limits of total damage) with respect to each relevant design failure mode scenario involving identified and/or unidentified hazard events should be provided by:

1. Verifying explicitly the structural performance with respect to the relevant design situations involving identified accidental hazard events and/or unidentified hazard events specified in design codes (e.g. BS EN 1991-1-7, see also Chapter 3).
   If the application of an action results in total damage that are not larger than a relevant acceptable limit, no particular consideration is necessary with respect to global structural safety for the considered design failure mode scenario. Since there is currently no design rules or guidance applicable to temporary structures, the recommended acceptable limit of total damage should be agreed with the client and/or the regulatory authority. The value of acceptable damage should be related to the Consequence Class of the design failure mode scenario.

In the case the condition presented above is not fulfilled for the considered design failure mode scenario, either the provisions given option 2 (defined below) are applicable and followed, or it should be demonstrated that an adequate level of global structural safety is provided by implementing one or a suitable combination of the design strategies specified in Sections 5.5.5 and 5.5.6, as relevant. Note that it should be demonstrated that the application of one or a suitable combination of the design strategies fulfils the requirements of global structural safety for temporary structures.

Or,

2. Implementing all the prescriptive design and detailing rules relevant for global structural safety specified in design codes (e.g. BS EN 1992 to BS EN 1999) that:
   a. Address explicitly and fully all the relevant identified accidental hazard events and unidentified hazard events;
   b. Are applicable for the type of structure under consideration;
   c. Are adequate with respect to the Consequence Class associated with the design failure mode scenarios of Design Approach 2 and Design Approach 3, as applicable (see Table 9).

Table 8. Design Approaches for temporary structures

<table>
<tr><th>Design Approaches</th><th>Analysis method</th><th>Design method</th></tr>
<tr><td>DA-1</td><td>Analyses can be based on simplified models of loads and structural behaviour.</td><td>Design according to design codes (e.g. BS EN 1990 to BS EN 1999, see Chapter 6) with respect to relevant design situations involving identified non-accidental hazard events as defined in design codes (e.g. BS EN 1991, see Chapter 3).
In cases where the expected value of the consequences associated with the complete collapse of a structure is acceptable, no specific consideration is necessary with respect to global structural safety.</td></tr>
<tr><td>DA-2</td><td>Assessment of relevant design failure mode scenarios.
Analyses can be based on simplified models of loads and structural behaviour.</td><td rowspan="2">Design according to design codes (e.g. BS EN 1990 to BS EN 1999, see Chapter 6) with respect to relevant design situations involving identified non-accidental hazard events as defined in design codes (e.g. BS EN 1991, see Chapter 3).
Design for global structural safety with respect to both identified and unidentified hazard events should be provided by:
design strategies for providing an adequate level of global structural safety,
or,
relevant and complete prescriptive design and detailing rules.</td></tr>
<tr><td>DA-3</td><td>Assessment of relevant design failure mode scenarios.
Analysis of structural performance can be based on simplified and idealised models, subject to justification.
Where appropriate, use of dynamic and/or nonlinear structural analysis models.</td></tr>
<tr><td>DA-4</td><td colspan="2">Risk-based design of structural safety may be carried out when relevant.
Identification and analyses of scenarios leading to structural collapses, utilizing risk screening meetings involving experts on all relevant subject matters.
Detailed assessments that, as appropriate, can involve dynamic and/or nonlinear structural analysis.
Risk analysis that addresses direct and indirect consequences of design failure mode scenarios.
Specific guidance is provided in Sections 5.3 to 5.5 of this Chapter but also in BS EN 1990 and in ISO 2394.</td></tr>
</table>

NOTE 1 Identified hazard events are associated with the nonabstract actions defined in design codes (e.g. permanent and variable loads but also accidental loads such as impacts, see Chapter 3).

NOTE 2 Unidentified hazard events supplement identified hazard events by simulating the effects of uncertainties and errors during the design and execution of the structure that are not accounted for in the design codes, or which exceed the limits considered in these documents. For example, events associated with the loss of elements due to persistent or transient design situations. Notional actions accounting for unidentified hazard events are provided in design codes, e.g. BS EN 1991-1-7, see Chapter 3.

Table 9. Design Approaches (DA), expected value of consequences, examples of temporary structures, related Consequence Class (CC) and global structural reliabilities

DA	Expected value of consequences of design failure mode scenarios	Related CC	Target global structural reliability, β_1 (1 year reference period)	Example of temporary structures
DA-1	Material damage where risk to human life, and economic, social or environmental consequences are small or negligible.	CC1	3.8	Standard solutions of scaffolding/falsework.
DA-2	Material damage where risk to human life, and economic, social or environmental consequences are limited.	CC2a Lower Risk Group	4.2	Simple designs of routine scaffolding/falseworks. Departures from catalogue design for standard components. Temporary structures involved in the construction of permanent structures belonging to CC1 (see examples in Table 1).
DA-3	Material damage where risk to human life, and economic, social or environmental consequences are moderate.	CC2b Higher Risk Group	4.7	Complex or innovative designs of scaffolding/falsework. Routine BCE. Temporary structures involved in the construction of permanent structures belonging to CC2 (see examples in Table 1).
DA-4	Material damage where risk to human life, and economic, social or environmental consequences are moderate.	CC3	5.2	Abnormal and highly innovative designs beyond the scope of normal design codes and practice. Long span BCE. Wind sensitive structures. Temporary structures involved in the construction of permanent structures belonging to CC3 (see examples in Table 1).

Table 10. Design supervision levels (DSL)

Design Supervision Levels	Characteristics	Minimum recommended requirements for checking of calculations, drawings and specifications
DSL1 Relating to CC1	Basic supervision	Self-checking: Checking performed by the person who has prepared the design
DSL2 relating to CC2	Normal supervision	Checking by different persons than those originally responsible and in accordance with the procedure of the organisation.
DSL3 relating to CC3	Extended supervision	Third party checking: Intensive supervision performed by an organisation different from that which has prepared the design

Table 11. Inspection levels (IL)

Inspection Levels	Characteristics	Requirements
IL1 Relating to CC1	Basic inspection	Self inspection Random sampling
IL2 Relating to CC2	Normal inspection	Inspection in accordance with the procedures of the organisation Increased effort with respect to supervision and inspection during the construction of the structural key elements.
IL3 Relating to CC3	Extended inspection	Third party inspection Intensive inspection by well-qualified people with an expert knowledge All elements to be inspected

If at least one of the above conditions is not fulfilled, the provisions specified in option 1 (above) should be implemented for the considered design failure mode scenario.

Very limited guidance is available from design codes regarding prescriptive design and detailing rules relevant for global structural safety of temporary structures. However, important guidance is given in Sections 5.5.5 and 5.5.6. In Section 5.6.2 an illustrative application example is provided, using a bridge falsework system. For BCEs, guidance is presented in Rosignoli (2007). In cases where no guidance is available or where appropriate, these prescriptive rules can be agreed with the client and/or the regulatory authority.

In all cases, the prescriptive design and detailing rules provided in design codes (e.g. BS EN 1992 to BS EN 1999) that represent general good design practice, including design for robustness, shall also be complied with, see Chapter 6. In cases where no guidance is available or where appropriate, these prescriptive rules can be agreed with the client and/or the regulatory authority.

In Design Approach 3, particularly when nonlinear analyses are performed, sensitivity studies of design assumptions may be required to assess the modelling uncertainty on the damage propagation (e.g. type and sequence of failure modes). This may be relevant to ensure that the intended hierarchy of resistance of the various structural components and configuration of failure modes are verified. For example, sensitivity studies may consist of structural analyses using different material properties, e.g. mean, characteristics or design values, to assess the influence of these variables in the results. To ease the use of this procedure, sensitivity studies should only concentrate on the assessing the influence of the most critical variables, varying the value of each variable one at a time.

5.6.2 Example of a Risk Analysis

5.6.2.1 Justification

Why is it necessary to apply risk management to temporary structures systems? The answers to this question are:

- An important percentage of collapses during construction relate to the failure of temporary structures. For example, almost 20% of the total number of bridge collapses, total or partial, are due to the failure of the bridge falsework (Scheer, 2010). Furthermore,
- The total costs of the consequences of a temporary structure collapse outweigh multiple times the direct costs of reconstruction of the permanent structure, namely because of the costs associated with loss of human lives, the additional user costs and sustainability costs, and
- The costs of upgrading temporary structures are usually very low when compared with the costs of the permanent structure, see Table 12.

Additionally, the risk framework presents several advantages over traditional engineering methods which make it a useful tool to properly consider the effects of uncertainties in the safety of future and in-service infrastructures, such as temporary structures:

- Risk management encompasses a holistic assessment of the system's structural performance and safety;

- Estimated risks are evaluated against established risk criteria, from which risks are classified as acceptable, as tolerable, as intolerable or as unacceptable;
- Risk management includes whole-life cycle considerations and encourages continuous improvements;
- Explicit and transparent consideration of uncertainties in the risk analysis process, with the chance of including newly gained information;
- Decisions are made based on a rational multivariable optimal decision-making process: individual and societal, economy and safety issues.

Yet, are these factors enough to justify undertaking a complex risk management methodology over the traditional design methods? Are the existing risks acceptable?

In order to answer these important questions, estimates of risks to individuals, and of structural risks, are presented, respectively the Individual Risk Per Annum (IRPA), i.e. the annual probability of a fatal accident, and the annual probability of structural failure of temporary structures will be calculated taking as an illustrative example the construction of a bridge using bridge falsework systems. The results which are presented in the following constitute an improvement from previously published information, namely André et al. (2012).

The above variables are calculated based on the results of a survey of bridge falsework failures since 1970 in presented in Chapter 7. The information concerning 16 countries will be used. Note that the two collapses recorded in the survey for the UAE occurred in the Emirate of Dubai. The values presented below correspond to notional estimates since they are based on a necessarily limited sample, and therefore are subject to uncertainties. The methodology adopted can only provide an estimate of the average of individual and structural risk, since it is determined from a sample of heterogeneous data in terms of: design standards used (e.g. target reliability levels), context and exposure characteristics, modes of failure and procedural, enabling and triggering events and types of bridge falsework systems. Therefore, they should only be interpreted in a comparative sense and not taken as the actual values.

In the absence of information, the variables were obtained considering the following assumptions (assumed conservative):

- 60% of the concrete bridges and viaducts were built using bridge falsework systems;
- 80% (in the case of developed countries) and 90% (in the case of developing countries) of the existing concrete bridges were built after 1970;
- The average number of persons exposed to the risk of collapse of the bridge falsework structure was determined based on the number of reported fatalities and injuries, considering a minimum number of 40 persons (in the case of developed countries) and a maximum number of 100 persons (in the case of developing countries) at risk in each falsework structure.

Table 12. Typical costs of temporary structures

Type of temporary structures	Cost
General access scaffolding	50€/m^2 for steel system, a fraction of it for bamboo scaffold
Shoring equipment	From 100€/m^2 up to 400€/m^2 depending on the floor height
Balance cantilever systems for bridge	Between 400,000 to 500,000 €/system

Using these assumptions and the data presented in Table 13, the value of IRPA is obtained by:

$$\text{IRPA} = \frac{\text{Number of fatalities}}{\text{Number of persons at risk}} \tag{52}$$

The data presented in Table 13 corresponds to the complete information collected in each country between 1970 and 2016. Since the data of the number of bridges built each year in each country analysed is not available, an average value of IRPA in the last 46 years was determined for each country considering the total number of reported fatalities and the total number of persons exposed to the risk of collapse of a bridge falsework structure. The latter is given by the product of the total number of concrete bridges built using these systems with the average number of persons exposed to the risk of collapse of the bridge falsework structure.

Figure 18 illustrates the IRPA values for 16 countries. It can be observed that in two countries (Brazil and Portugal), there is an estimated chance close to 100 in 10^6 of a fatal accident per year for this bridge construction method, which is much higher than the one registered in the UK for the construction sector which is 24 in 10^6, 2008/2009 figures (HSE, 2009a) – which represents a significant improvement following the high rate of 59 in 10^6 registered in 2000/2001 (HSE, 2001).

Table 13. Summary of data used to calculate risk estimates for bridge falsework systems in 16 countries since 1970

Country	Accidents	Fatalities	Injuries	Number of bridges [a]	Persons at risk [b]
Australia	1	0	15	22424	50
Austria	1	2	0	14871	40
Brazil	2	32	40	4073	100
Canada	3	7	16	17897	40
Czech Republic	1	7	67	7872	60
China	8	98	118	67719	100
Denmark	2	1	5	7062	40
Germany	19	19	42	53085	40
India	3	53	24	58542	100
Japan	1	4	14	49493	60
New Zealand	1	0	0	7829	40
Portugal	7	10	38	2457	40
South Africa	1	2	20	4215	80
UAE	2	7	29	2819	100
UK	1	3	10	28244	40
USA	17	24	72	168190	60
Total	**98**	**296**	**642**		

(a) Number of concrete bridges built after 1970 using the bridge falsework construction method
(b) Average number of workers exposed to the risk of collapse of the bridge falsework structure.

Figure 18. IRPA values for 16 countries considered in the survey

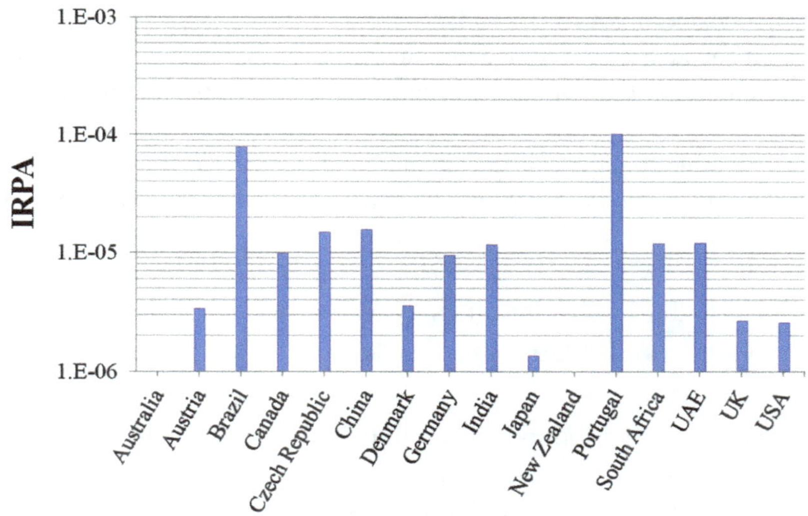

The results obtained from the survey carried out can be considered conservative because it is very likely that there are a number of unreported accidents with bridge falsework systems. This fact makes the recorded number of collapses and, possibly, the number of fatalities, a lower boundary.

It can be considered, with confidence, that the relative effect, in the IRPA values, of the uncertainties associated with the assumptions used to determine the total number of workers at risk (involving the number of bridges built each year and the number of workers involved in the casting operations), is lower than the relative effect of the uncertainties associated with the total number of accidents that happened in each of the 16 countries for the time period considered in the analysis.

Comparing the value for the individual risk obtained for bridge falsework systems with the limits presented in Section 5.4.4 for the acceptable and unacceptable annual risk levels, it can be concluded that in all countries included in the analysis, except Australia and New Zealand where no fatal injury was reported, the individual risk is higher than the broadly acceptable risk level (taken as 1 in 10^6 fatalities per year). Even if a less stringent criterion was used, equal to 10 in 10^6 fatalities per year, the risk would still be higher than broadly acceptable risk level in the seven countries. However, in all countries the individual risk is lower than the unacceptable risk level (taken as 1000 in 10^6 fatalities per year). Therefore, the individual risk for bridge falsework systems is in general within the risk tolerability zone and must be ALARP.

The data presented in Table 13 can also be used to estimate the annual probability of failure, $P_{f,1}$, of a bridge falsework system, which can be obtained by Eq. 53. The results are presented in Figure 19. As for the IRPA, the annual probability of failure was determined considering the total number of failures and the total number of concrete bridges built using bridge falsework systems in each country since 1970.

$$P_{f,1} = \frac{\text{Number of failures}}{\text{Number of concrete bridges}} \tag{53}$$

Using Eq. 27, a value for the acceptable annual probability of failure of the bridge falsework equal to 1×10^{-6} is obtained, considering $K_s = 0.5$ and $n_r = 50$, or 2.5×10^{-6}, considering $K_s = 0.5$ and $n_r = 20$. Observing Figure 19 it can also be concluded that this criterion is not satisfied. Additionally, using Eq. 28 the acceptable annual probability of failure is equal to 18×10^{-6} (considering $\eta_i = 0.1$), which is also not satisfied.

Finally, it is also possible to calculate the annual conditional probability of a person being killed given the failure of the bridge falsework system, which is obtained by Eq. 54. The results are presented in Figure 20.

$$\mathrm{P}\left(d \cap f\right) = \frac{\text{Number of failures with fatalities}}{\text{Number of bridges}} \tag{54}$$

It can be observed that the criterion specified in ISO 2394 for accepting the annual probability of a person being killed in a structural accident, equal to 1 in 10^6 fatalities (see Eq. 26), is not satisfied in all countries, except Australia and New Zealand for the reason already mentioned.

In conclusion, the estimated annual probabilities of a fatality and of a failure of a bridge falsework system are higher than the acceptable risk levels and, therefore, the development of a risk informed decision-making framework for bridge falsework systems is fully justified.

Comparing the $P_{f,1}$ values presented in Figure 19 with the structural risk of other temporary structures, the UK Health and Safety Executive (HSE) investigated 471 reported scaffold collapses during the period between 1986-1993, see Beale & Godley (2003). Considering an estimated 7.5 million scaffold erections it gives a failure rate of 63 collapses per 10^6 erections (i.e. 63×10^{-6} per year), which in the UK compares with an estimated probability of 46 collapses per 10^6 bridge projects for bridge falsework systems, a value close to the one observed for scaffold systems. Therefore, the same conclusion can

Figure 19. $P_{f,1}$ values for 16 countries considered in the survey

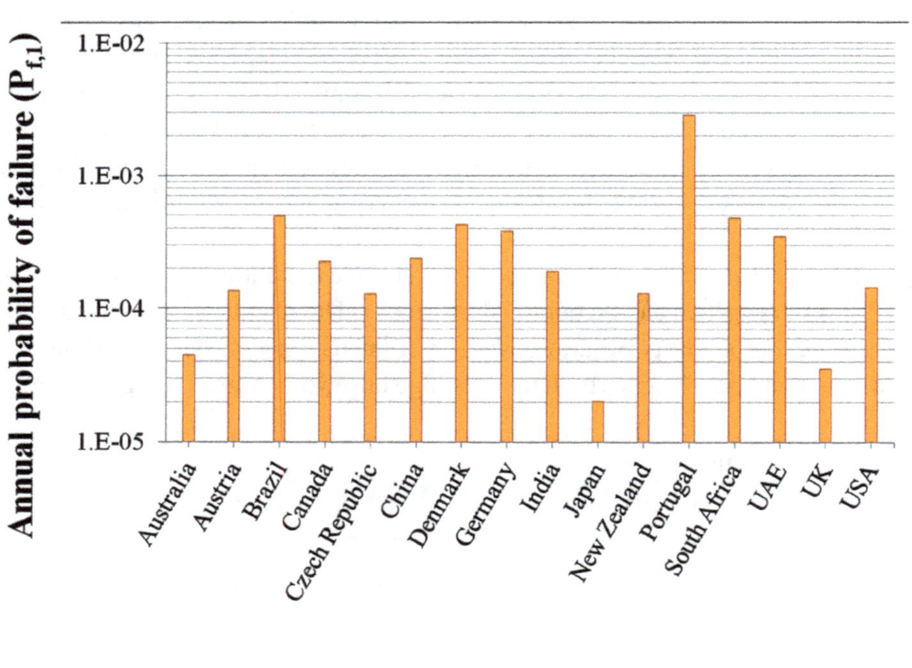

be drawn for this type of temporary structures. For BCE, available information about failures is very scarce, but given the sensibility of their *modus operandi* to human errors, their design requirements and their importance to the overall success of the bridge project, the use of a risk informed decision-making framework is advisable.

5.6.2.2 Bridge Falsework Example

5.6.2.2.1 Basis

Figure 22. presents the selected risk informed decision-making framework considered in this application example.

This example will include identification of the relevant risks for the safety of the structure and the critical failure modes during assembly, operation and QC/QA (quality control/quality assurance) phases of a bridge falsework project. The risks to workers and users not deriving from structural damage will not be included in the analysis. Therefore, risks to workers from falling at height or risks to workers or users from falling objects will not be considered. Risks related exclusively with the formwork system and with the superstructure (bridge) will also not be considered.

This investigation will focus on Cuplok® bridge falsework. The falsework system A2 (hereon labelled as Model A2) tested in the University of Sydney, see Chandrangsu & Rasmussen (2011) for the complete details, will be considered in the illustrative examples. Figure 21 illustrates the numerical representation of Model A2. Complete details of the risk analysis example are provided in André, Beale, & Baptista (2017).

5.6.2.2.2 Risk Identification

Hazards exist everywhere, in particular at the interfaces of system activities. The basis for the hazard identification is primarily based on reported information concerning accidents but also on expert judgement. In this Section the failure modes, failure effects and failure consequences will be presented in

Figure 20. P(d∩f) values for 16 countries considered in the survey

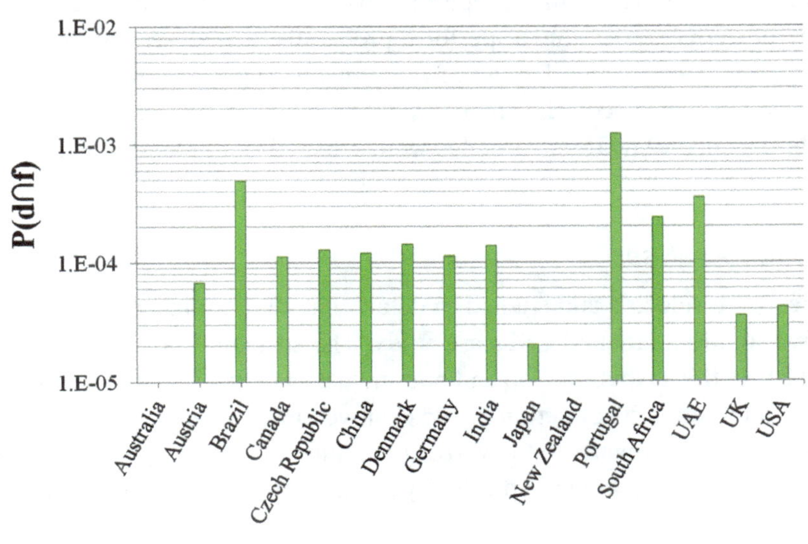

Figure 21.. Overview of Model A2 numerical model

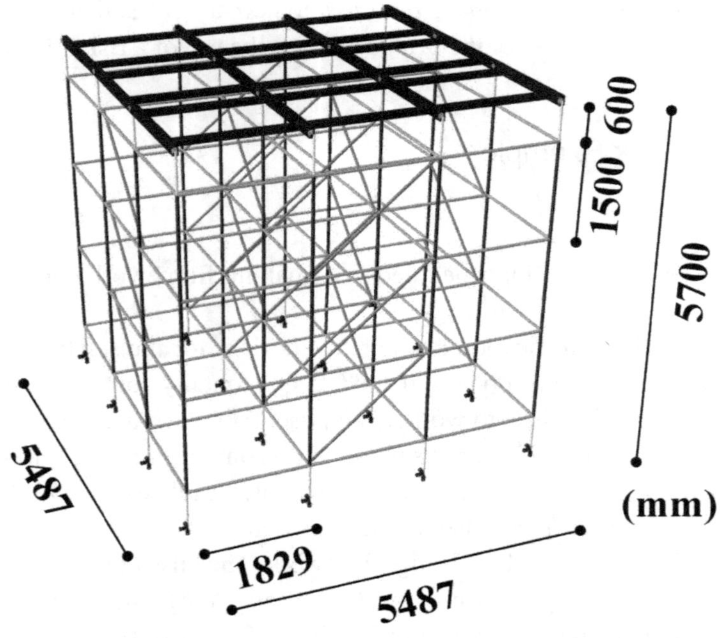

graphic terms using logic trees and hazard tables, see Figure 23 and Table 14. In Chapter 7 some of the listed hazards will be discussed.

Furthermore, from the information given in Figure 23 it is possible to identify opportunities to include suitable barriers to manage the failure modes and the failure effects. This is illustrated in Figure 24.

5.6.2.2.3 Choice of Probabilistic Variables

Early in any risk analysis it is critical to carry out a probabilistic sensitivity analysis in order to understand the influence of the variability of the input values in the variability of the output results and to determine which input random variables are the most important to explain the probabilistic structural behaviour.

There are many procedures available to perform probabilistic analyses (André, 2014; Benjamin & Cornell, 1970; Melchers, 1999). Here, Design of Experiments (DoE) followed by surrogate modelling and Monte Carlo analyses were used. The procedure is explained in detail elsewhere (André, 2014; André et al., 2017). In the end, 20 random variables were selected from the total 34, see Table 15 (André et al., 2017).

5.6.2.2.4 Case Study Analyses

Several case studies were selected and studied in detail. The results presented in this Section concern only one alternative solution, labelled CS2 case study. The structural layout of the CS2 case study is depicted in Figure 25. In this layout, the top and bottom jacks are braced by a continuous brace element placed in every bay, alternating its direction in consecutive bays, along two orthogonal directions.

In the CS2 case study, additionally to the vertical pressure applied on top of the formwork, the wind pressure corresponding to the working wind velocity and a localised differential ground settlement were

Figure 22. Selected risk informed decision-making framework

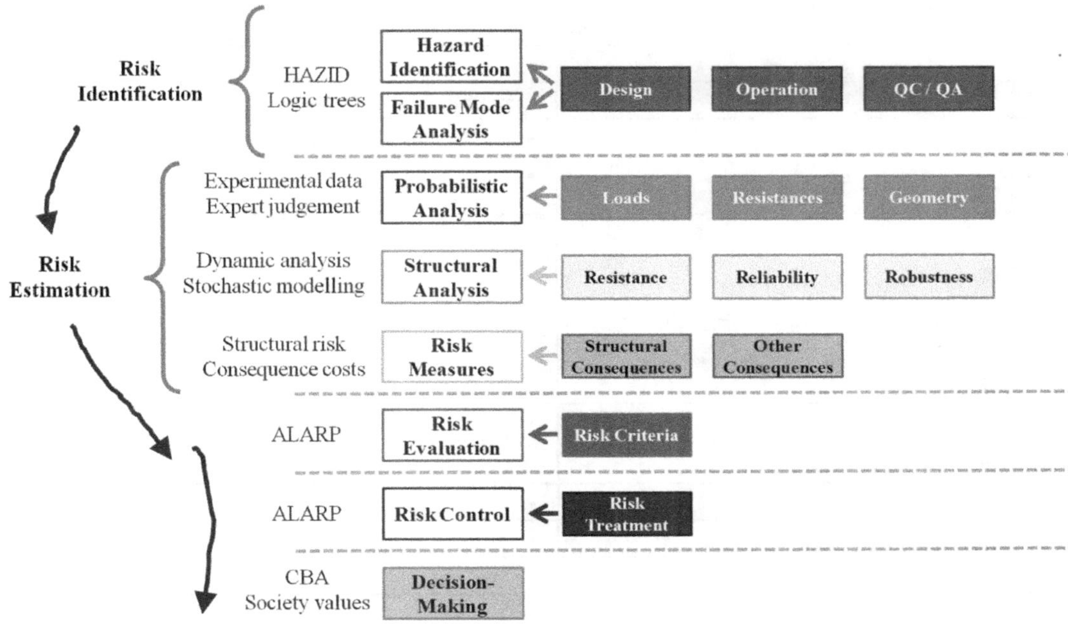

Figure 23. Decomposition of failure effects, failure modes and failure effects

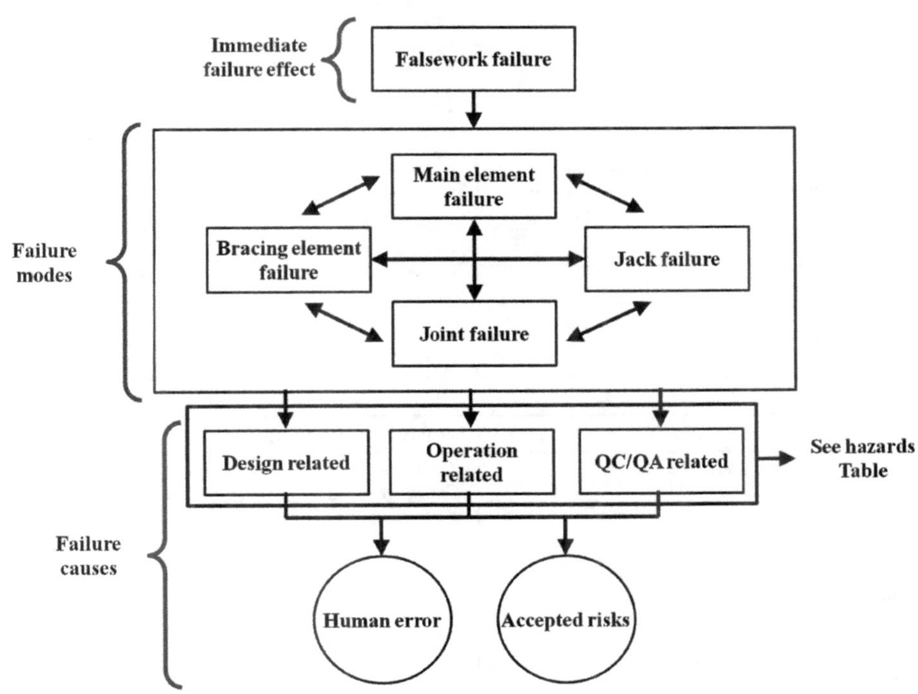

Table 14. List of primary hazard events

(A) Design	
4. Actions	
Activities	**Primary hazard events (Enabling events)**
1.4. Action cases selection	• Forgetting to consider action cases
Permanent Loads: • Self-weight	•
Construction Loads: • Concrete weight • Reinforcing steel weight • Precast units weight • Dynamic effects (concrete casting, bridge launching, etc.) • Storage of materials and equipment • Personnel • Post-tensioning	•
Settlements: • Ground • Foundation elements	•
Wind	•
Flood	•
Temperature	•
Snow	•
Ice	•
Earthquake	•
Impact Loads	•
1.5. Estimation of actions values	• Underestimation of the action values • Dynamic effects • Storage loads • Wind loads (use of reduction factors) • Incorrect assessment of the ground conditions • Underestimation of the period of exposure • Unaccounted changes in the design of the bridge project and/or of the method of construction
Directly specified in design standards. However, some loads values are project specific.	•
1.6. Load combinations	• Forgetting to consider concomitant action cases • Underestimation of concomitant action values
During assembly	
During operation	
During dismantling	
5. Resistances	
Activities	**Primary hazard events (Enabling events)**
Determination of mechanical properties of materials, ground characteristics: • Testing • Design standards • Expert judgment	• Deficient estimation or overestimation of resistance values determined by testing or obtained from theoretical or empirical models • Forgetting to consider actions cases
Determination of behaviour and resistance of elements and joints under static and dynamic actions • Testing • Design standards • Expert judgment	

continued on following page

Table 14. Continued

6. Modelling	
Activities	**Primary hazard events (Enabling events)**
Determination of actions effects: • Loads • Displacements • Rotations • Vibrations • Static/dynamic loads	• Forgetting to consider actions cases • Underestimation of actions effects • Not accounting for all actions effects • Forgetting to consider resistances models • Inadequate modelling of resistances models • Inadequate modelling of geometry • Forgetting to consider imperfections, damage, deterioration mechanisms • Underestimation of effect of imperfections, damage, deterioration mechanisms
Resistance modelling: • Ground • Materials • Elements • Joints • Deterioration mechanisms	
Geometry modelling: • Elements • Joints • Ground • Imperfections • Damage	
(A) Design	
9. Structural Analysis	
Activities	**Primary hazard events (Enabling events)**
Selection of analysis type: • Design tables • Analytical methods • Numerical methods • Linear or nonlinear analysis • Static or dynamic analysis	• Inaccuracy of analysis results • Incomplete analysis • Inadequate analysis
10. Structural Design	
Activities	**Primary hazard events (Enabling events)**
Serviceability and safety verification: • Sections • Main elements, brace elements and foundations • Joints • System • Ground	• Incorrect use of serviceability and safety design procedures • Incomplete serviceability and safety design checks • Under-designed element, joint, foundation, structure • Incomplete or wrong documentation
Documentation: • Drawings • Design justification • Method statement: o Assembly procedure o Ground investigation o Minimum ground characteristics o Maximum loads for various stages o Loading sequence o Monitoring requirements o Inspection and testing requirements o Imperfections considered o Dismantling procedure	

continued on following page

Table 14. Continued

<table>
<tr><th colspan="2">(B) Operation</th></tr>
<tr><td colspan="2">3. Assembly</td></tr>
<tr><td>Activities</td><td>Primary hazard events (Enabling events)</td></tr>
<tr><td>Carry out ground investigation</td><td rowspan="5">• Inadequate or incomplete ground investigation
• Incorrect analysis of the results of ground investigation
• System assembled over ground with weaker characteristics than the ones considered in the design
• Errors in the execution of the foundation elements
• Assembly procedure different from the one considered in the design
• System's configuration different from the one specified in the design (overextended jacks, spacing of lacing members, bracing configuration, deficient joints between formwork and falsework)
• Assembly with weather conditions not in accordance with the specified in the design
• System's imperfections larger than the design tolerances
• Use of incorrect or damaged elements, joints</td></tr>
<tr><td>Execute foundations (sole plates, concrete pads, piles, ground improvement)</td></tr>
<tr><td>Assembly baseplate, jack, vertical, lacing and bracing elements, jacks and forkheads, bolts and tie rods, formwork
Lock joints between members</td></tr>
<tr><td>Check extension lengths of jacks, tightness of joints, correct execution of joints between falsework and formwork, spacing of lacing and configuration of bracing</td></tr>
<tr><td>Check member and system imperfections, element defects</td></tr>
<tr><td colspan="2">4. Operation</td></tr>
<tr><td>Activities</td><td>Primary hazard events (Triggering events)</td></tr>
<tr><td>Check loading sequence and allowed weather conditions</td><td rowspan="3">• Loading sequence not as considered in the design
• Operation under weather conditions outside the maximum design limits
• Loading method not as specified in the design (by bucket or by pump, e.g.: casting concrete from large heights)
• Impact of the crane with the falsework system
• Falling precast units from crane into falsework
• Impact of a vehicle with the falsework
• Overload falsework by adopting an inadequate post-tensioning plan or different from the one considered in the design
• Overload falsework with equipment and storage materials more than considered in the design
• Occurrence of actions with intensities larger than the ones considered in the design
• Occurrence of actions not considered in the design</td></tr>
<tr><td>Concrete casting of bridge deck or placement of precast units from cranes</td></tr>
<tr><td>Apply partial post-tensioning</td></tr>
<tr><th colspan="2">(B) Operation</th></tr>
<tr><td colspan="2">6. Dismantling</td></tr>
<tr><td>Activities</td><td>Primary hazard events (Triggering events)</td></tr>
<tr><td>Check if the superstructure is already self-supporting</td><td rowspan="2">• Superstructure is still not self-supporting
• Early dismantling or improper dismantling procedure</td></tr>
<tr><td>Follow dismantling procedure</td></tr>
<tr><th colspan="2">(C) Quality management</th></tr>
<tr><td>Activities</td><td>Primary hazard events (Procedural causes)</td></tr>
<tr><td>Selection of skilled staff and workers</td><td rowspan="12">• Insufficient communication and cooperation between stakeholders (e.g. unreported changes in the bridge design)
• Deficient assignment of responsibilities of supervision
• Selection of unskilled, untrained staff and workers
• Undetected, uncorrected errors, damage, imperfections
• Improperly corrected errors, damage, imperfections
• Selection of inadequate methods of inspection and maintenance
• Under-designed structure</td></tr>
<tr><td>Training programmes</td></tr>
<tr><td>Appointment of a health and safety team</td></tr>
<tr><td>Appointment of a temporary structure supervision team</td></tr>
<tr><td>Cooperation and communication between stakeholders</td></tr>
<tr><td>Self-checking</td></tr>
<tr><td>Internal and external reviews of the project procedures and documents</td></tr>
<tr><td>Preparation of inspection plans</td></tr>
<tr><td>Preparation of maintenance plans</td></tr>
<tr><td>Definition of damage and imperfections limits</td></tr>
<tr><td>Definition of criteria for the selection of methods, methods of appraisal and review procedures</td></tr>
<tr><td>Approval requirements to start assembly, operation and dismantling</td></tr>
</table>

Figure 24. Possible barriers to manage the failure modes and the failure effects

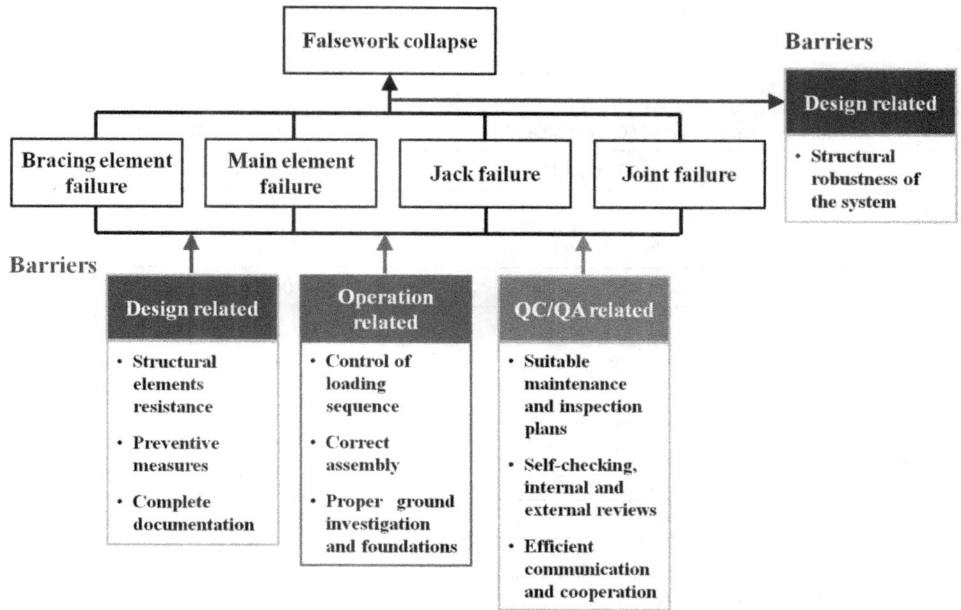

Figure 25. Case studies structural layout (André et al., 2017)

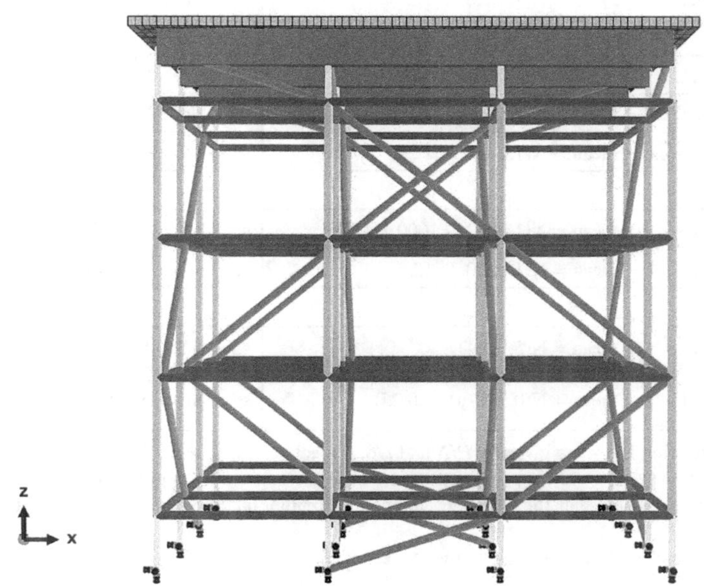

Table 15. Final set of random variables to be considered in the probabilistic analyses

Random variables	Prob. Dist.	Distribution parameters***		Min. value	Max. value
Initial geometrical imperfections					
Bow imp. factor*	Lognormal	mean = N(750,100,650,850)	sd = N(1000,250,500,1500)	50	2000
Sway imp. factor*	Normal	mean = N(625,100,500,700)	sd = N(1800,250,1000,2500)	50	2000
Material properties					
f_y, MPa	Lognormal	mean = N(419,10,400,440)	sd = N(20,5,15,25)	355	500
ε_u	Lognormal	mean = N(0.26,0.1,0.2,0.3)	sd = N(0.06,0.02,0.1,0.3)	0.1	0.3
Cuplok joint, strong bending axis					
Looseness, rad	Normal	mean = N(-0.008,0.005,-0.03,0.01)	sd = N(0.012,0.007,0.006,0.019)	0	0.04
k_{22Lc}, kN.m/rad	Weibull	shape = N(8,2,6,10)	scale = N(75,5,71,78)	30	90
k_{23Lc}, kN.m/rad	Weibull	shape = N(4.8,1.5,3.2,6.5)	scale = N(90,10,80,99)	30	120
k_{24Lc}, kN.m/rad	Weibull	shape = N(6.2,2.5,3.7,8.6)	scale = N(92,7,84,100)	30	140
d_{fc}**	Normal	mean = N(1,0.2, 0.75, 1.25)	sd = N(0.25, 0.1, 0.1, 0.5)	0.5	2
Spigot joint					
Looseness, rad	Normal	mean = N(0.005,0.005,0.001,0.01)	sd = N(0.007, 0.002, 0.005, 0.01)	0	0,04
k_{22s}, kN.m/rad	Normal	mean = N(128,10,110,140)	sd = N(35,15,20,50)	100	150
k_{23s}, kN.m/rad	Normal	mean = N(28,10,20,40)	sd = N(7,4,2,10)	20	50
d_{fs}**	Normal	mean = N(1,0.2,0.75,1.25)	sd = N(0.25,0.1,0.1,0.5)	0.5	2
Forkhead joint					
Looseness, rad	Normal	mean = N(0.005,0.005,0.001,0.01)	sd = N(0.007, 0.002, 0.005, 0.01)	0	0.04
k_{2f}, kN.m/rad	Normal	mean = N(29,10,20,40)	sd = N(5.5,5,2,15)	20	50
d_{ff}**	Normal	mean = N(1,0.2,0.75,1.25)	sd = N(0.25,0.1,0.1,0.5)	0.5	2
Brace joint					
k_{2b}, kN/m	Normal	mean = N(1360,250,1000,1500)	sd = N(322,150,100,500)	1000	2000
d_{fb}**	Normal	mean = N(1,0,2,0,75,1,25)	sd = N(0,25,0,1,0,1,0,5)	0,5	2
Baseplate joint					
Looseness, °	Normal	mean = N(5,1,3,4)	sd = N(2,1,1,3)	0	20
Maximum rotation, rad	Normal	mean = N(0.2,0.1,0.15,0.25)	sd = N(0.1,0.1,0.05,0.2)	0.1	0.5

*The local and global imperfection factor are given by H/Δ and L/δ, respectively, with H, L, Δ and δ representing the total system height, the element's length, the maximum system sway imperfection and maximum element bow imperfection, respectively.

**Deformation capacity factor (d_f) represents the ratio between the maximum joint deformation and the joint deformation at maximum force.

***N(a,b,c,d) represents a truncated Normal distribution with mean equal to a, standard deviation (sd) equal to b, minimum value equal to c and maximum valued equal to d.

also considered. The working wind velocity was considered equal to a wind pressure equal to 200 N/m^2 (BSI, 2010). The differential ground settlement was applied under a central column with a value equal to 100 mm.

The pressure applied to the formwork was selected as the leading action and was increased until structural collapse occurred.

The fragility and consequently the structural risk were analysed by means of predictive models. The procedure to validate, verify and select the predictive models is provided in André (2014) and André et al. (2017). In all cases the boosted trees family provided the best predictive models, either by the Stochastic Gradient Boosting (SGB) or by the Cubist model.

Using a surrogate model to foresee the actual behaviour under unknown and uncertain conditions introduces a component to the model uncertainty, besides the uncertainty of the numerical results. Both have been estimated and considered in the analysis.

5.6.2.2.5 Probabilistic Analyses

Twenty variables associated with the system's resistance were modelled as random variables, while the rest of the variables were considered deterministic with values equal to the mean values of the parent probabilistic distribution, with the exception of the stiffness (k_l) associated with joint looseness which was considered equal to 1 kNm.rad.

Regarding actions, only the value of the pressure load applied on top of the formwork surface was considered random. For the reliability analyses, the performance of the system was compared against a vertical pressure action modelled by a Normal distribution with mean value equal to 24.0 kN/m^2 and a COV equal to 0.075. On the contrary, the action distribution was considered uniform with a minimum value equal to 20 kN/m^2 and a maximum value equal to 26 kN/m^2 for the structural robustness and structural fragility analyses.

Figure 26. Example of the accuracy of SGB models for maximum resistance (André et al., 2017)

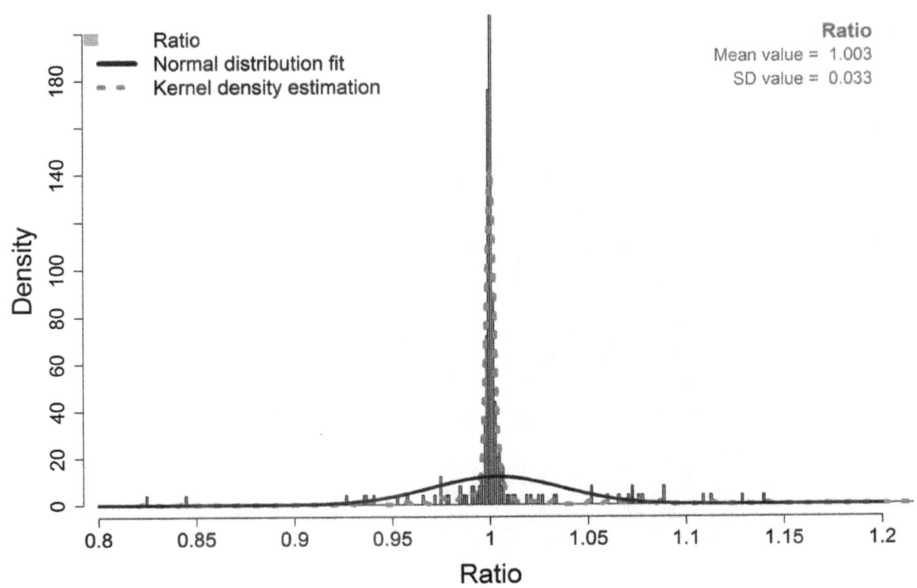

Figure 27. Accuracy of numerical models used (André et al., 2017)

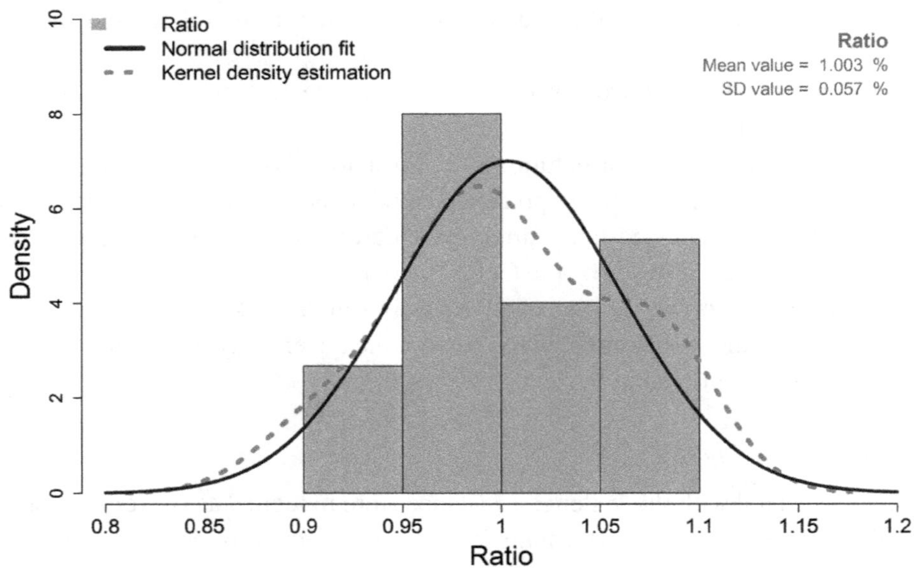

The results for case study CS2 are presented in Figure 28 to Figure 30. Observing the results in terms of reliability, robustness index and fragility index, it is possible to conclude that the value of the probability of failure is unacceptably high and the results for fragility demonstrate that it is very likely that significant damage will occur. The mean value of robustness index is not extremely low but it may not be enough to avoid collapse without warning. The effect of propagating uncertainty is also shown and it is seen that the variability value is considerable (between 10%-15% of variation of fragility values), see Figure 28 and Figure 30.

Figure 28. Histogram of the probability of failure, P_f, CS2 model (André et al., 2017)

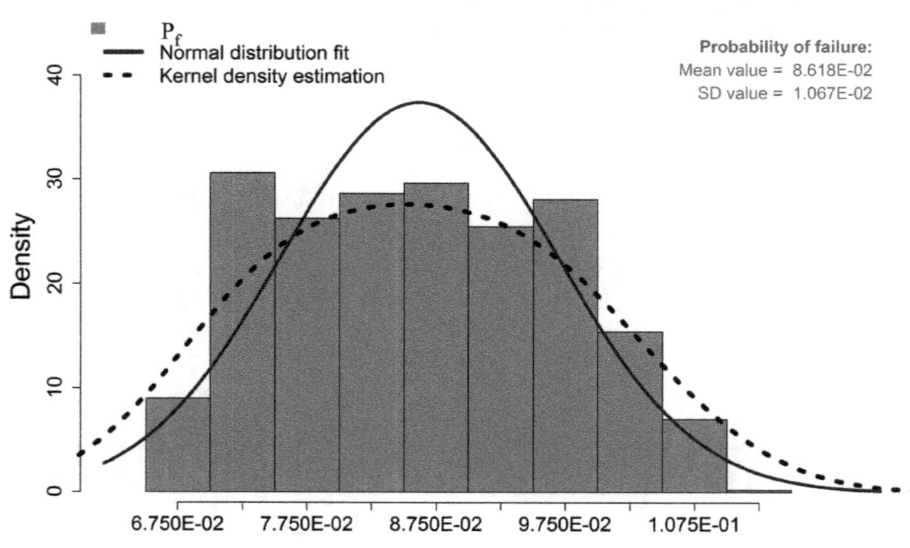

Figure 29. Histogram of the robustness index, I_R, CS2 model (André et al., 2017)

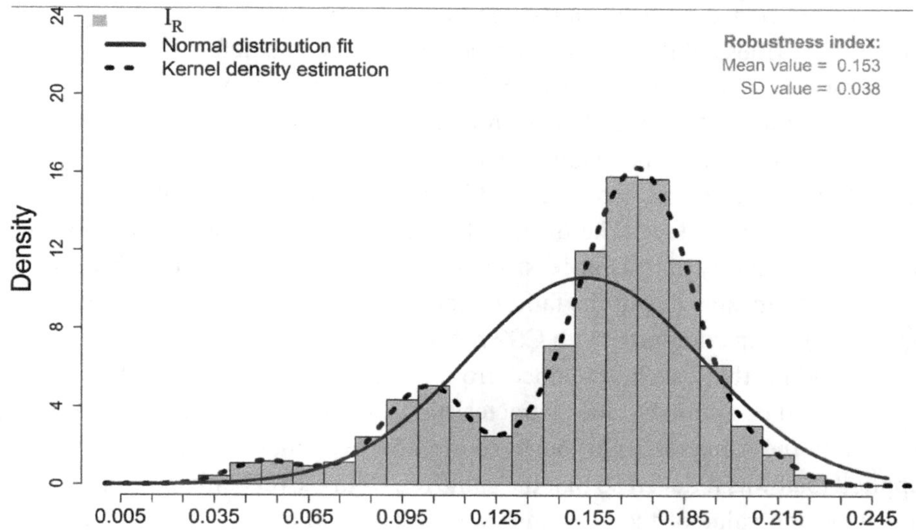

Figure 30. Empirical cdf of the fragility index, F_R, with the dispersion due to uncertainty propagation and highlighting the average curve, CS2 model (André et al., 2017)

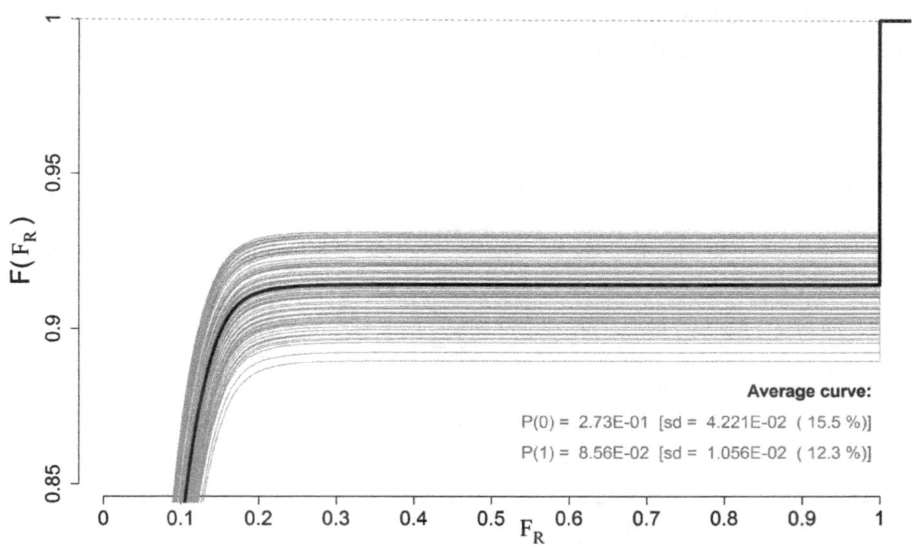

5.6.2.2.6 Additional Probabilistic Analyses

Based on the results presented in the previous Section it is necessary to discuss alternative strategies to increase the structural robustness and decrease the structural fragility of the bridge falsework Cuplok® system under analysis.

As detailed in Section 5.5.6, there are several possible strategies to increase the robustness of a structure. Nevertheless, some of these will be more or less efficient depending on the type of structure. For bridge falsework systems the following strategies seem more appropriate: increase resistance, increase structural integrity and increase ductility.

Applying the above concepts and guidance to the case study at hand, the values of the following random variables were modified: decreasing initial geometrical imperfections, decreasing looseness rotation (θ_1), increasing the k_2 stiffness of the cuplok joints and increasing the deformation capacity of the joints (d_f). The changes form an alternative (improved) scenario to the reference (baseline) case study (CS2) discussed previously. In practice, these changes reflect simple controls related to better quality checks, inspection and maintenance plans (i.e. quality management), see Table 16

The results of these changes are given in the Figure 31 to Figure 33 for case study CS2 improved model, CS2a. Results are expressive. The mean value of resistance and structural robustness increased also over 10% and the variability of resistance, structural robustness and structural fragility also decreased considerably. In fact, the mean value of the failure probability decreased four orders of magnitude (from 80000×10^{-6} to 7×10^{-6}) when compared with CS2 model results.

An interesting finding that can be obtained from the analyses results concerns with value of the partial factor that should be applied to the characteristic value of the resistance in order for the baseline model (CS2) to attain the same reliability level of the improved model (CS2a). To this end, the mean value of the applied load was reduced so that the same reliability level could be attained in both models. As a result, a new design value of the resistance was determined for each model. The modified partial factor was obtained by dividing the characteristic value of the CS2a model with the new design value CS2 model, for the same reliability level. The results are presented in Figure 34 for various values of the reliability index. For the conditions under analysis, the value of the modified partial factor is equal to 2.90. This value compares with 1.16 and 1.10 which are the resistance partial factor values obtained for the improved model and specified in BS EN 12812 for Class B1 structures, respectively. Such a

Table 16. Improved random variables values (changes highlighted in bold)

Initial geometrical imperfections	Minimum value	Maximum value
Local bow imperfection factor (l_{impf})	**1000**	2000
Global sway imperfection factor (g_{impf})	**1000**	2000
Cuplok joint, strong bending axis	Minimum value	Maximum value
Looseness (θ_{1c}), rad	0	**0.01**
Stiffness after looseness, 2 ledgers (k_{22Lc}), kN.m/rad	**60**	90
Stiffness after looseness, 3 ledgers (k_{23Lc}), kN.m/rad	**60**	120
Stiffness after looseness, 4 ledgers (k_{24Lc}), kN.m/rad	**60**	140
Deformation capacity factor (d_{fc})	**1**	2
Spigot joint	Minimum value	Maximum value
Looseness (θ_{1s}), rad	0	**0.01**
Deformation capacity factor (d_{fs})	**1**	2
Forkhead joint	Minimum value	Maximum value
Looseness (θ_{1f}), rad	0	**0.01**
Deformation capacity factor (d_{ff})	**1**	2
Brace joint	Minimum value	Maximum value
Deformation capacity factor (d_{fb})	**1**	2

Figure 31. Histogram of the probability of failure, P_f, CS2a model (André et al., 2017)

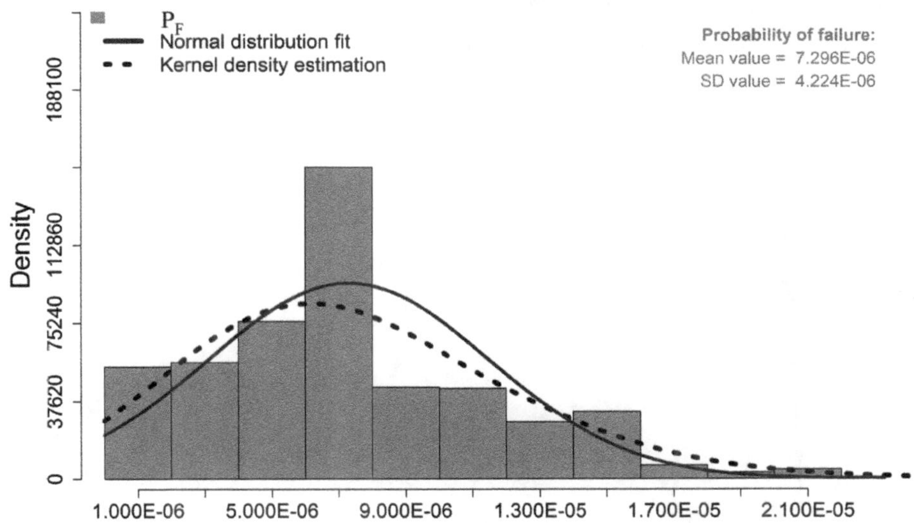

Figure 32. Histogram of the robustness index, I_R, CS2a model (André et al., 2017)

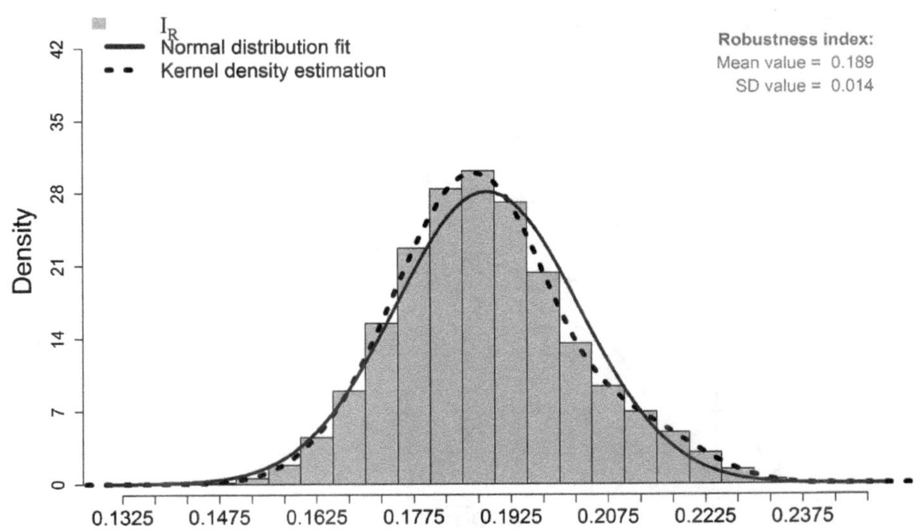

large difference between values clearly illustrates the significant adverse effects on the performance of falsework of poor quality management.

As the CS2a model consists of the CS2 model plus the effects of improved quality management provisions (presented in Table 16), the modified partial factor can be considered to be an estimate of the resistance partial factor value that should be applied to the characteristic value of the resistance of falsework systems that exhibit geometrical imperfections and joint looseness values of the same order of magnitude as showed in Table 15, so that adequate margins of safety are maintained.

Figure 33. Empirical cdf of the fragility index, F_R, with the dispersion due to uncertainty propagation and highlighting the average curve, CS2a model (André et al., 2017)

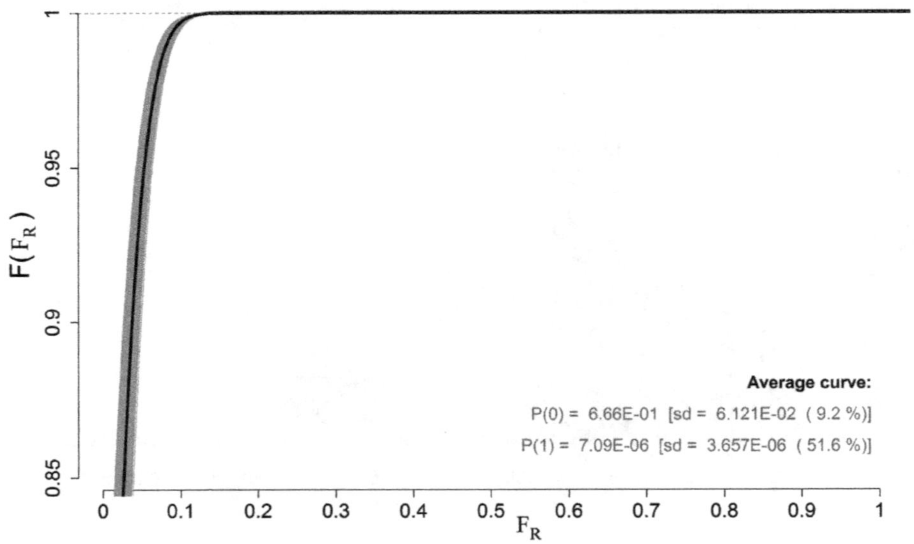

Figure 34. Partial factor as a function of the reliability index

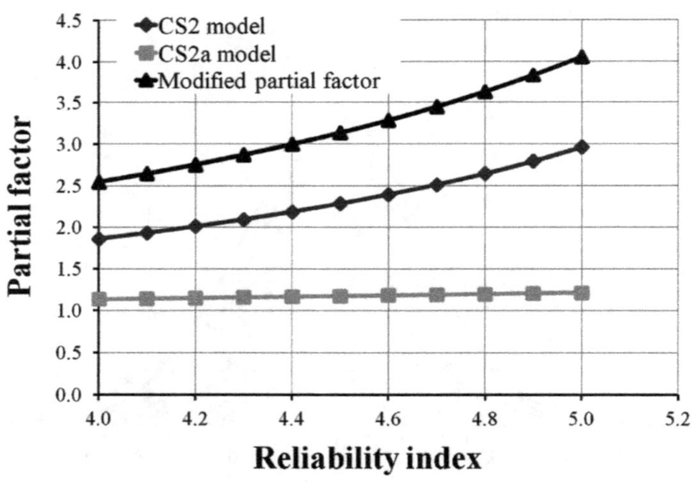

5.6.2.2.7 Risk Evaluation

Assuming an acceptable probability of failure equal to 1×10^{-5}, and comparing it with the results presented in previous Sections it is possible to conclude that in the original (reference) scenario the risk is exceedingly high and cannot be accepted or tolerated. Therefore, corrective measures need to be implemented to lower the risk level to acceptable, or tolerable, levels. By applying better quality management, it can be seen that it was possible to decrease considerably the failure probability to a level within the range of acceptability.

5.6.2.2.8 Risk Control and Risk Informed Decision-Making

In this Section an economical justification for adopting the improved (alternative) scenario, instead of the reference (baseline) scenario, will be analysed. In order to perform this analysis, a cost function must be derived, such as the one suggested in André (2014).

It was assumed that the sum of the cost of the structure supported by the bridge falsework with the cost of the bridge falsework, C_{max}, was equal to £200,000.00. Additionally, the function between fragility index and damage costs was considered to be linear, see Figure 35.

It was assumed that implementing the improved quality management represents a fraction, e.g. 20%, of the total cost of a new bridge falsework system, per use. Fixing this latter value at £20,000.00 (based on material's cost and labour cost), the extra costs associated with the alternative scenario are estimated to be equal to £4,000.00 (2014 prices), per use.

The benefits are calculated using the Value of Preventing a Fatality (VPF) concept, which is fixed annually in the UK by the Department for Transport (DfT). The 2014 number is equal to £1,700,000.00. Benefits are calculated by the improvements relative to the worst case scenario: the collapse of the structure, i.e. when structural fragility equals one. As a simplification it was considered that benefits decrease linearly with the fragility index, see Figure 36. The maximum benefits value (B_{max}) was considered equal to 50% of the VPF. This value was estimated taking into account the possible differences between the probabilities of various injury levels when fragility is equal to zero (due to falls from height or being struck by an object during assembly of the falsework for example) and equal to 1.0 (due to structural collapse of the falsework).

Considering CS2 and CS2a models as two independent bridge falsework structures subject to uncertain actions, a single use per year of each structure and that only one person is at risk per use, the cdf of the relative Net Value (equal to the Net Value of the CS2a model minus the Net Value of the CS2 model) between choosing the improved scenario (CS2a model) and the reference scenario (CS2 model) is presented in Figure 37 (light grey curve). In this analysis, structural fragility was calculated considering a vertical pressure action modelled by a Normal distribution with mean value equal to 24.0 kN/m^2 and a COV equal to 0.075.

Figure 35. Functions between Costs and Fragility

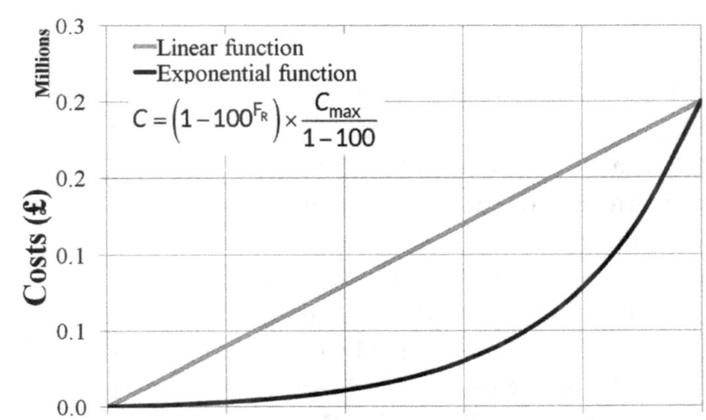

Figure 36. Functions between Benefits and Fragility

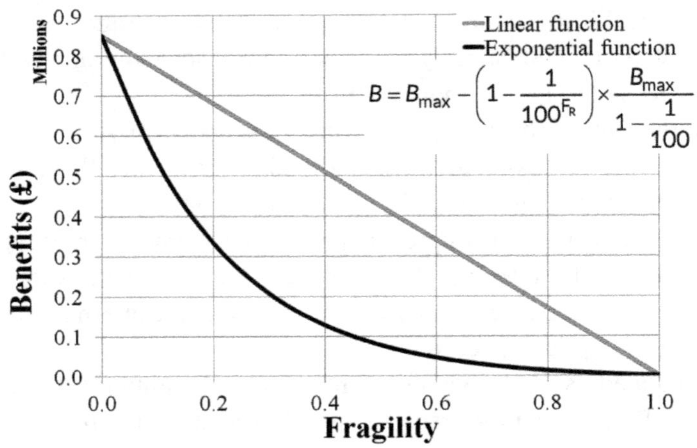

Figure 37. Cdf of relative Net Value (for CS2 type models), linear and exponential functions

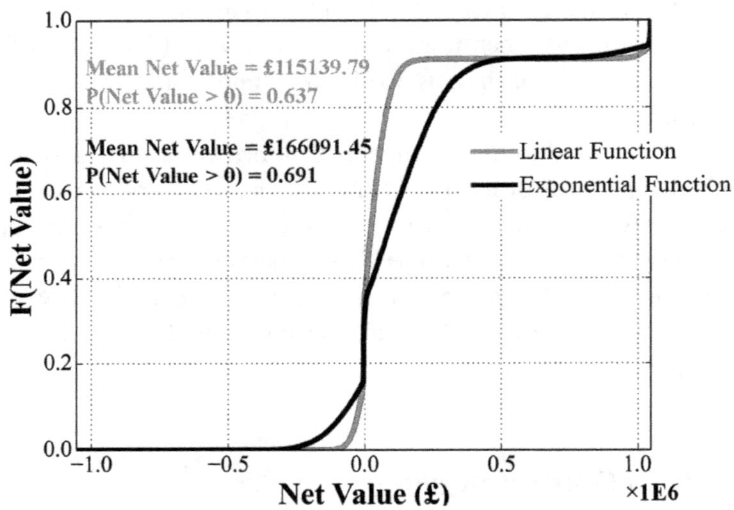

It can be observed that there is approximately 64% probability that a positive relative Net Value is obtained, with a mean relative Net Value of more than £100,000.00. It can be concluded that the choice of selecting the improved scenario, CS2a model, over the reference scenario, CS2 model, is justified since the additional costs incurred by adopting better quality management are outweighed by the dramatic reduction in individual and structural risks.

It is of interest to study how the relative Net Value varies for instance with the function between costs and structural fragility, and between benefits and structural fragility. Choosing an exponential law instead of a linear law leads to the dark curve shown also in Figure 37. It can be observed that with this modification the cdf of the relative Net Value is considerably shifted with the mean relative Net Value increasing significantly and the probability that a positive relative Net Value is obtained also increasing. This occurs because of the different configuration of the fragility curves of the reference and improved scenarios.

It can be concluded that if the cost of the permanent structure significantly exceeds (about one order higher) the cost of the temporary structure, the extent of improvements in terms of structural and economical risks completely justified the small extra costs incurred by adopting better quality management in the modified scenario.

5.6.3 Discussion of the Use of Reduction Factors and Target Reliabilities

5.6.3.1 Basis

According to the design codes approaches are design strategies, temporary structures are designed for both serviceability limit states (SLS) and ultimate limit states (ULS) using reduced return periods for environmental loads, such as the 10 or the 15 year return period for ULS wind action. It can be argued that this methodology is justified due to the smaller exposure time of temporary structures to hazards such as extreme wind gusts. However, the choice of the return periods to be considered is often quite arbitrary and tries to balance empirically the optimal use of resources at an acceptable safety level.

One should not forget that return period is just an alternative statement of annual risk of exceedance, e.g. a 50 year return period is equivalent to say a probability of exceedance of 0.02 in one year and a 10 year return period is equivalent to say a probability of exceedance of 0.1 in one year. Therefore, using lower return periods in the design of temporary structures is equivalent to accepting a higher risk of annual load exceedance than the one considered for permanent structures, see also Blackmore & Freathy (2004).

For example, assuming that the maximum wind velocities recorded in each year are discrete, identically distributed and statistically independent events, the risk of exceedance, over 10 years, of the 10 years and 50 years return periods wind velocities is given by Eq. 55, where m represents the number of years. Accordingly, the probability that the 10 years return period wind velocity will be exceeded at least once during a 10 years period is greater (64%) than the one estimated for the 50 years return period wind velocity (18%).

$$P_m = 1 - \left(1 - P_1\right)^m \tag{55}$$

Rosowsky (1995) argues that the use of design loads as specified in structural codes, which can in most cases correspond to maximum lifetime loads, to temporary structures may be excessively conservative. Rosowsky then proposed a method based on the concept of maintaining comparable load exceedance probabilities to modify the partial factors of loads to take into account reduced reference periods. This concept is based on the philosophy that the probability of exceedance of a given nominal load (for a given reference period) should be the same for every reference period considered. Accordingly, for an exposure time of less than one year a reduction factor of 0.85 is suggested to be applied to the 50 years return period wind velocity. A somewhat similar analysis is presented by Boggs & Peterka (1992) and Willford & Allsop (1990). In the latter document, it is proposed that the wind velocity to be considered during the construction of buildings, including the design of temporary structures, could just represent between 77% to as low as 55% of the design value specified in the design code for the permanent structure (for an exposure of less than two years). According to Mohammadi & Heydari (2008) the method of the reduced load level has become popular among designers of temporary structures, such as bridge falsework systems.

Calgaro, Tschumi, & Gulvanessian (2010) and CEN (1996) justify the use of return periods shorter than those agreed for persistent design situations for temporary structures by the important hidden additional reliability margins included in the climatic action models. As will be shown in the following, this is obviously not the case at least for wind action.

It is also argued that if a full independence of failure probabilities during transient and persistent design situations is assumed, by reducing for transient situations the mean return periods proportionally to the duration of the situations one gets the same failure probability during one transient design situation and one persistent design situation. Furthermore, the number of expected failures during transient situations is said to be obviously very low by comparison with what is accepted for persistent situations. This is obviously an incorrect conclusion, since if the same target annual reliability is used in both transient and persistent situations, the same probability of failure is achieved (assuming stationary processes for actions and resistances) at the end of the period corresponding to the transient situation.

Finally, it is argued that a reduced return period is acceptable for the design of temporary structures since this situation may be considered to be analogous to the choice of combination values for persistent situations. However, the straightforward uninformed application of this analogy is flawed by the fact that in general the design of temporary structures is controlled by a single dominant action.

In the following, a simple exercise is presented to analyse the adequacy of the use of reduction factors to derive the design wind load for temporary structures, including bridge falsework systems. It should be mentioned that wind loads have been chosen to illustrate existing relations between exposure time, return period and reliability. However, in some cases it may not necessarily be the most important action for the design of temporary structures. See André et al. (2013b) for complete details.

5.6.3.2 Probabilistic Models of Wind Velocity

One crucial point when analysing the use of reduction factors relates to the accuracy of the probabilistic models used to derive the design loads of climatic actions, for instance wind. In Europe, the existing design codes use the Gumbel distribution (one of the three Generalised Extreme Value (GEV) distributions) to obtain extreme value predictions of the maximum wind velocities for high values of return periods. This distribution predicts unlimited values of the wind velocity as the return period increases, which can overestimate the actual maximum wind velocity physically possible that can be generated in earth's atmosphere. However, this might be counterbalanced by the uncertainty that will always exist by using a finite size sample of the data to determine the distribution parameters, see Coles (2001) and Leadbetter, Bailey, & Rootzén (1983). According to Holmes (2007), "*the approach of extracting a single maximum value of wind velocity from each year of historical data obviously has limitations in that there may be many storms during any year and only one value from all these storms is being used*", but also only data values that are statistically independent are used, meaning that only the maximum wind velocity value per storm is used. If there are multiple similar events in each year the GEV approach might underestimate the load values for small return periods.

Additionally, Fawcett (2005) by analysing recorded wind velocities and including the seasonal variations in each year, showed that the results obtained with the cluster peaks approach were less accurate and less precise than the results obtained when all the data, above a properly chosen wind velocity threshold, are used, according to the Peaks Over Threshold approach (POT). The study revealed that, *at levels of temporal dependence often encountered in real-life data, the cluster peaks analyses were constantly underestimating return levels relative to the analyses making use of all threshold excesses. This suggests that designing to the maximum likelihood estimates which use cluster peak excesses – an approach currently employed by most practitioners – would result in considerable underprotection. However, further investigation has also revealed that this underestimation becomes even more pronounced as the strength of temporal dependence is increased, but for low temporal dependence the cluster peaks analyses actually overestimate return levels. (Fawcett, 2005)*

There is also the question of by how much is a given return period wind gust velocity exceeded. In the example presented in André, 2014) and André et al. (2013b) it was found that the 2 and 10 years return periods wind gust velocities are exceeded at maximum by 47% and 27%, respectively, and on average by 9% and 10%, respectively. Owing to the squared relationship between wind velocity and wind pressure, the magnitudes of exceedance of the latter variable are even larger. Taking into account the statistical uncertainties presented in the previous two paragraphs, together with the design uncertainties related to wind load effects on structures and to the structural response, the risk of failure in some structures under wind loading might be considered unacceptable.

5.6.3.3 Partial Factors for Wind Loads

Therefore, it is important to check whether the usual wind load partial factor specified by the most recent structural codes, e.g. BS EN 1991-1-4, accounts for the variability observed in the wind measurements and for the size of exceedance over a given wind velocity threshold. In general, the wind load partial factor, γ_W, is separated into two factors: γ_f that covers the uncertainty on the value of the action itself and γ_{Sd} that covers the uncertainty in modelling the effects of actions.

The conventional values of these two factors are 1.35 and 1.10 (Gulvanessian, Calgaro, & Holický, 2002). If the latter value has been derived from comparison studies between results obtained by design models and measurements of wind effects on structures, the former *"has no scientific justification, and results from engineering judgement"* (Gulvanessian et al., 2002). Nevertheless, it is understood that this value leads in general to safe design load values. However, in some cases the value of 1.35 may not suffice to account for the variability of wind velocity and wind pressure, such as in the example presented in the previous Section (André, 2014; André et al., 2013b).

It is also important to verify if the notional reliability indices associated with using the specified partial factor to be applied to the wind actions satisfy the specified target reliability indexes. According to André (2014) and André et al. (2013b), the partial factor, γ_W, to be applied to the wind action effects can be obtained by Eq. 56.

$$\gamma_W = 1.10 \cdot \frac{1 + \left\{-0.45 - 0.78 \cdot \ln\left[-\ln\left[\Phi\left(0.7 \cdot \left[\Phi\left(\beta_n\right)\right]^{\frac{1}{n}}\right)\right]\right]\right\} \cdot V_{W[1]}}{1 + \left\{-0.45 - 0.78 \cdot \ln\left[-\ln\left(1 - \frac{1}{R}\right)\right]\right\} \cdot V_{W[1]}} \tag{56}$$

where R represents the return period and n represents the exposure time.

According to JCSS (2001), the value of the coefficient of variation, V, of the annual maximum wind velocities, $V_{W[1]}$, depend on the climate and usually assumes values between 0.10 and 0.35. Here a value of 0.26 will be considered, the same value considered in BS EN 1991-1-4 (BSI, 2010).

It would be very interesting to understand how the value of partial factor, γ_W, varies with different R values and also with the design value of the maximum wind velocities (function of the reliability index value and of the exposure time, n; note that the latter values may be different from the values of R). The results of varying the exposure time (n: 2, 10, 20 and 50 years), the return period (R: 2 to 100 years) and the reliability index (β: 1 to 5) are illustrated in Figure 38.

It may be observed that for the same return period, the annual reliability index achieved with $\gamma_W = 1.5$ does not change with the exposure time, as expected. For instance for $R = 50$ years return period $\beta_{50} \approx 3.5$ (i.e. $\beta_1 \approx 4.4$) for $n = 50$ years, and $\beta_2 \approx 4.3$ (i.e. $\beta_1 \approx 4.4$) for $n = 2$ years. An important observation is that the notional structural risk level for the wind action achieved following the existing structural codes can be higher than the target probability of failure. For instance, the notional annual probability of failure($P_{f,1}$) achieved for $R = 50$, $\gamma_W = 1.5$ and $n = 50$ is larger than the specified annual probability of failure for a structure whose collapse would have high or medium consequences in terms of human life, economy, society or environment (Consequence Class CC3 or CC2 in BS EN 1990 (BSI, 2005), 1×10^{-7} and 1×10^{-6}, respectively), although smaller than the specified annual probability of failure (1×10^{-5}) for a structure whose collapse would have low consequences (Consequence Class CC1).

If in the case of the design of a permanent structure against wind loads, a lower target reliability than Consequence Class CC2 ($\beta_1 = 4.7$) could be accepted since the relative cost of safety measures is high;

Figure 38. Wind load partial factor (red curve represents values for γ_W = const. = 1.5).

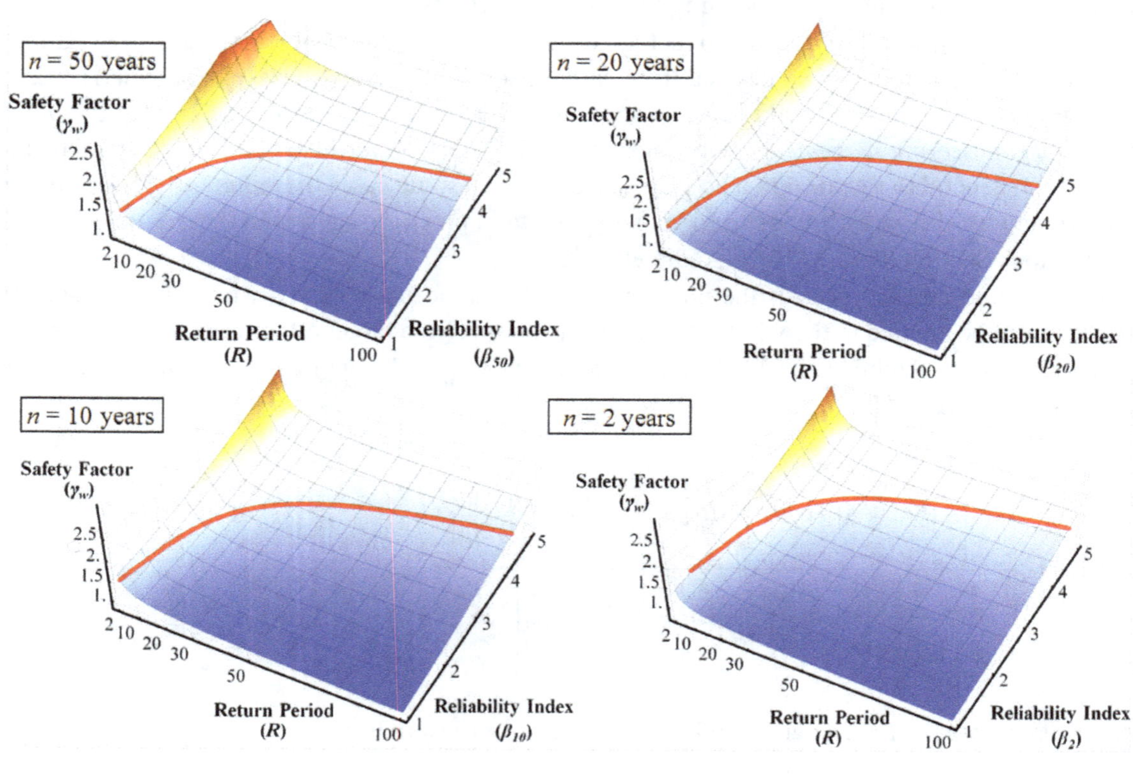

in the case of a temporary structure an annual target reliability at least equal to 4.7 may be used due to the low relative cost of safety measures.

It may also be observed that for reduced exposure times, for example 10 years, a structure designed considering a load return period of 10 years and a partial factor equal to 1.5, has an annual probability of failure of 4.0×10^{-4} ($\beta_1 \approx 3.4$) which is even higher than the Consequence Class CC1 target reliability value ($\beta_1 = 4.2$). In this case, in order to achieve the CC2 annual target reliability it would be necessary to adopt a partial factor equal to 2.0 (instead of 1.5), or a return period of 80 years (instead of 10 years). Therefore, the standard use of reduction factor for short exposure times is further placed in question.

The high number of collapses of temporary structures resulting from the January and February 1990 wind storms in the UK, see Buller (1993), along with the severe storms of 1987, 2002, 2004, 2010 and 2011, and the trend for an increase of their frequency in the future due to the global warming (Fawcett, 2005), can be used as evidence to support the use of an improved approach to analyse the wind data and to determine more accurate reduction factors to determine design wind loads for short return periods. An improved methodology should take into account not only the definition of the exceedance probability percentile, i.e. from which return periods (*R*) of loads are obtained, but all wind velocity values higher than a certain threshold should also be included in the assessment of design wind velocities.

This approach can be further enhanced if the assumption of independence between successive extremes within seasons is broken and the short-term temporal dependence between consecutive extremes is taken into account. In this context, the Markovian dependence structure lends great appeal, either using first-order Markov chains or higher order if greater precision is necessary. The distribution of consecutive wind extreme events can be modelled using a bivariate distribution or a multivariate logistic distribution, constructed in a way such that the marginals have a Generalised Pareto Distribution (GPD). See Fawcett (2005) for details and an example. To solve this complex problem, simulation techniques such as the Markov Chain Monte Carlo (MCMC) method can be used, see Fawcett (2005). Additionally, Bayes' Theory can be used to update the probability functions of the distribution parameters to properly account the model uncertainties in the results. Furthermore, Bayes' Theory can be used to explicitly estimate the wind loads for a given return period, based on the posterior distribution, thus avoiding the issues involved when using the MLE method to obtain estimates of the distribution parameters (Coles, 2001).

A further and important issue related to the design of temporary structures is the multiple reuse cycles during their design working life. For example, Rosignoli (2007) argues that because these structures are reused many times the meteorological loads should therefore be determined without reductions in relation to the work duration.

This problem is clearly explained in an excellent paper by Hill (2004). Hill gives an example of two cities, Constantown and Fickleville, where the first city maintains public buildings in service for 100 years while the second demolishes and rebuilds public buildings with more than five years (which therefore are designed for five year service life). Hill questions: which buildings are safer, Constantown's or Fickleville's? The answer is simple: it is likely that Constantown's buildings will stay in service for more than 100 years, therefore the probability that one building of Constantown is damaged after 100 years in service is greater than the probability of each one of Fickleville's buildings over the same period. However, for the same exposure time, the Fickleville's population is more likely to suffer injuries than Constantown's. Therefore Hill concludes that establishing design loads on the basis of service-life assumptions may result in significant safety inequities.

The information presented in the previous paragraphs has an immediate application to temporary structures. For example, it may be decided that a given bridge falsework structure should have the same probability of failure at the end of its design working life, 20 years for instance, than a building at the end of the usually considered design working life, 50 years, for a Consequence Class CC2 ($\beta_{20,\text{falsework}} = \beta_{50,\text{building}} = 3.8$). The partial factor to be applied to the wind action is plotted in Figure 39 (André, 2014; André et al., 2013b). It may be observed that considering return periods less than the design working life for the bridge falsework, for instance 10 years, would imply the use of a higher partial factor, 11% in this example. However, this is only true if the yearly probability of failure of the bridge falsework system is constant in all the reusage cycles during their design working life. This may not be the case: bridge falsework elements can be reused several times in various bridge projects with their own design particularities (design teams, ground characteristics, workers skills, maintenance plans, etc.) in different places with different types of weather.

It may be concluded that the choice of the appropriate reduction factors to determine the design wind velocities for short return periods is influenced by many uncertainties: from the validity of the assumptions regarding the stationarity and the temporal independence of the measured data, to the methods used to fit probabilistic distributions to the wind velocity data records and to obtain the distribution parameters and moments, ending in considerations about the design working life of bridge falsework systems. Furthermore, it was demonstrated that if load values are derived from return periods less than the design working life of bridge falsework structures than in order to achieve an acceptable risk at the end of this period it is necessary to use larger partial factors applied to the loads, which is often not the case, see also Sexsmith (1998).

Therefore, it appears reasonable to recommend not using any reduction factor and adopt return periods equal to the design working life when designing temporary structures. At least, for cases where meteorological data and site specific information is not available that can be used to reduce the uncertainty

Figure 39. Wind load partial factor, considering a consequence class CC2 ($\beta_{20,falsework} = 3.8$)

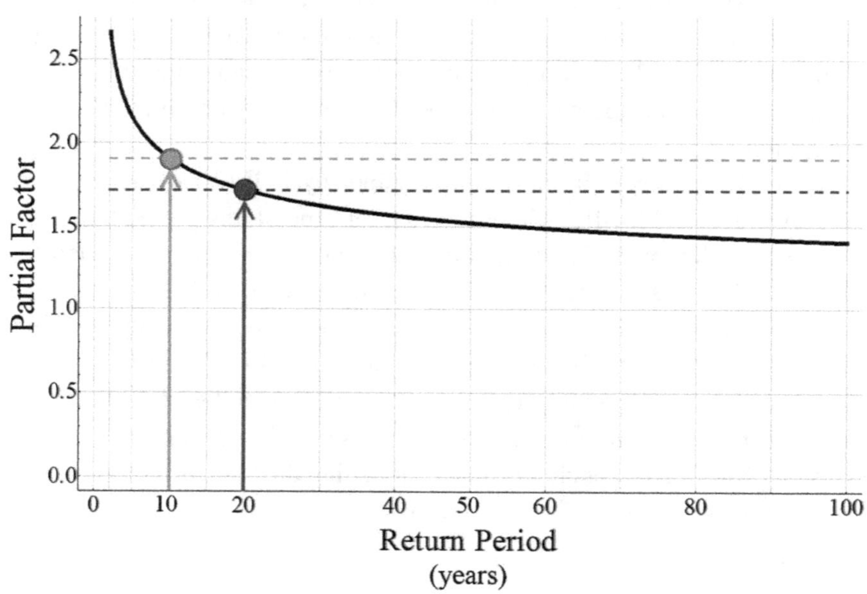

levels. In particular for short term usage periods of temporary structures. This recommendation is also well supported by the findings published by Sexsmith & Reid (2003) and Tamura, Wu, & Yang (2015) which found that the use of reduced load return periods, such as a ten year wind load, can be seriously irrational. For example, if the cost of a falsework failure is 500 times the cost of providing additional safety, the load factor to be applied to the ten year wind load in the Vancouver region, in Canada, would be higher than 2.5 in order to minimise the total expected costs (Sexsmith & Reid, 2003). Blackmore & Freathy (2004) only recommend using reduced return periods for climatic actions when the temporary structure is placed at a safe distance from public access to minimise the risk of injury in case of a failure, or where an adequate monitoring system of the wind velocity is used and effective evacuation procedures and exclusion areas are planned, see IStructE (2007) and also Chapter 8 for more information on site management.

5.7 CONCLUSION

In this Chapter a complete and detailed review of structural reliability and structural risk methods applied to structural engineering was presented.

At the start, a comprehensive list of the most relevant definitions was given. Next, the basis for the calibration of modern design codes was provided, allowing the reader to correctly interpret the safety format philosophy of the existing design codes.

Afterwards, the classical and advanced structural reliability methods were presented, contributing to the correct use of the latter methods to demonstrate structural adequacy.

Following, the risk management framework for structural design was detailed, including risk analysis and risk control. The basis for risk informed decision-making was also presented.

A novel framework for fragility and robustness was presented. The relative advantages of this new framework when compared with the classical approach were extensively detailed. Design strategies to reduce risk and enhance robustness were presented.

This Chapter also proposed a new design philosophy for temporary structures where several consequences classes were presented, each one with its own design requirements, design methods, inspection and supervision classes. The suggested design philosophy accounts for the safety with respect to identified and unidentified hazard events.

An application example of the novel framework for fragility and robustness was presented, using a bridge falsework system. Based on the results, the Chapter recommended that, as the cost of improving temporary structures against collapse and management procedures is a relatively small amount, and that the costs of temporary structures failures, both financially and in terms of potential life losses, can have a significant impact on companies and society, that these structures should be designed and maintained to higher standards than currently adopted in practice.

Wind is a hazard with the potential to cause the collapse of temporary structures under extreme conditions. At the end of this Chapter, the statistical procedures behind the determination of wind pressures are discussed. Based on the results obtained and contrary to commonly adopted design principles, the authors recommend not using any reduction factor and adopt return periods equal to the design working life, or longer, when designing temporary structures, for cases where meteorological data and site specific information are not available.

REFERENCES

ANCOLD. (2003). Guidelines on risk assessment. The Australian National Committee on Large Dams (ANCOLD).

Ando, T. (2010). *Bayesian model selection and statistical modeling*. Boca Raton, FL: Chapman & Hall/CRC. doi:10.1201/EBK1439836149

André, J. (2014). *Determination of the main parameters affecting the performance of bridge falsework systems* (PhD Thesis). Oxford Brookes University, Oxford, UK.

André, J., Beale, R., & Baptista, A. (2012). A survey of failures of bridge falsework systems since 1970. *Proceedings of the ICE - Forensic Engineering, 65*(4), 161–172. doi:10.1680/feng.12.00012

André, J., Beale, R., & Baptista, A. (2013a). Experimental behaviour of bridge falsework joints. In *Proceedings of the IX Portuguese National Conference of Steel and Composite Construction* (p. 10). Guimarães, Portugal: The Portuguese Steelwork Association (cmm).

André, J., Beale, R., & Baptista, A. (2013b). Recent advances and existing challenges in the design of bridge falsework systems. *Civil Engineering and Environmental Systems, 30*(2), 130–145. doi:10.1080/10286608.2012.733374

André, J., Beale, R., & Baptista, A. (2014a). Numerical investigation of bridge falsework structures. In *Proceedings of the 5th Portuguese Structural Engineering Conference (JPEE 2014)* (p. 15). Lisbon, Portugal: Portuguese National Laboratory for Civil Engineering (LNEC).

André, J., Beale, R., & Baptista, A. (2014b). Robustness and risk evaluation of bridge falsework structures. In *Proceedings of the 5th Portuguese Structural Engineering Conference (JPEE 2014)* (p. 15). Lisbon, Portugal: Portuguese National Laboratory for Civil Engineering (LNEC).

André, J., Beale, R., & Baptista, A. (2015). New indices of structural robustness and structural fragility. *Structural Engineering & Mechanics, 56*(6), 1063–1093. doi:10.12989/sem.2015.56.6.1063

André, J., Beale, R., & Baptista, A. (2017). *Risk analysis of bridge falsework Cuplok systems*. Accepted for Publication in Structure and Infrastructure Engineering.

Ang, A., & Tang, W. (1975). *Probability concepts in engineering planning and design: Basic principles* (1st ed.; Vol. 1). Wiley.

Ang, A., & Tang, W. (2007). *Probability concepts in engineering: Emphasis on applications in civil and environmental engineering* (2nd ed.). London, UK: Wiley.

Apeland, S., Aven, T., & Nilsen, T. (2002). Quantifying uncertainty under a predictive, epistemic approach to risk analysis. *Reliability Engineering & System Safety, 75*(1), 93–102. doi:10.1016/S0951-8320(01)00122-3

Artamonov, A., Bendel, C., Blueckert, C., Landrin, H., Spencer, M., & Wallace, P., … Wannick, H. (2008). Bridges – Construction, insurance and risk management. In *Proceedings of the 41st annual Conference of The International Association of Engineering Insurers (IMIA)* (p. 44). Gleneagles, Scotland: The International Association of Engineering Insurers (IMIA).

ASCE. (2010). ASCE 7–10 – Minimum design loads for buildings and other structures. American Society of Civil Engineers (ASCE).

Aven, T. (2004). On how to approach risk and uncertainty to support decision-making. *Risk Management, 6*(4), 27–39. doi:10.1057/palgrave.rm.8240196

Ayyub, B., & McCuen, R. (2011). *Probability, statistics, and reliability for engineers and scientists.* London, UK: Taylor and Francis.

Baker, J., Schubert, M., & Faber, M. (2008). On the assessment of robustness. *Structural Safety, 30*(3), 253–267. doi:10.1016/j.strusafe.2006.11.004

Bažant, Z., Le, J., Greening, F., & Benson, D. (2008). What did and did not cause collapse of World Trade Center Twin Towers in New York? *Journal of Engineering Mechanics, 134*(10), 892–906. doi:10.1061/(ASCE)0733-9399(2008)134:10(892)

Bažant, Z., & Verdure, M. (2007). Mechanics of progressive collapse: Learning from World Trade Center and building demolitions. *Journal of Engineering Mechanics, 133*(3), 308–319. doi:10.1061/(ASCE)0733-9399(2007)133:3(308)

Beale, R. (2014). Scaffold research — A review. *Journal of Constructional Steel Research, 98*, 188–200. doi:10.1016/j.jcsr.2014.01.016

Beale, R., & Godley, M. (2003). Common causes of failure of metal scaffolds in the UK and stability checks on scaffolds. In *Proceedings of the Technical Seminar on Metal Scaffolding (Falsework) – Design, Construction and Safety* (pp. 62–79). Hong Kong, China: Hong Kong Institute of Steel Construction.

Bedford, T., & Cooke, R. (2001). *Probabilistic risk analysis: Foundations and methods.* Cambridge, UK: Cambridge University Press. doi:10.1017/CBO9780511813597

Bellinger, W. (2007). *The economic analysis of public policy.* Routledge. doi:10.4324/9780203946480

Benjamin, J., & Cornell, C. (1970). *Probability, statistics, and decision–making for civil engineers* (3rd ed.). New York: McGraw-Hill.

Berry, L., Lindsay, J., Salomon, E., & Veale, S. (2006). Risk, responsibility and regulation – Whose risk is it anyway?. London, UK: UK's Better Regulation Commission.

Birnbaum, M. (2008). New paradoxes of risky decision making. *Psychological Review, 115*(2), 463–501. doi:10.1037/0033-295X.115.2.463 PMID:18426300

Blackmore, P., & Freathy, P. (2004). Wind loads on temporary stage decks (No. Special Digest 483). Bracknell, UK: Building Research Establishment (BRE).

Blockley, D. (1992). *Engineering safety.* London, UK: McGraw-Hill Book Company.

Boggs, D., & Peterka, J. (1992). Wind speeds for design of temporary structures. In *Proceedings of the 10th ASCE Structures Congress.* San Antonio, TX: American Society of Civil Engineers (ASCE).

Bowles, D. (2003). ALARP evaluation: Using cost effectiveness and disproportionality to justify risk reduction. In *Proceedings of ANCOLD 2003 Conference on Dams* (p. 17). Hobart, Australia: The Australian National Committee on Large Dams (ANCOLD).

Bowles, D. (2007). Tolerable risk for dams: How safe is safe enough? In *Proceedings of the US Society on Dams Annual Conference* (p. 24). Philadelphia: United States Society on Dams (USSD).

Box, G., & Draper, N. (1987). *Empirical model–building and response surfaces*. Wiley.

BSI. (2005). BS EN 1990:2002+A1:2005 – Basis of structural design. London, UK: British Standards Institution (BSI).

BSI. (2010). BS EN 1991–1–4:2005+A1:2010 – Actions on structures – General actions – Part 1–4: Wind actions. London, UK: British Standards Institution (BSI).

BSI. (2011). BS 5975:2008+A1:2011 – Code of practice for temporary works procedures and the permissible stress design of falsework (3rd ed.). London, UK: British Standards Institution (BSI).

BSI. (2014). BS EN 1991–1–7:2006+A1:2014 – Actions on structures – Part 1–7: General actions – Accidental actions. London, UK: British Standards Institution (BSI).

Buller, P. (1993). Gales of January and February 1990: Damage to buildings and structures (No. BR 248) (p. 24). Bracknell, UK: Building and Research Establishment (BRE).

Bunni, N. (2003). *Risk and insurance in construction* (2nd ed.). London, UK: Taylor & Francis.

Cajot, L., Haller, M., Conan, Y., Sedlacek, G., Kraus, O., Rondla, J., … Johansson, B. (2003). Probabilistic quantification of safety of a steel structure highlighting the potential of steel versus other materials. Brussels, Belgium: European Commission.

Calgaro, J., Tschumi, M., & Gulvanessian, H. (2010). *Designers' Guide to Eurocode 1: Actions on bridges*. London, UK: Thomas Telford Ltd. doi:10.1680/dgeab.31586

CEN. (1996). Background document to ENV 1991–2–6:1996 – Basis of design and actions on structures – Part 2–6: Actions during execution. Brussels, Belgium: European Committee for Standardization (CEN).

Chandrangsu, T., & Rasmussen, K. (2011). Structural modelling of support scaffold systems. *Journal of Constructional Steel Research*, *67*(5), 866–875. doi:10.1016/j.jcsr.2010.12.007

Chilton, S., Jones-Lee, M., Loomes, G., Robinson, A., Hopkins, L., & Cookson, R. … Beattie, J. (1998). Valuing health and safety controls. A literature review (No. CRR 171). London, UK: Health and Safety Executive (HSE).

CIRIA. (1977). Rationalization of safety and serviceability factors in structural codes (No. 63). London, UK: Construction Industry Research and Information Association (CIRIA).

CIRIA. (1996). Control of risk: A Guide to the systematic management of risk from construction (No. SP125). London, UK: Construction Industry Research and Information Association (CIRIA).

CIRIA. (2009). Whole-life infrastructure asset management: Good practice guide for civil infrastructure (No. C677). London, UK: Construction Industry Research and Information Association (CIRIA).

Claeskens, G., & Hjort, N. (2008). *Model selection and model averaging*. Cambridge, UK: Cambridge University Press. doi:10.1017/CBO9780511790485

Coles, S. (2001). *An introduction to statistical modeling of extreme values*. London, UK: Springer-Verlag. doi:10.1007/978-1-4471-3675-0

Comité Euro-International du Béton. (1993). *CEB–FIP Model code 90*. Thomas Telford Ltd.

Congdon, P. (2006). *Bayesian statistical modelling* (2nd ed.). Chichester, UK: John Wiley & Sons. doi:10.1002/9780470035948

Cooke, R. (1991). Experts in Uncertainty. Opinion and Subjective Probability in Science. Oxford, UK: Oxford University Press. Retrieved from https://global.oup.com/academic/product/experts-in-uncertainty-9780195064650

Cooke, R., & Goossens, L. (2004). Expert judgement elicitation for risk assessments of critical infrastructures. *Journal of Risk Research*, *7*(6), 643–656. doi:10.1080/1366987042000192237

Cropper, M., & Sussman, F. (1990). Valuing future risks to life. *Journal of Environmental Economics and Management*, *19*(2), 160–174. doi:10.1016/0095-0696(90)90066-8

Das, P. (1997). *Safety of bridges*. Thomas Telford Ltd.

Det Norske Veritas. (2002). Marine risk asessment. London, UK: The Health and Safety Executive (HSE).

Ditlevsen, O., & Friis-Hansen, P. (2005). Life quality time allocation index – an equilibrium economy consistent version of the current life quality index. *Structural Safety*, *27*(3), 262–275. doi:10.1016/j.strusafe.2004.12.001

Dusenberry, D., & Hamburger, R. (2006). Practical means for energy-based analyses of disproportionate collapse potential. *Journal of Performance of Constructed Facilities*, *20*(4), 336–348. doi:10.1061/(ASCE)0887-3828(2006)20:4(336)

Eales, R., Smith, S., Twigger-Ross, C., Sheate, W., Özdemiroglu, E., Fry, C., & Tomlinson, P. (2003). *Integrated appraisal methods (No. E2–044/TR)*. Bristol, UK: UK Environment Agency.

Efron, B. (1979). Bootstrap methods: Another look at the jackknife. *Annals of Statistics*, *7*(1), 1–26. doi:10.1214/aos/1176344552

Ellingwood, B. (2008). Reliability based codes in the United States. Challenges to professional engineering practice. *Proceedings of the 4th International ASRANet Colloquium on Integrating for Structural Analysis, Risk & Reliability*.

Ellingwood, B., Smilowitz, R., Dusenberry, D., Duthinh, D., Lew, H., & Carino, N. (2007). Best practices for reducing the potential for progressive collapse in buildings. National Institute of Standards and Technology (NIST). doi:10.6028/NIST.IR.7396

Faber, M. (2009). Risk and safety in engineering course lecture notes. Swiss Federal Institute of Technology Zurich (ETHZ).

Faber, M., & Stewart, M. (2003). Risk assessment for civil engineering facilities: Critical overview and discussion. *Reliability Engineering & System Safety*, *80*(2), 173–184. doi:10.1016/S0951-8320(03)00027-9

Fawcett, L. (2005). *Statistical methodology for the estimation of environmental extremes* (PhD Thesis). University of Newcastle, Newcastle upon Tyne, UK.

Ferry Borges, J., & Castanheta, M. (1968). Structural safety (1st ed.). Lisbon, Portugal: Laboratorio National de Engenharia Civil (LNEC).

Ferry Borges, J., & Castanheta, M. (1985). Structural safety (3rd ed.). Lisbon, Portugal: Laboratorio National de Engenharia Civil (LNEC).

Fischhoff, B., Lichtenstein, S., Slovic, P., Derby, S., & Keeney, R. (1981). *Acceptable risk*. Cambridge University Press.

Frangopol, D., & Curley, J. (1987). Effects of damage and redundancy on structural reliability. *Journal of Structural Engineering*, *113*(7), 1533–1549. doi:10.1061/(ASCE)0733-9445(1987)113:7(1533)

Fu, G., & Frangopol, D. (1990). Balancing weight, system reliability and redundancy in a multiobjective optimization framework. *Structural Safety*, *7*(2–4), 165–175. doi:10.1016/0167-4730(90)90066-X

Goda, K., & Hong, H. (2008). Application of cumulative prospect theory: Implied seismic design preference. *Structural Safety*, *30*(6), 506–516. doi:10.1016/j.strusafe.2007.09.007

Goodwin, P., & Wright, G. (2010). *Decision analysis for management judgment* (4th ed.). Wiley.

Gulvanessian, H., Calgaro, J., & Holický, M. (2002). *Designer's guide to Eurocode: Basis of structural design* (1st ed.). London, UK: Thomas Telford Ltd.

Haimes, Y. (2009). *Risk modeling, assessment and management* (3rd ed.). New York: John Wiley & Sons.

Haldar, A., & Mahadevan, S. (2000). *Probability, reliability, and statistical methods in engineering design*. New York: John Wiley & Sons.

Hartford, D., & Baecher, G. (2004). *Risk and uncertainty in dam safety*. London, UK: Thomas Telford Ltd. doi:10.1680/rauids.32705

Hastie, T., Tibshirani, R., & Friedman, J. (2009). *The elements of statistical learning – Data mining, inference, and prediction* (2nd ed.). New York: Springer.

Hayes, K. (2011). *Uncertainty and uncertainty analysis methods (No. EP102467)*. Dickson, Australia: CSIRO Mathematics, Informatics and Statistics.

Hill, H. (2004). Rational and irrational design loads for temporary' structures. *Practice Periodical on Structural Design and Construction*, *9*(3), 125–129. doi:10.1061/(ASCE)1084-0680(2004)9:3(125)

Holmes, J. (2007). *Wind loading of structures* (2nd ed.). Oxford, UK: Taylor & Francis.

House of Lords. (2006). *Government policy on the management of risk* (No. 5th Report of Session 2005–06). London, UK: The Stationery Office Limited.

HSE. (1992). The tolerability of risk from nuclear power stations. London, UK: The Health and Safety Executive (HSE).

HSE. (1995). Generic terms and concepts in the assessment and regulation of industrial risks. London, UK: The Health and Safety Executive (HSE).

HSE. (2001). Reducing risks, protecting people: HSE's decision making process (No. R2P2). London, UK: The Health and Safety Executive (HSE).

HSE. (2009a). Construction intelligence report: Analysis of construction injury and ill health intelligence 1997/98 - 2008/09. London, UK: The Health and Safety Executive (HSE).

HSE. (2009b). ND guidance on the demonstration of ALARP (as low as reasonably practicable) (No. T/AST/005-Issue 4-Rev 1). London, UK: The Health and Safety Executive (HSE).

ISO. (1998). ISO 2394:1998 – General principles on reliability for structures (3rd ed.). Geneva, Switzerland: International Organization for Standardization (ISO).

ISO. (2009a). ISO 13824:2009 – Bases for design of structures – General principles on risk assessment of systems involving structures. Geneva, Switzerland: International Organization for Standardization (ISO).

ISO. (2009b). ISO 31000:2009 – Risk management – Principles and guidelines. Geneva, Switzerland: International Organization for Standardization (ISO).

ISO. (2009c). ISO 31010:2009 – Risk management – Risk assessment techniques. Geneva, Switzerland: International Organization for Standardization (ISO).

ISO. (2015). ISO 2394 :2015 – General principles on reliability for structures (4th ed.). Geneva, Switzerland: International Organization for Standardization (ISO).

IStructE. (2007). Temporary demountable structures: Guidance on procurement, design and use (3rd ed.). London, UK: The Institution of Structural Engineers.

JCSS. (2001). *JCSS Probabilistic model code*. Joint Committee on Structural Safety.

Jones-Lee, M., Loomes, G., Yates, D., Mitchell, E., Spackman, M., McQuaid, J., & Thomson, T. (2008). T440: Fatalities and weighted injuries. London, UK: Rail Safety and Standards Board (RSSB).

Kahneman, D., & Tversky, A. (1979). Prospect theory: An analysis of decision under risk. *Econometrica*, *47*(2), 263–291. doi:10.2307/1914185

Kaplan, S., & Garrick, B. (1981). On the quantitative definition of risk. *Risk Analysis*, *1*(1), 11–27. doi:10.1111/j.1539-6924.1981.tb01350.x PMID:11798118

Khun, M., & Johnson, K. (2013). *Applied predictive modeling*. New York: Springer. doi:10.1007/978-1-4614-6849-3

Knoll, F., & Vogel, T. (2009). Design for robustness. Zurich: International Association for Bridge and Structural Engineering (IABSE).

Kolmogoroff, A. (1933). *Grundbegriffe der wahrscheinlichkeitsrechnung*. Berlin, Germany: Springer Berlin Heidelberg. doi:10.1007/978-3-642-49888-6

Konakli, K., & Faber, M. (2014). Value of information analysis in structural safety. In *Proceedings of 2nd International Conference on Vulnerability and Risk Analysis and Management (ICVRAM) and the 6th International Symposium on Uncertainty, Modeling, and Analysis (ISUMA)* (pp. 1605–1614). Liverpool, UK: American Society of Civil Engineers (ASCE). doi:10.1061/9780784413609.161

Korol, R., & Sivakumaran, K. (2014). Reassessing the plastic hinge model for energy dissipation of axially loaded columns. *Journal of Structures, 2014*(1), 7.

Kumamoto, H. (2007). *Satisfying safety goals by probabilistic risk assessment*. London, UK: Springer-Verlag.

Le Guen, J. (2008). *Tolerability of risk: UK principles & practice for controlling work activities*. Presented at the Workshop on Tolerable Risk Evaluation sponsored by US Army Corps of Engineers. U.S. Department of the Interior: Bureau of Reclamation.

Leadbetter, M., Bailey, G., & Rootzén, H. (1983). *Extremes and related properties of random sequences and processes*. New York: Springer-Verlag. doi:10.1007/978-1-4612-5449-2

Levy, H. (2006). *Stochastic dominance: Investment decision making under uncertainty* (2nd ed.). New York: Springer.

Marsden, J., McDonald, L., Bowles, D., Davidson, R., & Nathan, R. (2007). Dam safety, economic regulation and society's need to prioritise health and safety expenditures.*Proceedings of the NZSOLD/ANCOLD Workshop on 'Promoting and Ensuring the Culture of Dam Safety'*, 12.

McDonald, L., Bowles, D., Hartford, D., van der Meer, J., & Vrijling, J. (2005). Risk assessment in dam safety management. International Commission of Large Dams (ICOLD), Bulletin 130.

McGill, W., & Ayyub, B. (2007). The meaning of vulnerability in the context of critical infrastructure protection. In *Critical infrastructure protection: Elements of risk* (pp. 25–48). George Mason University School of Law.

Melchers, R. (1999). *Structural reliability analysis and prediction* (2nd ed.). Wiley.

Mohammadi, J., & Heydari, A. (2008). Seismic and wind load considerations for temporary structures. *Practice Periodical on Structural Design and Construction, 13*(3), 128–134. doi:10.1061/(ASCE)1084-0680(2008)13:3(128)

Muhanna, R., Rao, M., & Mullen, R. (2013). Advances in interval finite element modelling of structures. *Life Cycle Reliability and Safety Engineering*, *2*(3), 15–22.

Narasimhan, H., & Faber, M. (2009). Categorisation and assessment of robustness related provisions in European standards.*Proceedings of the Joint Workshop of COST Actions TU0601 and E55*, 147–164.

Nathwani, J., Lind, N., & Pandey, M. (1997). *Affordable safety by choice: The life quality method*. Waterloo, Canada: Institute of Risk Research, University of Waterloo.

Nishijima, K. (2009). *Issues of sustainability in engineering decision analysis* (PhD Thesis). ETH Zurich, Zurich, Switzerland.

Nwogugu, M. (2006). A further critique of cumulative prospect theory and related approaches. *Applied Mathematics and Computation*, *179*(2), 451–465. doi:10.1016/j.amc.2005.11.102

Oakley, J., & OHagan, A. (2004). Probabilistic sensitivity analysis of complex models: A Bayesian approach. *Journal of the Royal Statistical Society. Series B, Statistical Methodology*, *66*(3), 751–769. doi:10.1111/j.1467-9868.2004.05304.x

Ouchi, F. (2004). *A literature review on the use of expert opinion in probabilistic risk analysis*. Washington, DC: The World Bank. doi:10.1596/1813-9450-3201

Rackwitz, R. (2004). Discounting for optimal and acceptable technical facilities involving risks. *HERON*, *49*(2), 139–170.

Rao, M., Mullen, R., & Muhanna, R. (2011). A new interval finite element formulation with the same accuracy in primary and derived variables. *International Journal of Reliability and Safety*, *5*(3/4), 336. doi:10.1504/IJRS.2011.041184

Rimington, J., McQuaid, J., & Trbojevic, V. (2003). *Application of risk-based strategies to workers' health and safety protection: UK experience*. Ministry of Social Affairs and Employment.

Rosignoli, M. (2007). Robustness and stability of launching gantries and movable shuttering systems - Lessons learned. *Structural Engineering International*, *17*(2), 133–140. doi:10.2749/101686607780680673

Rosowsky, D. (1995). Estimation of design loads for reduced reference periods. *Structural Safety*, *17*(1), 17–32. doi:10.1016/0167-4730(94)00037-Q

Savage, S. (2009). *The flaw of averages: Why we underestimate risk in the face of uncertainty*. John Wiley & Sons.

Scheer, J. (2010). *Failed bridges: Case studies, causes and consequences*. London, UK: Wiley VCH. doi:10.1002/9783433600634

Schneider, J. (2006a). *Introduction to safety and reliability of structures* (2nd ed.). IABSE.

Schneider, J. (2006b). Structural & Construction Safety. Specially recorded for IABSE E-Learning 2006, Switzerland.

Schweiger, H., & Peschl, G. (2005). Reliability analysis in geotechnics with the random set finite element method. *Computers and Geotechnics*, *32*(6), 422–435. doi:10.1016/j.compgeo.2005.07.002

Sexsmith, R. (1998). Reliability during temporary erection phases. *Engineering Structures*, *20*(11), 999–1003. doi:10.1016/S0141-0296(97)00195-8

Sexsmith, R., & Reid, S. (2003). Safety factors for bridge falsework by risk management. *Structural Safety*, *25*(2), 227–243. doi:10.1016/S0167-4730(02)00058-9

Smith, J. (2006). Structural robustness analysis and the fast fracture analogy. *Structural Engineering International*, *16*(2), 118–123. doi:10.2749/101686606777962521

Sommer, H., Seethaler, R., Chanel, O., Herry, M., Masson, S., & Vergnaud, J. (1999). Health costs due to road traffic-related air pollution. An impact assessment project of Austria, France and Switzerland. Economic evaluation. Bern, Switzerland: World Health Organization (WHO).

Spackman, M. (2009). Review of the J–value literature – Final report. London, UK: The Health and Safety Executive (HSE).

Spackman, M., Evans, A., Jones-Lee, M., Loomes, G., Holder, S., & Webb, H. (2011). *Updating the VPF and VPIs: Phase 1: Final report.* UK Department for Transport, Local Government and the Regions.

Spackman, M., & Holder, S. (2007). *Discount rates for rail safety scheme appraisals. Final report for the office of rail regulation.* NERA Economic Consulting.

Starossek, U. (2006). Progressive collapse of structures: Nomenclature and procedures. *Structural Engineering International, 16*(2), 113–117. doi:10.2749/101686606777962477

Starossek, U. (2009). *Progressive collapse of structures.* Thomas Telford Ltd. doi:10.1680/pcos.36109

Starossek, U., & Haberland, M. (2008). Measures of structural robustness — Requirements & applications.*Proceedings of the ASCE SEI 2008 Structures Congress – Crossing Borders*, 10.

Starossek, U., & Wolff, M. (2005). Progressive collapse: design strategies. In *Proceedings of IABSE Symposium, Structures and Extreme Events* (p. 8). Lisbon, Portugal: International Association for Bridge and Structural Engineering (IABSE).

Steven, S. (2010). The different phases in construction – Design in health and safety to the project life cycle. In *ICE Manual of health and safety in construction* (pp. 51–69). London: Thomas Telford Ltd.

Sudret, B. (2007). *Uncertainty propagation and sensitivity analysis in mechanical models. Contributions to structural reliability and stochastic spectral methods.* Paris, France: Ecole Nationale des Ponts et Chauss'ees.

Tamura, Y., Wu, D., & Yang, Q. (2015). Relation between design load level and lifetime of individual building and its elements. In *Proceedings of the 2nd International Conference on Performance-based and Lifecycle Structural Engineering (PLSE 2015)* (p. 10). Brisbane, Australia: The University of Queensland.

Thoft-Christensen, P., & Baker, M. (1982). *Structural reliability theory and its applications.* Springer. doi:10.1007/978-3-642-68697-9

Todinov, M. (2007). *Risk–based reliability analysis and generic principles for risk reduction.* Amsterdam, The Netherlands: Elsevier.

HM Treasury. (2003). *Appraisal and evaluation in central government (The green book).* Controller of Her Majesty's Stationery Office.

HM Treasury. (2004). *Management of risk – Principles and concepts (The orange book).* Controller of Her Majesty's Stationery Office.

Turkstra, C., & Madsen, H. (1980). Load combinations in codified structural design. *Journal of the Structural Division, 106*(12), 2527–2543.

Tversky, A., & Kahneman, D. (1992). Advances in prospect theory: Cumulative representation of uncertainty. *Journal of Risk and Uncertainty*, *5*(4), 297–323. doi:10.1007/BF00122574

TWf. (2014). Clients' guide to temporary works (No. TWf2014:02). London, UK: Temporary Works forum.

UK DfT. (1999). *Guidance on robustness and provision against accidental actions. The current application of requirements A3 of the building regulations 1991. Final report including background documentation.* London, UK: UK Department for Transport, Local Government and the Regions.

UK DfT. (2011). *Guidance documents – Expert. Tag unit 3.4: The safety objective*. London, UK: UK Department for Transport.

UKOOA. (1999). Industry guidelines on a framework for risk related decision support (No. HS006). London, UK: UK Offshore Operators Association (UKOOA).

U.S. DOT. (2009). *Treatment of the economic value of a statistical life in departmental analyses – 2009 annual revision*. Washington, DC: U.S. Department of Transportation.

USDOD. (2010). Unified facilities criteria (UFC) – Design of buildings to resist progressive collapse. Washington, DC: United States of America Defence Department, Report UFC 4-023-03.

van Gelder, P. (2000). *Statistical methods for the risk-based design of civil structures* (PhD Thesis). Delft University of Technology, Delft, The Netherlands.

Viscusi, W. (1993). The value of risks to life and health. *Journal of Economic Literature*, *31*(4), 1912–1946.

Viscusi, W., & Aldy, J. (2003). The value of a statistical life: A critical review of market estimates throughout the world. *Journal of Risk and Uncertainty*, *27*(1), 5–76. doi:10.1023/A:1025598106257

Viscusi, W., & Gayer, T. (2002). Safety at any price? *Regulation*, *25*(3), 54–63.

von Neumann, J., & Morgenstern, O. (1944). *Theory of games and economic behavior*. Princeton University Press.

Vrijling, J., van Gelder, P., Goossens, L., Voortman, H., & Mandey, P. (2004). A framework for risk criteria for critical infrastructures: Fundamentals and case studies in the Netherlands. *Journal of Risk Research*, *7*(6), 569–579. doi:10.1080/1366987032000081178

Vrijling, J., van Hengel, W., & Houben, R. (1998). Acceptable risk as a basis for design. *Reliability Engineering & System Safety*, *59*(1), 141–150. doi:10.1016/S0951-8320(97)00135-X

Willford, M., & Allsop, A. (1990). Design guide for wind loads on unclad framed building structures during construction (No. BR 173). Bracknell, UK: Building and Research Establishment (BRE).

Xiao, N., Huang, H., Wang, Z., Pang, Y., & He, L. (2011). Reliability sensitivity analysis for structural systems in interval probability form. *Structural and Multidisciplinary Optimization*, *44*(5), 691–705. doi:10.1007/s00158-011-0652-9

Xie, Q., Lu, S., Cóstola, D., & Hensen, J. (2014). An arbitrary polynomial chaos-based approach to analyzing the impacts of design parameters on evacuation time under uncertainty. *Fire Safety Science*, *11*, 14. doi:10.3801/IAFSS.FSS.11-1077

Zhang, H., Mullen, R., & Muhanna, R. (2010). Finite element structural analysis using imprecise probabilities based on p-box representation. In *Proceedings of the 4th International Workshop on Reliable Engineering Computing (REC2010)* (pp. 211–225). University of Singapore: Research Publishing Services.

Zhang, H., Mullen, R., & Muhanna, R. (2012). Structural analysis with probability boxes. *International Journal of Reliability and Safety*, *6*(1), 110–129.

Chapter 6
Design Codes and General Design Guidance

ABSTRACT

The philosophies behind design codes with particular reference to the use of modern limit state design are presented in this chapter. Comments are made on the design life of temporary structures which vary considerably between different countries. Design codes of the USA, Europe and Australia/New Zealand for temporary structures are compared with particular reference to the loads combinations and the partial factors applied. It is noted that whilst the European design codes do not specify how construction, use and disassembly of the temporary structures are to be executed the USA code for scaffolding includes such specification. The Hong Kong code for bamboo scaffolds is described showing the similarities and differences between bamboo and metal scaffolds. The chapter concludes with design examples for selected temporary structures based on design codes.

6.1 INTRODUCTION

Every temporary structures project is a unique endeavour, given a particular set of challenges and a specific context. The planning, design, execution and operation processes of a temporary structure vary with the intended application, the site where it will be used and the role in the construction process. Therefore, temporary structures are exposed to a multiplicity of external and/or internal hazard events. Many of the hazards have been appropriately researched and design rules have been incorporated in existing codes of practice and guidance documents. However, there are still gaps of knowledge that need to be filled, with emphasis on the risks originating from human interaction, namely human errors during all phases of temporary structures life cycle.

The aim of this Chapter is to introduce the framework of structural design in the context of temporary structures, to discuss the associated challenges, and to provide an understanding concerning the design rules specified in modern design codes. The Chapter also identifies aspects that are not covered by existing structural design codes and presents state-of-the-art methods that help to overcome these limitations.

DOI: 10.4018/978-1-5225-2199-0.ch006

Copyright © 2017, IGI Global. Copying or distributing in print or electronic forms without written permission of IGI Global is prohibited.

On the basis of this Chapter it is expected that the reader will acquire knowledge on the following topics:

1. Difference between designs for temporary structures and permanent structures.
2. Differences between design codes in various countries including the use of different materials.
3. Understanding of bridge falsework and access scaffolds.
4. Design of elements of temporary structures including bearings, jacks and elements made of bamboo.
5. An understanding of the problems involved in bridge falsework and the design and use of Bridge Construction Equipment.

6.2 CURRENT DESIGN CODES

6.2.1 Basis

Temporary structures must provide structural safety and safety to workers. Furthermore, the interaction between the temporary and the permanent structures must guarantee the specified geometry, durability, quality and, ultimately, the safety of the permanent structure. However, temporary structures do not receive the same research attention as do permanent structures, and as a result this sector of civil engineering lacks the existence of strong regulation and standardization.

Until recently, the available guidance to designers was summed up in a few general statements in permanent structures design codes such as "proper provisions shall be made" and "adequate temporary bracing shall be provided" (Ratay, 2009). Therefore, the majority of temporary structures were designed at the discretion of the sub-contractor, usually based on in-house design procedures prepared by the producers of the proprietary equipment and often without supervision or in some cases with inadequate supervision. As a result, the design rules applied to temporary structures were not uniform and therefore the actual reliability levels were often smaller and exhibited a greater variation than the corresponding reliability levels of permanent structures.

In recent years there has been an effort in Europe, in the USA and in other developed countries to diminish the uncertainty and risk involved with the use of these structures. Nowadays, the importance of temporary structures has increased to the point where their design, assembly and maintenance are ruled by specifically prepared standards and codes of practice. As expected, existing temporary structures codes profited from the available knowledge accumulated over the years, the majority of which was empirical in nature. Lessons from past mistakes led to changes in all phases of temporary structures, from design to use and from maintenance to inspection. This was further enriched with added value from recent research studies, most of them numerical investigations. However, the improvement of design and construction codes for temporary structures was made at a slower pace than those codes for permanent structures. Until recently, existing standards were based on the allowable stress design concept – which was replaced by the Limit State Design (LSD) philosophy in the 1980s in permanent structures' design codes. Nonetheless, at present most of the design codes adopt the LSD philosophy with the exception of some USA and UK design codes.

However, since many of the temporary structures codes were developed based on existing codes for permanent structures, more often than not guidance is found to be missing: "Is this structure as safe, as wind-resistant, and as cost-effective as we can make it?" (Gorlin, 2009). Additionally, some deep rooted axioms shared by designers and other relevant stakeholders were transmitted to the existing temporary structures codes. The majority of these codes allow that in the design of temporary structures, lower design loads may be used than those required for completed permanent structures. This is partly based on the idea that short-term loads are more closely predictable and can be more effectively controlled than the long-term variable loads during the decades of use of permanent structures. Additionally, because the design and use of temporary structures is considered to be not as important as for permanent structures, the specified partial factors applied on the resistance side are sometimes smaller than the ones specified for permanent structures. In Chapter 5, the claim for the universal application of both ideas is rebutted.

One must always have in mind that "codes codify safety; good designs provide it" (fib, 2000). Structural safety comes from a coherent and comprehensive global structural concept. That is, proportionate structural forms with logically defined load paths together with simple construction procedures, not from the fulfilment of code requirements which by themselves cannot assure the required safety (fib, 2000). This is obviously applicable to all temporary structures.

According to Ratay (2009) there are several factors inherently involved in the establishment of design loads, resistance models and reliability levels that are used in the design of any structure. These include the following:

1. "Intended function of the structure
2. Nature of loads
3. Predictability of occurrence of loads
4. Certainty in the magnitudes of loads
5. Possibility of simultaneous occurrence of loads
6. Possible secondary stresses, redundancy, and instability.
7. Condition of the member and its material (new, used, damaged, deformed)
8. Acceptable behaviour of the structure (such as tolerable deflections, and vibrations)
9. Allowable degree of unacceptable behaviour
10. Acceptable probability of total failure
11. Consequences of failure
12. Construction tolerances
13. Workmanship in the construction
14. Inspection standards
15. Protection of the structure against damage, deterioration, and extremities of weather
16. Intended life-span of the structure with increasing probability of occurrence of maximum loads, abnormal loads, damage, and deterioration with time".

It can be observed there are many topics that separate the design of temporary structures from the design of permanent structures. However, because of the constraints listed above, research is still needed to improve the design codes for temporary structures in order to properly address important aspects such as evaluation of loads for reduced exposure periods, consideration in the design process of multiple usage cycles, structural performance requirements (including robustness), acceptable and unacceptable risk levels.

A brief summary of the principles underlying the USA, European and Australian standards related to the design of temporary structures is now presented. It is notable that the European design codes, with the exception of BS 5975 (BSI, 2011a) are purely design codes with no specification of how the design is to be executed safely. This is in contrast to other regions' codes which include procedures to manage the erection, use and disassembly of temporary structures. BS 5975, although primarily an old code still using elastic analysis, is still current because it defines the process of management control of these structures. This process is constantly under development and no European strategy has been agreed. Details on the management aspects included in BS 5975 are found in Chapter 8.

In the USA, the most complete and up-to-date standard outlining a design philosophy and specifying minimum design loads and load combinations for temporary structures is the ASCE/SEI 37 (ASCE, 2014). Since the scope of this document concerns mainly the load requirements during construction of buildings, other standards exist, for example the AASHTO standard GSBTW-1-M (AASHTO, 2008b) which covers the design of bridge falsework systems and the ACI standard 347-04 (ACI, 2004) which covers the design of formwork for concrete construction. Design guidance about bridge falsework can also be found in the State of California Falsework Manual (Department of Transportation, Division of Structures, 2001). Scaffolding is covered by ANSI/ASSE A10.8 (ANSI, 2011).

In Europe reference should be made to BS EN 12812 (BSI, 2011d) and BS EN 12813 (BSI, 2004h) for falsework, and to BS EN 12810 (BSI, 2003a, 2003b) and BS EN 12811 (BSI, 2002, 2003c, 2004g) for scaffolding. BS EN 12812 gives performance requirements and provides methods to design falsework to meet those requirements. The information on structural design is supplementary to the relevant Eurocodes. The standard also describes different design classes: Class A and Class B (further subdivided into Class B1 and Class B2).

Class A falsework is one where the structural performance can be individually evaluated and where simplified and traditional design methods (determined by experience and established good practice) can be safely used. Examples of Class A falsework are adjustable telescopic steel props and formwork. BS EN 12812 does not provide guidance for the structural design of Class A falsework, which may be found in BS 5975 (BSI, 2011a).

Class B falsework is one where simplified design methods cannot be applied and more comprehensive approaches are needed, for instance in the case of bridge falsework. Therefore, the general design philosophy of the Eurocodes is followed. BS EN 12812 gives the falsework designer two options: Classes B1 and B2. For Class B1, one has to adopt the Eurocodes with some additional requirements concerning information that must be included in the drawings, inspection and method statements. For Class B2, BS EN 12812 specifies an alternative design procedure which differs from the one given in the Eurocodes on the amplitude of the initial geometrical imperfections (global and local), the construction loads values, the loads combinations and the partial factors to be used. Nevertheless, the underlying design philosophy is that the resulting designs for both classes have equivalent reliability levels.

The structural Eurocodes shown in Table 1, consist of a number of parts. They are intended for the design of new construction works using traditional materials (reinforced and post-tensioned concrete, steel and composite construction, timber, masonry, and aluminium). However, BS EN 1990, "Basis of Structural Design", is also applicable for the structural evaluation of existing constructions, in developing designs for repairs and rehabilitation or in assessing changes of use.

Still in Europe, the British Standard BS 5975 (BSI, 2011a) establishes a design framework based on the Allowable Stress Design (ASD) method for falsework, namely Class A falsework according to BS EN 12812. Note that this standard, although European, is not part of the Eurocode series. BS 5975 has not been withdrawn by the UK British Standards Institution because, unlike the Eurocodes, not only does it have design procedures, it also gives recommendations and guidance on the procedural controls

Table 1. List of existing Eurocodes

EN 1990	Eurocode 0	Basis of Structural Design
EN 1991	Eurocode 1	Actions on Structures
EN 1992	Eurocode 2	Design of Concrete Structures
EN 1993	Eurocode 3	Design of Steel Structures
EN 1994	Eurocode 4	Design of Composite Steel and Concrete Structures
EN 1995	Eurocode 5	Design of Timber Structures
EN 1996	Eurocode 6	Design of Masonry Structures
EN 1997	Eurocode 7	Geotechnical Design
EN 1998	Eurocode 8	Design of Structures for Earthquake Resistance
EN 1999	Eurocode 9	Design of Aluminium Structures

to be applied to specification, construction, use and dismantling of falsework, which are absent from BS EN 12812.

This Chapter will also compare the Australian standard AS 3610 (SAA, 1995) to the USA and European design codes. The Canadian standard CSA 269.1 (CSA, 1975), reaffirmed in 2003, will not be considered since it is a basic document which implied that all analyses will be conducted using other Canadian codes. It has some similarities to the USA codes but it is in SI units whereas the USA codes are usually in modified Imperial units.

With respect to telescopic props, only in Europe do product and design standards exist. BS EN 1065 (BSI, 1999) for steel props and BS EN 16031 (BSI, 2012) for aluminium props.

Nevertheless, there are no specific standards for the design of Bridge Construction Equipment (BCE), contrary to bridge falsework (such as built-up towers used as temporary supports during bridge launching) and formwork systems for which standards exist (AASHTO, 2008a, 2008b; ACI, 2004; BSI, 1982, 2011d, 2011a; CSA, 1975; SAA, 1995). Therefore, the designer of BCE systems must use first structural engineering principles and best practices.

To this end, BCE designers often resort to structural codes for permanent structures, for instance the structural Eurocodes in Europe or the American Society of Civil Engineers (ASCE) and the American Association of State Highway and Transportation Officials (AASHTO) standards, for the design of steel members, connections and bearings. Therefore, BCE are designed using the same principles and methodologies which are applied to permanent structures such as buildings and bridges.

As BCE are also highly mechanised systems, mechanical engineering codes have to be followed for the design of hydraulic jacks and lifting systems, for example. Additionally, materials used such as steel, bolts, anchors, wires, post-tension cables and rods should be certified or else evidence should be made available of compliance with specified project requirements.

The design of cranes often used during bridge construction is not discussed here as guidance can be found in the BS EN 13001 series, such as BS EN 13001-2 (BSI, 2014a), or in the BS 7121 series, such as BS 7121-1 (BSI, 2016), or in the ISO standards published by the ISO TC 96/SC 10, such as ISO 8686-1 (ISO, 2012). Additionally, guidance on bridge design and safety verification during construction which is very important for the safety of BCE (especially the post-tensioning solution), can be found elsewhere (AASHTO, 2016a, 2016b, BSI, 2005d, 2006c, 2007c, 2008e; CSA, 2014; fib, 2000; SAA, 2007).

The main USA design standard dealing with access scaffolds is ANSI A10.8 (ANSI, 2011). In Europe, the standards applicable to scaffolding are BS EN 12810 and BS EN 12811. The ANSI standard covers all types of scaffold from façade scaffolds made of steel, aluminium or timber to supported and suspended scaffolds and tower scaffolds. Metal scaffolds constructed in Hong Kong must conform to the Eurocodes or to other International Standards (Hong Kong Buildings Department, 2006). Tower scaffolds are covered in the Europe by BS EN 1004 (BSI, 2004a).

6.2.2 Design Philosophies

Almost all the existing structural codes for the design and analysis of civil engineering infrastructures have abandoned the Permissible Stress Design philosophy (PSD), also called the Allowable Stress Design (ASD) in the USA, and are now based on the Limit State Design (LSD) principles. In the USA, Limit State Design is termed Load and Resistance Factor Design (LRFD). Nevertheless, in the USA codes still allow the use of ASD.

The LSD principles are semi-probabilistic, see Chapter 5 for a complete presentation. In this methodology, the format for structural design verification is expressed by a simple comparison between factored resistances and factored actions (or actions effects) without explicitly assessing the reliability or the risks.

Due to the fact that resistances and actions are subject to uncertainties, probabilistic analyses were performed to derive statistically representative values (called characteristic values) taking into account the design working life of the structure and the uncertainty of different physical properties and conditions.

To ensure that the basis for design provides an appropriate level of structural reliability (or probability of failure), partial factors are introduced to take into account the effects of uncertainties in the methods used to assess the characteristic values but also in the specified analysis and verification procedures. Therefore, design values for resistances are determined by dividing the characteristic values by a partial factor, γ_R, (larger than or equal to 1.0) and design values for load effects are obtained by multiplying the characteristic values by a partial factor, γ_E, (typically larger than 1.0).

LSD specifies the verification of the structural reliability for several limit states, i.e. states beyond which the structure no longer fulfils the relevant design criteria: Ultimate Limit States (ULS) in which all possible failure modes must be evaluated and Serviceability Limit States (SLS) in which it is verified that specified service requirements are met, and other limit states such as fatigue resistance.

Finally, several load combinations must be checked to guarantee that all reasonable possible sets of physical conditions that can occur during a certain time interval, also known as design situations, are taken into account. This time interval is dependent on the design working life of the structure and is associated with a limit value for the annual probability of exceedance of the loads.

In general, three different design situations are defined. Each one represents a certain time interval with associated hazards, conditions and relevant structural limit states: persistent, transient and accidental situations which refer to normal, temporary and exceptional situations. Each load combination is formed by the permanent loads, a leading variable action and the relevant accompanying variable actions which are multiplied by combination factors (smaller than 1.0) in order to obtain concomitant actions values taking into consideration the unlikely event of the simultaneous occurrence of the different actions at their maximum design values (Turkstra & Madsen, 1980).

For the case of earthquakes and accidental actions, the design requirements may be established as a function of the probability of exceedance of the design action values calibrated so that an acceptable balance between safety, economy and feasibility is achieved. This is called performance-based design philosophy. For example, a structure should resist undamaged to moderate earthquakes, whereas for higher magnitude earthquakes the structure should be able to sustain the imposed action effects through plastic energy dissipation of selected structural elements and thus maintain adequate safety margins against overall collapse. However, the performance-based design rules specified in design codes may not be directly applicable to all temporary structures since the threshold values defined for the control design variables (e.g. joint hysteretic energy dissipation capacity or inter-story drifts) were not calibrated for every temporary structures structural solution.

LSD has been established in the Australian standard series AS 5400 (SAA, 2007), in the Canadian standard S6 (CSA, 2014), in the European Structural Eurocodes, namely BS EN 1990 (BSI, 2005b), BS EN 1991 series, such BS EN 1991-1-6 (BSI, 2005c), and in the USA standards ASCE/SEI 37 (ASCE, 2014) and AASHTO LRFD bridge code (AASHTO, 2016b), for example. The ASD philosophy is still the basis for the AASHTO design guide for bridge falsework (AASHTO, 2008b), and its use is still allowed by BS 5975 (BSI, 2011a).

The USA standard ANSI A10.8 (ANSI, 2011) does not define the analysis procedure underlying the standard, only giving prescriptive definitions of allowable loading and structural arrangements.

It should not be forgotten that the Eurocodes are only valid if used together with the corresponding National Annexes published by every European Union member state. These contain the national choices for the Nationally Determined Parameters (NDPs). In this book the UK National Annexes will be used as an example.

It should be noted, that most codes allow the designer to use alternative design rules different from the ones specified if it can be demonstrated analytically or experimentally that the structural safety, serviceability and durability achieved will be at least equal to the ones expected when using the codes.

In both the AASTHO bridge code (AASHTO, 2016b) and in BS EN 1990 it is possible for the stakeholders to manage the target reliability level of the structure. In the former code this is dependent on the structural characteristics (ductility, redundancy, etc.) and in the latter code of the consequences of failure, types of quality control and inspection procedures implemented. Temporary structures may in general be considered as standard structures based on the AASTHO and BS EN 1990 criteria. However, since the relative cost of safety measures is usually low compared to the large consequences of their collapse it may be rational to design for increased reliability levels (ISO, 2015; JCSS, 2001).

The target reliability index (see Chapter 5 for definition) used in the AASHTO bridge code calibration for normal structures is equal to 3.5 (unfortunately there is no indication of the associated reference period), whereas in BS EN 1990 (BSI, 2005b) for the CC2 consequence class (i.e. limited to moderate risks to human lives), an annual target reliability index equal to 4.7 is specified.

Considering that the AASHTO bridge code specifies 75 years as the expected design working life, and assuming that the reference period associated with the specified reliability index is equal to the expected design working life, a target reliability index of 4.5 is obtained for a one year reference period, which is lower than the one used in BS EN 1990. It is also lower than the target reliability index suggested in ASCE 7 (ASCE, 2010), of 4.8, for structures whose failure is sudden, which results in wide spread progression of damage and for which 5 to 500 persons are exposed (Risk Category II).

No information is given in the USA standards regarding the design working life of temporary structures. In BS EN 1990, a minimum of ten years for the design working life of temporary structures is indicated. However, for temporary structures that can be dismantled with a view to being re-used (most of the cases), this code demands they should not be considered as temporary. For example, for gantry elements such as girders and bearings a design working life of 25 years is recommended (BSI, 2007b,

2009e). In general, it is thought acceptable to consider a design working life between 15 to 30 years for temporary structures (BSI, 2005b, 2009d). In the Australian standard AS 1170.0 (SAA, 2002), the minimum design working life for ultimate limit state consideration of any structure is 25 years. In the UK, if a given temporary structure is on the same site for over five years it needs planning permission and is considered a permanent structure and designed accordingly.

6.2.3 Actions, Partial Factors and Load Combinations

Actions relevant for temporary structures have been introduced and discussed extensively in Chapter 3.

The general format for load combinations specified in the various codes for ULS verification (applicable both to persistent and transient design situations; not applicable to accidental and seismic design situations) can be expressed by Eq. 1 (BSI, 2005b).

$$\left.\begin{matrix}\gamma_{G,\max}\\ \gamma_{G,\min}\end{matrix}\right| \cdot \sum G_{k,j} + \gamma_{Q,1} \cdot Q_{k,1} + \sum \left.\begin{matrix}1.0\\ 0\end{matrix}\right| \cdot \gamma_{Q,i} \cdot \psi_{0,i} \cdot Q_{k,i} \tag{1}$$

where $\gamma_{G,max}$ and $\gamma_{G,min}$ represent the maximum and minimum partial factors to be applied to the characteristic value of permanent loads $G_{k,j}$, respectively; $\gamma_{Q,1}$ represents the partial factor to be applied to the characteristic value of the leading variable action $Q_{k,1}$; $\gamma_{Q,i}$ represents the partial factors to be applied to the characteristic value of the accompanying variable loads $Q_{k,i}$ and $\psi_{0,i}$ represents the combination factors of each $Q_{k,i}$.

The favourable effects from the accompanying variable actions should not be considered (i.e. they should be multiplied by combination factors equal to zero). The treatment of the favourable effects from permanent actions needs a careful analysis. If both favourable and unfavourable effects are correlated, safety must be verified considering separately the cases of $\gamma_{G,max}$ and $\gamma_{G,min}$ multiplied by all the permanent actions effects. If the favourable and unfavourable effects are not correlated, then $\gamma_{G,max}$ should be applied to the unfavourable effects and $\gamma_{G,min}$ to the favourable effects.

Specific load combinations can be found in ASCE/SEI 37, AS 3610, BS EN 12811 and BS EN 12812. BS EN 1991-1-6 does not specify load combinations, leaving this duty to the temporary structures designer which increases his responsibilities but also allows him to optimise the design in view of performance requirements and limitations specific to each project. However, the envisaged load combinations must in all cases follow the philosophy defined in BS EN 1990. Load combinations that will not occur do not need to be considered. For instance, often a maximum working wind velocity is set. This therefore means it is not necessary to consider simultaneously the design wind load and all construction loads. For example, provided that there are no personnel on the structure or any concrete operations occurring under maximum working wind conditions, the former actions need not be combined during design with the latter action.

A comparison of the load partial factors specified in the AS 3610 and in the ASCE/SEI 37 is given in Table 2 and in the European standards is given in Table 3. Regarding the latter Table, the information included concerning the Eurocodes relates to the CC2 consequence class as defined in BS EN 1990 (BSI, 2005b), see Chapter 5 for details.

In the case of scaffolding/falsework, BS EN 12810 to BS EN 12813 provide simplified loading combinations, compared with the Eurocodes, see Table 4. All the combination factors are 1.0 and there is no use of leading and other variable actions. Contrary to BS EN 12812 and the Eurocodes, BS EN 12811-1 does not differentiate between permanent and variable loads and uses a single value of the load partial factor equal to 1.5.

For BCEs used in cast in-place or precast segmental construction of concrete bridges, partial factor and load combinations additional to the ones presented in this Section are specified in the AASHTO bridge code (AASHTO, 2016b).

Detailed comparison of the several codes shows different partial factors for the same conditions and it may be difficult to determine the reasons for the differences. For example, considering equilibrium limit states (e.g. sliding, overturning and uplift), AS 3610 specifies $\gamma_{G,min} = 0.8$, BS EN 12812 $\gamma_{G,min} = 0.9$, while for bridges BS EN 1990 indicates $\gamma_{G,sup} = 1.05$ and $\gamma_{G,min} = 0.95$. In particular, when counterweights are used to control the stability of the temporary structure, as is often used in launched BCE, the variability of the counterweights load value may be taken into account. In these cases, according to BS EN 1990, either a value of $\gamma_{G,min} = 0.8$ should be applied to the counterweights load value or alternatively, a value of $\gamma_{G,min} = 1.0$ may be used provided that the counterweights position is varied ± 1 m from its nominal position so to produce the most unfavourable load case.

Regarding accidental design situations, such as vehicle or crane impact, the ULS load combinations format can be expressed by (BSI, 2005b):

$$\sum G_{k,j} + A_d + \sum \left.{}^{1.0}_{0}\right| \cdot \psi_i \cdot Q_{k,i} \tag{2}$$

where A_d represents the design value of the accidental action and ψ_i represents the combination factor for the accompanying variable load $Q_{k,i}$. The values of the combination factors ψ_i are project specific but can be considered smaller than the values of the combination factors used in the persistent and transient design situations.

The favourable effects from the accompanying variable actions should not be considered (i.e. they should be multiplied by combination factors equal to zero).

Interpreting the values, it can be observed that in general the partial factors to be applied to the leading variable action specified in ASCE/SEI 37 are slightly higher than the corresponding values given in the Eurocodes and higher than the Australian ones. However, the combination factors specified in the USA code are much lower than the ones recommended in the Eurocodes and in the AS 3610, in particular for construction loads.

The fact that partial factors for loads are smaller in one code in comparison with others, does not necessarily imply that the design loads determined using that code will also be smaller. Design loads are obtained in various codes by different methods and using different initial representative values, all of which can be more or less conservative. Additionally, during code calibration different target reliabilities can be selected by the code committees (see Chapter 5). Therefore, code comparison has to be carried out carefully and not by a simple comparison of partial factor values. Examples of some published studies about code comparison exercises are referenced in André, Beale, & Baptista (2012b). It is also noted that when all the combinations of actions are taken into account that the overall safety of the structures achieved with different design codes is comparable.

Finally, regarding partial factors for resistance there are no major differences between the values specified in the AASHTO bridge code for the verification of SLS and ULS and the corresponding values

Table 2. Comparison of load partial factors for falsework and BCE design, between the Australian and USA standards

Load type	AS 3610		Load type	ASCE/SEI 37		
	ID	**Load partial factor γ_l**		**ID**	**Load partial factor γ_l**	$\psi_{0,l} \cdot \gamma_l$
Permanent loads (Q_l)	Steel: sections, wires, cables, etc.	γ_{max} = 1.25 γ_{min} =1.00	Permanent loads	Steel (CD)	γ_{max} = 0.9; 1.2 or 1.4 γ_{min} =0.85	N/A
	Wood: formwork					
	Concrete weight					
Construction loads	Concrete casting loads (Q_c)	1.0	Construction loads	Fixed (CFML)	1.2	0
	Load from stacked materials (M)	1.5		Variable (CVML)	1.4	Analysis dependent
	Lateral pressure of concrete (P)	1.5		Lateral pressure of concrete (CC)	1.2 or 1.6	0
	Personnel and equipment (Q_{uv})	1.5		Personnel and equipment (CP)	1.6	0.5
				Erection and fitting (CF)	2.0	Analysis dependent
				Equipment reactions (CR)	1.6 or 2.0 (Rated or Unrated)	0
Variable loads	Horizontal (Q_{uh})	1.5		Horizontal (CH)	1.6	0.5
	Wind (W_u)	1.0	Variable loads	Wind (W)	1.0	0.3
	Thermal (T)	1.5		Thermal (T)	1.2	0
				Snow (S)	1.6	0.5
	Earthquake (E_u)	1.0		Earthquake (E)	1.0	0
Other loads	Loads caused by post-tensioning or other actions (X_m)	1.5	Other loads	Loads caused by post-tensioning or other actions (O)	1.6	0.5
Basic combinations	Unloaded: $\gamma_G \cdot G+\gamma_{Quv} \cdot Q_{uv}+\gamma_{Quh} \cdot Q_{uh}+\gamma_M \cdot M$ EQU: $\gamma_G \cdot G+\gamma_{Wu} \cdot W_u$		Examples (Basic combinations)	1.4·D + 1.4·CD + 1.2·CFML + 1.4·CVML		
				1.2·D + 1.2·CD + 1.2·CFML + 1.4·CVML + 1.6·CP + 1.6·CH		
	During loading: $\gamma_G \cdot G+\gamma_{Qc} \cdot Q_c$			1.2·D + 1.2·CD + 1.2·CFML + 1.4·CVML + 1.0·W + 0.5·CP		
	After loading: $\gamma_G \cdot G+\gamma_{Quv} \cdot Q_{uv}+\gamma_{Quh} \cdot Q_{uh}+\gamma_M \cdot M+\gamma_{Xm} \cdot X_m$ $\gamma_G \cdot G+\gamma_M \cdot M+\gamma_{Xm} \cdot X_m+\gamma_{Wu} \cdot W_u$			1.2·D + 1.2·CD + 1.2·CFML + 1.4·CVML + 1.0·E + 0.5·CP		
	Seismic action: $G+E_u$			0.9·D + 0.9·CD + (1.0·W or 1.0·E)		

Table 3. Comparison of load partial factors for falsework and BCE design, between the different European standards

Load type	BS EN 12812		Load type	Eurocodes (Medium consequence class)		
	ID	Load partial factor γ_i		ID	Load partial factor γ_i (STR/GEO)	$\psi_{0,i}$ (Road bridges)
Permanent loads (Q_1)	Steel: sections, wires, cables, etc.	$\gamma_{max} = 1.35$ $\gamma_{min} = 1.00$	Permanent loads	Steel (G)	$\gamma_{max} = 1.35$ $\gamma_{min} = 1.00$	N/A
	Wood: formwork		Construction loads (Qc)	Formwork system (Qcc)	1.5	1.0
Construction loads	Fresh concrete weight, precast units weight (Q_2)	1.5		Concrete casting loads, precast units weight (Qcf)	1.5	1.0
	Concrete casting loads (Q_4)	1.5				
	Construction loads due to working personnel (Q_2)	1.5		Construction loads due to working personnel (Qca)	1.5	1.0
	Horizontal (Q_3)	1.5		Construction loads due to moveable heavy machinery and equipment, lifting, hoisting (Qcd)	1.5	1.0
	Construction loads due to storage of materials (Q_2)	1.5		Construction loads due to storage of moveable items (Qcb)	1.5	1.0
Variable loads	Wind actions (Q_5)	1.5	Variable loads	Wind actions (W)	1.5	0.8
	Thermal (Q_8)	1.5		Thermal (T)	1.5	0.6
	Snow (Q_2)	1.5		Snow (S)	1.5	0.8
	Earthquake (Q_7)	1.0		Earthquake (E)	EN 1998-2 + NA	0.0
Other loads	Loads caused by post-tensioning or other actions (Q_9)	1.5	Other loads	Loads caused by post-tensioning or other actions (O)	EN 1990, EN 1992, EN 1993 + NAs	EN 1990
Basic combinations	Unloaded: $\gamma_1 \cdot Q_1 + \gamma_5 \cdot Q_5$		Analysis dependent, refer to EN 1990			
	During loading: $\gamma_1 \cdot Q_1 + \gamma_2 \cdot Q_2 + \gamma_3 \cdot Q_3 + \gamma_4 \cdot Q_4 + \gamma_5 \cdot Q_5 + \gamma_8 \cdot Q_8$					
	After loading: $\gamma_1 \cdot Q_1 + \gamma_2 \cdot Q_2 + \gamma_3 \cdot Q_3 + \gamma_5 \cdot Q_5 + \gamma_8 \cdot Q_8 + \gamma_9 \cdot Q_9$					
	Seismic action: $Q_1 + Q_2 + Q_4 + Q_7 + Q_8 + Q_9$					

STR/GEO: Limit states where the resistance is governed by the failure or excessive deformation of structural elements or of the supporting ground. Additional limit states are defined in BS EN 1990. For example, EQU consist in the verification of the loss of equilibrium of the structure or any part of it, considered as a rigid body.

Table 4. Actions combinations for scaffolding (European standards)

Provision	See Table 6 for actions legend	
	ID	Value
Basic combinations	Service condition	$Q_1 + Q_2$ (working level) $+0.5 \cdot Q_2$ (Level below) $+ Q_{3a}$
	Out-of-service condition	$Q_1 + c \cdot Q_2 + Q_{3b}$ where c = 0, class 1; c =0.25, classes 2 & 3 c=0.50, classes 4, 5 & 6

given in the Eurocodes. For SLS, they are set to 1.0 and for ULS (persistent and transient design situations) they vary but are always larger than 1.0 and in general not greater than 1.25.

In BS EN 12810 to BS EN 12813, a constant value of the partial factor for resistance equal to 1.1 is specified for steel and aluminium (BS EN 12812 also requires the application of an additional resistance partial factor of 1.15 for Class B2 falsework designs), whereas in the Eurocodes the values are a function of the type of material, design verification (element or joint) and expected failure mode (ductile or brittle). However, for ULS verification of seismic and accidental design situations, the Eurocodes require the use of the same values as for the other design situations whereas the AASHTO code specifies unit resistance partial factors. Additionally, in the Eurocodes, resistance partial factors can alternatively be derived by testing, updated using new information by Bayesian methods and can be reduced depending on the efficiency of the control and inspection measures implemented.

The AASHTO guide for bridge falsework (AASHTO, 2008b) by still being based on the Allowable Stress design philosophy, specifies design values for the actions and resistance variables, meaning that unit safety factors are used, unless a global safety factor is provided. The same occurs in BS 5975, where global safety factors of 1.65 on yield condition and 2.0 on failure condition are specified.

6.2.4 Structural Analysis and Design

6.2.4.1 Steel elements analysis

Basis

The designer of temporary structures can use many types of structural analysis (see Chapter 4): from the simple first-order elastic analysis which can only approximately simulate the "real" behaviour of the structure, to the complex but more accurate geometrical and material nonlinear imperfect analysis (GMNIA).

The calculation model and basic assumptions for the calculations should reflect the structural behaviour at the relevant limit state with appropriate accuracy and reflect the anticipated type of behaviour of the cross-sections, members, joints and bearings. Analyses should be based upon calculation models of the structure that are appropriate for the limit state under consideration. The method used for the analyses should be consistent with the design assumptions.

For example, there are limits for the validity of application of linear analyses: they return accurate results only if the design loads are sufficiently smaller than the critical buckling load of the structures. In some cases, amplification factors based on the reduction of load capacity based on buckling loads can be used. In all other cases, second-order analyses should be performed. See Chapter 4 for guidance.

Additionally, the analyses can be either static or dynamic. In the former case, and when the resonant part of the response is not significant, equivalent static loads may be used to conservatively take into account the dynamic effects. The latter is usually performed by multiplying the static loads by notional dynamic factors.

Finite element analyses (FEA) of structures can use beam, shell or solid elements, each with their own advantages and disadvantages. If beam elements are used, shear lag and local buckling effects could be simply simulated by considering an effective cross-section, or by other equivalent methods. Allowance should be made for distortional, warping and torsional stresses as appropriate (e.g. in the case of open or of closed cross-sections subject to eccentric transverse loading) and for the effect of large transverse loads generated for example by vehicle wheels, lifting equipment operations or launching elements over rigid supports.

It appears that the most common methods of structural analysis in the USA involve the use of approximate analysis methods and tables. In Europe, load tables such as those produced by the UK National Access and Scaffolding Association (NASC, 2013) or manufacturers of proprietary equipment for scaffolds are used in standard configurations. However, for BCE, falsework systems and non-standard scaffolds it is commonly accepted that as a minimum requirement beam FEA will be undertaken. For more complex cases shell or even solid FEA should be undertaken.

The AASHTO LRFD specifications are less informative than the Eurocodes regarding criteria for judging the adequacy of using simplified analysis methods and do not provide criteria for performing advanced structural analysis (Hida et al., 2010). It also appears that the USA codes are more prescriptive-based than the Eurocodes, which support FEA. The Eurocodes allow for departure from traditional design approach through performance-based design and full probabilistic methods for the verification of the structural safety and serviceability together with a rational definition of minimum reliability levels. In the Eurocodes, the use of advanced structural analysis methods, such as GMNIA, is well supported by providing information not only regarding initial imperfections, material, connections and load modelling but also the safety format to be used.

For nonlinear analyses, the partial factor γ_E must be applied considering the following general simplified rule (valid in the case of a single predominant action) (BSI, 2005b, 2009d):

- When the action effect increases more than the action, the partial factor γ_E should be applied to the representative value of the action;
- When the action effect increases less than the action, the partial factor γ_E should be applied to the action effect of the representative value of the action.

When applying the above rule, the nature of the action should be taken into account so to prevent unrealistic action values being considered. This is relevant in intrinsically dynamic actions such as wind and waves, but also in geotechnical related problems.

In terms of resistance variables, it is overly conservative and inadequate to use design values of resistance properties in nonlinear analyses. It is more appropriate to calculate the resistance of the structure directly using the characteristic values (or the mean values) of the resistance properties.

From the information of the two previous paragraphs, it is clear that the use of a global safety factor is especially suitable for design based on nonlinear analysis. For example, when characteristic values of the resistance properties are used, the global safety factor can be determined by $\gamma_{R1}\gamma_{E1}$ in the case of a single design dominant resistance property ($R1$) and a single action effect ($E1$), or when mean values of the resistance properties are used, the global safety factor is equal to $\gamma_{R1*}\gamma_{E1}$ (for the simple case considered). The value of γ_{R1} and γ_{R1*} can be calculated from (assuming that resistance variables follow a Lognormal distribution):

$$\gamma_{R1} = \frac{e^{\left[\left(\alpha_{R1}\cdot\beta-1.64\right)\cdot V_{R1}\right]}}{\theta_{m}} \quad , \quad \gamma_{R1*} = \frac{e^{\left[\alpha_{R1}\cdot\beta\cdot V_{R1}\right]}}{\theta_{m}} \tag{3}$$

where θ_m represents the average value of the ratio between the actual resistance and the predicted resistance for the modelling approach chosen, α_{R1} represents the sensitivity coefficient of variables $R1$, β represents the reliability index and V_{R1} represents the coefficient of variation of the resistance variables, accounting for various types of uncertainties (see Chapter 5) given by:

$$V_{R1} = \sqrt{V_{a}^{2} + V_{.}^{2} + V_{m}^{2}} \tag{4}$$

where V_a, V_θ, V_m represent the coefficient of variation of the geometry, model and material, respectively.

Nevertheless, the bulk content of most modern design codes is devoted to the use of simple structural analysis methods, such as linear elastic analyses, in combination with correction factors in order to get "accurate" solutions. For seismic design, simple analysis methods based on elastic static analysis have been developed as an alternative to the more complex nonlinear time history analysis, such as equivalent-static loads and response spectrum analysis methods. More recently displacement-based analysis methods, such as nonlinear static (pushover) analysis, have also been introduced as a compromise between elastic static analysis and time-dependent nonlinear analysis (BSI, 2004e, 2008f, 2008f, 2009b).

6.2.4.2 Material Properties

Regarding material requirements, AASHTO CHBTW-1-M (AASHTO, 2008a) does not recommend the use of steel grades greater than ASTM A 36/A 36M for falsework construction. The minimum value of the yield strength of this steel grade is 250 MPa (250 N/mm^2). BS 5975 also includes a general recommendation to limit the use of steel grades higher than S 275JR (BSI, 2004f). However, AASHTO bridge temporary structures code (AASHTO, 2008b) allows the use of steel grades with higher resistance properties as long as they conform with AISC 360 (AISC, 2010) and BS EN 12811-1 (BSI, 2003c) for scaffolds requires the use of steels with a minimum nominal yield strength of 235 MPa.

Table 5 presents the design requirements of the steel grades complying with BS EN 10025-2 (BSI, 2004f) for hot rolled structural steel.

The design values to be considered in analyses are provided directly from the design codes which in general adopt conservative simplifications of the values specified in product standards. To take into account products that are not covered by product standards, design codes establish minimum material characteristics. For example, BS EN 1993-1-1 requires the ratio between f_u/f_y to at least equal to 1.10, the strain at failure be at least the larger of 15% and 15 ε_y, where ε_y is the yield strain. These provisions are extremely important since they lay the foundation for the safe use of simplified analysis and design procedures (see Chapter 4).

Table 5. Material properties for hot-rolled steel grades and qualities according to BS EN 10025-2

Steel grades and qualities	Minimum yield strength f_y (MPa)				Tensile strength f_u (MPa)		Minimum percentage elongation after fracture (proportional test specimens)		
	Nominal thickness (mm)				Nominal thickness (mm)		Nominal thickness (mm)		
	≤16	>16≤40	>40≤63	>63≤80	<3	≥3≤100	≥3≤40	>40≤63	>63≤100
S235JR	235	225	215	215	360–510	360–510	26	25	24
S235J0	235	225	215	215	360–510	360–510			
S235J2	235	225	215	215	360–510	360–510	24	23	22
S275JR	275	265	255	245	430–580	410–560	23	22	21
S275J0	275	265	255	245	430–580	410–560			
S275J2	275	265	255	245	430–580	410–560	21	20	19
S355JR	355	345	335	325	510–680	470–630	22	21	20
S355J0	355	345	335	325	510–680	470–630			
S355J2	355	345	335	325	510–680	470–630			
S355K2	335	345	335	325	510–680	470–630	20	19	18

6.2.4.3 Initial Imperfections

The idealised structural system considered during analysis and design should consider material and structural deviations from the perfect state. The design values should allow conservative simulations of the representative values of the imperfections that are generated during fabrication, maintenance and assembly activities.

There are many types of imperfections, from variability of material properties to deviations of the geometrical characteristics (see Chapter 4). Traditionally, product standards specify limiting values for the geometrical tolerances (e.g. out-of-straightness) and representative values for some properties of the materials (e.g. steel's yield stress). For rolled steel elements values are provided for example in BS EN 39, BS EN 10210-2 or BS EN 10219-2 (BSI, 2001, 2006d, 2006e) for tubular sections and in BS EN 10034 (BSI, 1993b) or in AASHTO GDBTW-1-M (AASHTO, 2008b) for I and H sections. For elements with welded profiles geometrical tolerances are provided for example in BS EN 1090-2 (BSI, 2011c).

Design codes take into account initial imperfections by grouping various types of imperfections in equivalent initial geometric imperfections calibrated against results of rigorous and accurate analysis studies. These equivalent initial geometric imperfections typically account for elements initial geometric imperfections, joint eccentricities, material properties variability as long as the individual values of each of these imperfections is not larger than tolerance limits specified in product standards.

With respect to temporary structures, in particular scaffolding and falsework, Birch, Booth, & Walker (1971), Burrows (1989), Chandrangsu & Rasmussen (2011a) and Pallett, Burrow, Clark, & Ward (2001) provide results of site surveys of initial geometric imperfections of actual systems, including measurement of local and global deformations and load eccentricities. From the findings, it seems that a considerable percentage (up to 50%) of the elements of scaffolding and falsework systems exhibited imperfections that were outside the design code's tolerances and that there does not appear to have been an overall improvement in this regard over time. This is a violation of the assumptions made during the design of temporary structures in accordance to design codes directly putting the lives of workers at risk.

Table 6. Types of equivalent initial geometric imperfections for falsework, BS 5975

<table>
<tr><th>Falsework Category</th><th colspan="2">Member imperfections</th><th>Eccentricity of loading</th><th>Eccentricity at supports</th></tr>
<tr><td>Adjustable steel props and forkheads.</td><td colspan="2">Lateral displacement due to 1.0° rotation at base, not exceeding 17 mm over a height of 1 m.</td><td colspan="2">No eccentricity in excess of that assumed by the designer</td></tr>
<tr><td>Tube and coupler systems</td><td colspan="2">Lateral displacement not exceeding 15 mm over a height of 2 m, subject to a maximum value of 25 mm at full height.</td><td>Eccentricity of equal to 25 mm.
The centrelines of tubes at a node point should be as close together as possible, and never more than 160 mm apart</td><td>Sole plates normally be set within a tolerance not exceeding 25 mm per 1 m.</td></tr>
<tr><td>Purposely fabricated steel work.</td><td>Columns:
Sway imperfection of max(L/600, 5 mm/m) but not larger than 25 mm.
Bow imperfection of a max(L/1000, 3 mm/m), but not larger than 25 mm.
L is the clear length of the strut or column (mm).</td><td>Beams:
Bow imperfection of a max(L/1000, 3 mm/m), but not larger than 25 mm.
L is the clear length of the beam (mm).</td><td>Eccentricity of equal to 5 mm</td><td></td></tr>
</table>

For scaffolding and falsework, BS 5975 (BSI, 2011a) specifies several types of equivalent initial geometric imperfections, see Table 6. Note that the method of substituting the sway imperfection with an equivalence horizontal load as commonly used in many design codes was shown by Prabhakaran to be not valid if the structure is unbraced as significant differences in behaviour and maximum load capacity occur. The alternative method was shown to be valid only for braced structures (Prabhakaran, 2009; Prabhakaran, Beale, & Godley, 2011). In the case of an unbraced structure, the analysis model must include all the relevant geometrical imperfections in order to achieve a safe design.

Table 7 shows the equivalent set of geometrical imperfections for steel and aluminium telescopic props, specified in the applicable European design codes.

For steel elements design according to BS EN 1993-1-1 (BSI, 2005e, 2008a), the equivalent initial geometrical imperfections specified in the design code are presented in Table 8 for unbraced systems:

Other types of imperfections need to be considered for example torsional imperfections for thin wall open sections.

The Eurocodes also allow for nonlinear FEM-based analysis. In these cases, the model should include both geometric and structural imperfections. The use of equivalent geometric imperfections is allowed applied in the direction that returns the lowest resistance. For steel structures, the types and values of the imperfections is presented in Table 9.

Table 7. Types of equivalent initial geometric imperfections for telescopic props

Material	Member imperfections	Eccentricity of loading	Eccentricity at supports		
Steel (BSI, 1999)	Bow imperfection equal to *L*/500 + lateral displacement due to gap between inner and outer tubes. *L* is the length of the prop (mm).	Eccentricity of equal to 10 mm	Eccentricity of equal to 0.4 (*D*+2 *t*) (mm) *D* is the tube's external diameter *t* is the baseplate thickness		
Aluminium (BSI, 2012)	Bow imperfection equal to *L*/375 (elastic analysis) + lateral displacement due to gap between inner and outer tubes. *L* is the length of the prop (mm).	Eccentricity of equal to 10 mm	Eccentricity of equal to 0.4 (*D*+2 *t*) (mm) *D* is the tube's external diameter *t* is the baseplate thickness		

Table 8. Types of equivalent initial geometric imperfections for unbraced systems, BS EN 1993-1-1

Sway imperfection	Bow imperfection
Rotation, φ, equal to: $\varphi = \varphi_0 \cdot \alpha_h \cdot \alpha_m$ φ_0 is equal to 1/200 $\alpha_h = \frac{2}{\sqrt{h}}$, $\frac{2}{3} \leq \alpha_h \leq 1.0$ $\alpha_m = \sqrt{\frac{1}{2} \cdot \left(1 + \frac{1}{m}\right)}$ *h* total height of the structure (m); *m* is the number of columns in a row, including only those columns which carry a design axial vertical load not less than 50% of the average value of the axial force in the columns in the vertical plane considered.	Mid-span lateral deflection equal to: e_0/L *L* is the length of the member e_0 depends on the geometry of the cross-section, material strength and of the type of analysis For example, e_0=L/300 for S355 steel closed cross-section and elastic analysis

Table 9. Equivalent geometric imperfections for nonlinear FEM-based analysis, Eurocode 3

Type of imperfection	Component	Shape	Magnitude
Global	Individual linear element	bow	As specified in BS EN 1993-1-1
Local	Individual plate	buckling shape	min (*L*/200, *h*/200) *L* is the length of the member; *h* is the height of the cross-section.
All types	Shells of revolution	Analysis dependent	See BS EN 1993-1-6 (BSI, 2007a).

For Class B2 scaffolding and falsework, BS EN 12812 (BSI, 2011d) specifies several types of equivalent initial geometric imperfections, see below. For Class B1, Eurocodes apply.

- Looseness at spigot joints
  - Angular imperfection, φ:

$$\tan\left(\varphi\right) = 1.25 \cdot \frac{d_i - d_0}{l_0} \cdot \min\left(\sqrt{0.5 + \frac{1}{n_v}}, 1.0\right) \tag{5}$$

  - Eccentricity, e:

$$e = 1.25 \cdot \frac{d_i - d_0}{2} \tag{6}$$

where d_i, d_0 and l_0 represent the internal diameter of the standards, the external diameter of the spigot and the overlap length, respectively, and n_v represents the number of standards arranged side by side in a row.

This imperfection can be applied in two ways: (i) sway-like imperfection or (ii) bow-like imperfection.

- Bow imperfections
  - System bow imperfection amplitude, e:

$$e = \frac{l}{250} \cdot \min\left(\sqrt{0.5 + \frac{1}{n_v}}, 1.0\right) \tag{7}$$

where l represents the overall length of the standards in each bay and n_v represents the number of standards arranged side by side in a row.

  - System bow imperfection amplitude, e_0:

$$\begin{aligned} e_0 &= \frac{l_e}{300} \text{ for elastic analysis and} \\ e_0 &= \frac{l_e}{250} \text{ for plastic analysis} \end{aligned} \tag{8}$$

where l_e represents the member length.

- Sway imperfections, base rotation, φ:

$$\tan\left(\varphi\right) = 0.01 \cdot \sqrt{\frac{10}{h}} \quad \text{but if } h \leq 10 \text{ m} \Rightarrow \tan\left(\varphi\right) = 0.01 \tag{9}$$

The overall sway imperfection, Eq. 9, and the sway for individual components, see Eqs. 5 and 6, need not be considered as simultaneous effects.

- Load eccentricities:

BS EN 12812 specifies that load eccentricity at load points shall be taken as a minimum of 5 mm where there is no centring device. Where there is a centring device the eccentricity taken may be reduced to a value consistent with the tolerances of the relevant components.

6.2.4.4 Steel Design Verifications

The main elements of most BCEs are made of steel: girders (I or box), frames or trusses for example. The safety and serviceability verifications must follow the rules specified in the codes applicable in each country and are the same as for permanent structures. For scaffolding and falsework, simplified alternative rules may apply.

As stated earlier the majority of rules found in existing structural codes assume that simple analysis methods are used together with practical assumptions for many parameters, such as the buckling length of elements, and thus the proposed safety verification procedures contain correction factors, additional to the ones mentioned previously. These procedures, generally performed at individual element level, were developed based on calibration studies against results obtained using advanced analysis for general and common applications. Code rules must always be analysed and interpreted having this information in mind.

6.2.4.5 Ultimate Limit States

Regarding the verifications of the ultimate limit states (ULS), code rules usually divide them in different parts:

1. Resistance of the steel cross-section against tensile, compressive, shear and torsion forces, individually or combined;
2. Resistance of the steel element: flexural buckling, torsional buckling, flexural-torsional buckling, coupled local and global buckling, shear buckling, flange-induced buckling and web resistance to high local transverse loads (web buckling, crippling and crushing);
3. Resistance of the connections against tensile, compressive, shear and torsion forces, individually considered or in combination;
4. Resistance to fatigue and brittle failure;
5. Loss of equilibrium of the structure or any part of it, considered as a rigid body.

The design procedures to verify the safety of steel elements to these ULS are given in the USA AASHTO bridge code (AASHTO, 2016b), AISC 360 (AISC, 2010), in the various parts of Eurocode 3, namely BS EN 1993-1-5 (BSI, 2006a, 2008b), BS EN 1993-1-11 (BSI, 2006b, 2008d), BS EN 1993-2 (BSI, 2006c, 2008e) and BS EN 1993-6 (BSI, 2007b, 2009e) and in BS 5975 (BSI, 2011a).

Regarding the design of compressed elements (e.g. columns), until recently, national and international design rules were based on simple design procedures, for example: the columns' effective length would be governed only by the vertical spacing of horizontal members, not considering the system's overall stability. The use of the effective length concept as a design procedure, although simple, is not accurate since it is based on an element level safety check and it assumes that the element's deformed shape is very similar to its first global elastic buckling mode. Therefore, the use of second-order nonlinear analysis and design procedures is recommended, see Chapter 4.

The design rules and geometrical requirements for stiffeners, longitudinal and transverse, which can be essential to verify safety of steel girders under high transverse loads or of steel girders with large unbraced lengths, are included in the AASHTO bridge code, AISC 360 and in BS EN 1993-1-5. These rules are particularly important in the design of connection panels of falsework systems commonly used in the USA consisting in top and bottom steel beams (termed cap and sill beams, respectively) connected by steel or timber columns (posts).

The strength verification rules specified in the AASHTO bridge code are generally taken from the AISC 360-10 (AISC, 2010) for building structures. In 2009, Sahin (2009) compared the AISC 360-05 (the 2010 edition contains some few changes but these are not fundamental) and the Eurocode 3 rules for ULS. Although the cross-section classifications used in the two codes are different, the number of classes, four, and the design concepts are the same. In terms of cross-section resistance it was found that in general, AISC-360 gives approximately 10 to 20% higher values for compact sections while the Eurocode 3 returns 15 to 35% higher values in slender sections. Regarding member resistance only a single curve for flexural buckling is given in AISC-360 whereas five separate curves are presented in Eurocode 3. The AISC code curve is always less conservative (up to 40%) than the Eurocode curves except for one curve. For lateral buckling, the concepts are different and the results obtained for slender sections using the AISC code are about 15% higher. For shear buckling the AISC code uses the tension field method whereas Eurocode 3 uses the rotated stress field method.

With respect to connections between elements of temporary structures, as the joint types in BCE resemble solutions already in use for heavy construction of buildings and bridges, guidance concerning the design of these joints is abundantly available (BSI, 2005f, 2008c; Jaspart & Weynard, 2016).

In scaffolding and falsework it also has to be verified that the design values of the loads acting on the couplers do not exceed the corresponding design values of the resistances. Some design models are available, exceptions are the values and equations provided in BS EN 12811-1 and in BS EN 12812 for some types of couplers used in scaffolding/falsework complying with BS EN-74-1 (BSI, 2005a), such as the right-angle coupler. For other types of couplers, guidance is presented in Chapter 4.

For the design of anchoring elements to concrete, reference is made to the rules given in design documents such as ACI (2011, 2014a), CEN (2009) and EOTA (2013a, 2013b) or to the design guidelines included in the technical manuals developed by the manufacturers of proprietary fastening solutions as long as they comply with the principles and requirements for structural design set in the applicable design codes. Safety against relevant anchor failure modes should be evaluated, namely:

- Tension resistance of the steel anchor;
- Shear resistance of the steel anchor;
- Tension resistance of the concrete;
- Shear resistance of the concrete;
- Bond resistance in the interface between the anchor and the surrounding concrete;
- Concrete splitting resistance.

In the UK, the National Access and Scaffolding Confederation (NASC) has developed a design guide concerning anchorage systems for scaffolding (NASC, 2011).

Elements that support cranes need additional verification in order to properly consider the detrimental effects of the torsional moments that arise from the eccentricities of vertical actions and of lateral horizontal actions relative to their shear centre. Additional procedures which cover this and other special safety verifications of these elements, such as web slenderness limits and maximum stresses in service (SLS design), can be found in BS EN 1993-6 (BSI, 2007b, 2009e).

Designing stability against overturning if no specific bracing is provided (AASHTO, 2008b) requires that the ratio between the resisting moment and the overturning moment shall be equal to or greater than 1.2 for all load combinations.

The AASHTO bridge code requires top or bottom lateral bracing for I-girders when the span exceeds 60 m. For horizontally curved bridges it may be necessary to consider both top and bottom lateral bracing. Truss girders must always have both top and bottom lateral bracing (AASHTO, 2016b).

Regarding the friction developed in PTFE (Teflon) bearings, the AASHTO bridge code requires considering a friction value on launching bearings between 0 and 4%, whichever is critical, plus the actual superstructure gradient. BS EN 1991-1-6 recommends the same lower and upper bounds of friction coefficients, but also highlights that they should be defined for each specific project.

Concerning design of geotechnical structures, such as foundation elements, reference is made in Europe to BS EN 1997-1 (BSI, 2013, 2014b). In particular, Annex D provides a simple analytical method for bearing resistance calculation which can be used for the basis of calculation.

Finally, with respect to design against seismic events, if simplified analysis methods that consider the linear elastic response of the structural system, design codes may allow for a reduction of seismic forces (dividing them by behaviour factors equal to or larger than 1.0) in order to account for the nonlinear response of the structure. The values of behaviour factors strongly depend on the structural system configuration, materials' elastoplastic behaviour and structural robustness. If values are not provided, behaviour factors can be determined by dividing the total elastoplastic displacement of a reference node by the elastic component of the displacement of that node or more accurately by dividing the base shear assuming linear elastic behaviour by the base shear assuming elastoplastic behaviour.

6.2.4.6 Serviceability Limit States

Serviceability limit states (SLS) are usually specified in the form of deflections and rotations limits, as well as vibrations and stress limits. The design loads are in general the same as those considered for ULS, although unit safety factors are applied to all loads. Regarding load combinations, the same guidance as for ULS is used although the loads involved can differ in order to take into account specific operation conditions. Combination factors lower than the ones specified for ULS can be used, allowing for higher probabilities of load exceedance, whose values can also vary between each SLS depending on the reversibility characteristics, or lack, of the considered SLS.

However, few design criteria for SLS of temporary structures are given in the USA and European codes. Listed below are some exceptions.

For incremental launching, BS EN 1991-1-6 recommends the following design values of vertical deflection, see Figure 1:

1. $\pm$ 10 mm longitudinally for one bearing, the other bearings being assumed to be at the theoretical level (Figure 1(a)) ;

Figure 1. Deflections of bearings during execution for bridges built by the incremental launching method, recommended in BS EN 1991-1-6. (a) Longitudinal section, (b) transverse section

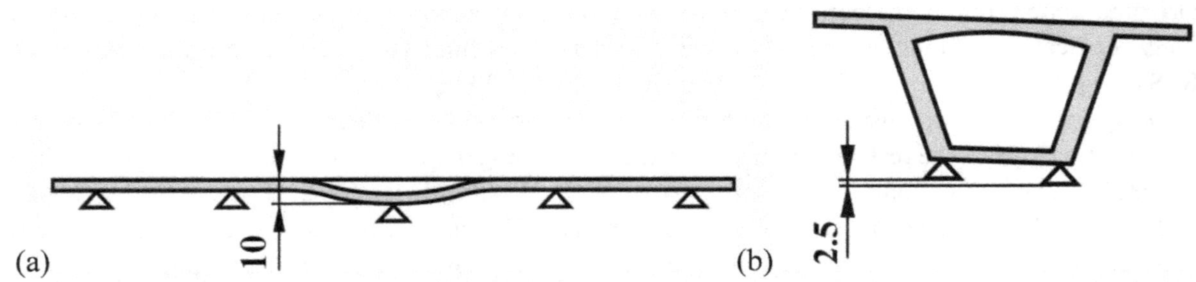

2. ± 2.5 mm in the transverse direction for one bearing, the other bearings being assumed to be at the theoretical level (Figure 1(b)).

Guidance on the maximum values for both horizontal and vertical displacements for elements which support cranes can be found in BS EN 1993-6 (BSI, 2007b, 2009e).

For scaffolding, SLS limits refer only to the deflection of working platform elements, see BS EN 12811-1 (BSI, 2003c) and Table 10, whereas for falsework almost no criteria are provided. Therefore, in the latter case, SLS criteria of falsework should satisfy, as a minimum, the provisions included in the execution standards of the permanent structures, for example BS EN 1090-2 (BSI, 2011c) and BS EN 13670 (BSI, 2009a). In the AASHTO standard GSBTW-1-M (AASHTO, 2008b) it is required that the vertical deflection for falsework members do not exceed 1/240 of their span under only the permanent load of the concrete.

For grandstands and stages, the maximum horizontal displacement should not exceed height/300 under the combined effect of the permanent loads, imposed loads, wind and notional horizontal loads. The wind loads are determined for the working wind velocity. The notional horizontal loads are specified in Chapter 3.

6.2.4.6.1 Design of Bearings

To allow movement of a BCE to a new position it is necessary to support the structure on a series of sliding bearings, which offer only a minimal resistance to movement, and to provide a jacking system capable of dealing with the combined effects of friction and longitudinal gradient.

Temporary sliding-guided bearings are provided at each of the piers and at any temporary intermediate supports. These bearings usually comprise a neoprene pad to facilitate rotation of the structure and a PTFE/stainless steel sliding surface to minimise friction. The design of these elements is specified in product standards (BSI, 2000, 2004b, 2004d, 2004c, 2006c, 2008e, 2009c).

Table 10. Serviceability criteria for scaffolding (European standards)

Deflections	Platform units	Bay span/100 and defection difference between loaded span and adjacent span ≤25 mm
	Side protection	Guardrail ≤35 mm, fencing structures < 100 mm

For segmental bridges built using the incremental launching method, the AASHTO bridge code specifies a minimum value of 1% of the vertical support reaction for the horizontal load acting on the lateral guides of the launching bearings. Additionally, bearings shall be designed in such a way that they can compensate for local deviations of the sliding surface of up to 2.0 mm by elastic deformation. Construction tolerances of launching bearings should not exceed 5.0 mm and 2.5 mm in the longitudinal and transverse direction, respectively, between two adjacent bearings.

6.2.4.7 Design of Hydraulic Jacks

For every type of BCE, a mechanical system of hydraulic jacks must be used and properly designed. Jacks must be securely anchored to a reaction structure in order to create the launching movement. Three different types of systems can be used to launch a BCE: push, pulling or friction systems.

The friction launching system is one of the most common methods to launch bridges. This system consists of vertical jacks which are used to slightly elevate the structure in order to generate sufficient friction to grip it, and horizontal jacks, which then push the structure forward (Rosignoli, 2000). A more ingenious system used to launch the deck of the Millau bridge in France consisted in using two wedge steel plates with PTFE layers in between and two horizontal jacks, one to produce the launching movement by raising one of the wedges along the contact area, and the other to lower the bridge again into its supports by applying the opposite action. Another solution was applied in the construction of the Normandie bridge, also in France, to suit large longitudinal gradients (6%) (see Virlogeux (1993)). This involved placing small steel rollers between the trapezoidal bearing and the pier, and using jacks to lift and lower the bridge deck to allow the sliding bearing to recover to its original launching position.

The design of a launching jack system must consider the force necessary to overcome the combined frictional resistance offered by the sliding bearings plus any resistance arising from the longitudinal slope of the bridge. A proper safety factor should also be considered. If no other information is available, it may be required to specify a jack capacity equal to twice the design launching load.

When braking devices are necessary they can consist of hydraulic jacks or high friction plates. The former solution, generally used only in push and pulling launching systems, can for instance comprise two outside braking jacks while the central jacks provide the launching force. The latter solution consists of rubber or high friction steel plates which are activated against the bottom surface of the superstructure after the stroke of the launching jacks is exhausted.

The design of the launching system should take into account accidental scenarios involving for instance the chance of uncontrolled sliding of the superstructure especially when the kinematic friction coefficient is significantly less than the slope of the bridge. For example, the braking system should have redundancies built in the system in order to safely accommodate failure of one of the primary elements or the occurrence of power shutdowns.

6.2.4.8 Design of Shallow Foundations

The purpose of shallow foundations (e.g. steel baseplate supported on the ground) is to transfer loads from the structure to the supporting ground. Traditionally design of shallow foundation elements has been carried out using simplified analyses or empirical approaches of which some are still currently used (BSI, 2013, 2014b). In these methods the verification of the bearing capacity and of the deformations is performed separately. More recently, the FEM has started to be applied extensively to geotechnical problems such as those involving shallow foundations (Potts & Zdravkovic, 1999, 2001).

There are essentially two main design objectives: to avoid lack of equilibrium of the structure and the ground, and to ensure sufficient bearing resistance capacity and stiffness to the foundation elements and to the supporting ground.

The former objective is concerned with the possibility of rigid body failure of the structure often by excessive base rotations and with global instabilities of the ground (only relevant when the foundation rests on a sloped soil surface or on soil sustained by a retaining wall). The latter objective addresses the possibility of insufficient bearing resistance capacity of the ground (or of the foundation element) as well as of excessive ground settlements.

Global instabilities of the ground are usually not applicable to most temporary structures and will not be considered herein.

A stable equilibrium state at the base level of a temporary structure can be guaranteed by adequate proportions of the foundation element so that there is always a region of the surface in contact with the soil that is in compression. Additionally, the magnitude of the base rotations should not exceed pre-established tolerance limits so to avoid triggering local collapses and the associated consequences.

With respect to soil resistance, the bearing resistance capacity, q_f, under pure axial compression of a shallow rectangular foundation over drained soil (e.g. sand) is governed by the following equation (BSI, 2013, 2014b):

$$q_f = c \cdot N_c \cdot s_c \cdot i_c + q \cdot N_q \cdot s_q \cdot i_q + \frac{1}{2} \cdot \gamma \cdot B \cdot N_\gamma \cdot s_\gamma \cdot i_\gamma \tag{10}$$

where:

c represents the cohesion of the soil;
q represents the magnitude of any imposed pressure on the ground surface adjacent to the foundation element;
γ' represents the effective unit weight of the soil;
B represents the width of the foundation element (lesser dimension in plan);
N_c, N_q and N_γ represent bearing capacity factors;
b_c, b_q and b_γ represent inclination factors to take into account the effect of inclined foundation base;
s_c, s_q and s_γ represent shape factors to allow applicability to shapes other than rectangular;
i_c, i_q and i_γ represent inclination factors to take into account the effect of inclined loading.

For eccentric loading, e, the width B should be replaced by an effective width B' equal $B - 2 \cdot e$. Comprehensive design examples are provided in Frank et al. (2013).

AASHTO (2008b) and BSI (2011a) documents state that in the absence of information relative to ground mechanical properties, the allowable bearing resistance values, i.e. q_f in Eq. 10, (to be used under the ASD philosophy) shown in Table 11 may be used as a guide, subject to site reassessment. The presumptive bearing values shown in Table 11 are for level and sloping ground where the slope is not greater than 1 vertical to 6 horizontal, with no water present at least at a depth smaller than the larger dimension of the foundation element.

Table 11. Allowable bearing pressure values (AASHTO, 2008b; BSI, 2011a)

Foundation support	Foundation material	Allowable bearing pressure
Excellent	Hard shales and soft sandstones Soft shales, soft claystones, and very soft sandstones Weak and fractured limestone Dense sands and gravels Very stiff to hard clays	2,100 kN/m² 600 to 1,000 kN/m² 600 kN/m² 400 kN/m² 300 to 400 kN/m²
Generally adequate	Medium dense sands and gravels Medium dense uniform-size sand Stiff clays	200 to 300 kN/m² 200 kN/m² 100 to150 kN/m²
Poor	Loose sand or loose sand and gravel Loose uniform-size sand Soft to medium clays Loose silts	100 kN/m² 75 kN/m² 50 kN/m² 50 kN/m²

In the design of the temporary structures, full account should be taken of the effects of settlement of the foundations, both uniform and differential. These settlements can be caused by various conditions, such as variation of ground properties (including those caused by the permanent structure both during construction and after completion) and variations in loading.

The short-term elastic settlement of shallow foundations over a layer of a single type of soil can be estimated by (BSI, 2013, 2014b):

$$\delta_v = \frac{q \cdot B}{E} \cdot \left(1 - \upsilon^2\right) \cdot I_s \tag{11}$$

where:

q represents the uniform pressure load at the base of the foundation;
B represents the width of the foundation element;
E and ν represent the soil's Young's Modulus and Poisson's coefficient, respectively, see Table 12;
I_s represents an influence factor depending on the shape of the loaded area.

The rotation, θ, of a shallow foundation over a layer of a single type of soil can be estimated by (Bowles, 1997):

$$\tan\theta = \frac{1-\nu^2}{E} \cdot \frac{M}{B^2 \cdot L} \cdot I_\theta \tag{12}$$

where:

M represents the bending moment at the base;
B and L represent the width and length of the foundation element, respectively;
E and ν represent the soil's Young's Modulus and Poisson coefficient, respectively, see Table 12;
I_θ represents an influence factor depending on the type of soil.

Table 12. Range of values for Young's Modulus and Poisson's coefficient of soils (Bowles, 1997)

Soil	Type	E (MPa)	ν (-)
Clay	Very soft	2 to 15	Clay, saturated: 0.4 to 0.5 Clay, unsaturated: 0.1 to 0.3
	Soft	5 to 25	
	Medium	15 to 50	
	Hard	50 to 100	
	Sandy	25 to 250	
Sand	Silty	5 to 20	0.3 to 0.4
	Loose	10 to 25	
	Dense	50 to 81	
Sand and gravel	Loose	50 to 150	
	Dense	100 to 200	
	Shale	150 to 5000	
	Silt	2 to 20	

Equations 11 and 12 should only be applied to determine the elastic settlement if for the imposed loading the constitutive model of the soil may be considered to be linear elastic. Great caution is required in the case of non-homogeneous ground.

The maximum acceptable relative rotations of $L/500$ is adequate for many structures, where L represents the larger of the dimensions of the foundation element, whereas a settlement up to 50 mm is often acceptable (BSI, 2013, 2014b).

6.3 EXAMPLES

6.3.1 Determination of Cantilever Bridge Construction Safety for Equilibrium Limit State

Figure 2 shows a section of a bridge during execution by the cantilevering construction method. The design is made according to BS EN 1990 (BSI, 2005b, 2009d). The bridge is supported on a single pier, and fully unsupported at both ends, cantilevering 18 m at the left side and over 20 m at the right side. The deck is loaded with the material's self-weight, wind loading, and construction loads. It is assumed that the coefficient of variation of the material's self-weight is very small and therefore, the mean value of the self-weight may be applied.

To verify safety with respect to a possible loss of equilibrium, equation 7 of BS EN 1990 states that:

$$E_{d,dst} \leq E_{d,stb} \tag{13}$$

where $E_{d,dst}$ and $E_{d,stb}$ are the design values of the destabilising and stabilising actions respectively.

Equation 10 of the Eurocode states that:

Figure 2. Bridge cantilever design against overturning

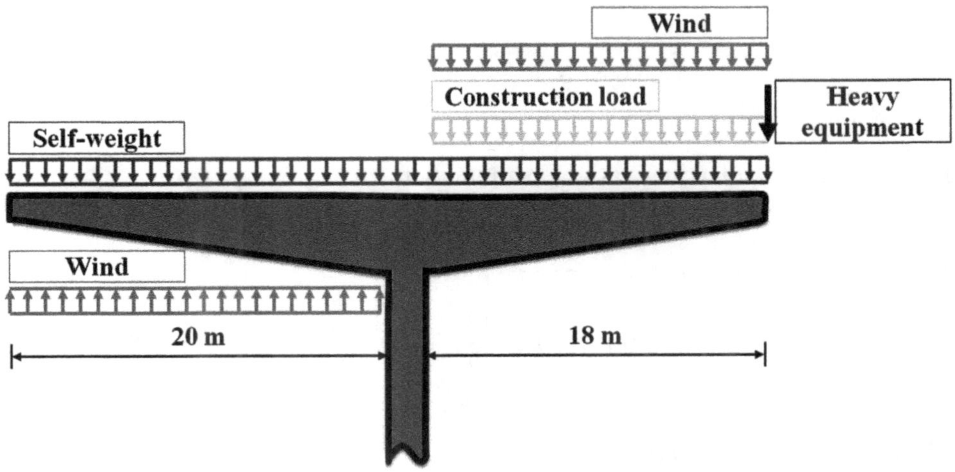

$$E_{d,dst} = \gamma_{G,sup} \cdot G_{k,sup} + \gamma_{Q,1} \cdot Q_{k,1} + \gamma_{Q,i} \cdot \psi_{0,i} \cdot Q_{k,i} \tag{14}$$

where:

$E_{d,dst}$ is the design destabilising effects from the actions combination;

γ_G and γ_Q are action partial factors. For unfavourable actions: $\gamma_{G,sup} = 1.05$ for permanent actions and $\gamma_{Q,i}$ = (1.35 for construction actions and 1.5 for all other actions during execution). For favourable actions: $\gamma_{G,sup} = 0.95$ for permanent actions and $\gamma_{Q,i} = 0$ for variable actions;

ψ_0 are combination factors, whose values are found in the UK National Annex to BS EN 1990. When the wind is the accompanying action, the value of the combination factor should be equal to $\psi_0 = 0.8$. For construction loads $\psi_0 = 1.0$.

In the design example, let us assume that the density of the post-tensioned reinforced concrete is 25 kN/m^3 and the bridge has a cross-section area of 8.3 m^2. The self-weight of the structure is therefore 25 8.3 = 207.5 kN/m. We can also assume that the weight of heavy machinery at the end of the cantilever is 60 kN. Using the UK National Annex to BS EN 1991-1-6 (BSI, 2009d) the construction load due to personnel $Q_{ca} = 1.0$ kN/m^2 and for moveable storage is $Q_{cb} = 0.2$ kN/m^2.

Using BS EN 1991-1-4 and its UK National Annex (BSI, 2010, 2011e) we can determine the maximum wind pressure on the structure. For simplicity, let us assume that the value is ±6.3 kN/m per metre of the bridge length. An analysis must assume the most destabilising direction of the wind, namely upwards on the left part (see Figure 2) and downwards on the right portion.

The destabilising calculation must be performed for two load cases: one where the construction load is the leading variable action and one where the wind is the leading variable action.

When the construction load is the leading variable action, one has:

$$E_{d,dst} = 1.05 \cdot 207.5 \cdot \frac{18^2}{2} + 1.35 \cdot \left(0.2 \cdot \frac{18^2}{2} + 1.0 \cdot \frac{18^2}{2} + 60 \cdot 18\right) + \ldots$$
$$\ldots + 1.5 \cdot 0.8 \cdot 6.3 \cdot \frac{18^2}{2} = 36.93 \text{ MN.m} \tag{14}$$

When the wind load is the leading variable action, then:

$$E_{d,dst} = 1.05 \cdot 207.5 \cdot \frac{18^2}{2} + 1.5 \cdot 1.0 \cdot 6.3 \cdot \frac{18^2}{2} + \ldots$$
$$\ldots + 1.35 \cdot 1.0 \cdot \left(0.2 \cdot \frac{18^2}{2} + 1.0 \cdot \frac{18^2}{2} + 60 \cdot 18\right) = 38.55 \text{ MNm} \tag{15}$$

The stabilising effects are given by:

$$E_{d,stb} = 0.95 \cdot 207.5 \cdot \frac{20^2}{2} - 1.5 \cdot 1.0 \cdot 6.3 \cdot \frac{20^2}{2} = 37.54 \text{ MNm} \tag{16}$$

In this example, the destabilising forces are greater than the stabilising forces and the design is unstable. Therefore, either temporary supports, tie-down elements or counterweights should be designed and added to the left hand side of the bridge to ensure the required stability.

6.3.2 Access Scaffold Design

As part of the verification process for the NASC design manual TG20 (NASC, 2013), models of different access scaffolds were constructed in 2-D and tested against a set of 3-D models constructed using a finite element model. Figure 3 shows a model of a 10 bay, 5 lift access scaffold with no ledger bracing on the bottom two lifts as it is common practice to remove ledger bracing at one level to enable workmen to work at that level without encountering an obstruction, and to have no ledger bracing at the lowest level to enable pedestrians to walk beneath the scaffold. The bottom level frequently has no ledger bracing as well to enable pedestrians to walk under the scaffold which, of course, must be adequately externally clad to prevent materials or tools falling.

The scaffold modelling is detailed in Chapter 4.

The loading from the scaffold boards, due to permanent and variable loads, was applied at nodal points. Wind loading was applied as a combination of point and distributed loads to all standards. The service wind load (called working wind) was 200 N/mm^2 in accordance with BS EN 12811-1 (BSI, 2003c). For the maximum (out-of-service) wind load, the pressure, q, at height h above the ground was given by:

$$q = a \cdot \ln(h) + b \tag{17}$$

where the values of the coefficients a and b were derived by regression from Table 4 of BS 6399-2 (BSI, 1997), for different town and country distances and wind velocity factors.

Figure 3. 3-D model of an access scaffold with two bottom lifts without ledger bracing

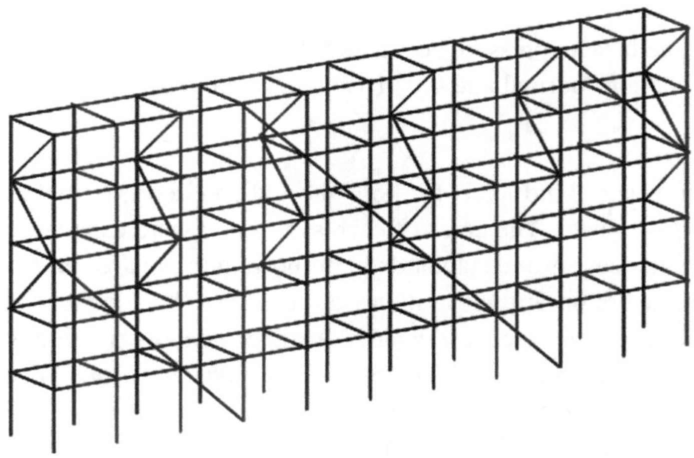

In accordance with a recommendation given in BS EN 12811-1, the out-of-service wind pressures were reduced to 70% of the values calculated by the above formula. Note that this is a simplification. The general method specified in BS EN 1991-1-4 should be used, see Chapter 3.

The direction of the wind load normal to the façade was such as to put the ties into tension and hence suction was applied to the scaffold structure. The compressive direction was not considered as it was assumed that the scaffold would deflect into bearing with the façade, thus improving its stability. For unsheeted scaffolds, the wind load on standards, ledgers and transoms was applied as distributed line loads. The loads on toeboards and guard-rails were applied as point loads on the standards at the appropriate levels. The guard-rail wind load on the top lift was exceptionally applied at the top lift due to the model not including standards above the top lift level. The scaffold was analysed as a "bare-poles" scaffold, as a fully sheeted scaffold and as a debris netted scaffold. The shielding effect of the building leeward façade was considered as specified in BS EN 12811-1 through the use of a site coefficient, see Chapter 3. The solidity ratio of the building façades were taken equal to 0.95, resulting in a site coefficient equal to 0.38. The force coefficients of the sheeting and netting face were based on wind tunnel test results. For example, for the netting, the force coefficient normal to wind direction was considered to be 0.65, and 0.25 parallel to wind direction.

For sheeted scaffolds, the following loading was applied (taken from BS EN 12811-1, Annex A, see Chapter 3):

- Winds normal to the façade: pressure coefficient equal to -0.5 normal to the façade (suction), no load on the end of the façade;
- Winds parallel to the façade: pressure coefficient 1.0 on the ends parallel to the façade (applied as pressure), -0.5 (suction) normal and 0.01 (friction) parallel to all front elements along the front of the façade.

The pressure coefficient was applied to either the working wind load pressure of 200 N/m^2 and to the unfactored out-of-service wind load pressure given by Eq. 17.

For debris netted scaffolds, the following was applied:

- Winds normal to the façade: pressure coefficient -0.25 normal to the façade (suction), no load on the end of the façade;
- Winds parallel to the façade: pressure coefficient 0.5 on the ends parallel to the façade (applied as pressure), -0.25 (suction) normal and 0.03 (friction) parallel to all front elements along the front of the façade.

To calculate the equivalent wind loads for both sheeted and debris netted scaffolds, the load was applied to each bay and lift panel. The areas of each subsection were calculated and the total load on the panel distributed to the ledgers and standards in proportion to the area of each subsection. A small correction was made to the bottom lift where it was assumed that the wind load was only applied to the ledger above the panel and to the adjacent standards. A trapezoidal distribution was applied to the largest element of either the standard or the ledger which was equivalent to total pressure over that element found by Eq. 17. The effect of wind loads acting on the guard-rails was ignored as it would have meant two different loading distributions for boarded and unboarded scaffolds.

For the in-service wind condition, the trapezoidal and triangular loads were modelled accurately as the analysis program used was able to apply distributed multi-linear constant loads. For the out-of-service load condition, the assumption was made that the wind pressure distribution given by 17 could be approximated to the exposed rectangle-shaped surface areas by applying the appropriate load at nodal points and using linear interpolation. For a 20 m high scaffold, the error in this assumption was shown to be less than 0.1%. In addition, as the analysis program used could not handle trapezoidal and triangular loads with variable magnitudes these loads were modelled by uniformly distributed loads with the same total external force on the sides of each panel. The resulting total load applied to the scaffold by the program was within 0.5% of the total load calculated by integration using a symbolic algebra package.

The variable loads were applied to the top two levels only as vertical loads in agreement with the European standard BS EN 12811-1. This meant that a service load of between 0.75 kN/m^2 to 3 kN/m^2 (Load classes 1-4, see Chapter 3) were applied to the platform at the top level and a load of 50% applied to the next level in the maximum load condition. The top level was considered to be the working level for the structure as it generated the maximum loads throughout the structure. In accordance with the code, a load factor of 1.5 was applied to the loads in the ultimate load condition.

The following load combinations were considered for every scaffold analysed:

Load Case 1: Self weight + service imposed load + equivalent frame imperfection load normal to the façade + a horizontal load on side protections equal to 0.3 kN in each bay, normal to the façade at the working lift.

Load Case 2: Self weight + service imposed load + equivalent frame imperfection load normal to the façade + service wind load normal to the façade.

Load Case 3: Self weight + out-of-service imposed load + equivalent frame imperfection load normal to the façade + out-of-service wind load normal to the façade.

Load Case 4: Self weight + service imposed load + equivalent frame imperfection load parallel to the façade + a horizontal load on side protections equal to 0.3 kN in each bay, parallel to the façade at the working lift.

Load Case 5: Self weight + service imposed load + equivalent frame imperfection load parallel to the façade + service wind load parallel to the façade.

Load Case 6: Self weight + out-of-service imposed load + equivalent frame imperfection load parallel to the façade + out-of-service wind load parallel to the façade.

Frame imperfections were ignored for the early buckling analyses which only considered point loads at the top of the scaffold. For non-linear analyses, these were represented by equivalent horizontal loads directly proportional to all vertical loads with the constant of proportionality, ϕ, given by the formula:

$$\phi = \phi_o \cdot \left(0.5 + \frac{1}{n_c}\right)^{0.5} \cdot \left(0.2 + \frac{1}{n_s}\right)^{0.5} \tag{18}$$

in which n_c = the number of fully loaded standards (= no of bays), n_s = the number of lifts and $\phi_o = 0.01$ (corresponding to the erection tolerance set in BS 5973 (BSI, 1993a). Note that this formula has been replaced in BS EN 12811-1 by Eq. 5, the same as indicated in BS EN 12812. However, the results being quoted do not vary when different models of imperfection are included as they only made differences of approximately1.5% on the results.

All the above loads were applied either normal or parallel to the façade depending upon the load case being considered. The non-linear analyses started from 10% of the design load and increments were made until either the structure was unable to carry further load or three times the design load was achieved. The following analyses were made in each case:

1. A linear analysis to check structural loads and structural geometry including restraints.
2. A buckling analysis.
3. A non-linear analysis including geometrical non-linearity. This started from 10% of the design load and increments were made until either the structure was not able to carry further load or three times the design load was achieved.
4. A program was written to process the output of each load increment in the non-linear analysis. The following checks were incorporated:

$$\left(\frac{N_{Sd}}{P_c} + \frac{M_{xSD} + M_{zSD}}{M_{Rd}}\right) \cdot 1.1 = k \text{ (compressive loads)} \tag{19}$$

$$\left(\frac{N_{Sd}}{A \cdot f_y} + \frac{M_{xSD} + M_{zSD}}{M_{Rd}}\right) \cdot 1.1 = k \text{ (tensile loads)} \tag{20}$$

$$\frac{M_{coupler} \cdot 1.35}{400} = m_{test} \tag{21}$$

where:

N_{Sd} is the design axial load;
M_{xSd} is the design bending moment about the x axis;
M_{zSd} is the design bending moment about the z axis;

P_c = 48 kN for 2 m lift and M_{Rd} =1.85 kN.m (plastic moment of resistance of a 4 mm thick scaffold tube with f_y = 235 N/mm^2);

k and m_{test} are limit state values with values less than 1.0 implying that the test is satisfied and values greater than 1.0 implying that the element of the structure has failed;

$M_{coupler}$ is the characteristic moment for the coupler.

Initial checks of coupler and joint slippage were made but as the internal forces in the connections were low these checks were not made for all analyses. It was found that the use of a partial factor of 1.35 for the coupler resistance was always conservative. Failure was deemed to have occurred when the limit state values exceeded 1.0 for any element in the scaffold.

The results of applying BS EN 12811-1 were that the scaffold with ledger bracing removed for the bottom two levels was capable of carrying the design load in both service and out-of-service conditions. Unfortunately, the analysis program used was incapable of applying loads in excess of the design loads due to convergence problems. More examples of this analysis can be found in Beale & Godley (2006).

6.3.3 Design of a Bamboo Scaffold Standard

Chung & Chan (2002) describe the material properties of bamboo in scaffolding and also give design procedures for scaffolds. Using their procedures let us consider the design of a bamboo standard in Mao Jue (called a post in Hong Kong – note Kao Jue standards are also called poles).

Let us consider a standard 2.0 m long with an external diameter equal to 90 mm at the top and 80 mm at the bottom, with corresponding internal diameters of 72 mm and 64 mm, respectively.

If the standard is placed on the ground and pinned at the top its length, L, is equal to 2.0 m and its effective length is also 2.0 m.

From the formulae in Chapter 4, Section 4.7.4 (note numbers used in this calculation are stored in the computer and used to reduce rounding errors in calculations):

$$A_1 = \left[\frac{\pi}{4}\cdot\left(D_o^{\ 2} - D_i^{\ 2}\right)\right]_1 = \left[\frac{\pi}{4}\cdot\left(80^2 - 64^2\right)\right] = 1809.6 \text{ mm}^2, A_2 = \left[\frac{\pi}{4}\cdot\left(90^2 - 72^2\right)\right] = 2290.2 \text{ mm}^2 \tag{22}$$

$$I_1 = \left[\frac{\pi}{64}\cdot\left(D_o^{\ 4} - D_i^{\ 4}\right)\right]_1 = 1187070 \text{ m}^4 \quad , \quad I_2 = 1901456 \text{ mm}^4 \tag{23}$$

$$\text{radii of gyration,} \quad r_1 = \sqrt{\frac{I_1}{A_1}} = \sqrt{\frac{1187070}{1809.6}} = 25.62 \text{ mm}, \quad r_2 = \sqrt{\frac{I_2}{A_2}} = \sqrt{\frac{1901456}{2290.2}} = 28.81 \text{ mm} \tag{24}$$

$$\text{slenderness ratios,} \ \lambda_1 = \frac{L}{r_1} = \frac{2000}{25.61} = 78.09, \ \lambda_2 = \frac{L}{r_2} = \frac{2000}{28.81} = 69.41 \tag{25}$$

$$\text{ratio of section change,} \ \rho = \frac{I_2 - I_1}{I_1} = 0.602 \tag{26}$$

Coefficients of the cubic equation $c_3 \cdot \alpha^3 + c_2 \cdot \alpha^2 + c_1 \cdot \alpha + c_0 = 0$ are:

$$\begin{aligned} c_3 &= -0.2880 \\ c_2 &= 2.016 \cdot (2+\rho) = 5.2452 \\ c_1 &= -\left(14.11 + 14.11 \cdot \rho + 3.098 \cdot \rho^2\right) = -23.7235 \\ c_0 &= 10.37 + 15.55 \cdot \rho + 7.047 \cdot \rho^2 + 0.932 \cdot \rho^3 = 22.483 \end{aligned} \tag{27}$$

Solving the equation numerically $\alpha = 1.289$.

The critical buckling strength, $f_{cr} = \alpha \cdot \dfrac{\pi^2 \cdot E_b}{\lambda_1^2} = 1.289 \cdot \dfrac{\pi^2 \cdot 11.4}{78.09^2} = 23.79\ \mathrm{kN/mm^2}$ (28)

where E_b is Young's modulus for bending and is given by 11.4 kN/mm^2 for the example considered.

The design compressive strength, $f_{c,d} = \dfrac{f_c}{\gamma_R} = \dfrac{80.5}{1.5} = 53.67\ \mathrm{N/mm^2}$ (29)

where the compressive strength of Mao Jue is 89.5 N/mm^2 for the example considered, and a resistance partial factor $\gamma_R = 1.5$ is used.

The limiting slenderness ratio, $\lambda_0 = 0.2 \cdot \pi \cdot \sqrt{\dfrac{E_b}{f_c}} = 0.2 \cdot \pi \cdot \sqrt{\dfrac{1.4 \cdot 1000}{53.67}} = 9.158$ (30)

Perry factor, $\eta = 0.001 \cdot a \cdot (\lambda_1 - \lambda_0) = 0.001 \cdot 15 \cdot \left(78.09 - 9.158\right) = 1.034$ (31)

$$\varphi = \frac{f_{c,d} + \left(1+\eta\right) \cdot f_{cr}}{2} = \frac{53.67 + (1+1.034) \cdot 23.79}{2} = 51.03\ \mathrm{N/mm^2} \tag{32}$$

buckling strength, $f_{cc,d} = \dfrac{f_{cr} \cdot f_{c,d}}{\varphi + \sqrt{\varphi^2 - f_{cr} \cdot f_{c,d}}} = \dfrac{23.79 \cdot 53.67}{51.03 - \sqrt{51.03^2 - 23.79 \cdot 53.67}} = 14.598\ \mathrm{N/mm^2}$

(33)

modified slenderness ratio, $\lambda_0 = 0.2 \cdot \pi \cdot \sqrt{\dfrac{E}{f_{c,d}}} = 0.2 \cdot \pi \cdot \sqrt{\dfrac{11.4 \cdot 1000}{53.67}} = 9.16$ (34)

Finally, the design maximum axial load is equal to $f_{cc,d} \cdot A_1 = \dfrac{14.598}{1000} \cdot 1.8096 = 26.42\ \mathrm{kN}$. (35)

6.4 DESIGN GUIDANCE

6.4.1 Steel Telescopic Props

Often, the design of steel telescopic props involves high levels of uncertainty, due to the inherent particularities of their behaviour and to the lack of sufficient quality control during their use. In order to give a contribution to the improvement of the efficiency and safety of these structural elements a campaign of experimental tests was carried out, as well as analytical analysis and numerical modelling. Based on the results presented in André (2008) and André, Baptista, & Camotim (2007, 2009b, 2009a), a proposal of a design curve for this specific type of structural elements is presented, thus contributing to overcome the limitations in existing standards dealing with their safety check.

The European standard BS EN 1065 (BSI, 1999) is the current product standard for adjustable telescopic steel props. This document specifies two alternative ways for the design of these structural elements: experimental and numerical simulations. Both options involve considerable work, either by requiring a large number of tests, or by imposing the use of a rather complicated numerical model (the exact formulae are given in Chapter 4). At the end of these two procedures one has to classify the prop in one of five classes defined in the standard (A to E). Each class has different resistance requirements, specified as a function of the props extension length and its maximum height, as well as of geometrical constraints, such as minimum values for the sizes of the endplates. These requirements become increasingly more stringent from Class A to Class E props.

Thus, the development of a design curve for these structural elements is a task of great practical utility, because this tool can overcome the limitations of the design methods set out in European and national documents, for example the complexity of the rules presented in BS EN 1065, or the restricted field of application of the design charts given in BS 5975 (BSI, 2011a). Note that the design methodologies given in these two standards can sometimes be unnecessarily conservative and cannot be used efficiently to optimise prop design.

The development of the proposed design curve was based on the results of experimental tests on 35 props, performed at Labatório National de Engenharia Civil, LNEC (Portugal), and of 79 numerical simulations (André, 2008; André et al., 2007, 2009b, 2009a).

The design curve is based on a simplified structural model of a prop, illustrated in Figure 4, and on the concept of the buckling length of an equivalent prop, i.e. the buckling length of an equivalent simply supported prop with a constant cross-section. Using energy methods one can determine an approximation of the equivalent moment of inertia of a stepped column (prop), and using a length/area relation the equivalent cross-sectional area can be determined. From these two variables the radius of gyration of an equivalent prop can be obtained:

$$i_{eq} = \sqrt{\frac{I_{eq}}{A_{eq}}} \quad (36)$$

with

$$A_{eq} = \frac{A_{n,inner} \cdot L^*_{inner} + A_{gr,outer} \cdot L_{outer}}{L} \tag{37}$$

$$I_{eq} = \frac{L \cdot I_{i,inner} \cdot I_{gr,outer}}{I_{i,inner} \cdot L_{outer} + I_{gr,outer} \cdot L^*_{inner}} \tag{38}$$

where:

L_{outer} and L represent the length of the outer tube and the total extension length of the prop, respectively;
L^*_{inner} is given by $L^*_{inner} = L - L_{outer}$;
$A_{n,inner}$ and $A_{gr,outer}$ represent the net cross-sectional area of the inner tube and the gross cross-sectional area of the outer tube, respectively. The net cross-sectional area is given by:

$$A_n = 2 \cdot \left(\varphi_R \cdot R^2 - \varphi_r \cdot r^2\right) - d_0 \cdot \left(R \cdot \sin\varphi_R - r \cdot \sin\varphi_r\right) \tag{39}$$

Figure 4. Simplified structural model of a prop with three different cross-sections

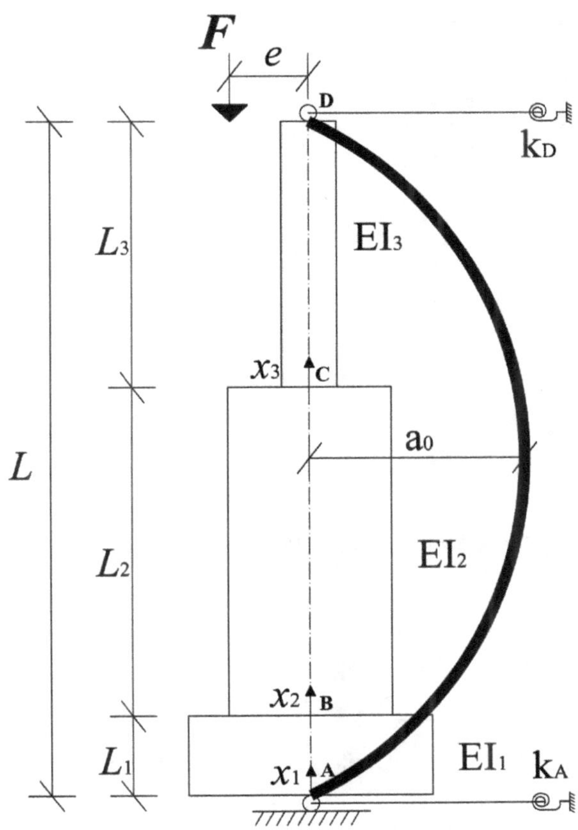

where $I_{\mathrm{i,inner}}$ and $I_{\mathrm{gr,outer}}$ represent the ideal moment of inertia of the inner tube, and the gross moment of inertia of the outer tube, respectively.

$$I_{\mathrm{gr}} = \frac{\pi}{4} \cdot \left(R^4 - r^4\right) \tag{40}$$

$$I_{\mathrm{i}} = I_{\mathrm{gr}} \cdot \frac{1}{1 + 2 \cdot \frac{d_0}{a} \cdot \left(\frac{I_{\mathrm{gr}}}{I_{\mathrm{n}}} - 1\right)} \tag{41}$$

$$I_{\mathrm{n}} = \frac{R^3}{2} \cdot \left[\varphi_{\mathrm{R}} \cdot R - \frac{d_0}{6} \cdot \sin\varphi_{\mathrm{R}} \cdot \left(3 + 2 \cdot \sin^2\varphi_{\mathrm{R}}\right)\right] - \frac{r^3}{2} \cdot \left[\varphi_{\mathrm{r}} \cdot r - \frac{d_0}{6} \cdot \sin\varphi_{\mathrm{r}} \cdot \left(3 + 2 \cdot \sin^2\varphi_{\mathrm{r}}\right)\right] \tag{42}$$

The equivalent buckling length can be evaluated by:

$$L_{\mathrm{eq}} = \sqrt{\frac{\pi^2 \cdot E \cdot I_{\mathrm{eq}}}{P_{\mathrm{cr}}}} \tag{43}$$

where the buckling load of the prop, P_{cr}, is given by:

$$P_{\mathrm{cr}} = \frac{\left(\varsigma_{\mathrm{A}} + 0.4\right) \cdot \left(\varsigma_{\mathrm{D}} + 0.4\right)}{\left(\varsigma_{\mathrm{A}} + 0.2\right) \cdot \left(\varsigma_{\mathrm{D}} + 0.2\right)} \cdot \frac{\pi^2 \cdot E \cdot I_{\mathrm{eq}}}{L^2} \tag{44}$$

with

$$\varsigma_{\mathrm{A}} = \frac{E \cdot I_{\mathrm{eq}}}{k_{\mathrm{A}} \cdot L}, \ \varsigma_{\mathrm{D}} = \frac{E \cdot I_{\mathrm{eq}}}{k_{\mathrm{D}} \cdot L} \tag{45}$$

Figure 5. Notation concerning the geometrical properties of the tubes

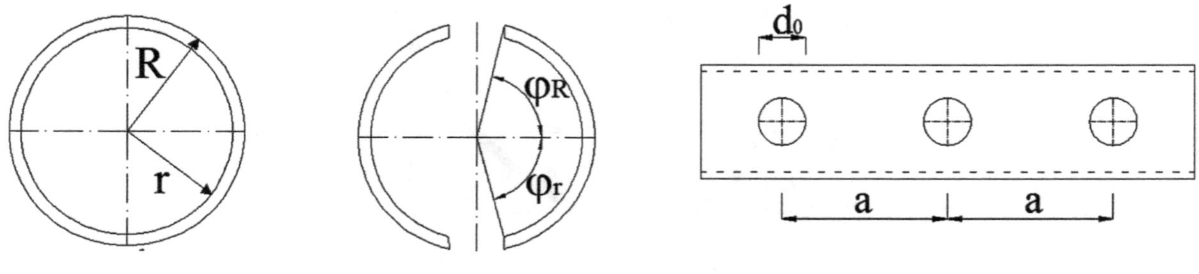

where k_A and k_D represent the elastic constant of the springs that simulate the stiffness of the prop's endplates, and L is the extension length of the prop. k_A and k_D are determined by considering that the rotational stiffness of the endplate fraction in contact with the top and bottom supports is simulated by the rotational stiffness of an element totally fixed on one side and with only free transverse displacements on the other side. Therefore k_A and k_D are given by $E\,I\,/\,l$, where E represents the Young's Modulus of the endplate's steel, I represents the moment of inertia of the endplate and l represents the free length of the endplate given by $l = (B - D_{ext}) / 2$, where B represents the side length of the endplate and D_{ext} the external diameter of the tube.

Based on the above variables it is possible to calculate the slenderness of the equivalent prop:

$$\lambda_{eq} = \frac{L_{eq}}{i_{eq}} \tag{46}$$

and the normalized slenderness:

$$\bar{\lambda}_{eq} = \frac{\lambda_{eq}}{\lambda_1} \tag{47}$$

with $\lambda_1 = \pi \cdot \sqrt{E / f_{y,min}}$, $\quad f_{y,min} = \min\left(f_{y,inner}, f_{y,outer}\right)$.

Finally, the design resistance of the prop under axial compression (P_{Rd}) is determined by:

$$P_{Rd} = \frac{\chi \cdot P_{y,min}}{\gamma_R} = \frac{\chi \cdot \left(A \cdot f_y\right)_{min}}{\gamma_R} \tag{48}$$

where χ represents the reduction factor due to flexural buckling, $P_{y,min}$ corresponds to the minimum plastic resistance to axial forces of the inner and outer tubes, and γ_R is the partial factor for resistance. The evaluation of the reduction factor is based on the formula proposed by Maquoi-Rondal (Maquoi & Rondal, 1978):

$$\chi = \frac{\beta_1}{\phi + \sqrt{\phi^2 - \bar{\lambda}_{eq}^{\,2}}}, \text{ but } \chi \leq 1,0 \tag{49}$$

with

$$\phi = \beta_2 \cdot \left(\bar{\lambda}_{eq}^{\,2} + \Theta + 1\right) \tag{50}$$

$$\Theta = \alpha_1 \cdot \left(\bar{\lambda}_{eq}^{\,2} - \alpha_2\right) \tag{51}$$

β_1, β_2, α_1 and α_2 being imperfection factors, see Figure 6.

Figure 6. Characteristic and design curve

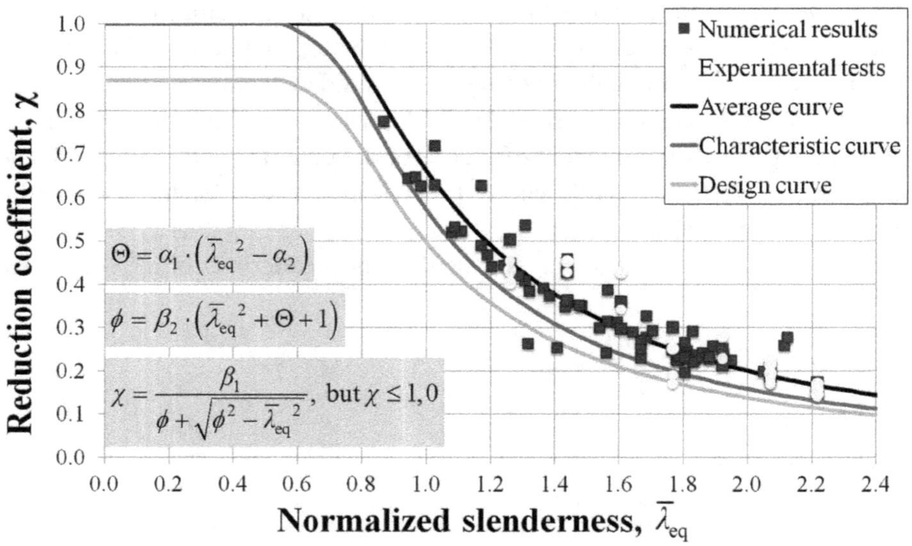

Based on the procedure specified in the Annex D of BS EN 1990, a value of γ_R equal to 1.15 was obtained.

It is important to note that the value of the resistance, P_{Rd}, given by Eq. 48 only accounts for the failure of the prop by elastic or elastoplastic instability, or by complete yielding of the cross-section of at least one tube. Therefore, additional verifications involving the calculation of the connection between the two tubes should be carried out to determine the maximum resistance of the props, see BS EN 1065 (BSI, 1999).

As an application example, consider a prop with the characteristics specified in Table 13 where:

$D_{ext,\,inner}$ is the external diameter of the inner tube;
$D_{ext,\,outer}$ is the external diameter of the outer tube;
t and b are the thickness and the side length of the square endplate, respectively.

The nominal value of the steel yield strength of the various components is 355 MPa for the inner tube, 275 MPa for the outer tube and 460 MPa for the pin. Additionally, the nominal value of the steel tensile strength of the pin, used to connect the two tubes, is 540 MPa.

Table 13. Geometrical characteristics of the prop (mm)

Prop		Tubes								Endplates		Pin
		Inner tube				Outer tube						
L_{max}	L_{min}	L_{inner}	$D_{ext,inner}$	t_{inner}	d_0	a	L_{outer}	$D_{ext,outer}$	t_{outer}	t	b	d
3500	2000	2000	48.3	3.2	14	80	1900	60	2.5	6	110	13

Using this data and applying the equations specified above for the prop resistance one gets:

Normal Configuration
Maximum extension length (L_{max})
Minimum extension length (L_{min})
$A_{gr.outer} = 451.60\ mm^2$
$A_{n.inner} = 355.52\ mm^2$
$I_{gr.outer} = 186{,}992.26\ mm^4$
$I_{i.inner} = 91{,}578.88\ mm^4$
$L_{outer} = 1{,}900\ mm$
$L_{inner} = 1{,}600\ mm$
$L_{inner} = 100\ mm$
$A_{eq} = 407.68\ mm^2$
$A_{eq} = 446.80\ mm^2$
$I_{eq} = 12{,}6664.16\ mm^4$
$I_{eq} = 177733.49\ mm^4$
$i_{eq} = 17.63\ mm$
$i_{eq} = 19.94\ mm$
$k_A = 16{,}236.00$ kN.mm/rad, $k_D = 131{,}157.21$ kN.mm/rad
$P_{cr} = 34.43\ kN$
$P_{cr} = 116.56\ kN$
$L_{eq} = 2728.07\ mm$
$L_{eq} = 1756.42\ mm$
$\lambda_{eq} = 154.77$
$\lambda_{eq} = 88.06$
$\bar{\lambda}_{eq} = 1.80$
$\bar{\lambda}_{eq} = 1.03$
$\chi = 0.19$
$\chi = 0.54$
$P_{Rd} = 20.78\ kN$
$P_{Rd} = 58.82\ kN$

Recent shoring procedures allow time to be saved and expenditure reduced on temporary structures equipment whilst maintaining adequate levels of safety for workers and for the permanent structure. These procedures consist in the early striking of formwork and shoring (i.e. when concrete is less than three days old and therefore the slabs are not fully self-resistant) and improved reshoring procedures (also known as backpropping). Reshoring consists in installing suitable propping at levels below the slab supporting the falsework to redistribute internal forces to adequate supporting elements.

In order to strike a slab, the concrete must be strong and stiff enough to avoid failure or excessive cracking and deformation of the slab. BCA (2001) and The Concrete Society (2003) documents provide the following guidance (see also ACI (2014b)):

- Determining backpropping loads during construction using empirical methods for simple repetitive structures and a three-dimensional approach for special structures. A report from The Concrete Society (2012) specifies four methods of calculating the loads in backprops and Other documents (ACI, 2014b; The Concrete Society, 2003) provide various calculation examples;
- Determining required striking concrete strength and stiffness from serviceability criteria;
- Determining early concrete strength and stiffness in situ, see The Concrete Society (2003, 2012) for guidance on in situ testing of concrete;
- Striking formwork and shoring in an agreed sequence after consideration of concrete pouring sequence. In general:
    - on large slab areas, comprising internal and edge panels, strike internal bays first, followed by edge and corner bays;
    - where soffit form is part of cantilever, start removal from tip and work towards wall, beam or columns.
- Controlling the loads on newly struck slabs until they have gained the required strength.

The compressive strength of concrete at early ages can be estimated by Eq. 52, provided that an adequate curing (at 20ºC) is verified.

$$f_{cm}\left(t\right) = \beta\left(t\right) \cdot f_{cm} \quad \text{with} \quad \beta\left(t\right) = e^{\left[s \cdot \left(1-\sqrt{\frac{28}{t}}\right)\right]} \tag{52}$$

where:

$f_{cm}(t)$ represents the mean compressive strength in MPa at age of t days;
f_{cm} represents the mean compressive strength in MPa at age of 28 days;
t represents the concrete age in days;
s represents a coefficient which depends on the type of cement, see Table 14.

The Young's Modulus of Elasticity of concrete at early ages can be estimated by Eq. 53, provided that an adequate curing (at 20ºC) is verified.

$$E_c\left(t\right) = \sqrt{\beta\left(t\right)} \cdot E_c \tag{53}$$

where:

$E_c(t)$ represents the Young's Modulus of Elasticity in MPa at age of t days;
f_{cm} represents the Young's Modulus of Elasticity in MPa at age of 28 days.

The number of days needed to strike the slab may be estimated associating the ratios $f_{cm}(t)/f_{cm}$ and $E_c(t)/E_c$ to the ratios of $S_d(t)/S_d$ and $C_d(t)/C_d$, respectively, where:

$S_d(t)$ and S_d represent the actions (or actions effects) design combination at time t and for the critical ULS, respectively;

Table 14. Coefficients to be used in Eq. 52

f_{cm} (MPa)	Type of cement (according to BS EN 197-1 (BSI, 2011b))	s
≤ 60	32.5 N, 32.5 R	0.38
	42.5 N	0.25
	42.5 R, 52.5 N, 52.5 R	0.20
> 60	all classes	0.20

$C_d(t)$ and C_d represent the actions (or actions effects) design combination at time t and for the critical SLS, respectively.

6.4.2 Scaffolding

6.4.2.1 Comments on Design Codes for Metal Scaffolding

Scaffolds systems, prefabricated or modular, are very common structures. Scaffold components are often erected according to producer's design guidelines which define the spacings of vertical and horizontal members. Similarly, these documents have been constructed for tube-and-fitting scaffolds (European terminology) or tube-and-coupler scaffolds (USA terminology).

In the UK, the National Access Scaffold Confederation (NASC) produced a set of design load tables (NASC, 2013). In the USA, ANSI 10.8 (ANSI, 2011) requires all access scaffolds to be erected according to this code or to a set of design guidelines issued by manufacturers of proprietary or system scaffolds.

It is to be noted that the ANSI standard does not require the same calculations as those required in Europe. However, the ANSI standard gives detailed charts showing allowable spacings for metal scaffolds up to 38.1 m (125 feet). The ANSI standard defines scaffolds made from metal as well as from wood.

The latter range from a height of 6 m (20 feet) for a single pole (standard) light duty scaffold made with timber standards of with cross-section dimensions 50 × 100 mm, up to 18 m (60 feet) for timber standards with dimensions 100 × 100 mm. The spacing of timber standards is either 1.8 m (6 feet) or 3.05 m (10 feet). The range of vertical spacings of horizontal members of wooden scaffolds is 1.2 m (4 feet) up to 2.4 m (8 feet).

The outside diameter of steel tubes for access scaffolds is in both the European and USA codes 48.3 mm (1.9 inch) for medium and light duty, see Chapter 2. However, ANSI standard specifies a larger tube for heavily loaded scaffolds of 60.3 mm (2.375 inch). This larger tube is also required for load bearing horizontal members in the USA. The working height between levels is at least 1.90 m (often 2.0 m) in Europe and typically 2.13 m (7 feet) in the USA. The design charts in TG 20 (NASC, 2013) allow for one level on each scaffold to have no internal bracing in order to enable workmen to work without hindrance on that level. This bracing must then be reinserted before work on the next level can be undertaken. Unfortunately, many scaffold collapses occur when workmen remove this internal bracing but do not reinstate.

The horizontal spacing between steel standards is often between 2.0 m to 3.0 m in Europe but specified as 2.44 m (8 feet) in the USA (it is 1.83 m or 2.44 m for timber). Transverse spacings in Europe are often 1.3 m but again can vary. In the USA, they are specified normally as 1.83 m. The European codes do not specify values but require the designer to choose appropriately and justify the analysis. Plan bracing is beneficial to safely transmit wind loads to the support structure and to triangulate the framework which might otherwise behave as parallelograms and have a tendency to distort.

6.4.2.2 Comments on Design Codes for Bamboo Scaffolding

Bamboo is a common choice of material for scaffolds in Asia. In Hong Kong, design procedures and guidelines are given in publications by the Buildings Department, Hong Kong Government and by the Labour Department, Hong Kong Government (Hong Kong Buildings Department, 2006; Hong Kong Labour Department, 2014). The material that can be used is Kao Jue (minimal external diameter 40 mm) or Mao Jue (minimum external diameter 75 mm with a minimum thickness of 10 mm). For design purposes, Kao Jue is taken to have a constant diameter of 40 mm over a 6 m length with a constant thickness of 5 mm, whereas for Mao Jue, the pole is said to have an external diameter of 90 mm at the bottom with thickness 9 mm to an external diameter of 60 mm with thickness 6 mm at the top over the 6 m length with a linear variation between the top and the bottom. The codes state that the overlap when joining two separate poles is between 1.5 m to 2.0 m (the latter as a minimum for braces) with the "head" of one pole being joined to the "tail" of the second pole. Nylon strips or bamboo strips are used with the nylon strips having a tensile strength of at least 0.5 kN and width 5.5 mm to 6.0 mm. The strips must be wrapped around at least five times before knotting and strips must be 300 mm apart.

On a typical two-layer façade scaffold the main standards and ledgers on the outer layer must be made of Mao Jue whilst the standards and ledgers on the inner layer are made from Kao Jue. Two main standards on the outer layer must be not more than 1.3 m apart. Midway between each main standard is a standard made of Kao Jue. Vertically ledgers must be placed with a spacing $\leq$ 1.2 m. The height of each boarded lift must be between 1.9 m to 2.0 m. Transoms, made of Kao Jue, must be spaced not less than 0.75 m apart and boards placed on them. Diagonal braces must be placed continuously from the ground to the top at angles between 45° and 60° to the vertical on both inner and outer faces and made from Kao Jue. All lines of standards must be attached to a diagonal brace in a checker-board pattern from the ground. Ties, with a tensile capacity of at least 7 kN must be attached with a maximum spacing of 4 m vertically and 7 m horizontally. These provisions apply to scaffolds not exceeding 15 m in height. For higher scaffolds more detailed calculations using the procedures given in Chapter 4 must be used to ensure safety. Note that a material factor $\gamma_R = 1.5$ is commonly used.

Guidelines to erection of bamboo scaffolds may be found in the INBAR report by Chung & Siu (2002).

6.4.2.3 Tie Types and Tie Configuration

Tying the scaffolding to an adjacent support structure is the most effective way to prevent buckling failures.

Ties are often placed vertically at every second platform level and horizontally at the same level between adjacent pairs of standards (Gylitoft & Mroz, 1995).

In the ANSI standard (ANSI, 2011), the spacings of ties are defined as required if the height of a scaffold is more than four times the narrowest horizontal dimension. In these cases, ties are required at least every 9.15 m (30 feet) horizontally and every 6.10 m (20 feet) or 7.92 m (26 feet) vertically, for scaffolds narrower than 0.91 m (3 feet) and for widths greater than 0.91 m, respectively. The European codes do not specify tie spacings.

The NASC TG20 guide (NASC, 2013) presents recommended patterns for three classes of ties loaded in pure tension: light duty ties (safe working load equal to 3.5 kN), standard duty ties (safe working load equal to 6.1 kN) and heavy duty ties (safe working load equal to 12.2 kN). In cases where the imposed design load exceeds the safe working load of a certain class of ties, the alternatives are to select ties from a class with higher resistance, and/or use additional tie points and/or reinforce the base material.

The authors in their design calculations have produced recommendations that ties are spaced on alternate rows of standards and spaced at ever second or third level vertically although the position of the first vertical tie can be staggered in the adjacent standards to give additional support. The authors also recommend that the top level should always be tied as this prevents a cantilever failure under wind load on sheeted or clad scaffolds. This is because the upper part of the scaffold pole could simply fail as a plastic hinge above the last tie, as the moment induced by maximum wind could exceed the allowable plastic moment in the tube. This is contrary to the ANSI standard specification which says that the top tie can be up to four times the minimum base dimension from the top.

Note that the authors have demonstrated that under high wind conditions, as occurs in North Scotland for example, a 50 m high scaffold requires tying at every level in every standard to ensure safety.

There are many types of ties available in the market but not all ties are suitable for use in all materials. The NASC TG 14 guide (NASC, 2011) provides ranges of adequate ties as a function of the base material that can be used for tie selection during preliminary design, see Table 15. This guide also contains guidance concerning tie characteristics, anchor installation and testing procedures.

Guidance with respect to shear resistance of ties can be found in the manufacturer's technical documentation, which should have been developed based on results of suitable tests from which records need to be available. Under combined shear and tensile loads, a possible safety verification of the ties is given by:

$$\frac{F_{v,Ed}}{F_{v,Rd}} + \frac{F_{t,Ed}}{F_{t,Rd}} \leq 1.2 \tag{54}$$

where $F_{v,Ed}$ and $F_{t,Ed}$ represent the shear and tensile design loads, respectively, and $F_{v,Rd}$ and $F_{t,Rd}$ represent the shear and tensile design resistances, respectively.

Table 15. Tie types and base materials

Base material	Suitable tie types
Concrete	Drop-in expansion anchor
	Self-tapping screws
	Nylon anchors with screw-in eyes
	Resin anchors
Brickwork and stonework	Self-tapping screws
	Self-tapping screws with resin
	Nylon anchors with screw-in eyes
	Resin anchors

Ties are typically designed as simple connections, not transmitting bending moments to the supporting structure. If in the event it is necessary for the ties to transmit bending moments, appropriate structural connections need to be designed (e.g. anchored baseplate).

6.4.2.4 Earthquakes

In general, structures less than two stories high are normally exempt from considering seismic action.

Limited research has been conducted into the effect of earthquakes on temporary structures. The major recommendation from the research is that these structures should not be tied to the ground but be allowed to move freely, thereby only small seismic loads are transmitted into the structure (Blair & Woods, 1990).

During the Christchurch, New Zealand Earthquake on 22 February 2011, subcontractors were working on scaffolding five floors above the ground. The subcontractors saved their lives by jumping through open windows into the main structure (Prestney, 2012). Following the earthquake changes were made to the scaffold by ensuring the revised scaffold was built on its own load bearing wall which was tied to the main structure. This illustrates the point that fixings must be designed to allow for earthquake loads. The latter ensures that if another earthquake occurred the scaffold would move with the structure and hence its ties would not fail in shear. Additional toeboards were applied to the original scaffold to ensure that objects did not leave the scaffold and fall to the ground.

6.4.3 Falsework

6.4.3.1 Basis

In this Section, guidance is provided for the design of falsework, but which also applicable to scaffolding/grandstands/stage structures, based on the analysis of falsework structures under critical external and internal hazards. Note that although the effect of individual hazards will be considered, it has been shown that most often the combined effects of several hazards are the triggering event of collapses, see André (2014), André, Beale, & Baptista (2015b) and Milojkovic, Beale, & Godley (2002).

Several types of joints exist in falsework systems, the most common being: beam-to-column joints (*aka* standard-to-ledger joints), column-to-column joint (*aka* spigot joints), brace-to-ledger joints, and top and bottom boundary joints (i.e. forkhead and baseplate joints). Different solutions exist for each type of joint, for example, cuplok or wedge joints for standard-to-ledger joints; hook or swivel joints for brace-to-ledger joints. Figure 7 shows types of joints that can be found in bridge falsework Cuplok® systems. For other systems see Chapter 2.

Different numerical modelling techniques are available to simulate these types of joints: from the more complete 3-D joint modelling using solid elements to the simple spring joint modelling, see Chapter 4.

In this Section, the phenomenological-based analytical joint models presented in Chapter 4 are used. The analytical models used for the cuplok joints, spigot joints and forkhead joints have been derived from the experimental tests conducted by André and presented in André (2014) and André, Beale, & Baptista (2013). The FEM formulation, verification and validation is presented and detailed in André (2014) and André, Beale, & Baptista (2014), but also in Chapter 4.

Different models were considered in this Section. Unless noted otherwise, the models considered resemble the structures 2 and 4 tested in the University of Sydney (referenced here as Models A2 and A4, respectively), see Chandrangsu & Rasmussen (2011b) and Figure 8. Both structures display a grid frame

Figure 7. Some types of joints of bridge falsework Cuplok® systems

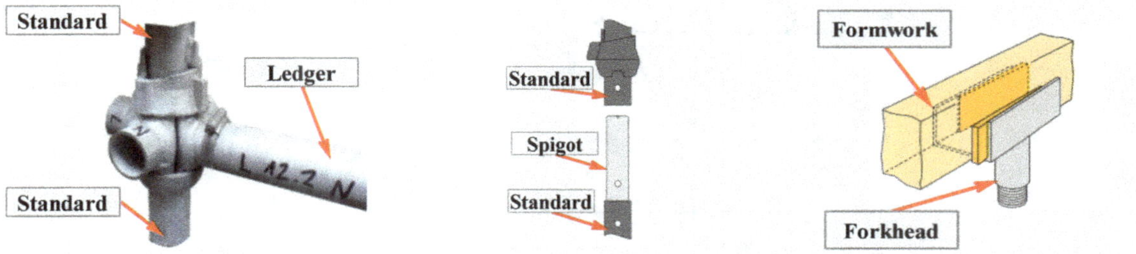

(a) Cuplok joint

(b) Spigot joint. ©2016 Brand Energy & Infrastructure Services. Used with permission

(c) Forkhead joint. ©2016 Brand Energy & Infrastructure Services. Used with permission

of three-by-three bays with a constant nominal bay width of 1829 mm in both directions, with three lifts with equal nominal height of 1.5 m and 600 mm of jack extension length. The bracing configuration of Model A2 is represented in Figure 9. The bracing arrangement is the same in each bay in each direction. Model A4 is unbraced. All structures were considered to be free-standing.

For the systems considered, the weak links are the over-extended unbraced jack elements or the spigot joints. The systems were defined so that the effect of localised internal and external actions on their safety and performance is more significant than the effect due to the same localised actions on more complex (larger) systems. As a consequence, the results presented in the following were obtained for severe scenarios and serve as a point of reference for other cases. In spite of constituting severe scenarios, it is possible that the analysis may not have encompassed all the potential failure modes. This is

Figure 8. Overview of the numerical model

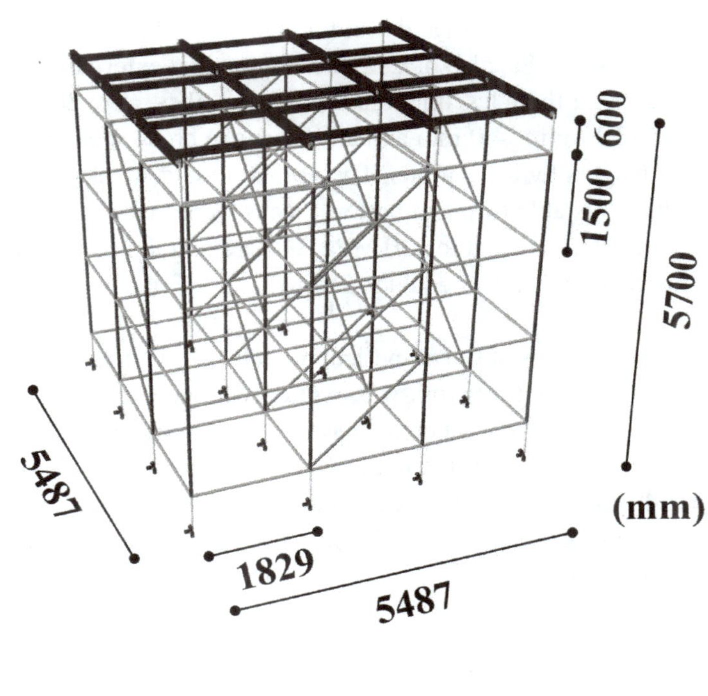

Figure 9. Structural layout of Model A2

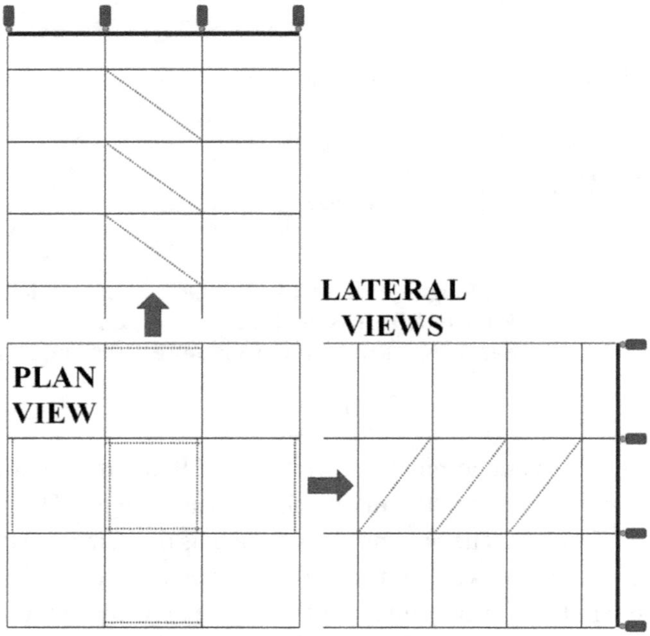

something important to keep in mind when extrapolating the results obtained, for example relative to the definition of the bracing requirements.

In all the cases considered, the cross-section geometrical characteristics, as well as the material properties of the various elements which make the falsework system, are identical to the ones used in the structures tested at the University of Sydney. The standards were made from cold-formed steel elements of circular hollow section (CHS) with a nominal yield stress of 450 MPa. The cross-section had a nominal external diameter of 48.3 mm and a wall thickness of 4 mm. Ledgers were made of steel with a nominal yield stress of 350 MPa, also of CHS with a nominal external diameter of 48.3 mm and thickness of 3.2 mm. Concerning the telescopic brace elements, they were also made of CHS with an outer tube cross-section of 48.3 mm × 4.0 mm and an inner tube cross-section of 38.2 mm × 3.2 mm, and were connected to the ledgers by hook joints. The nominal yield stress of the brace elements steel was equal to 400 MPa. The adjustable (top and bottom) jacks were made of 36 mm diameter threaded steel rods with a nominal yield stress equal to 430 MPa. The rectangular baseplates were 180 mm wide and 10 mm thick with a nominal yield stress equal to 250 MPa. See Chandrangsu & Rasmussen (2011b) for supplementary details.

The finite element mesh properties of the numerical models are given in André (2014) and André et al. (2014, 2015b) and Chapter 4. The formwork was explicitly modelled in all models, with an equivalent thickness equal to 100 mm, and the joint characteristics considered in all models, unless otherwise noted, were taken as the average values of the experimental test results, see André (2014) and André et al. (2013) and Chapter 4. Finally, the reference initial geometrical imperfections were taken as the values measured in situ at the beginning of the full-scale tests performed at University of Sydney, see Chandrangsu & Rasmussen (2011b).

6.4.3.2 External Hazards

6.4.3.2.1 Concrete Casting Action

Concrete can be placed either by skips or by pumps. The latter is nowadays the most used method for placing concrete. Concrete casting loads consist in a combination of different types of variable loads, see Chapter 3.

Most falsework collapses have been found to be triggered by concrete casting actions (André, Beale, & Baptista, 2012a; Hadipriono & Wang, 1987), see Chapter 7. As the influence of the dynamic effects and of different concrete placing methods have only been analysed in past studies using simple numerical methods, e.g. using the tributary area method or beam models, that can perform poorly in some cases, it was decided to develop several advanced numerical models to determine the different scenarios in which the concrete casting could be a critical hazard event to the safety and performance of falsework structures (André, 2014).

Therefore, the formwork was explicitly modelled and the concrete load applied directly over it. Two different concrete placing methods were analysed combined with various local concrete heaping values (from zero to two times the slab thickness), different reference slab thickness (0.25 m to 1.5 m) and different numbers of casting layers (one to ten).

In all the models, pumps were considered as the method used to place concrete and a 0.5 kN equivalent dynamic load was therefore applied, see Chapter 3, associated to concrete blocks (1 to 36 in Figure 10) representing 1 m^2 formwork area. In addition, the stiffness of the poured fresh concrete was considered negligible, thus not contributing to the load distribution to the formwork system.

The only loads considered were the ones associated with the concrete casting action itself: weight of the fresh concrete; local concrete heaping and equivalent dynamic loads. If the collapse of the falsework system had not been attained until the end of the concrete casting up to the reference slab thickness, the latter was uniformly increased until the collapse was reached.

Comparing the numerical results for the columns axial force with the ones registered in situ during the casting of concrete in a real bridge, see Figure 11, it is possible to observe that the numerical models are able to satisfactorily represent the behaviour of a falsework system during this construction phase of a concrete structure.

Based on the results obtained, see André (2014) and André, Beale, & Baptista (2015c), the factor related with the concrete casting action which has the highest influence is the local concrete heaping height. However, only high values (e.g. two times the slab thickness) lead to an important degradation of the maximum concrete pressure value that the system can resist. In general, these are unrealistically high values, but stress the importance of planning suitable operational procedures during the risk management. All other variables, i.e. concrete placing method, slab thickness and dynamic effects, seem to have a very small influence on the resistance of the falsework system. This is in accordance with findings published by other authors using simpler models, e.g. Peng et al. (1994).

In conclusion, in general, the collapse of correctly designed, assembled and operated falsework systems may not be attributed to the concrete casting actions acting in isolation. However, it is a fact that collapses of these systems have often been found to be triggered by concrete casting actions. This occurs because the concrete casting actions interact with other loads and defects of the system caused by material and geometrical imperfections and human errors. The accumulation of the detrimental effects of all these variables is responsible for most of the collapses of falsework systems.

Figure 10. Numbering of the concrete casting blocks on formwork surface

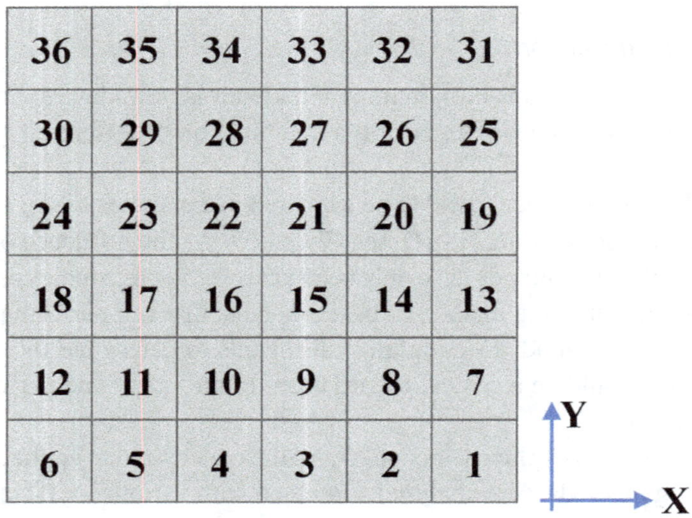

Figure 11. Comparison between column axial force numerical results and in situ results during concrete casting

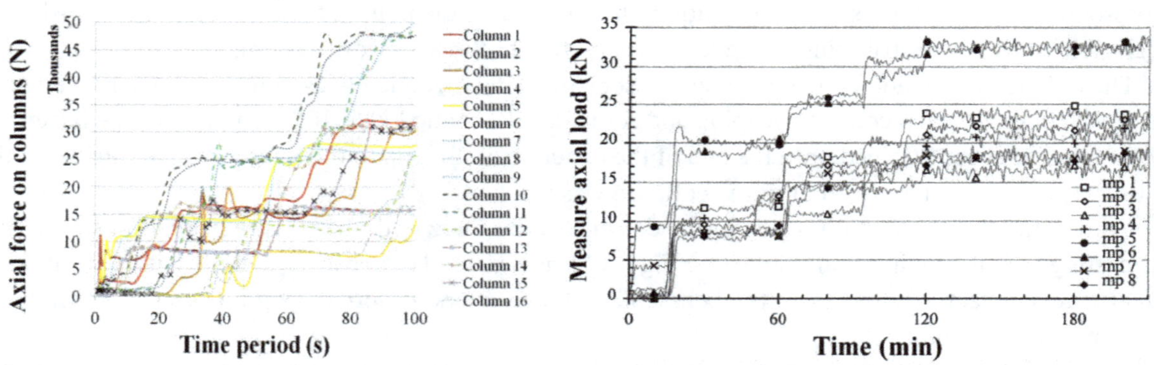

(a) Numerical results

(b) In situ results (Ikäheimonen, 1997)

6.4.3.2.2 Wind Action

Wind must also be considered in the design of falsework structures. Based on the findings in André et al. (2012a), see Chapter 7, it is not common for wind to trigger the collapse of falsework. This may be directly related to the fact that falsework structures are only used for periods of time comparatively smaller than the return periods usually considered for determining the design wind action. However, there is a trend to consider smaller return period values and it is important to assess whether this might have an impact on the risk of falsework structures. The latter has been discussed from the wind action point of view in detail in Chapter 5.

As a single uncoupled action, wind might not be decisive to the safety of falsework structures, but for analogous structures used in stages and grandstands it is the action that most commonly causes damage.

During the operation of falsework structures, wind may play a critical role in any of the following phases: during assembly of the falsework system; during the casting of the concrete and after concrete has been placed but before the fresh concrete has hardened to a degree where it can resist the applied actions by itself. Traditionally, wind action is specified in design standards, see Chapter 3.

Several numerical models were developed to test if under a number of different scenarios the wind action could be a critical hazard to the safety and performance of falsework structures, see André (2014) and André et al. (2015c) for details. Wind action was only considered in one direction: the direction of the collapse mode of Models A2 and A4 under vertical loads (determined by the results of the models presented in the previous Section).

The potential resistance against uplift loads conferred by incorporating anchor bolts at the baseplates and pins at the spigot joints was also analysed.

Based on the results obtained, see André (2014) and André et al. (2015c), the most influential factor related with the wind action is the occurrence of the maximum design wind velocities.

High values of wind action led to a significant degradation of the resistance of the system when compared with the resistance obtained when only vertical loads are applied. This is particularly true for unbraced falsework systems. Even the occurrence of working wind velocities had an impact on the system resistance; in particular, for braced falsework systems. This is justified because wind action subjects column elements, in particular the existing spigot joints, to larger rotations, thus larger bending moments. As spigot joints are a weak link in falsework, the collapse occurs for lower concrete pressures than the ones obtained when wind action is not considered.

It was also possible to conclude that including pins at the spigot joints and anchor bolts at the baseplates had a significant beneficial effect on the system's resistance when compared with the option of not using these components. Therefore, one option to increase the structural resistance if high wind velocities are forecasted to occur is to use brace elements and anchor bolts at the baseplates.

6.4.3.2.3 Ground Settlements

Ground settlements were often found to contribute to the collapse of bridge falsework (André et al., 2012a). Due to the low robustness of falsework systems, see André, Beale, & Baptista (2015a) and Chapter 5, any required imposed load redistribution may not be able to be achieved, driving the system to collapse.

Ground settlements are a function of the ground characteristics and applied actions. The ground upon which falsework structures foundations are placed typically exhibits poor resistance and rigidity characteristics since it consist of top ground layers, e.g. soft and loose soils. Without proper care, large ground settlements can result from the applied loads transmitted to the falsework system and from this to the ground via the foundation elements. Differences between displacements of the foundation ground can create differential settlements at the foundation level of the falsework system with potential negative structural consequences.

Several numerical models were developed using the already presented Models A2 and A4. In all scenarios, the ground settlement action was applied as imposed displacements (of 10 mm or 100 mm magnitude) at bottom node(s) of bridge falsework models – a limit case scenario considering the ground has no stiffness. In particular, localised and widespread ground settlements were considered, see Figure 12 for a plan distribution of the various cases studied.

Based on the results obtained, see André, 2014; André et al. (2015c), it can be concluded that there is a noticeable (negative) sensitivity of the considered falsework structures resistance, and of its variability,

Figure 12. Examples of differential ground settlements configurations considered in this study

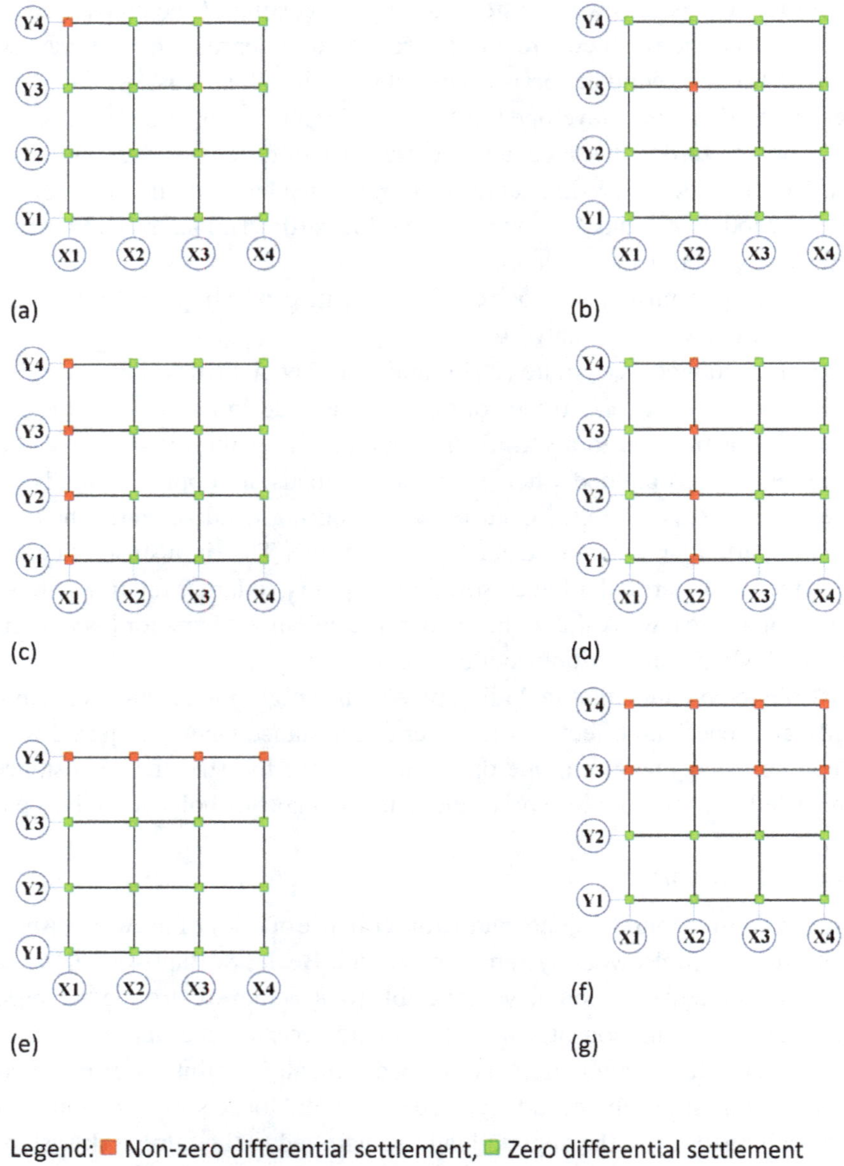

to the possibility of differential ground settlements. Even for residual differential ground settlements (e.g. 10 mm) it was found that there is a critical scenario where a localised residual differential ground settlement can generate a 20% reduction of the resistance capacity of the studied falsework structures. For higher settlement values, the reduction in resistance can attain 50% of the resistance of the reference system. BS EN 12812 (BSI, 2011d) reflects this sensitivity by imposing a maximum admissible differential ground settlement equal to 5 mm but this must be ensured on site, although frequently it is not.

It was also possible to conclude that the resistance of stiffer falsework systems seems to be more sensitive to differential ground settlements. This can be justified because the presence of significant looseness at the cuplok joints helps the system to accommodate differential ground settlements with less induced strains than the ones that occur in a system with smaller looseness at the cuplok joints.

6.4.3.3 Internal Hazards

6.4.3.3.1 Bracing Elements Configuration

Bracing is an essential part of any falsework since it increases the system's lateral stability. However, there are many possible bracing configurations and it is of interest to analyse which of these is more beneficial in terms of safety and performance of falsework structures. The minimum bracing recommended is

"one complete brace from the top to the bottom ledger level in a continuous diagonal line, on each row of standards, one in seven bays in each direction" (SGB, 2009),

see Figure 13. It is however common to find falsework structures with bracing only at the extreme faces, or just in one direction. In addition, jack bracing is usually avoided because it is not easy to assemble, but often it is necessary.

In order to assess the outcome of different bracing arrangements on the performance of falsework structures several models were developed, see Figure 14 (André, 2014). The only action considered was the one due to the concrete weight placed on top of the formwork. This load was increased until collapse was attained.

Based on the results obtained, see André (2014) and André et al. (2015b), it was observed that by adding bracing to over-extended jack elements the resistance of the structure can increase significantly. In structures with over-extended jack elements they are typically critical to the collapse resistance. Therefore, in these systems introducing additional brace elements but not in the over-extended jack elements is an extremely inefficient design strategy.

Reducing the number of connection points between the brace elements and column or ledger elements decreases the resistance of the system. The same conclusion can be drawn for the cases where bracing is only placed at external faces or just in one direction.

Figure 13. Recommended bracing layout. Adapted from (SGB, 2009)

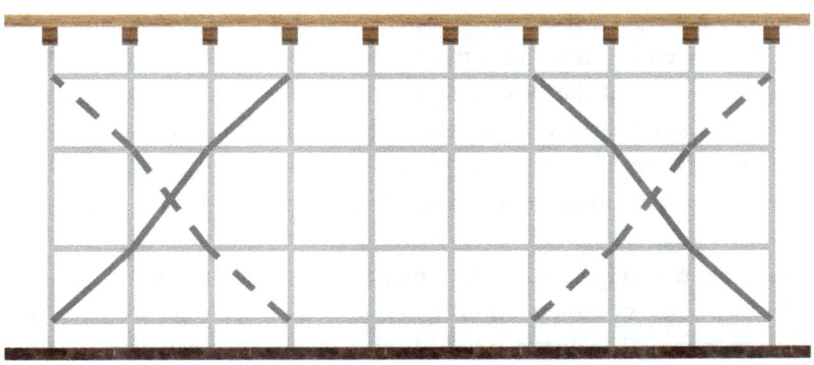

Figure 14. Different bracing arrangements

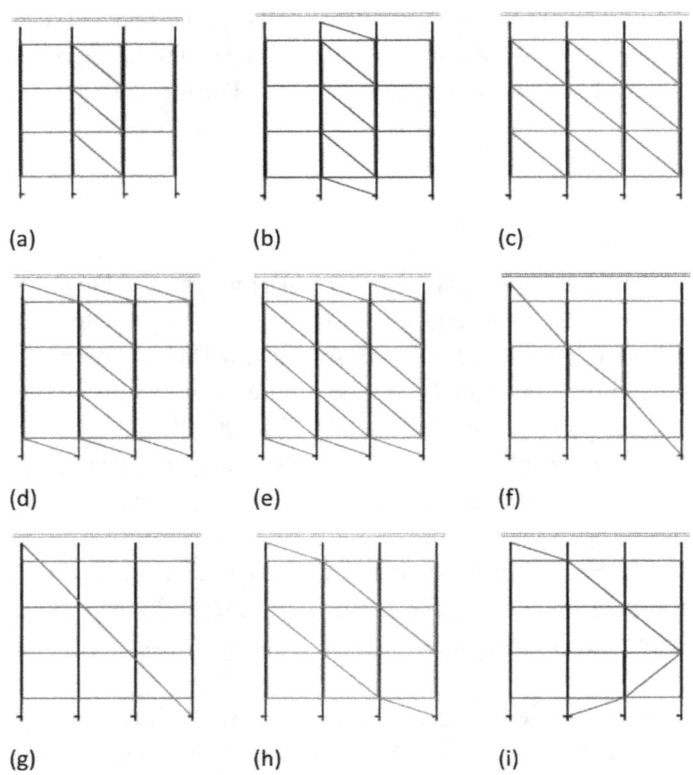

As given in the producer recommendations, it is more efficient to have one complete brace from the top to the bottom falsework levels in a continuous diagonal line, alternated in direction in adjacent faces, than parallel or than having multiple diagonal brace lines that are discontinuous.

However, for long and/or tall falsework, the recommendation of

one complete brace from the top to the bottom ledger level in a continuous diagonal line, on each row of standards, one in seven bays in each direction

is often insufficient. This occurs, since in terms of effectiveness of the bracing, its free length is also important (by free length is meant the piecewise length where the slope of the bracing diagonal line segments have the same orientation). A piecewise diagonal line from top to bottom levels of the false-work loses its effectiveness in providing lateral stability, if its free length is large when compared with a bracing arrangement with the same total bracing length, but where the orientation of the bracing diagonal alternates.

As a result, for long and/or tall falsework, the authors recommend numerical modelling the proposed falsework system since design recommendations given in documents released by system producers only contain minimum requirements that may not be sufficient for the specific use.

6.4.3.3.2 Spigot Joints Configuration

Spigot joints are a weak link of any falsework structure (André, 2014; Chandrangsu & Rasmussen, 2009). If spigot joints can be avoided they should. In a spigotless falsework the resistance increases, as full plastic hinges can be formed anywhere along the columns, in some cases considerably.

In Cuplok® systems, spigot joints can only be positioned at discrete locations along the column element between two consecutive storeys, but they can be positioned near the bottom, the middle or the top of the storey, depending on the length of the column.

The behaviour of spigots joints depends not only of its characteristics but also on the bending-moment distribution at the column elements. In an elastic regime, the bending-moment distribution at the column elements depends mainly on the rotational restraint provided by the cuplok and spigot joints, the lateral restraint at the top, the initial column imperfection shape and magnitude and the value and pattern of the applied actions. In general, the stiffness of the cuplok and spigot joints is not large and resembles a semi-rigid joint. The location of maximum bending-moments can either be located near the column ends or around mid-height of the column. The deciding factor is the relative influence of the second-order bending-moments distribution due to the combination of local and global initial imperfections of the column against the first-order bending-moments distribution. As the actual initial imperfection configurations are unknown variables during design, the actual effect of the spigot joints positioning on the falsework performance is uncertain. It is recommended to position the spigot joints at different heights from the ground in adjacent columns in order to avoid having a weak plane in a tall falsework/scaffold. This is achieved by varying the height of the columns in the lowest row.

In order to investigate what is the influence on the resistance of falsework of the positioning of the spigot joints, several models were developed, see André (2014) for details. Based on the results obtained, the collapse of unbraced models (Models A4) involves large sway displacements which favour the location of maximum bending-moments near the column ends. On the contrary, in braced models (Models A2) the collapse mode is dominated by column buckling which favours the location of maximum bending-moments near the column mid-height. As spigot joints are a weak link, positioning them near the maximum bending-moments leads to a reduction of the resistance of the system. The inverse case is also true.

6.4.3.3.3 Steel Girders Configuration

Where a bridge crosses waterways or roads, or the soil properties are weak, steel trusses or steel girders can be used to sustain the formwork over these areas, transmitting the loads to falsework towers placed at the ends of the span, in order to avoid the obstacles. This system can also be used if the height of the bridge piers is high.

If a simple falsework structure can be a complex design problem, designing falsework with steel girders sustaining a part of the falsework presents some further challenges.

One challenge is to quantify the influence that using steel girders has on the internal forces distribution on all falsework elements. The vertical elements resting on top of the steel girders will behave as if they were placed over a "soft ground" with the centre elements experiencing larger "settlements" due to the deformation of the steel girder.

Another challenge is to design properly the falsework towers that support the steel girders. These towers have to bear very large concentrated forces that are transmitted by the steel girders. Therefore, bracing has to be explicitly designed and properly assembled. Failure to properly consider them during design could lead to disaster as many collapse examples so clearly illustrate, see Chapter 7.

To investigate some of these problems, various numerical models were developed, see André (2014) and André et al. (2015b) for details. It was found that by not explicitly analysing and designing the tower elements, e.g. by using tabulated bracing layouts, the resistance of the falsework may represent a fraction (50% or less) of the required resistance.

Regarding internal forces distribution within the elements of the falsework that are supported by the steel girders, it can be concluded that the axial force distribution is not uniform across the vertical elements of the falsework for a given steel girder alignment, see Figure 15.

In fact, the maximum axial force value occurs at the vertical elements which are directly above the supports of the steel girders (i.e. falsework towers). This physically makes sense since the falsework towers act as a restraint to the displacements of the steel girders. Thus, the falsework vertical elements located near to the girder supports have a rigid support in contrast with the flexible support of the falsework vertical elements located near to the mid-span of the girders. As a result, the falsework vertical elements located near to the falsework towers draw higher internal forces than the rest of the vertical elements supported above the steel girders.

Consequently, if the falsework design is carried out based on the traditional and simple influence area method, where the internal forces transmitted from the formwork to the falsework vertical elements are determined based on the influence (formwork) area of each vertical element, a gross error is made with possibly catastrophic consequences. In the example under evaluation, based on the influence area method the axial forces of the vertical elements in the central girders alignments equals 90 kN, whereas the actual values range between 80 kN for the central vertical elements and 140 kN for the exterior vertical elements. For the latter level of forces, design using producer safe load values is not possible and an explicit model must be developed.

6.4.4 Temporary Grandstands and Stages

Most of the recommendations detailed in the previous Section are also applicable to temporary grandstands and stage structures.

These structures should possess sufficient transverse stiffness, longitudinal stiffness and strength to resist wind loads, notional horizontal loads and other dynamic loads induced by spectator movements

Figure 15. Axial force distribution within the falsework elements supported by the steel girders

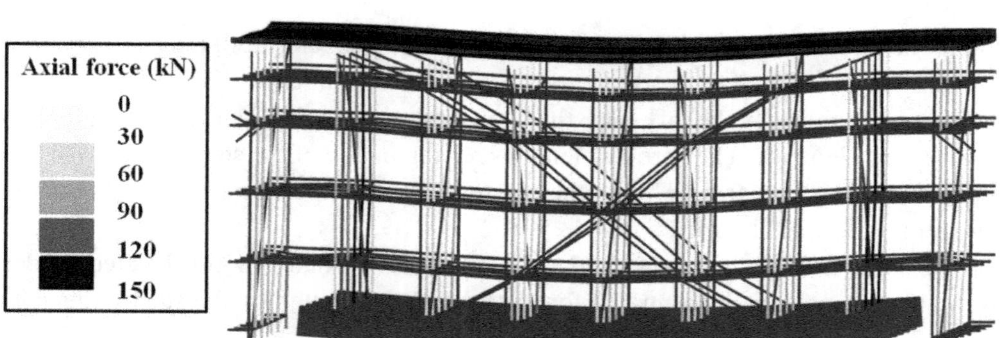

(Ellis, Ji, & Littler, 2000). Design of the bracing systems should also take into account the stiffness of the structure under dynamic loading, e.g. from spectator movements. Bracing or stiff frames should be provided regularly on transverse and longitudinal planes and should extend over the full height of the structure. As the loading is not fixed, the effects of local uplifts of the foundation should be considered in the design and if necessary to prescribe a connection of the foundation elements to the supporting ground must be provided.

The design of bracing systems should take account of robustness requirements. The structure should be designed with sufficient bracing so that removing up to two adjacent bracing members would not initiate a collapse (IStructE, 2007).

As dynamic crowd loading is usually confined to a narrow frequency range, see Chapter 3, it is possible to define the minimum frequencies for structures which should avoid safety problems from dynamic actions, otherwise the structure should be explicitly designed for the dynamic action (see Chapter 4) or crowd limits must be enforced.

The structural frequency above which vertical vibration should not pose a safety problem for jumping on floors is 8.4 Hz. For grandstands where concerts may be held, a frequency of 6 Hz has been given as an interim measure. These frequencies are for the appropriate mode of vibration of an empty structure (Ellis & Ji, 2004).

6.4.5 Bridge Construction Equipment

6.4.5.1 General Guidance

Despite the importance of BCEs, their current extensive use and the existence of multiple research areas not fully covered in design codes, the available published research is very limited. Apart from the limited guidance presented in some well known bridge design books, Leonhardt (1979) and Podolny & Muller (1982) for example, only a small number of research papers, technical reports and guidance documents exist directly addressing specific issues routinely encountered when designing BCEs. Detailed assembly and operational recommendations and guidelines for BCEs can be found elsewhere (Bakhoum, 2014; Chen & Duan, 2013; fib, 2009; Rosignoli, 2013).

During the planning, design and operation of BCEs there is a need for a close relationship between the teams responsible for the permanent and the temporary structures. In fact, some cases of BCEs will govern the design of the bridge structure. As the structural system and load cases change with each construction stage, and for time-dependent materials also the mechanical material properties (e.g. strength and stiffness), it is of critical importance to correctly perform a risk analysis, including defining and assessing the requirements for risk identification and risk control, planning, communication and cooperation between parties, quality management, structural analysis, construction sequence and schedule and on site operation, geometry and load monitoring and control for example.

In this Section, a brief review of the available literature organised by construction method will be presented. Additionally, failure examples will be presented to highlight some of the enabling and triggering events that can lead to accidents, see also Chapter 7.

In terms of design situations to be considered it is extremely important to identify all the relevant hazard scenarios. Consultation with other stakeholders should take place in order to assess the importance of each load case, the influence factors involved and possible risk treatment and risk control solutions available. In some particular applications, it may be desirable to modify the load safety factors on the basis of an iterative process in which risk levels and cost of measures are analysed in parallel (see Sexsmith & Reid (2003) for example).

In general, the following three different design situations must be checked (comparable cases are specified in Rosignoli (2007)):

1. Work situation
    a. **In Situ Construction**: The main action comes from the concreting procedure. Due allowance should be made to consider dynamic effects and unusual concrete concentrations. Other actions must be taken into account, such as post-tensioning, wind and construction loads from equipment and materials.
    b. **Pre-Cast Construction**: The main action comes from the hoisting procedure as the BCE structure receives the full load at once in contrast with cast in situ construction. Due allowance should be considered for dynamic effects and the BCE design should also include accidental design situations such as the possibility of asymmetrical loads occurring. Use of longitudinal and transverse devices in order to restrain movements is recommended. Other actions must be taken into account, such as post-tensioning and wind loads.
    In order to improve the robustness of a BCE during hoisting, a double safeguard system can be implemented by connecting two wires to each hook (Uemura, Kanda, Sakamoto, & Ito, 2000).
2. Movement situation

During the launching process, the structural scheme changes and the stability of the BCE main elements must be checked, especially in curved configurations. It is necessary to check the worst situation in all the phases of movement. The main actions are those due to wind and to movement operations.

Hastings, Zhao, & Burdette (2010) presented closed-form equations following the AASHTO bridge code for the maximum unbraced length (L) to compression flange width (b) ratios (L/b) for straight cantilever steel girders, beyond which lateral torsional buckling failure can occur under the combined effect of self-weight and wind action. The equations were derived for the AASHTO load combination strength III and considered 1/8 of the design wind pressure for transient design situations. It was concluded that cantilever girders with L/b ratios less than 30 should not require a stability checks but for L/b ratios higher than 50 temporary braces would be needed for any girder section.

The effects of local high transverse loads at the bearings should be considered in the design of BCE main members. Alignment deviations, geometric imperfections of the bearings and construction tolerances should be taken into account (see Rosignoli (2002)). All these factors can lead to high unbalanced stress concentrations in the flanges and high stresses in the bracing elements of I-girders or in the webs of box-girders caused by distortion deformations. The effect of the latter can be reduced by installing hydraulic jacks in order to be able to level the BCE elements over the piers.

Local overloading may also result from higher than anticipated friction forces over the sliding bearing elements. It is recommended to monitor the launching forces at the rear section but also to observe the structural behaviour over the supports.

Adequate longitudinal restraint at the supports must be assured, especially in bridges with a longitudinal gradient. The use of buffers, for example to stop the movement of the systems at a certain position, is recommended.

To analyse safety in the longitudinal direction during launching, the horizontal friction forces at every pier should be determined considering a range of values of the friction coefficients, μ_{min} and μ_{max}. BS EN 1991-1-6 (BSI, 2005c) recommends $\mu_{min} = 0.00$ and $\mu_{max} = 0.04$ for PTFE bearings.

If temporary supports are necessary in order to reduce deflections, stresses and improve stability, their design should consider not only the vertical but also the horizontal loads introduced by the movement of the BCE. Guidance on the assessment of these loads can be found in Rosignoli (2002). The installation of a lateral guide system on the top section of the piers to resist the lateral horizontal loads (e.g. locking plates on the sliding bearings) is recommended (Rosignoli, 2000). In LaViolette, Wipf, Lee, Bigelow, & Phares (2007) it is suggested that the guidance system should provide lateral resistance of at least 10% of the design vertical reaction at a given pier during the entire launching process.

3. Assembly, dismantling and transport situations.

Due to the size and weight of elements of BCEs special attention should be made during handling. Often this requires detailed structural analysis and design verification for all stages of the assembly, and it is common to use temporary stiffening and bracing elements.

The safety verification procedures to be used depend on the type of structural analysis employed. The simple verification rules provided by existing codes might be used in the case of elastic analysis. Alternative methods can be used when these rules prove to be inadequate. Deterministic GMNIA is a good solution. However, careful analysis should be given to the choice of material properties values to be considered, as the structural performance (brittle or ductile) returned by the numerical model can be quite different depending on whether the design, the characteristic or the mean values of the material mechanical properties were used.

For all design situations, Rosignoli (2007) recommends the use of a load safety factor of 2.5 for verifications related to global buckling phenomena. The use of this high value, much larger than the ones specified in structural codes, is justified as the geometrical imperfections present in BCE members can be significantly higher than the ones assumed during the development of the code rules.

It is also necessary to clearly define and properly state maximum wind speed velocities during the working and moving operations and also the design load combinations used for each design situation. In particular for cast in situ concrete construction, when the concrete is still fresh, the mass of the BCE increases without a corresponding increase in stiffness, and as result the aerodynamic characteristics of the BCE change (i.e. the fundamental natural frequency of the structure decreases) making it more susceptible to resonant dynamic phenomena, such as vortex shedding (see Chapter 3).

To reduce the wind-induced vibrations of the BCE and the bridge deck to acceptable levels, different alternative solutions can be adopted, such as to increase the stiffness of the system or the use of tune-mass dampers.

For precast construction, a critical step concerns the geometrical control of each segment during casting. The geometric tolerances should be minimal and a common solution is to use a stiff casting cell and to cast each segment directly against each other (Rosignoli, 2001).

Bridge erection machines are typically load tested upon completion of the first assembly. New load tests and comparisons with previous tests should also follow every major reassembly of the unit. Launching gantries are subjected to static and dynamic tests. The static test load is 10% to 40% (25%) higher than the design load of the unit. The dynamic test load is typically 10% to 20% higher than the design load. Requirements for these tests are usually defined amongst the relevant parties involved in the bridge project, but guidance is provided in Rosignoli (2010).

6.4.5.2 Incremental Launching Method

Complete and detailed design aids for the incremental launching method are given in Rosignoli (2002, 2013). Furthermore, examples of construction of bridges using this method are given in Ahmadi-Kashani, Brun, & Papanikolas (2007), Beavor & Cai (2006) and Liddle (2010).

A steel launching nose is rigidly connected to the bridge deck in order to reduce the bending moments and the amount of deflection experienced by the bridge deck during the construction phase. For the nose to be efficient it has to be a light structure (to reduce the cantilever bending moments) but also very stiff (to be able to transmit large loads directly to the supporting pier thus alleviating the contribution of the bridge deck resistance during launching).

The suggested design criteria for the steel launching nose is: the nose length between 0.60 and 0.70 of the longest span of the bridge and the nose bending stiffness between 0.10 and 0.20 of the bridge deck bending stiffness. Typically, a launching nose is made of two steel I-shaped tapered girders with the webs stiffened transversely. The bottom flanges are braced together horizontally whilst diagonal vertical brace elements connect the top flanges to the bottom flanges of the two beams.

The nose should be designed for all situations associated during the launching procedure, including the cantilever structural system when the nose is in-between piers, the structural system when it is simply supported at the pier, but also for the situation where there is a need for adjusting the level of the nose elements when they diverge (possibly asymmetrically) from the position of the pier support (i.e. upwards or downwards).

For long spans it may be necessary to use temporary cable-stayed systems to reduce the bending moments. In these cases, it is very important to properly assess the safety of the cables and of the mast. Dynamic analyses should be done in order to assess the vibration amplitudes. Guidance is found in. Caetano (2007). This also applies to launching gantries using temporary cable-stayed systems.

A monitoring system consisting of load cells and strain, displacement and tilt sensors should be implemented in the most critical sections of the BCE, bridge and other elements. For example:

- Horizontal launching jack forces and displacements;
- Vertical and horizontal displacements of the launching nose elements;
- Vertical and horizontal force applied to piers;
- Top horizontal and vertical displacements of the piers;
- Stresses at the BCE, bridge (e.g. contact and shear stresses over pier bearings and lateral guide system).

It is important to verify the measurements with the ones obtained in the numerical model, in particular the vertical displacements at the front end of the launching nose. The use of tapered girders for the launching nose and vertical jacks are suitable ways to correct possible excessive deflections. The stability of these jacks as well as of other vertical adjustment devices must be evaluated.

The BCE designer must consider in the analysis all possible hazards: from loads to equipment failures (in particular for variable slope launching and for sensitive exposed environments). Not doing so can lead to disaster. Take the case of a project in which the Teflon layers became damaged during the launching sequence. The design of the jacking system did not consider the scenario of having to change these layers during launching. The bridge was therefore left precariously exposed to high wind loads. Fortunately, in this case a solution could be found rapidly and the launching sequence safely continued.

Proper training of the site personnel and adequate inspection plans (before and after critical operations) are mandatory. These factors can reduce significantly the risks to the safety of the infrastructure during the construction phase. In one instance, the temporary support of the bridge suffered severe damage caused by excessive friction while launching of the superstructure. This occurred because waste materials accumulated at the sliding bearings and this was not detected in time to prevent the damage occurring.

The BCE designer should strive to include in the analysis provisions against critical human errors. If considered to be unfeasible the designer should make clear in the method statement that these scenarios have not been incorporated in the design. A practical example is given in which this procedure could have been beneficial. At the end of the launching of a major new bridge over a railway line it was realised that the launched end was out of position. The contractor relied on his experience and without informing the BCE designer installed horizontal jacks together with bearings and layers of PTFE and applied force to correct the bridge alignment. In the documentation provided by the BCE designer no information or guidance was given concerning this operation. Unfortunately the temporary bearings became unstable and the bridge slipped onto the railway line (RAIB, 2009). Another similar example is presented in Rosignoli (2007) where improper anchoring of a support leg to a pier cap caused the unit to collapse killing four workers in 2004 in Ohio, the USA.

6.4.5.3 Launching Gantries

Launching gantries are especially useful for high bridges or bridges located in complex or congested landscapes due to their ability to self-launch forward. They can be used in both precast and cast in situ structures. The launching procedure is multi-stage process and because many different proprietary systems exist in the market, or are tailored designed for each project, the exact procedure to be followed will vary from project to project and must be clearly defined in the respective method statement.

Launching gantries comprise of a light modular structural system made of two twin parallel steel girders, generally trusses, generally with a length twice the length of the standard bridge span unless intermediate temporary support systems are to be used (Rosignoli, 2010).

Movable scaffolding system (MSS) are used for cast in situ concrete bridges. The design of MSS must properly consider the interaction between the BCE and the permanent structure, especially if the concreting of the deck cross-section is carried out in several stages and in multi-span structures. Additionally, the design of the temporary elements and its connections to the piers must take into account not only vertical loads but also horizontal loads due to wind and other dynamic actions, such as sudden braking. In all cases, the MSS structure should be equipped with pressure cells to monitor the jacks and the support reactions during launching and stress measurement equipment to measure the structural response in real-time at the most stressed elements. Examples of measuring systems are discussed under Structural Health Monitoring in Chapter 8, Section 8.2.4.

Launching gantries for the segmental construction of bridges are similar to MSS. The main differences are the removal of the auxiliary structure supporting the formwork and the addition of the lifting equipment. Application examples of such equipment are presented in Collings (2001) and Mizon & Kitchener (1997). The design of these structures must take into account operational design situations such as asymmetrical lifting and the placement of segments but also accidental design situations such as when a failure of a winch cable during lifting a segment occurs, or when the braking system fails to operate properly during launching of the BCE or lifting of segments.

Alternatively, launching gantries can be used to place segments span-by-span, the full span method (FSM). With the increasing trend for faster construction of bridges with greater span lengths the outcome is a BCE with a massive weight. Therefore, in these structures, the main elements of the equipment used to carry the span and to launch the BCE consist in single or twin girders of closed cross-sections. Due to the massive weight and the fact that the construction loads are transferred instantaneously to the equipment, not gradually as in other launching BCE, it is very important to check the shear resistance of the connections between the BCE elements and the supports (piers and bridge deck). As a precaution it is recommended to implement a monitoring system to check reaction values (e.g. jacks or load cells for measurement of reactions).

For all types of equipment, it is very important to consider and design appropriate provisions for longitudinal fixity, lateral guidance and verify safety against overturning specially during launching of gantry. The adoption of redundant solutions is recommended to lower the operational risks.

Concerning wind action, a report issued by the International Federation for Structural Concrete (fib, 2009) suggested that the maximum wind speed in which any movement can occur should be limited to 35 km/h.

6.4.5.4 Balanced Cantilever Form-Travellers

A comprehensive set of technical guidance rules is provided in report by Sétra (2007). Furthermore, examples of construction of bridges using this method are given in Lucko & Garza (2003), Sétra (2007) and Wang, Tang, & Zheng (2004).

Form-travellers should be light weight, stiff structures with redundant anchorages during segment casting and repositioning, that can be easily and quickly fastened to the supporting permanent structure.

As the stability of this type of BCE depends largely on the resistance of the rear anchoring elements, the latter must be designed carefully. Nitschke (2010) suggested the use of a global safety factor of 2.0 in order to improve reliability. Construction design situations can be converted to have a resistance safety factor of 1.35 as for this construction method variable construction loads (with a load safety factor of 1.5) dominate over permanent loads. It is also considered good practice to check the safety of the BCE for the case where one of the connections fails. A sufficiently robust system must be designed which is capable of redistributing the loads to other elements in a controlled fashion, preventing the collapse of the entire structure (fib, 2009). The same design concept should be applied for the suspended tie bars that sustain the formwork.

Even though the weight of concrete is in general the dominant design action, the stage at which the traveller is moved requires special attention. All possible situations during the movement of the traveller must be carefully studied, with respect to resistance and stability, in particular under high wind loads. As for other BCEs, fib (2009) recommends monitoring the deflections of the traveller during all operations.

At the final stage of the construction of each span the safety of both cantilevers must be checked against possible differential displacements between the two segments due to differential thermal gradients, asymmetric wind action, concreting castings or accidental actions that result in imbalance loads.

To reduce the seismic vulnerability during construction of cable-stayed bridges using the balanced cantilever method, Wilson & Holmes (2007) suggest a seismic mitigation strategy based on the use of tie-down cables. In Bokan, Janjic, & Heiden (2006) an optimal tensioning strategy for the construction of cable-stayed bridges using the balanced cantilever method is proposed.

Finally, in Morgenthal, Sham, & Schwarz (2008) an example is presented of the design requirements and erection procedures applied to a lifting system used during the construction of the steel deck of a large cable-stayed bridge by the cantilever method.

The construction of the cantilevers is especially based on precamber design and in situ measurement and control of the deflection deformations. Precamber is computed based on specific assumptions as to the materials and their behaviour, such as creep and shrinkage, the environmental conditions and the construction time schedule. These assumptions must be continuously checked during construction and the design adjusted for any changes, see report by Sétra (2007) for detailed design and construction monitoring guidance. Al-Qarra (1999) gives a practical example.

The vertical deflection under maximum serviceability load is $L/400$ (where L is the span length) of the but should not be larger than 25 mm (Nitschke, 2010).

6.5 CONCLUSION

This Chapter has given guidance in the use of the design codes of Europe, the USA and Canada, Australia and Hong Kong and enabled readers to appreciate the differences in philosophy behind the codes. In general, design codes are based on limit state principles but some such as the AASHTO (USA) and BS 5975 (UK) are still based on allowable stress design philosophy.

The detailed review of the design codes showed that partial factors for actions which are time dependent such as wind and concrete casting loads are often reduced in temporary structures compared with those adopted for permanent structures. This has been questioned, particularly for wind, where some consultants have suggested that as the structures often have lower factors of safety for imposed loads and are often not constructed to the same standards as permanent structures, that wind effect require higher loads to maintain safety. It is also noted that different partial factors for the same action in the codes of different countries are prevalent but that when all the combinations of actions are taken into account that the overall safety of the structures is comparable.

Whilst simple temporary structures, such as access scaffolds, are often dealt with by applying load tables or elastic linear analyses, more complex structures, such as those occurring in bridge falsework or bridge construction equipment (BCE), require finite element analyses. In order to handle complex interactions between elements of the structure with external actions, and include geometrical and material nonlinearity, the authors recommended the use of second-order nonlinear analyses.

Many design codes used to refer to elements' effective lengths but this was shown to lead to inaccurate models for complex geometries.

The design philosophies and methodology of determining characteristic strengths of structural components such as telescopic props, bearings, hydraulic jacks and the design of shallow foundations, is discussed with differences between codes emphasised.

Detailed examples of the application of codes to the assessment of global stability of a cantilever bridge during construction, of the design resistance of a five level access metal scaffold and of a bamboo standard using the appropriate codes and analysis procedures described in Chapter 4 are then presented.

The Chapter then proceeds to give detailed design guidance for telescopic props and scaffolding (both constructed from metallic and bamboo components), emphasising for the latter case the importance of tying patterns. The authors noted that the traditional tying pattern of alternate levels being tied may either lead to inefficient use of ties as larger spacings may work or to failure due to an insufficient number of ties being used.

In the design of falsework, it was concluded that the most important factor in determining the maximum loads during concrete casting operations is the height of local concrete heaping during pouring as it can significantly degrade structural performance.

When considering the external loads acting on a falsework structure, wind was found to also degrade performance unless bracing is applied with appropriate tying patterns included.

Finally, the Chapter considers the special requirements of BCE which should be designed according to the rules applicable to permanent structures provided that adequate consideration is taken into account with respect to the intrinsic evolving and dynamic nature of the operation process and that these structures can be reused in different projects.

REFERENCES

AASHTO. (2008a). AASHTO CHBTW–1–M – Construction handbook for bridge temporary works, with 2008 interim revisions (1st ed.). Washington, DC: American Association of State Highway and Transportation Officials (AASHTO).

AASHTO. (2008b). AASHTO GSBTW–1–M – Guide design specifications for bridge temporary works, with 2008 interim revisions (1st ed.). Washington, DC: American Association of State Highway and Transportation Officials (AASHTO).

AASHTO. (2016a). AASHTO LRFDCONS–3 – LRFD Bridge construction specifications, with 2016 interim revisions (3rd ed.). Washington, DC: American Association of State Highway and Transportation Officials (AASHTO).

AASHTO. (2016b). AASHTO LRFDUS–7 – LRFD Bridge design specifications, with 2015 and 2016 interim revisions (7th ed.). Washington, DC: American Association of State Highway and Transportation Officials (AASHTO).

ACI. (2004). ACI 347–04 – Guide to formwork for concrete. Washington, DC: American Concrete Institute (ACI).

ACI. (2011). ACI 355.3R–11 – Guide for design of anchorage to concrete: Examples using ACI 318 Appendix D. Washington, DC: American Concrete Institute (ACI).

ACI. (2014a). ACI 318–14 – Building code requirements for structural concrete and commentary. Washington, DC: American Concrete Institute (ACI).

ACI. (2014b). ACI 347R–14 – Guide to formwork for concrete. Washington, DC: American Concrete Institute (ACI).

Ahmadi-Kashani, K., Brun, C., & Papanikolas, P. (2007). Design and construction of Megalorema Viaduct, Greece. *Proceedings of the ICE - Bridge Engineering, 160*(3), 117–127. doi:10.1680/bren.2007.160.3.117

AISC. (2010). AISC 360–10 – Specification for structural steel buildings. American Institute of Steel Construction (AISC).

Al-Qarra, B. (1999). The Dee Estuary Bridge: Control of geometry during construction. *Proceedings of the ICE -. Civil Engineering (New York, N.Y.), 132*(1), 31–39.

André, J. (2008). *Behaviour and quality of steel telescopic props. Analytical, numerical and experimental investigation* (M.Sc. Thesis). Instituto Superior Técnico, Technical University of Lisbon, Lisbon, Portugal. (in Portuguese)

André, J. (2014). *Determination of the main parameters affecting the performance of bridge falsework systems* (PhD Thesis). Oxford Brookes University, Oxford, UK.

André, J., Baptista, A., & Camotim, D. (2007). Experimental investigation of steel telescopic props (in Portuguese). In *Proceedings of the VI Portuguese National Conference of Steel and Composite Construction* (p. 12). Porto, Portugal: The Portuguese Steelwork Association (cmm).

André, J., Baptista, A., & Camotim, D. (2009a). Design buckling curve for steel telescopic props under compression (in Portuguese). In *Proceedings of the VII Portuguese National Conference of Steel and Composite Construction* (p. 10). Lisbon, Portugal: The Portuguese Steelwork Association (cmm).

André, J., Baptista, A., & Camotim, D. (2009b). Numerical investigation of adjustable telescopic steel props. In *Proceedings of the 12th International Conference on Civil, Structural and Environmental Engineering Computing* (p. 25). Madeira, Portugal: Civil-Comp Press.

André, J., Beale, R., & Baptista, A. (2012a). A survey of failures of bridge falsework systems since 1970. *Proceedings of the ICE - Forensic Engineering, 65*(4), 161–172. doi:10.1680/feng.12.00012

André, J., Beale, R., & Baptista, A. (2012b). Bridge construction equipment: An overview of the existing design guidance. *Structural Engineering International, 22*(3), 15. doi:10.2749/10168661 2X13363869853419

André, J., Beale, R., & Baptista, A. (2013). Experimental behaviour of bridge falsework joints. In *Proceedings of the IX Portuguese National Conference of Steel and Composite Construction* (p. 10). Guimarães, Portugal: The Portuguese Steelwork Association (cmm).

André, J., Beale, R., & Baptista, A. (2014). Numerical investigation of bridge falsework structures. In *Proceedings of the 5th Portuguese Structural Engineering Conference (JPEE 2014)* (p. 15). Lisbon, Portugal: Portuguese National Laboratory for Civil Engineering (LNEC).

André, J., Beale, R., & Baptista, A. (2015a). New indices of structural robustness and structural fragility. *Structural Engineering & Mechanics, 56*(6), 1063–1093. doi:10.12989/sem.2015.56.6.1063

André, J., Beale, R., & Baptista, A. (2015b). Numerical analysis of bridge falsework. In *Proceedings of the 8th International Conference on Advances in Steel Structures (ICASS 2015)* (p. 20). Lisbon, Portugal: Elsevier.

André, J., Beale, R., & Baptista, A. (2015c). Performance of bridge falsework structures under external hazards. In *Proceedings of the X Portuguese National Conference of Steel and Composite Construction* (p. 10). Coimbra, Portugal: The Portuguese Steelwork Association (cmm).

ANSI. (2011). ANSI/ASSE A10.8:2011 – Scaffolding safety requirements. American National Standards Institute (ANSI).

ASCE. (2010). ASCE 7–10 – Minimum design loads for buildings and other structures. American Society of Civil Engineers (ASCE).

ASCE. (2014). ASCE/SEI 37–14 – Design loads on structures during construction. American Society of Civil Engineers (ASCE).

Bakhoum, M. (2014). Bridge construction methods. In *Bridge engineering handbook: Construction and maintenance* (2nd ed.; Vol. 2, p. 60). CRC Press.

BCA. (2001). *Early striking and improved backpropping for efficient flat slab construction*. Crowthorne, UK: British Cement Association.

Beale, R., & Godley, M. (2006). Numerical modelling of tube and fitting access scaffold systems. *Advanced Steel Construction*, *2*(3), 199–223.

Beavor, N., & Cai, J. (2006). Design and construction of two launched bridges.*Proceedings of the ICE - Bridge Engineering, 159*(3), 137–145. doi:10.1680/bren.2006.159.3.137

Birch, N., Booth, J., & Walker, M. (1971). Effect of site factors on the load capacities of adjustable steel props (No. R27). London, UK: Construction Industry Research and Information Association (CIRIA).

Blair, T., & Woods, K. (1990). Seismic scaffolding concerns? Let it 'walk'. *Nuclear Plant Journal*, *8*(3), 76–77.

Bokan, H., Janjic, D., & Heiden, M. (2006). Optimisation of the tensioning schedule for cable-stayed bridges using dynamic software. In *Proceedings of the IABSE Symposium 'Responding to Tomorrow's Challenges in Structural Engineering'* (p. 9). Budapest, Hungary: International Association for Bridge and Structural Engineering (IABSE). doi:10.2749/222137806796169182

Bowles, J. (1997). *Foundation analysis and design* (5th ed.). New York: McGraw-Hill.

BSI. (1982). BS 5975:1982 – Code of practice for falsework (1st ed.). London, UK: British Standards Institution (BSI).

BSI. (1993a). BS 5973:1993 – Code of practice for access and working scaffolds and special scaffold structures in steel (3rd ed.). London, UK: British Standards Institution (BSI).

BSI. (1993b). *Structural steel I and H sections. Tolerances on shape and dimensions*. London, UK: British Standards Institution.

BSI. (1997). BS 6399–2:1997 – Loading for buildings. Code of practice for wind loads. London, UK: British Standards Institution (BSI).

BSI. (1999). *Adjustable telescopic steel props – Product specifications, design and assessment by calculation and tests*. London, UK: British Standards Institution.

BSI. (2000). BS EN 1337–1:2000 – Structural bearings – Part 1: General design rules. London, UK: British Standards Institution (BSI).

BSI. (2001). *Loose steel tubes for tube and coupler scaffolds – Technical delivery conditions*. London, UK: British Standards Institution.

BSI. (2002). BS EN 12811–3:2002 – Temporary works equipment – Part 3: Load testing. London, UK: British Standards Institution (BSI).

BSI. (2003a). BS EN 12810–1:2003 – Facade scaffolds made of prefabricated components – Part 1: Product specifications. London, UK: British Standards Institution (BSI).

BSI. (2003b). BS EN 12810–2:2003 – Facade scaffolds made of prefabricated components – Part 2: Particular methods of structural design. London, UK: British Standards Institution (BSI).

BSI. (2003c). BS EN 12811–1:2003 – Temporary works equipment – Part 1: Scaffolds – Performance requirements and general design. London, UK: British Standards Institution (BSI).

BSI. (2004a).*Mobile access and working towers made of prefabricated elements. Materials, dimensions, design loads, safety and performance requirements*. London, UK: British Standards Institution.

BSI. (2004b). BS EN 1337–2:2004 – Structural bearings – Part 2: Sliding elements. London, UK: British Standards Institution (BSI).

BSI. (2004c). BS EN 1337–4:2004 – Structural bearings – Part 4: Roller bearings. London, UK: British Standards Institution (BSI).

BSI. (2004d). BS EN 1337–7:2004 – Structural bearings – Part 7: Spherical and cylindrical ptfe bearings. London, UK: British Standards Institution (BSI).

BSI. (2004e). BS EN 1998–1:2004 – Design of structures for earthquake resistance – Part 1: General rules, seismic actions and rules for buildings. London, UK: British Standards Institution (BSI).

BSI. (2004f). BS EN 10025–2:2004 – Hot rolled products of structural steels - Part 2: Technical delivery conditions for non-alloy structural steels. London, UK: British Standards Institution (BSI).

BSI. (2004g). BS EN 12811–2:2004 – Temporary works equipment – Part 2: Information on materials. London, UK: British Standards Institution (BSI).

BSI. (2004h). *Temporary works equipment. Load bearing towers of prefabricated components. Particular methods of structural design*. London, UK: British Standards Institution.

BSI. (2005a). BS EN 74–1:2005 – Couplers, spigot pins and baseplates for use in falsework and scaffolds – Part 1: Couplers for tubes – Requirements and test procedures. London, UK: British Standards Institution (BSI).

BSI. (2005b). BS EN 1990:2002+A1:2005 – Basis of structural design. London, UK: British Standards Institution (BSI).

BSI. (2005c). BS EN 1991–1–6:2005 – Actions on structures – Part 1–6: General actions – Actions during execution. London, UK: British Standards Institution (BSI).

BSI. (2005d). BS EN 1992–2:2005 – Design of concrete structures – Part 2: Concrete bridges – Design and detailing rules. London, UK: British Standards Institution (BSI).

BSI. (2005e). BS EN 1993–1–1:2005 – Design of steel structures – Part 1–1: General rules and rules for buildings. London, UK: British Standards Institution (BSI).

BSI. (2005f). BS EN 1993–1–8:2005 – Design of steel structures – Part 1–8: Design of joints. London, UK: British Standards Institution (BSI).

BSI. (2006a). BS EN 1993–1–5:2006 – Design of steel structures – Part 1–5: Plated structural elements. London, UK: British Standards Institution (BSI).

BSI. (2006b). BS EN 1993–1–11:2006 – Design of steel structures – Part 1–11: Design of structures with tension components. London, UK: British Standards Institution (BSI).

BSI. (2006c). BS EN 1993–2:2006 – Design of steel structures – Part 2: Steel bridges. London, UK: British Standards Institution (BSI).

BSI. (2006d). BS EN 10210–2:2006 – Hot finished structural hollow sections of non–alloy and fine grain steels – Part 2: Tolerances, dimensions and sectional properties. London, UK: British Standards Institution (BSI).

BSI. (2006e). BS EN 10219–2:2006 – Cold formed welded structural hollow sections of non–alloy and fine grain steels – Part 2 : Tolerances, dimensions and sectional properties. London, UK: British Standards Institution (BSI).

BSI. (2007a). BS EN 1993–1–6:2007 – Design of steel structures – Part 1–5: Strength and stability of shell structures. London, UK: British Standards Institution (BSI).

BSI. (2007b). BS EN 1993–6:2007 – Design of steel structures – Part 6: Crane supporting structures. London, UK: British Standards Institution (BSI).

BSI. (2007c). UK National annex to BS EN 1992–2:2005. London, UK: British Standards Institution (BSI).

BSI. (2008a). UK National annex to BS EN 1993–1–1:2005. London, UK: British Standards Institution (BSI).

BSI. (2008b). UK National annex to BS EN 1993–1–5:2006. London, UK: British Standards Institution (BSI).

BSI. (2008c). UK National annex to BS EN 1993–1–8:2005. London, UK: British Standards Institution (BSI).

BSI. (2008d). UK National annex to BS EN 1993–1–11:2006. London, UK: British Standards Institution (BSI).

BSI. (2008e). UK National annex to BS EN 1993–2:2006. London, UK: British Standards Institution (BSI).

BSI. (2008f). UK National annex to BS EN 1998–1:2004. London, UK: British Standards Institution (BSI).

BSI. (2009a). *Execution of concrete structures*. London, UK: British Standards Institution.

BSI. (2009b). PD 6698:2009 – Recommendations for the design of structures for earthquake resistance to BS EN 1998. London, UK: British Standards Institution (BSI).

BSI. (2009c). PD 6703:2009 – Structural bearings. Guidance on the use of structural bearings. London, UK: British Standards Institution (BSI).

BSI. (2009d). UK National annex to BS EN 1990:2002+A1:2005. London, UK: British Standards Institution (BSI).

BSI. (2009e). UK National annex to BS EN 1993–6:2007. London, UK: British Standards Institution (BSI).

BSI. (2010). BS EN 1991–1–4:2005+A1:2010 – Actions on structures – General actions – Part 1–4: Wind actions. London, UK: British Standards Institution (BSI).

BSI. (2011a). BS 5975:2008+A1:2011 – Code of practice for temporary works procedures and the permissible stress design of falsework (3rd ed.). London, UK: British Standards Institution (BSI).

BSI. (2011b). BS EN 197-1:2011 – Cement. Composition, specifications and conformity criteria for common cements. London, UK: British Standards Institution (BSI).

BSI. (2011c). BS EN 1090–2:2008+A1:2011 – Execution of steel structures and aluminium structures – Part 2: Technical requirements for steel structures. London, UK: British Standards Institution (BSI).

BSI. (2011d). *Falsework. Performance requirements and general design*. London, UK: British Standards Institution.

BSI. (2011e). UK National annex to BS EN 1991–1–4:2005+A1:2010. London, UK: British Standards Institution (BSI).

BSI. (2012).*Adjustable telescopic aluminium props – Product specifications, design and assessment by calculation and tests*. London, UK: British Standards Institution.

BSI. (2013). BS EN 1997–1:2004+A1:2013 – Geotechnical design – Part 1: General rules. London, UK: British Standards Institution (BSI).

BSI. (2014a). BS EN 13001–2:2014 – Crane safety. General design – Part 2: Load actions. London, UK: British Standards Institution (BSI).

BSI. (2014b). UK National annex to BS EN 1997–1:2004+A1:2013. London, UK: British Standards Institution (BSI).

BSI. (2016). BS 7121–1:2016 – Code of practice for safe use of cranes – Part 1: General. London, UK: British Standards Institution (BSI).

Burrows, B. (1989). *Organisation and quality of falsework construction (PhD.)*. Warwick University.

Caetano, E. (2007). Cable vibrations in cable–stayed bridges. Zurich, Switzerland: International Association for Bridge and Structural Engineering (IABSE).

CEN. (2009). CEN/TS 1992–4:2009 – Design of fastenings for use in concrete – Parts 1–5. Brussels, Belgium: European Committee for Standardization (CEN).

Chandrangsu, T., & Rasmussen, K. (2009). Full-scale tests and advanced structural analysis of formwork subassemblies. In *Proceedings of the 6th International Conference on Advances in Steel Structures* (pp. 1083–1090). Hong Kong, China: Hong Kong Institute of Steel Construction.

Chandrangsu, T., & Rasmussen, K. (2011a). Investigation of geometric imperfections and joint stiffness of support scaffold systems. *Journal of Constructional Steel Research*, *67*(4), 576–584. doi:10.1016/j.jcsr.2010.12.004

Chandrangsu, T., & Rasmussen, K. (2011b). Structural modelling of support scaffold systems. *Journal of Constructional Steel Research*, *67*(5), 866–875. doi:10.1016/j.jcsr.2010.12.007

Chen, W., & Duan, L. (2013). *Handbook of international bridge engineering*. CRC Press. doi:10.1201/b15520

Chung, K., & Chan, S. (2002). *Design of bamboo scaffolds*. Hong Kong, China: INBAR and Hong Kong Polytechnic University.

Chung, K., & Siu, Y. (2002). *Erection of bamboo scaffolds*. Hong Kong, China: INBAR and Hong Kong Polytechnic University.

Collings, D. (2001). The A13 viaduct: construction of a large monolithic concrete bridge deck.*Proceedings of the ICE - Structures and Buildings, 146*(1), 85–91. doi:10.1680/stbu.2001.146.1.85

CSA. (1975). S269.1–1975 R2003 – Falsework for construction purposes – Part 1: Design. Canada: Canadian Standards Association (CSA).

CSA. (2014). S6–14 – Canadian highway bridge design code. Canada: Canadian Standards Association (CSA).

Department of Transportation, Division of Structures. (2001). *California falsework manual* (32nd ed.). State of California.

Ellis, B., & Ji, T. (2004). The response of structures to dynamic crowd loads (No. Digest 426). Bracknell, UK: Building Research Establishment (BRE).

Ellis, B., Ji, T., & Littler, J. (2000). The response of grandstands to dynamic crowd loads.*Proceedings of the ICE - Structures and Buildings, 140*, 355–365. doi:10.1680/stbu.2000.140.4.355

EOTA. (2013a). ETAG 001: Metal anchors for use in concrete – Parts 1–6. Brussels, Belgium: European Organisation for Technical Assessment (EOTA).

EOTA. (2013b). ETAG 029: Metal injection anchors for use in masonry. Brussels, Belgium: European Organisation for Technical Assessment (EOTA).

fib. (2000). *Guidance for good bridge design*. Lausanne, Switzerland: International Federation for Structural Concrete (fib), Bulletin No. 9.

fib. (2009). *Formwork and falsework for heavy construction. Guide to good practice*. Lausanne, Switzerland: International Federation for Structural Concrete (fib), Bulletin No. 48.

Frank, R., Bauduini, C., Driscoll, R., Kavvadas, M., Ovesen, N., Orr, T., & Schuppener, B. (2013). *Designers' Guide to Eurocode 7: Geotechnical design – General rules*. London, UK: Thomas Telford Ltd.

Gorlin, W. (2009). Wind loads for temporary structures: Making the case for industrywide standards. *Journal of Architectural Engineering*, *15*(2), 35–36. doi:10.1061/(ASCE)1076-0431(2009)15:2(35)

Gylitoft, K., & Mroz, K. (1995). Load-carrying capacity of scaffolds. *Structural Engineering International*, *5*(1), 37–42. doi:10.2749/101686695780601619

Hadipriono, F., & Wang, H. (1987). Causes of falsework collapses during construction. *Structural Safety*, *4*(3), 179–195. doi:10.1016/0167-4730(87)90012-9

Hastings, J., Zhao, Q., & Burdette, E. (2010). Steel girder stability during bridge erection: AASHTO LRFD check on L/b ratios. *Journal of Bridge Engineering*, *15*(6), 759–762. doi:10.1061/(ASCE)BE.1943-5592.0000102

Hida, S., Ibrahim, F., Capers, H., Bailey, G., Friedland, I., Kapur, J., … Sivakumar, B. (2010). Assuring bridge safety and serviceability in Europe (No. FHWA-PL-10-01). Washington, DC: U.S. Department of Transportation.

Hong Kong Buildings Department. (2006). *Guidelines for the design and construction of bamboo scaffolds*. Hong Kong, China: Occupational Safety and Health Branch.

Hong Kong Labour Department. (2014). *Code of practice for bamboo scaffolding safety* (3rd ed.). Hong Kong, China: Occupational Safety and Health Branch.

Ikäheimonen, J. (1997). *Construction loads on shores and stability of horizontal formworks* (PhD Thesis). Royal Institute of Technology (KTH), Stockholm, Sweden.

ISO. (2012). ISO 8686–1:2012 – Cranes – Design principles for loads and load combinations – Part 1: General. Geneva, Switzerland: International Organization for Standardization (ISO).

ISO. (2015). ISO 2394 :2015 – General principles on reliability for structures (4th ed.). Geneva, Switzerland: International Organization for Standardization (ISO).

IStructE. (2007). Temporary demountable structures: Guidance on procurement, design and use (3rd ed.). London, UK: The Institution of Structural Engineers.

Jaspart, J., & Weynard, K. (2016). *Design of joints in steel and composite structures*. Berlin, Germany: John Wiley & Sons. doi:10.1002/9783433604762

JCSS. (2001). *JCSS Probabilistic model code*. Joint Committee on Structural Safety.

LaViolette, M., Wipf, T., Lee, Y., Bigelow, J., & Phares, B. (2007). Bridge construction practices using incremental launching. Washington, DC: American Association of State Highway and Transportation Officials (AASHTO).

Leonhardt, F. (1979). *Vorlesungen über massivbau – Teil 6 – Grundlagen des massivbrückenbaues*. Berlin, Germany: Springer-Verlag. doi:10.1007/978-3-642-61863-5

Liddle, R. (2010). The launching of the River Esk bridge, UK.*Proceedings of the ICE - Bridge Engineering, 163*(3), 137–146. doi:10.1680/bren.2010.163.3.137

Lucko, G., & Garza, J. (2003). Constructability considerations for balanced cantilever construction. *Practice Periodical on Structural Design and Construction, 8*(1), 47–56. doi:10.1061/(ASCE)1084-0680(2003)8:1(47)

Maquoi, R., & Rondal, J. (1978). Mise en équations des nouvelles courbes européennes de flambement. *Revue Construction Métallique, 1*, 17–29.

Milojkovic, B., Beale, R., & Godley, M. (2002). Determination of the factors of safety of standard scaffold structures. In *Proceedings of the 3rd International Conference on Advances in Steel Structures (ICASS 2002)* (pp. 303–310). Hong Kong, China: Elsevier. doi:10.1016/B978-008044017-0/50035-4

Mizon, B., & Kitchener, J. (1997). Second Severn crossing: Viaduct superstructure and piers. *Proceedings of the ICE -. Civil Engineering (New York, N.Y.), 120*(2), 35–48.

Morgenthal, G., Sham, R., & Schwarz, A. (2008). The 4000 tonne lift of Stonecutters bridge steel back-spans, H.K.*Proceedings of the ICE - Bridge Engineering, 161*(4), 197–208. doi:10.1680/bren.2008.161.4.197

NASC. (2011). *Anchorage systems for scaffolding*. London, UK: National Access and Scaffolding Confederation.

NASC. (2013). *Guide to good practice for scaffolding with tube and fittings*. London, UK: National Access and Scaffolding Confederation.

Nitschke, J. (2010). *Form-traveller systems*. Presented at the IABSE WG 6 Seminar: State-of-the-art Bridge Deck Erection; Safe and Efficient use of Special Equipment, Singapore: International Association for Bridge and Structural Engineering (IABSE).

Pallett, P., Burrow, M., Clark, L., & Ward, R. (2001). *Investigation into aspects of falsework (No. crr01394)*. Birmingham, UK: The University of Birmingham.

Peng, J., Rosowsky, D., Pan, A., Chen, W., Chan, S., & Yen, T. (1994). *Concrete placement load effects during construction - Part I: Structural modeling and analysis (No. CE-STR-94-15)*. West Lafayette, IN: Purdue University.

Podolny, W., & Muller, J. (1982). *Construction and design of prestressed concrete segmental bridges*. New York: Wiley.

Potts, D. M., & Zdravkovic, L. (1999). Finite element analysis in geotechnical engineering.: Vol. 1. *Theory*. London, UK: Thomas Telford Publishing.

Potts, D. M., & Zdravkovic, L. (2001). Finite element analysis in geotechnical engineering.: Vol. 2. *Application*. London, UK: Thomas Telford Publishing.

Prabhakaran, U. (2009). *Nonlinear analysis of scaffolds with semirigid connections* (PhD Thesis). Oxford Brookes University, Oxford, UK.

Prabhakaran, U., Beale, R., & Godley, M. (2011). Analysis of scaffolds with connections containing looseness. *Computers & Structures*, *89*(21–22), 1944–1955. doi:10.1016/j.compstruc.2011.03.016

Prestney, L. (2012). Quake lessons for scaffolding safety. *Build Magazine, New Zealand*, *130*, 1.

RAIB. (2009). Collision with debris from bridge GE19 near London Liverpool Street 28 May 2008. London, UK: Rail Accident Investigation Branch (RAIB), Department for Transport.

Ratay, R. (2009). *Forensic structural engineering handbook*. McGraw Hill Professional.

Rosignoli, M. (2000). Thrust and guide devices for launched bridges. *Journal of Bridge Engineering*, *5*(1), 75–83. doi:10.1061/(ASCE)1084-0702(2000)5:1(75)

Rosignoli, M. (2001). Deck segmentation and yard optimisation for launched bridges. *Concrete International*, *23*(2), 64–73.

Rosignoli, M. (2002). *Bridge launching*. London, UK: Thomas Telford Ltd.

Rosignoli, M. (2007). Robustness and stability of launching gantries and movable shuttering systems - Lessons learned. *Structural Engineering International*, *17*(2), 133–140. doi:10.2749/101686607780680673

Rosignoli, M. (2010). Self-launching erection machines for precast concrete bridges. *PCI Journal*, *55*(1), 36–57. doi:10.15554/pcij.01012010.36.57

Rosignoli, M. (2013). *Bridge construction equipment*. London, UK: ICE Publishing. doi:10.1680/bce.58088

SAA. (1995). *3610:1995 – Formwork for concrete. Standards Association of Australia.*

SAA. (2002). *1170.0:2002 – Structural design actions – Part 0: General principles. Standards Association of Australia.*

SAA. (2007). AS 5100 – Bridge design – Parts 1–7. Standards Association of Australia (SAA).

Sahin, S. (2009). *A comparative study of AISC-360 and Eurocode 3 strength limit states (M.Sc.). Middle East Technical University.*

Sétra. (2007). Prestressed concrete bridges built using the cantilever method. Design guide. Paris, France: Service d'études sur les transports, les routes et leurs aménagements (Sétra).

Sexsmith, R., & Reid, S. (2003). Safety factors for bridge falsework by risk management. *Structural Safety*, *25*(2), 227–243. doi:10.1016/S0167-4730(02)00058-9

SGB. (2009). *Cuplok: Datasheet*. SGB Group (now HARSCO).

The Concrete Society. (2003). *Guide to flat slab formwork and falsework*. London, UK: The Concrete Society, Report CS140.

The Concrete Society. (2012). *Formwork: A guide to good practice* (3rd ed.). London, UK: The Concrete Society, Report CS30.

Turkstra, C., & Madsen, H. (1980). Load combinations in codified structural design. *Journal of the Structural Division*, *106*(12), 2527–2543.

Uemura, A., Kanda, K., Sakamoto, T., & Ito, S. (2000). Design, fabrication and erection of the stiffening girder of 3rd Kurushima Kaikyo bridge. *Kawasaki Steel Giho*, *32*(2), 95–102.

Virlogeux, M. (1993). Normandie bridge: Design and construction.*Proceedings of the ICE - Structures and Buildings, 99*(3), 281–302.

Wang, P., Tang, T., & Zheng, H. (2004). Analysis of cable-stayed bridges during construction by cantilever methods. *Computers & Structures*, *82*(4–5), 329–346. doi:10.1016/j.compstruc.2003.11.003

Wilson, J., & Holmes, C. (2007). Seismic vulnerability and mitigation during construction of cable-stayed bridges. *Journal of Bridge Engineering*, *12*(3), 364–372. doi:10.1061/(ASCE)1084-0702(2007)12:3(364)

Chapter 7
Analysis of Collapses

ABSTRACT

This chapter presents the procedural, enabling and triggering causes of temporary structures collapses, with an emphasis on falsework and scaffolding. The review into collapses described in Chapter 2 is extended by investigating in detail the causes of collapse in temporary structures and providing comprehensive lists of faults which can occur during design, erection, use and disassembly of these structures. As bridge falsework collapses are more commonly reported, with usually greater financial implications and greater risks to life, a survey, conducted by André, is summarised showing that these collapses occur regularly throughout the world. The chapter concludes with the presentation of two examples of forensic analyses, namely of a scaffold collapse and of a bridge falsework collapse.

7.1 INTRODUCTION

With the industrial revolution came novel challenges for civil engineers. New infrastructures (such as bridges, commercial, residential and industrial buildings), in larger scales, carrying more and heavier loads, had to be built at a fast pace. Temporary structures also experienced this novel complexity. However, little attention was drawn to this subject and as a result a series of collapses of major significance involving temporary structures occurred in the industrialised countries throughout the 20th century.

In 1970, as a response to the public outcry following a collapse with severe consequences, the UK construction industry established a committee under the chairmanship of S.L. Bragg to investigate the use of falsework. The result was the Bragg report (Bragg, 1975), a pioneer document which established the basis for the subsequent publication of the first UK standards concerning falsework (BSI, 1982).

Since the Bragg report, there have been a number of fundamental changes to the construction industry (HSE, 2003):

- "Clients and designers give insufficient consideration to health and safety, despite their obligations under the CDM regulations.
- Price competition among contractors gives advantage to companies less diligent with health and safety.

DOI: 10.4018/978-1-5225-2199-0.ch007

Copyright © 2017, IGI Global. Copying or distributing in print or electronic forms without written permission of IGI Global is prohibited.

- Key documentation, such as the health and safety plan, method statements and risk assessments are treated as a paper exercise, having little practical benefit.
- Lengthy sub-contractor chains result in elements of the construction team being distanced from responsibility, inadequately supervised, and with low commitment to projects.
- Frequent revision of work schedules leads to problems with project management and undesirable time pressure.
- A long hours culture in the industry results in fatigue, compromised decision-making, productivity and safety.
- Bonus payments act as a strong incentive, but encourage productivity over safety.
- A skills shortage in the industry is leading to increased reliance on inexperience workers, coupled with difficulties verifying competency.
- Problems exist with the availability, performance and comfort of PPE.
- Training is seen as a solution to all problems, but with content often superficial."

The above changes had a profound effect upon the manner temporary structures, in particular falsework, is dealt with by all relevant stakeholders (SCOSS, 2002):

- "Falsework design is no longer a task of the main contractor but the responsibility is passed to a sub contractor or a specialized supplier;
- The structural concept of the falsework is no longer arbitrary; proprietary systems and more often modular ones are widely used nowadays in order to optimise costs and operational efficiency. Additionally the number of usage cycles of falsework components has increased dramatically;
- The paradigm of the construction industry has changed: intense competition in a profit orientated environment has produced a reduction of technical competence and responsibility at the design, construction and quality assurance stages of a construction project".

As detailed in Chapter 1, the design and use of temporary structures places very complex and different challenges from the ones associated with permanent structures, such as:

- BCE systems have the capacity of moving. This is different from common permanent structures which are normally considered as static.
- Generally, most temporary structures are subject to load values close to, and sometimes even above, the assumed design values during almost their entire service period, whereas the design of permanent structures is often controlled by load cases that will only occur for a brief period of time, or that have a small probability of occurring, during their design working lifetime.
- Temporary structures are used for short periods of time, although due to multiple re-use cycles their design working life can sum up to 15 years or more.
- The ratio between the cost of temporary structures and the cost associated with their collapse is much lower than for permanent structures. Note that temporary structures collapses may also cause human fatalities and lead to company collapses due to legal procedures.
- Temporary structures are assembled, (re)used for short periods and dismantled several times in repetitive cycles. Permanent structures are generally assembled only once and are used for large periods of time.

- Finally, temporary structures due to their purpose are generally constituted by slender elements, and therefore their performance is more sensitive than permanent structures to errors during their erection and operation, and to the use of damaged elements due to inadequate maintenance and quality control. Permanent structures exhibit a much higher degree of inherent robustness against human errors.

Therefore, it is not surprising that several authors have concluded that the construction period of infrastructures is the most vulnerable phase during the infrastructures' lifetime (CIRIA, 2011; Hadipriono & Wang, 1987; Wardhana & Hadipriono, 2003). From the various possible causes leading to structural failure, Sexsmith (1998) states that

Human error, such as mistakes in the design concept or the calculations, especially for falsework and its supporting system, is a primary cause of many failures.

In particular, human errors occurring during the design phase of temporary structures (Bennett, 2004).

Accidents involving temporary structures often have vast and various negative impacts: on the project profitability, on the competence credibility of the involved companies, on the increase of the insurance premiums, on economic, financial and political costs due to postponed benefits. Additionally, when human victims occur many of the abovementioned effects are scaled-up by the media and public attention.

Taking as an example bridge falsework, in 2005, Wong, Onof, & Hobbs (2005) studied the possible failure consequences of bridges in service by a cost-evaluation method. The major costs involved, of a total of more than 25 million pounds (2009 prices) for the studied cases, sorted in a descending order, are the ones related to rebuilding costs, traffic delay costs, access and traffic management costs, casualty costs, repair costs and finally some other indirect costs. These results can easily be extrapolated to the failure of bridges during construction, by reclassifying traffic delay and traffic management costs as postponed benefits (loss of service and associated loss of revenue) due to the delayed bridge opening date.

Failures of temporary structures, contrary to permanent structures, often occur due to a combination of "low strengths" and "high loads". These are usually originated by departure from commonly accepted competent professional practice, i.e. human errors. "Low strengths" can arise from multiple human error sources: design, construction and quality control errors, (current knowledge gaps with respect to the real structural behaviour of temporary structures, namely the influence of human errors. Regarding the definition of "high loads", two different situations must be distinguished: action values, or actions effects, considered in the design phase but which become higher than their design values, or design action effects, due to construction and quality control errors; and actions, or action effects, not considered, either partially or entirely, in the design phase due to ignorance or oversight, such as foundation settlements, load redistribution or second-order effects, for example. In the definition of "high loads", catastrophic unexpected events such as floods, earthquakes, terrorism acts, vehicle impacts or other extraordinary actions are not included.

The collapse of temporary structures caused by an unaccounted or under-evaluated hazard event cannot be acceptable. According to Steven (2010), the

mind set within the construction industry over the decades from the 1900s has changed from accepting (...) 13 deaths for the construction of the major viaduct, to a mindset unacceptable of any level of injury.

The various concerns outlined in the above paragraphs illustrate the need for a holistic approach applied to temporary structures, e.g. a risk management framework, and adequate competency to undertake the task. It is essential that those involved with temporary structures realise the importance of this change, the relevant statutory need to consider whole working life risks to structural safety, and the commercial benefits that will accrue by doing so (SCOSS, 2005). Detailed guidance concerning the risk management of temporary structures is provided in Chapter 5.

It is clear that there is a need for scientific progress in the field of temporary structures. A measure of the accomplishment of this task is given by our ability to reduce the uncertainties associated with the use of temporary structures. This is precisely the aim of the present book. However, it must be realised that there will always exist a certain level of uncertainty that cannot be eliminated completely.

On the basis of this Chapter it is expected that the reader will acquire knowledge on the following topics:

1. History of temporary structures failures.
2. Typology of temporary structures collapses.
3. Common causes of temporary structures collapses, including procedural, enabling and triggering events, with emphasis on scaffolding and falsework.
4. Understanding by examples of how the joint effect of the above causes has been behind the collapses of various types of temporary structures.

7.2 BACKGROUND ON TEMPORARY STRUCTURES FAILURES

7.2.1 Learning from Failures

Failures are rarely caused by one reason only, but rather by the accumulation of the detrimental effects caused by a series of small events, each of which might be considered not critical, but the total effect exceeds the structure safety margin. This reasoning is clearly expressed by the "Swiss cheese" model of Reason (1990), see Figure 1, which shows that small errors may cumulate and have unexpected and disproportionate consequences. In this model, various protective barriers exist that keep a system from failing, such as: following good practice design recommendations; self-checking, internal and external reviews; adequate quality control, inspection and maintenance procedures, etc. However, holes exist in these safety barriers, originated by uncertainties, human errors and accepted risks. Failures will happen when these holes are aligned and the errors are not detected or properly corrected. The willingness and capacity to search for these errors is a characteristic of an organisation with a good safety culture (Blockley, 2011).

Uncertainties and human errors (errors, lapses or omissions) will always exist. These factors are present top to bottom in the decision-making process: from the limited knowledge of known risks and the existence of unknown risks, proper consideration of the known risks, to the competence in technical, organisational and management matters. It is a naive believe that errors can be avoided. What is important is to reduce as reasonably as possible the size, in particular the size of its effects, and the lifetime of errors to avoid serious consequences.

Failures of temporary structures are not uncommon events. Studying the most common causes of accidents is one of the available tools to assist in identifying significant risks in the construction industry

Figure 1. Swiss-cheese model of Reason, adapted from Reason (1990)

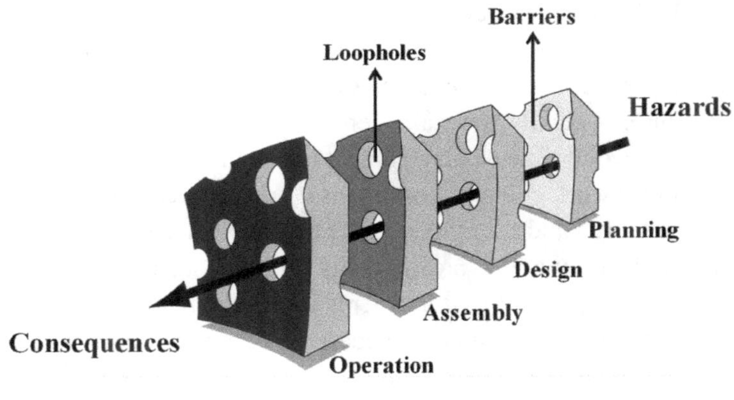

(Steven, 2010), so there is a need to understand the conditions giving rise to past failures and ways to avoid such failures so that loss of life and property can be minimised.

The term failure can be associated to two conditions, collapse and distress (Wardhana & Hadipriono, 2003). Failure can be defined as the incapacity of a constructed facility or its components to perform as specified in the design and construction requirements. Distress is the unserviceability of a structure or its components, representing the loss of ability of the structure to function as planned. Collapse of a structure happens when all or a substantial part of the structure loses its structural integration and comes down.

According to Bragg (1975):

Failures arise from many different causes. Each one has two elements: the technical cause which led to the collapse; and procedural errors which allowed the faults to occur and go undetected and uncorrected.

According to Ratay (2009):

construction failures caused by defective performance or complete absence of components of temporary structures in construction is an almost daily occurrence. Just about every step along the design-construction process includes hidden risks and has been shown to be prone to errors or omissions that result in subsequent construction failure. Failures of excavation supports, scaffolding, falsework, formwork, and temporary shoring, bracing, and guying (in approximately this order) are the most frequent occurrences of temporary structure failures.

However, with very few exceptions that involve a considerable number of fatalities, examples of failures of temporary structures do not fill the media headlines as much as a collapse of a building and a bridge do. They usually happen away from the public eye, at an isolated construction site and the knowledge of their occurrence is often kept limited to very few people. One should note that even when a disaster of a permanent structure during its construction is covered by the media, reported in ENR (Engineering News-Record, http://enr.ecnext.com/), in books, or in other technical publications, it is nearly always the permanent structure that is described, with little or no discussion of the details of the temporary structure even if it was the cause of the collapse (Ratay, 2009). In fact, in past decades several bridge disasters were catalogued as "bridge collapses" when actually, in various accidents, only the falsework collapsed resulting in injuries and fatalities, construction delays and cumulative economic costs.

According to Hadipriono, Lim, & Wong (1986), many researchers have discussed and evaluated failures of temporary structures, in particular falsework. However, it is often difficult or impossible to determine the precise cause of the failure. Typically, the main members of a falsework structure are slender elements which are used and re-used several times, being subject to rough usage and poor maintenance. A usual ground zero scenario of a falsework collapse comprises a pile of wreckage of bent tubes, in which the initial failure is probably obscured and the evidences that could be used to discover the causes of failure might have been destroyed. Additionally, typically many different scenarios exist that could explain the failure. Finally, when studying reports of past accidents it must be taken into account that many failure investigations are carried out by private companies generally recruited by a party involved in a legal action related to the failure, and therefore could be biased. Furthermore, the reports of accident investigations are generally sealed by court order as part of the resolution of the case and become unavailable to those not directly involved, but who wish to understand causes in order to avoid repetitions.

However difficult it may be, it is extremely important to carry out failure investigations so as to ascertain the likely causes of the temporary structures failures, and to be able to develop a failure database from which risk analysis of temporary structures can be developed. In more detail, the importance of failure investigations has both technical and public dimensions.

Technically, it is important to understand the physical causes of a failure in order to have an overview of the safety and reliability of the studied structure, and determine whether existing standards are adequate to prevent such failures or whether the design and construction standards require revision to be improved, and to disseminate these findings to the profession to avoid repetitions of the failure. It is never too important to emphasise the importance of trying to identify and characterise the hazard scenarios, i.e. the actual load combinations which lead to the collapse and the failure mechanisms involved in the collapse. In parallel, public and media attention are part of the aftermath of major failures, as the media, political leaders, concerned groups such as construction labour unions, and the general public become concerned about the safety of the class of structures involved in the failure.

Recently, CIRIA (2011) and HSE (2011) conducted an extensive study on what are the major hazard events in construction. They found out that failures in planning, design and management of temporary structures was a significant factor in about half of the case studies examined. Additionally, the main causal factors to accidents were identified: failure to recognise hazardous scenarios and influencing factors, poor teamwork and lack of experience and competence. Particularly, the following casual factors were highlighted with respect to the design of temporary structures: inadequate design or (late) design changes of permanent structures, underlying lack of robustness and incorrect as-built drawings and incomplete documentation.

7.2.2 Typology of Temporary Structures Collapses

Temporary structures usually fail in a disproportionate collapse fashion, where initially one or more critical elements (vertical, lacing and brace elements or connections) fails leading to a redistribution of the forces initially carried by these elements to the remaining structure, which may not have adequate resistance to them. As a consequence, equilibrium with the external forces can no longer be achieved and the system becomes unstable which is expressed by the consecutive failure of other critical elements in a domino-type collapse mode, see Figure 2.

Due to the diversity of temporary structures solutions (ranging from light and slender steel framed structures to the bulk and stiff elements used in BCEs), and of the nature of actions to which they are exposed to, the damage propagation varies substantially. In general, the collapse of scaffolds and falsework systems are disproportionate although they might not involve the collapse of the complete structure. These structures are constituted by several sub-systems sparsely connected to each other by weak joints. Thus, the existing continuity conditions usually do not offer redundancy between sub-systems: the joints may break before a total transfer of dynamic forces caused by the collapse of a sub-system to adjacent sub-systems takes place. In some cases the latter sub-systems may not have been subject to the design loads and they might be able to resist the partial transfer of dynamic forces, and therefore only partial collapse occurs. However, most often the stability of these sub-systems is strongly interdependent and a local spatially limited failure may spread to large areas of the structure (see Chapter 5).

This type of disproportionate collapse was studied by Starossek (2009): The domino-type collapse is characterised by a failure of one or more elements which are connected to other similar elements in a repetitive display, and because of the force redistribution, dynamic horizontal pushing forces develop causing overturning movements of the structure increasing the second-order effects in the adjacent vertical elements (for which the structure was not designed – lack of bracing) which then become unstable bringing the structure to the onset of collapse.

For BCEs, complete collapses of the structure are more likely to occur. This happens because these systems are less hyperstatic (statically indeterminate) than other types of temporary structures. As a result, modes of collapse involving loss of global equilibrium by sliding or overturning are more likely to occur. In addition, in BCEs fewer elements carry the loads to the supporting structures, thus they are more critical to safety. Consequently, the complete collapse of a BCE is more likely to be unavoidable when of one these elements fails.

7.2.3 Past Investigations

In the literature review in Chapter 2, a small Section was devoted to temporary structures collapses. From the literature there are several surveys of such failures. Examples of the surveys are those by Matousek & Schneider (1976) who surveyed 800 cases from European insurance files and by Walker (1981) who

Figure 2. Typical scenario after a bridge falsework collapse. ©2016 LUSA. Used with permission

summarised the results of 120 failures predominantly in the UK with approximately 30% of the failures taking place during construction. Schneider summarised Matousek's research (Schneider, 2006) which found that 37% of the failures occurred through ignorance, carelessness and negligence; 27% through insufficient knowledge; 14% through underestimating influences on the structure; 10% through forgetfulness, errors and mistakes; 6% for unjustifiably trusting in others and 6% through objectively unknown influences. In addition, Matousek found that the errors underlying the failures could have been detected in time in 32% of the cases by a careful review of the documents by a second person and 55% by additional checks on the correct strategy.

The faults that cause temporary structures to fail were not only defined by Matousek and Schneider but also by Lew (1984), Hadipriono & Wang (1986, 1987), Wardhana & Hadipriono (2003) and Maitra (1997) amongst others.

The UK Standing Committee on Structural Safety (SCOSS) has a system, similar to airlines, where correspondents can anonymously write small reports on all hazard incidents with which they have been involved and which might cause or have caused structural failures. Reporters can also make recommendations on methods of preventing future occurrences of similar incidents. These reports are made available to anyone via the website http://www.structural-safety.org/ (accessed 24 June 2016).

Examples of recent reports are:

1. An access scaffold on a refurbishment contract in a city centre contract collapsed. The following comments were made by the reporter – no clear briefing was given to highlight the particular use of the scaffold and the need to provide support for all the design loads; there was a need for a competent engineer to design and detail the scheme and that the designer should visit the site regularly; the user should confirm that he/she understands all points in the design brief and that drawings and design calculations should be available on site; that pull-out tests on ties be conducted and test results available on site; modifications should be confirmed in writing; the scaffolder must sign and issue a handover certificate; weekly checks should be made on site.
2. A scaffold collapsed onto parking area damaging 11 cars, fortunately without injuring anyone. The reporter commented that front-to-back diagonals and horizontals were offset by half bay and the horizontal members were only held by putlog connections (see Figure 2. in Chapter 2 for a picture of a putlog connection); the diagonals and some standards were wrapped in polythene before the putlogs were attached; these then slipped. Proper right-angled couplers should have been used and the polythene removed.
3. On a multi-lift access scaffold in a coastal area it was found that several of the ties had failed, even though they had been subjected to pull-out tests. Failure occurred because the bolts used were only 6 mm diameter with fixings supplied from abroad made from substandard material which failed partially due to corrosion and due to defects in the manufacturing process which in turn allowed hydrogen embrittlement to occur. The anchors to which the ties were connected did not have any reserve resistance capacity to allow for corrosion. The reporter stated that fortunately regular inspections had identified the faults in approximately 5% of the 440 fittings and that designers must allow for deterioration if a scaffold is to be in use for a significant period time in an exposed high risk situation. The reporter also stated that the source of components must be clearly defined and the technical data recorded.
4. A mobile scaffold tower fell seven storeys after a high wind occurred. The reporter felt that the tower had not been properly secured and its wheels had not been locked. The comment from SCOSS was

that this failure illustrated the generic risk of falls from height and that flying debris from roofs and unsecured scaffold boards are able to fly over 2.5 m high sheeted scaffolds.

5. A reporter who has worked on scaffolds erected on Public Highways stated that on many occasions he/she has seen insufficient consideration regarding the impact of vehicles with the scaffolds. The reporter stated that many codes do not emphasise the importance of considering impact on scaffold and falsework structures. Note that information with regard to the importance of guarding against impact assessment can be found in TG20 (NASC, 2013), CIRIA Report C579 (CIRIA, 2003), BS 5975 Clause 19.2.8 (BSI, 2011), Chapters 3 and 5.
6. A bridge was being "jacked down" into its final position and left unattended during the night. The temporary supports failed at one end resulting in the bridge dropping approximately 200 mm off the temporary support plates onto the permanent bearings. An enquiry showed that the main factor in the slippage was due to incorrect positioning of slipper pads covered with PTFE between the base of the bridge and the top of the sloping surface of a plate of taper plates. A reporter for SCOSS stated that he/she had observed a similar failure when a slab of concrete being cast sequentially and being jacked along a braced-pair of welded plate girders. There was an abutment at one end and a downward slope away from the abutment. The slab slipped and gained momentum as it slid. The reporter stated that advice should be given whenever low friction bearings are used to ensure that uncontrolled movement cannot occur.
7. A proprietary formwork/falsework system which was being used to construct an in situ, reinforced concrete, circular retaining structure for a sewage treatment works collapsed whilst the concrete was being poured. It transpired that the formwork calculations were for a different diameter tank with smaller loads and that the limits for the calculation had been exceeded. The calculations required the use of high strength bolts with an ultimate tensile strength of 700 N/mm^2 whereas those used, which sheared, only had a capacity of 250 N/mm^2. The reporter commentated that there was obviously insufficient supervision and highlighted the importance of a Temporary Works Coordinator being employed.
8. Several temporary structures such as stages and large video screens have collapsed recently such as the Sugarland Concert Indiana in August 2011 where the main stage roof collapsed causing seven fatalities and 44 people being injured. The Pukkelpop Music Festival Belgium, also in August 2011, is another example, where a video screen collapsed with five fatalities and 140 injured. In both cases the designers had not allowed for severe wind conditions and a plan of action if the design wind speeds were exceeded. SCOSS commented that the design of these temporary structures should be no less rigorous than those for ordinary structures with the same overview. They also recommend that licensing should be introduced.
9. During an 80 m^3 concrete pour on top of a birdcage falsework structure, it almost collapsed. It was noticed after erection that some standards had buckled. An inspection of the detailed design drawings revealed numerous deviations from the specification; in particular that transverse diagonal bracing was only attached to every fourth bay and not to every bay as specified. Collapse was fortunately prevented by the ledger elements becoming wedged and providing moment capacity and the decking was locked between two reinforced concrete walls. The reporter commented that adequate supervision by competent personnel would have prevented the construction errors and that regular safety inspections would have identified the difference between the erected and design specifications. SCOSS commented that photographs should be taken regularly and that a log of inspections be made.

10. A partial collapse of an area of approximately 300 m^2 of falsework occurred on a Southeast Asian project during the concrete slab pour of a post-tensioned slab. A post collapse forensic examination showed that a multitude of errors had occurred with respect to both design and construction of the falsework, any one of which could have caused collapse. A partial cause was due to the contractor being unable to obtain good quality falsework and hence an inadequate falsework was used. This failure once again emphasises the importance of correct design, construction and supervision of falsework.

7.2.4 Procedural Causes

Procedural causes are related to the context, and to organisational and management deficiencies. These can be expressed by improper and unclear attribution of responsibilities and work priorities, inadequate communication channels, incorrect information management and presentation, appointment of inexperienced (unqualified) or incompetent staff, insufficient internal (including self-checking) and external review and quality control policies.

In terms of context, the construction industry has been increasingly suffering from over-optimistic programmes and deadlines coupled with shrinking budgets and de-leveraging of responsibilities by multiple subcontracts. Temporary structures projects are highly sensitive to the above procedural causes. From many case studies in the UK, it was possible to conclude that insufficient consideration was being given to the management of temporary structures and that effective management of temporary structures is crucial to success and to avoid catastrophic events in construction (HSE, 2011).

The design and use of temporary structures involves reaching structural equilibrium by technical expertise and achieving the required levels of performance by management expertise. These two approaches, the technical and the managerial, the "hard" and the "soft" systems, although intimately linked are very different and their coexistence is not always straightforward and peaceful, especially at their interfaces (Blockley, 2011). This aspect embodies one of the first challenges of most temporary structures projects: to come up with a feasible solution for the temporary structures that is compatible with the project objectives (e.g. time and cost) and the permanent structure design. During the construction phase the stability and resistance of the latter depends on the stability and resistance of the former, but frequently the inverse is also true (in particular for BCEs).

There are several stakeholders directly or indirectly concerned with temporary structures: researchers, designers, producers, clients, consultants, insurers, contractors, sub-contractors and workers. In this context, the assemblage, use and dismantling of temporary structures is usually done by a specialised sub-contractor, in accordance with a standard project or with a special developed project depending on the work complexity.

The framework of construction industry consists of complex interactions between all the above mentioned stakeholders who have different backgrounds and can have different priorities, perceptions and goals, some of which can even be contradictory. Despite the construction phase being the most critical stage of a structures' lifetime, some stakeholders still do not recognise the importance of these systems: they are "temporary" and, therefore, their role is considered to be minor compared to that of the permanent structures. Consequently, the design and use of temporary structures are not usually treated as carefully as in the case of permanent structures and do not receive the same level of research attention and research funding.

Temporary structures projects are most of the times performed without interaction, consultation and planning with other relevant stakeholders such as the permanent structure's designer, the principal contractor and the supervision team. Frequently, changes in the permanent structures' design or in the construction sequence, which often occur, with a direct impact on the performance of the temporary structures are not properly addressed and communicated. Furthermore, specific temporary structures activities are not given the correct priority (e.g. ground investigation). Additionally, decision criteria regarding approval of temporary structures prior to use, geometrical and material quality requirements are sometimes set without consultation with the designer.

It is not uncommon for temporary structures projects to be made of "standard" solutions taken from the system's producer guide without ensuring that they are appropriate and consistent with the project specific design requirements. The same can be said regarding the design specifications or method statements which often are a copy of the system's producer standard recommendations. Often lack of information is found in the detailed design report regarding site investigation, foundation testing, assembly tolerances, material requirements (important because various material grades could be used), load cases considered, construction and loading sequence, maintenance and inspection procedures and priorities.

It must be acknowledged that most of the problems not dealt with during the planning and design phases will have to be handled on the site. However, the lack of expertise in the field and tight project deadlines have a tendency to make construction workers, often unskilled for the task at hand, behave unsafely, take unnecessary chances, and endanger both themselves and the structures (both temporary and permanent). Long sub-contractor chains lead inevitably to loss of communication between the various agents and to loss of responsibility for the supervision, inspection and dismounting procedures.

Not only errors during design are frequent but also errors during assembly and operation are also quite common. Therefore, controls (training, inspections and quality assurance plans) should be designed and enforced during assembly and at regular intervals during the operation to ensure that the system is built and used as it was designed. Chapter 8 presents a quality management guide for temporary structures.

7.3 ANALYSIS OF SCAFFOLD COLLAPSES

Maitra (1994) reported that of the scaffolds which collapsed in windy conditions, a disproportionate number were sheeted and that most of these collapses could have been prevented by better design and site control, concluding that the effects of sheeting a scaffold are not fully understood. He went on to draw attention to the fact that the practice of sheeting scaffolds was on the increase and, consequently, that there is a need to ensure that information is available to designers and users of such scaffolds to ensure that they are of adequate strength, when supplied and throughout their intended life. Guidance on this topic is provided in Chapters 3 and 4.

Note that sheeted scaffolds may collapse under windy conditions particularly in the cases where the top section of an access scaffold is not tied. Under the latter circumstances, the scaffold when subjected to wind pressures can simply collapse by a plastic hinge forming in the standard just above the highest tie, provided the spacing of the standards is inadequate (in most cases standards spaced over 2 m apart is an enabling cause). The authors recommendations are that the top of all standards on sheeted scaffolds should be tied or braced back to the supporting façade (see Chapter 6).

In a report using data from the UK MARCODE HSE database, it was found that when wind velocity was responsible for scaffold damage, 29% of these were catastrophic collapses (Maitra, 1997). Furthermore, of this number of scaffold collapses, 54% were sheeted of which nearly two thirds collapsed because they were inadequately tied, 35% were reported as never having had sufficient ties while in 25% of the cases ties had been removed by operatives and never replaced. This number of collapses is disproportionate to the number of scaffolds that were sheeted during the period in question (April 1986 to December 1993), and gives cause for concern.

Maitra (1997), determined that 28% of the failures were caused by missing ties, 25% by overload, 13% by use of faulty components and 9% missing bracing elements. These faults still occur as will be shown later in Section 7.5.2 where a discussion of the Milton Keynes access scaffold failure is analysed. Note that this list is not exhaustive. For example, other potential causes of failure are props being out-of-plumb. Burrows (1989), in his doctoral thesis examined 11 sites and found that over 50% of standards were erected with initial geometrical imperfections outside the UK and European codes allowable limits. Burrows also found that structures were not being inspected at the recommended weekly intervals.

Milojkovic, in her doctoral dissertation Milojkovic (1999) and in a paper with Beale and Godley (Milojkovic, Beale, & Godley, 2002) conducted a finite element model of a domestic scaffold (See Figure 1 in Chapter 2) and subjected it to various faults to determine the effects of these faults on the scaffold performance. Three load cases were considered:

Load Case 1: Permanent load + Working load of 1.5 kN/m^2 in every bay at the top level.
Load Case 2: Permanent load + Working load of 1.5 kN/m^2 in alternate bays at the top level.
Load Case 3: Permanent load + Overload on the top platform comprising two 300 kg loads acting on an area of 0.96 m × 0.46 m at the centre of each and every span.

In addition to the above, wind loading was considered acting normal to and parallel to the façade. The wind load pressure was considered uniform over the whole scaffold with a pressure corresponding to a wind speed of 25 m/s giving a pressure of 400 N/m^2. Note that this is in excess of the pressure used for working wind action, equal to 200 N/m^2, and represents an ultimate limit state condition.

Each of the load cases was considered in the following combinations:

Combination 1: Vertical load condition with no wind.
Combination 2: Vertical load condition with wind parallel to the façade, blowing from left side to right.
Combination 3: Vertical load condition with wind perpendicular to and away from the façade, assuming that the façade itself offered no wind resistance.

There were therefore initially a total of nine load case combinations.

However, when *Load Case 2* was analysed it was discovered that all standards had a higher overall load factor multiplier than occurred in *Load Case 1*. This is because at the top level the load applied in *Load Case 2* was only half (applied on alternate bays) of that in *Load Case 1*. The ledgers, however, have a lower overall load factor multiplier in *Load Case 2*, because bending moments are greater. This increase in bending moment caused local failure of the structure but did not lead to an overall collapse. In addition, the faults which were introduced to the structure in this research did not have any influence on this local failure of the top ledgers. As the aims of Milojkovic's research were to investigate overall failure and not local failure *Load Case 2* was neglected. This reduced the number of combinations to be analysed to six.

For each load case and combination, Milojkovic considered the structural faults which Maitra (1997) had identified as the commonest. These are presented in Table 1 along with the simulation methods used to include them in the numerical models of the scaffold.

Milojkovic conducted nonlinear analyses for each of the six load cases analysed and determined the lowest overall load factor multipliers for each load case (γ_{cr}) and for the whole structure. Table 2 gives the individual results. The reduction in γ_{cr} is taken to be the percentage drop of the minimum value of γ_{cr} for the fault case being considered against the minimum value of γ_{cr} for the perfect structure, obtained for all load case combinations. For example, the minimum γ_{cr} for the perfect structure is taken from the first row of Table 2 and equals 5.25. For fault 1, i.e. the second row of Table 2, the minimum γ_{cr} is 4.78. Hence the percentage reduction of γ_{cr} for this fault is 8.95%. Full details are found in Milojkovic (1999).

The lowest overall load factor multipliers for *Load Case 1* and *Load Case 3*, individually and in combinations 1 to 3, are given. These always occurred for *Load Case 3*. It can clearly be seen that the most significant reductions occurred for the cases of Gross Settlement (Fault 2), excessive curvature in the lowest standards (Fault 10), inadequate ties (Faults 17 and 18) and the incorrect use of putlog couplers instead of right angled couplers (Fault 13). It is interesting to note that Fault 3, irregular standard position, actually had a slightly higher buckling load factor. However, even the worst fault would not cause collapse (overall load factor multipliers less than 1.0) and hence combinations of faults were introduced.

In practice when scaffolds are designed and erected single faults do not arise on their own. Analyses were therefore undertaken with combinations of faults. In most cases the reduction in capacity of a set of

Table 1. Structural faults case studies

ID	Structural faults
1	A partial settlement caused by the baseplate of the scaffold not being level. This fault induces a bending moment into the bottom leg of the scaffold. This was modelled by inclining the bottom leg of the scaffold by 2%.
2	A gross settlement caused by a member not being supported by the ground. For maximum effect this was modelled by removing the support at the most heavily loaded base.
3	Initially the standards were regularly placed at 2.4 m centres. As alternative scenarios, the overall width of the scaffold was unaltered but the spacing of the central two standards was changed to 2.1 m and 2.7 m. The effect of this fault was to increase the strength of the scaffold. This is due to the fact that for regularly spaced scaffolds the flexural rigidity provided by the ledgers is a minimum.
4-7	The middle standards were assumed to be out-of-plumb by 1% (fault 4), 2% (fault 5) parallel to the façade and by 1% (fault 6) and 2% (fault 7) normal to the façade.
8-10	Curvature was applied to the standards below the bottom lift of a maximum of 3 mm (fault 8), 6 mm (fault 9) and 12 mm (fault 10).
11	The height to the first lift was increased from the original 2.1 m to 2.7 m. In order to keep the whole scaffold to the same height the top lift was reduced to 1.15 m. (Note, the model is shown in Figure 2.1, Chapter 2)
12	To gauge the effects of corrosion one standard was reduced from 4 mm thickness to 3.25 mm.
13	Connections between transoms and ledgers adjacent to standards were made with right-angle couplers. In order to model the common fault of using putlog connections, pinned joints were inserted at all these positions.
14	Ledger bracing was initially placed as close to standards as possible. In practice this is not possible and in faulty structures the ledger bracing was placed 300 mm vertically away from the correct position. In addition, swivel connectors instead of right-angled couplers were also used.
15	The perfect structure had ledger braces at every level. Commonly, however, the bottom diagonal is omitted.
16-18	The most common cause of failure is inadequate tying. The perfect structure had two ties on the middle standard. Fault 16 – top tie omitted; fault 17 – bottom tie omitted, fault 18 – no ties.

Table 2. Overall load factor multipliers

Fault ID	Description	Overall load factor multiplier, γ_{cr}								Reduction in γ_{cr} (%)
		Load Case 1	Load Case 3	Load Case 1 + Combination			Load Case 3 + Combination			
				1	2	3	1	2	3	
–	Perfect structure	11.79	6.45	10.91	9.88	9.88	6.02	5.39	5.25	–
1	Partial settlement	11.82	6.49	10.34	10.32	9.19	5.66	5.98	4.78	-8.95
2	Gross settlement	11.54	6.37	7.00	7.00	6.63	3.36	3.10	3.24	-40.95
3	Irregular standards	11.78	6.46	10.81	9.87	9.83	6.03	5.45	5.27	+0.38
4	1% out-of-plumb parallel to façade	11.92	6.44	10.41	9.35	9.31	5.60	5.10	4.80	-8.57
5	2% out-of-plumb parallel to façade	11.72	6.43	9.76	8.90	9.23	5.39	4.93	5.05	-6.10
6	1% out-of-plumb normal to façade	11.92	6.49	9.96	8.95	8.74	5.30	4.84	4.47	-14.86
7	2% out-of-plumb normal to façade	11.80	6.42	9.24	8.45	8.85	4.90	4.40	4.60	-16.19
8	3 mm curvature	11.71	6.34	10.24	9.37	9.11	5.52	4.96	4.56	-13.14
9	6 mm curvature	11.72	6.37	9.61	9.15	8.03	5.28	4.94	4.13	-21.33
10	12 mm curvature	11.74	6.39	8.61	8.23	6.74	4.71	4.45	3.35	-36.19
11	2.7 m bottom lift	10.26	5.52	9.61	8.69	8.60	5.32	4.79	4.48	-14.67
12	Corrosion	11.73	6.37	10.84	9.83	9.83	5.97	5.41	5.19	-1.14
13	Putlog couplers	9.68	5.01	9.23	8.19	7.98	4.87	4.35	4.00	-23.81
14	Eccentric bracing	11.42	6.10	10.55	9.55	9.15	5.59	5.12	4.76	-9.33
15	No bottom brace	11.63	6.21	10.48	9.49	9.29	5.40	4.91	5.02	-6.48
16	No top tie	11.21	5.89	10.45	9.37	9.11	5.52	4.96	4.56	-13.14
17	No bottom tie	9.69	5.16	9.07	7.66	8.03	4.92	4.16	4.14	-21.14
18	No ties	9.48	5.06	9.02	7.62	7.87	4.86	3.53	3.70	-32.76

faults could approximately be estimated by adding the reductions in load factor multiplier given by the single faults in Table 7.2 above. For example, a combination of putlog couplers (instead of right angled couplers) in combination with no ties (i.e. faults 13 + 18) gave a reduction of 59%. The sum of the two faults is 54.5%. The major exception occurred in the case of putlog couplers in combination with gross settlement. In this case the analysis produced a reduction of 28% instead of the combined total of 65%. The effect of putlog couplers is to change the structure from one containing semi-rigid connections to a structure with pinned joints only. This therefore prevents moment transfer between adjacent elements in a frame. When the scaffold is near to collapse additional moments transferred from the frame farthest from the façade precipitate failure in the most heavily loaded member. When putlog couplers are used these moments are not transferred. The most heavily loaded members in this case are those adjacent to the "settled" support in the same plane frame. They have higher axial loads than members with right angle couplers but much lower moments. The use of the interaction formulae produces a higher collapse load.

A typical scaffold is influenced by two important stages in its lifetime. The first of these stages is the design and construction stage. In this stage, decisions are first made on paper about the specification required to meet a defined set of purposes. Then the scaffold is erected in accordance (or not) with this specification. The design and erection stage of the process may be carried out by experts, and would then perform as intended, or it may be carried out less well, so that if the full demands specified for the structure were actually asked of it, the scaffold would not perform to specification. The second stage in the life of the scaffold occurs once it has been erected on site, when its performance is influenced by the quality of the site control. Site control is good if the scaffold is correctly used, so that the imposed loading never exceeds that specified for it, and if it is conscientiously inspected and serviced at regular intervals so that no significant deterioration takes places during its life.

Both stages can be subdivided into three categories: good, average and poor according to the faults which are present. Some faults may be generated in both stages of the life of the structure while others may be generated in one or the other. For instance, ties may be missed by the designer or the erector, or they may be removed while the scaffold is in use and, because of poor site control, not replaced. This is a fault possible in both stages. Table 7.3 summarises the combinations of faults assigned to each category. Note that overload, *Load Case 3* (LC3), should not occur in practice and has hence only been assigned to faults in site control. Common geometric faults such as standards out-of-plumb by 1% or with 6 mm curvature and partial settlement have been assigned to average categories. Gross settlement, or extreme errors of curvature (12 mm), or large out-of-plumb (2%) have been assigned into the category of poor site control. The use of putlog couplers was considered to be an element of poor design or poor construction. Joint eccentricity usually arises due to poor design. The absence of ledger braces was considered to belong to poor erection/site control as it should be noticed and corrected.

The resulting overall load factor multipliers are given in Table 7.4. The table indicates that in general poor site control is potentially more detrimental to the safety of scaffolds than poor design. Each combination of faults has been analysed in the same way as the single faults. It is interesting to note that the extreme case of poor design in conjunction with poor site control led to a negative safety factor. In this case, if all the faults were present, it would not be possible to erect the scaffold. In practice, not all the faults occur and so such scaffolds are constructed but with very little margin of safety. Although the analyses undertaken in this research have dealt with only one scaffold it is thought that the principles and results obtained are applicable to scaffolds in general; particularly, the absence of ties and the use of putlog connectors.

Table 3. Fault combinations

	Site Control		
Design/Erection	**Good**	**Average**	**Poor**
Good	-	17+LC3	15+18+LC3
Average	1+4/5+9+11+13+15	1+4/5+9+11+13+17+LC3	1+4/5+9+11+13+15+18+LC3
Poor	2+3+6/7+10+11+12+ 13+14+18	2+3+6/7+10+11+12+ 13+14+18+LC3	2+3+6/7+10+11+12+ 13+14+15+18+LC3

Table 4. Load factors for different site design/site control combinations

	Site Control		
Design/Erection	Good	Average	Poor
Good	9.88	4.14	0.98
Average	3.88	1.79	0.01
Poor	1.19	0.51	-0.18

7.4 ANALYSES OF FALSEWORK COLLAPSES

7.4.1 Survey of Collapses of Bridge Falsework

The reality shows that more accidents involving scaffold structures are reported than those concerning falsework systems. This is easy to understand, since the number of potential situations requiring the use of scaffolding is significantly larger than for falsework which are typically only used during the construction phase of buildings and bridges. However, consequences of a collapse of a falsework structure are generally far more severe than the ones due to a scaffold collapse, since the former ones are generally associated with loss of human lives, loss of considerable equipment and partial or total collapse of the permanent structure being built.

Additionally, the forensic work carried out to investigate why has the falsework collapsed, not only but mainly to account for responsibilities, involves a considerably larger time span, human and technical resources. All in all, a failure of a bridge falsework structure represents a heavy burden in social and economical terms.

In this Section, extensive research over the available literature and media information has been performed concerning the numbers and causes of bridge falsework incidents and accidents. The framework of the survey was based on Bragg (1975) and was divided into three major components, following Hadipriono & Wang (1987): (information on the occurrences of failure, details on the enabling and triggering causes of failure, and (information on the consequences of failure.

From past investigations, various cases of bridge falsework failures have been reported in the last 20 years, including accidents in European countries, USA and more recently in China, India and in the United Arab Emirates (UAE). Wardhana & Hadipriono (2003) revealed that in the period between 1989 and 2000 more than 500 bridge failures were reported in USA. According to these authors less than 2% of the failures occurred during bridge construction. However, this finding can be attributed to lack of official reports describing bridge collapses during construction. Scheer (2010) in his book reported 440 bridge failures, of which 125 (28%) occurred during construction and 74 (17%) were related to bridge falsework. Similar results can be obtained using the data made available in the website www.bridgeforum.org developed and maintained by the "Bridge Research Group" at Cambridge University, UK.

The survey carried out in this book found that since 1970 up to 2016, 107 major accidents occurred involving the collapse of bridge falsework structures in 25 countries. The results which are presented in the following constitute a significant evolution from previously published information, namely André, Beale, & Baptista (2012).

No evidence was found that any collapse happened because of accepted risks related to deficiencies in structural codes, or related to extraordinary severe external hazards like earthquakes, floods, landslides

Figure 3. Origins of errors leading to bridge falsework collapses since 1970

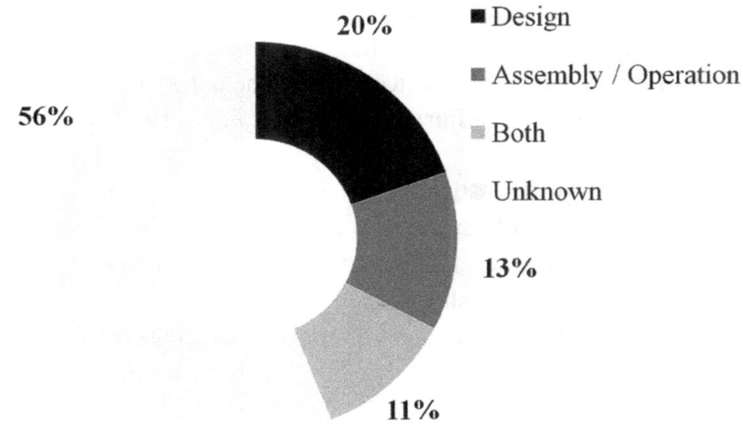

and hurricanes or tornados. All the collapses resulted from human errors, and the main cause of failure were design errors (20%), see Figure 3. However, in 56% of the accidents the causes were unknown.

Figure 4 presents the evolution with time of the total number of collapses in the countries where three or more collapses have been registered. It can be observed that in developed countries such as Germany and the USA most of accidents occurred during the beginning of the concrete construction revolution, while in developing countries (e.g. China and India) the majority of the accidents occurred during the last decade in parallel with the progressive industrialisation and globalisation of these countries.

Looking in detail into the available information, it was possible to distinguish between procedural causes, enabling events and triggering events. The procedural causes are related to management issues and the interrelationship between parties involved in a project. The enabling events are related to the

Figure 4. Evolution with time of the registered bridge falsework collapses since 1970 (countries with three or more collapses)

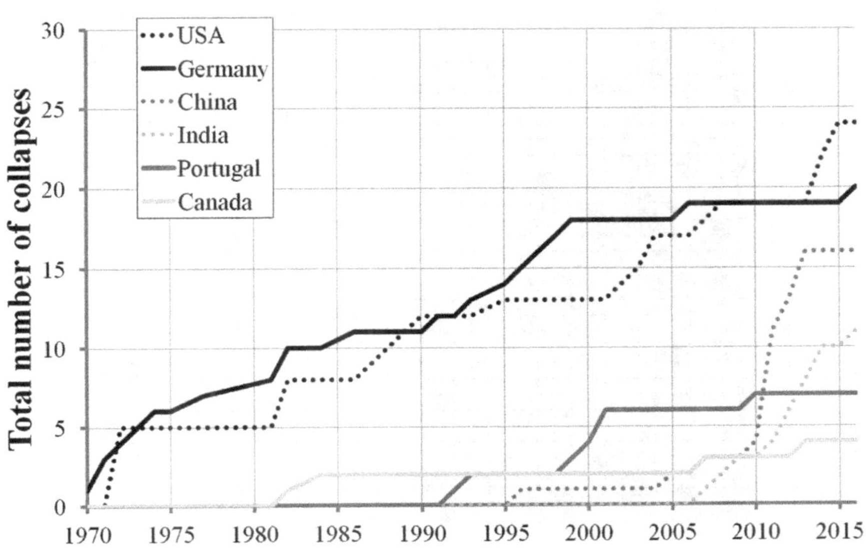

internal condition or performance of the bridge or its components that contribute to failure. The triggering events are external events that could initiate failure of a structure. It is considered that every collapse occurs due to a series of events that involve deficiencies in management framework, errors in design, assembly and operation and a hazard scenario which triggers the collapse. The findings are in line with results of previous research studies, e.g. Hadipriono & Wang (1986, 1987) which means that progress over time has been limited.

The insight achieved by this deeper investigation is considered to be extremely valuable information for the identification of the major hazards and of the critical paths of events which could lead to the collapse of a bridge falsework structure. In addition, it makes it easier to setup effective and efficient barriers to reduce and control the existing risk levels.

It should be mentioned that in a high percentage of reported accidents no detailed information about the factors that explained the collapse was found. Nonetheless, it is assumed that the results presented below are representative.

Procedural causes are related to human behaviour, organisational, planning and supervision issues. They are important in all project life management phases: from the more general asset management philosophy to the more particular conceptual design, detailed design, information and site management. These areas can be further sub-divided. In the present book, six possible areas were considered, see also Figure 5:

- Inadequate and/or insufficient communication and collaboration between parties;
- Inadequate and/or insufficient inspection and checking of falsework/formwork;
- Inadequate and/or insufficient review of falsework design / assembly / operation methods;
- Inadequate and/or insufficient site supervision, monitoring and control of construction methods;
- Inadequate documentation;
- Other or unknown causes.

Figure 5. Procedural causes of bridge falsework collapses since 1970

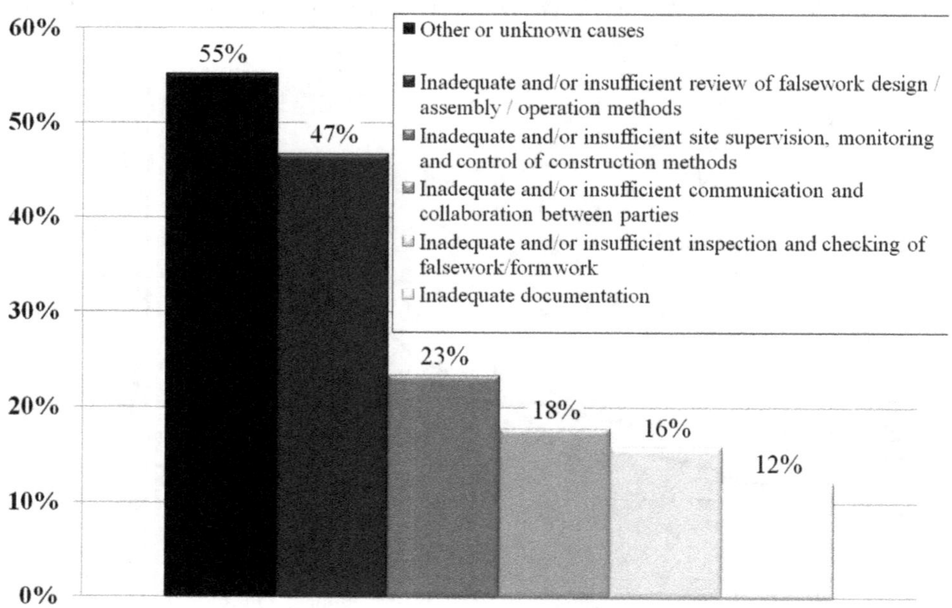

It was found that the main contributors to procedural causes are inadequate and/or insufficient review of falsework design/assembly/operation methods, including falsework dismantling, (47%), and four more specific procedural causes which occurred in more than 10% of the collapses. However, in 55% of the accidents the procedural causes were still unknown. It can also be concluded that in general several procedural causes coincide in a given accident, meaning that accidents are caused by the occurrence of multiple errors in the various phases of the project.

Six different enabling events were considered, see also Figure 6:

- Inadequate and/or insufficient falsework bracing;
- Inadequate falsework foundation;
- Inadequate falsework main element;
- Improper assembly procedure;
- Other or unknown design related causes;
- Other or unknown causes.

It was found that the most important ones are inadequate falsework bracing (19%), inadequate falsework main element (14%) and inadequate falsework foundation (9%). The survey showed that the primary enabling event associated with bridge falsework collapses is insufficient or missing bracing elements. This can be justified by the lack of awareness in the design and in the construction stage of the stability requirements of each bridge falsework solution. The second most important enabling event was found to be under-designed components such as jacks, couplers, standards or ledgers, but also support steel girders used to span open traffic areas. This in turn can in part be justified due to the reuse of falsework elements which are subjected to heavy loads and improper maintenance and thus can accumulate damage leading to a reduced load bearing capacity. Incorrect assembly procedures of the falsework system were reported to have been involved in only 2% of the collapses. However small this percentage is, it must be noted that before or after the collapse of the system, it is not very easy to determine if it was erected as planned, so this number should be read taken this into account. Finally, in a great number of accidents (51%) the enabling events are still unknown. Additionally, 21% of the accidents were caused by unknown design related errors.

Finally, six triggering events were analysed, see also Figure 7:

- Heavy rain;
- Strong winds;
- Construction material loads;
- Improper/premature falsework or formwork assembly/removal;
- Other loads;
- Unknown causes.

Three events emerged as the most critical ones: construction material loads (50%), unknown events (24%) and effects of improper/premature falsework or formwork assembly/removal (8%). It can be seen that expected loads during design of the falsework are responsible for 50% of collapses by triggering a local failure which then generally develops as a progressive and disproportionate collapse of part of the bridge falsework structure. These loads are mainly due to concreting operations.

Figure 6. Enabling events of bridge falsework collapses since 1970

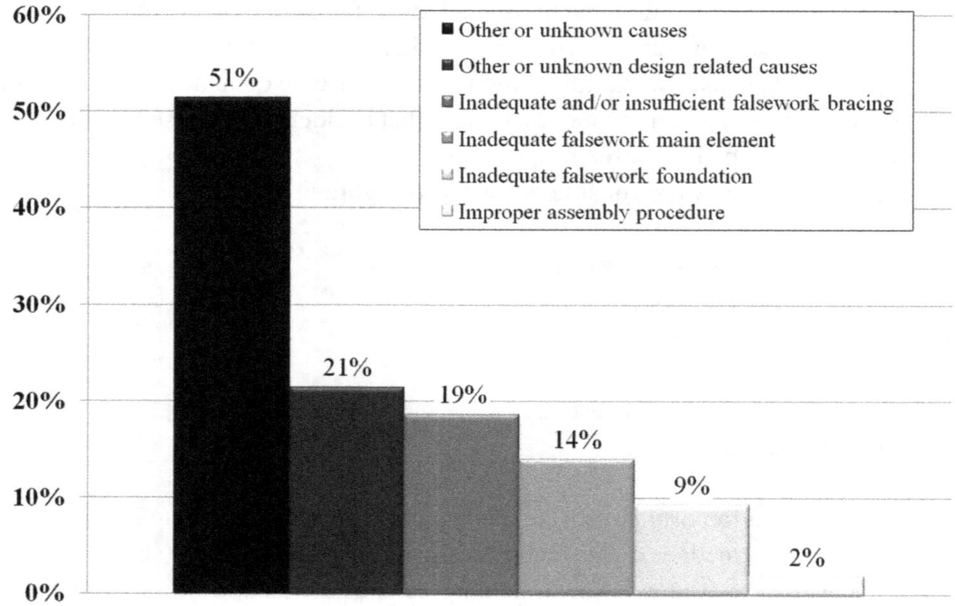

Figure 7. Triggering events of bridge falsework collapses since 1970

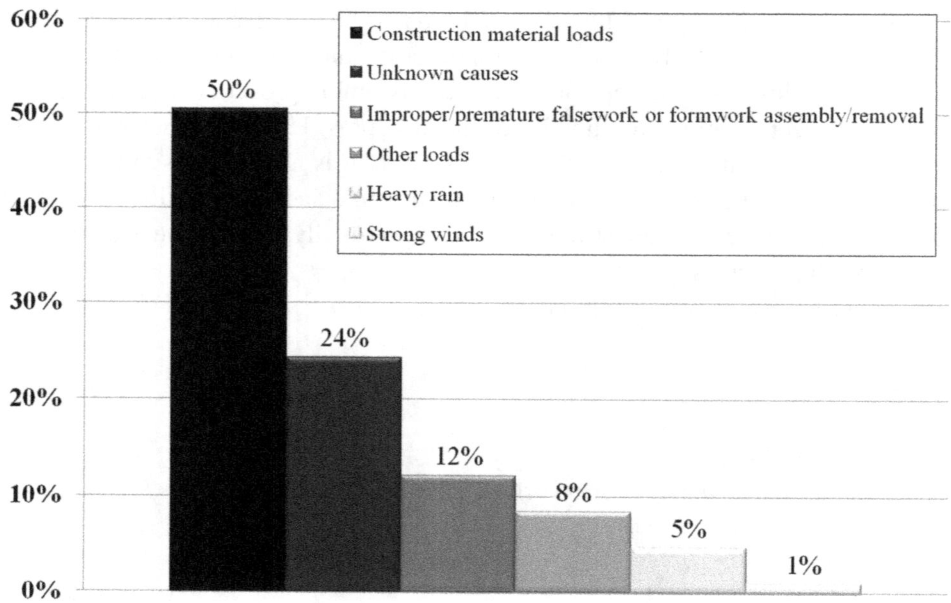

Looking at the data it could be concluded that until the year 2000, the reported accidents occurred mainly in developed countries such as Germany and USA, and that after the year 2000 there has been an increasing number of reported bridge falsework failures in the developing world such as China, India and Dubai. The numbers also indicate a growing trend in the number of reported collapses, injuries

Figure 8. Number of bridge falsework collapses since 1970

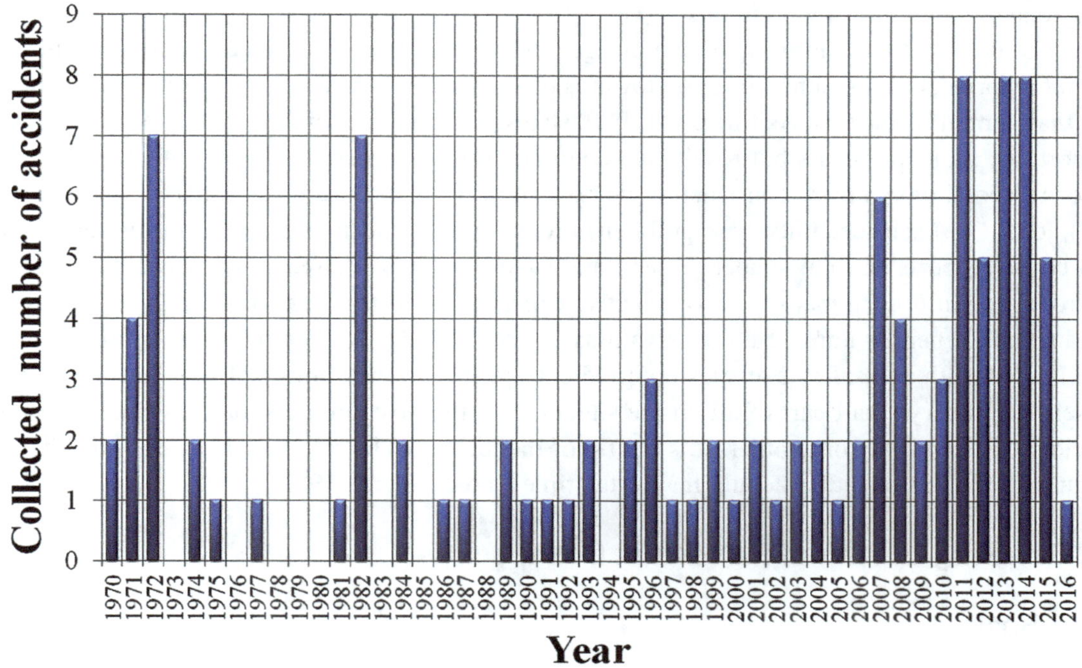

Figure 9. Number of victims due to bridge falsework collapses since 1970

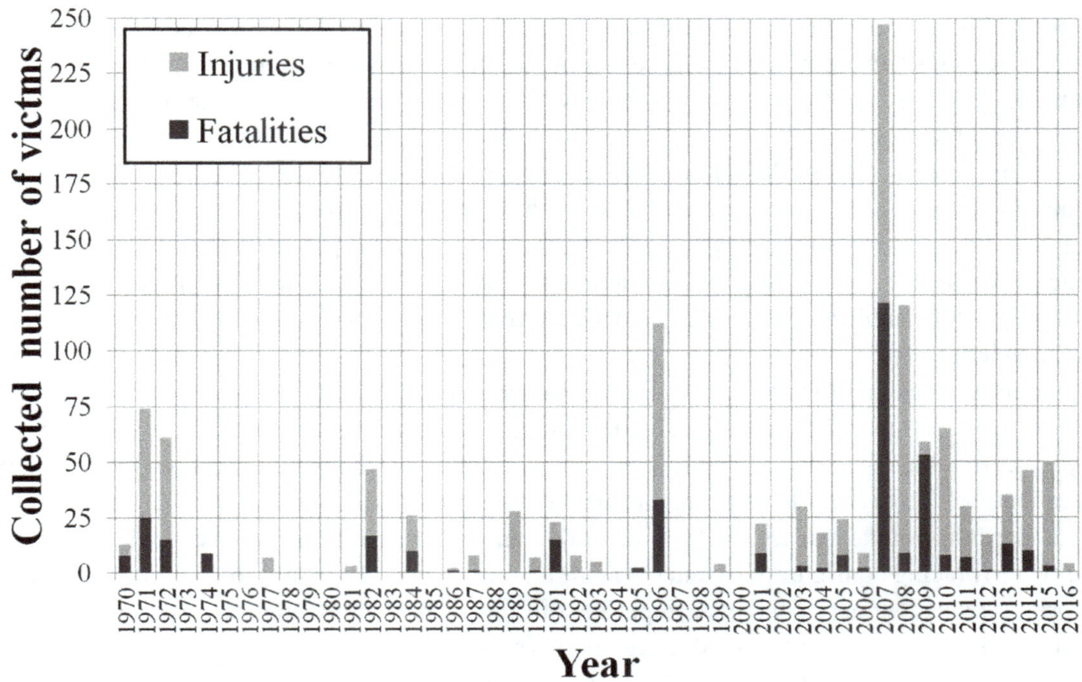

and fatalities since 2000, see Figure 8 and Figure 9. A total of 369 fatalities and 807 injuries happened directly due to the failures found in this survey.

One should note that failures involving bridge falsework structures are much more frequent but are often not reported, because attention is set towards major accidents which result in severe consequences rather than "small" accidents. As Burrows (1989) states: "The number of failures that occur daily where no reportable accident occurs but results in economic loss for the contractor or sub-contractor in the form of remedial works or re-construction works can only be surmised". Melchers, Baker, & Moses (1983) adds: "(...) serviceability-type problems are extremely under-represented in formal enquiries, in "in-house" reports and newspaper reports, and even in technical papers, but constitute a considerable proportion, if not the major part, of negative experiences in structural engineering, as assessed by individual and generally unreported observations". Also, Sikkel (1982) emphasises an important point which is the health and safety statistics reported each year with the number of accidents and fatalities represent just a very small part of all unsafe situations: only those unsafe situations which brought us an accident in some way or another. It is the tip of the iceberg. The total number of unsafe situations, including all the near-accidents could maybe ten times more (Sikkel, 1982).

7.4.2 Enabling Events of Falsework Collapses

7.4.2.1 Basis

Enabling events of falsework failures are related to design and operation issues, most often, both of them. Take for example a partial bridge collapse that occurred in 2001 in Portugal (the report is not publicly available). The collapse happened when concreting operations of the deck were being carried out. According to the failure investigation report, the accident was caused by the collapse of the bridge falsework which was found to be under-designed but also the material quality used failed to meet the design requirements and the structure was not assembled correctly – in particular some bracing elements were missing and the way the formwork beams were positioned in the system forkhead plates lead to high load eccentricities.

According to the results of the failure survey, design errors were found to be one of the most common enabling events in bridge falsework failures. This finding is expected to encompass falsework in general (i.e. extensible to shoring systems, including telescopic props). No matter what bullet proof construction controls are put in place the structure will be likely to fail if it is not properly designed for the actions it will be subjected to. Design errors stem from errors in judgment, wrong assumptions, and lack of knowledge.

In practice the design of falsework is usually an oversimplified process, based on "safe load values" taken from a design load table developed by the system's producer. In these cases, care should be taken to fully understand the hypothesis, requirements and limitations of the methods behind the design rules. This frequently involves considerations regarding the system configuration, load cases considered, length of the standards and of the ledgers, the location of the standards (face or inner elements), lateral restraints requirements (bracing configurations and top restraint provided by the plate and membrane action of the formwork), type and values of member imperfections (sway, bow and load eccentricities) and material grade of the elements. Some are included, clearly highlighted and easy to understand. However, others are not obvious, are not given or have not been accounted for.

In addition, often errors also occur during site activities, and in most cases the actual circumstances of use differ somewhat from those envisaged in the design. Typical errors involve assembling the falsework system not in agreement with the design drawings and using unfit elements/materials.

For standard cases, all of the above shortcomings might be compensated by the large partial factor included, but for more complex structures with particular load requirements or complex system configurations it may not be enough and lead to under-designed structures.

In the following, the most important enabling events will be presented.

7.4.2.2 System Stiffness Considerations

A very common enabling event is related with the failure to properly consider in the structural analysis the stiffness of the falsework system and the interaction of the falsework system with the permanent structure. Stiffness is an important characteristic of any structure because it not only controls its deformation but also the forces distribution, in the case of continuous structures. Interaction between adjacent structures must be assessed correctly in order to get an unbiased estimate of the values of the forces and the loads path from the formwork to the foundations. Therefore, in special projects where the general design hypothesis of the design load tables developed by the falsework producer are not valid, the designer must always strive to develop as accurately as possible the structural model of the falsework system including the permanent structure.

This is especially critical in complex systems such as falsework systems used to span an obstacle, such as a river or a road, where steel girders or trusses are used as flexible supports to a 3-D falsework steel structure, but also when large post-tensioning loads must be applied with the permanent structure still supported on the falsework structure, or when soil settlements are important.

In the former case, as shown in Chapter 6, the load distributions between the standards over the steel girders will not be uniform, with the outer standards receiving larger axial forces due to the higher system stiffness over the supports. The same is applicable to the standards of the support towers where the outer elements will be more stressed than the inner ones.

These effects cannot be determined by the usual analysis hypothesis and methods of calculation of the load distribution between standards, such as the tributary area method and assuming the formwork and the standards as rigid elements. Therefore, in special projects these effects should be considered explicitly in the design analysis as they will control the design of the main elements of the falsework and the bracing and lacing configurations.

7.4.2.3 Main Elements

The safety of falsework steel members should be verified by design rules specified in structural codes or by testing. However, what is not commonly taken into account in both cases is the fact that these elements are reused many times and therefore their resistance will be reduced overtime (fib, 2009). Also, the use of non-conform materials and elements or particularly corroded or damaged elements (e.g. bent tubes out of tolerance limits) is not infrequent.

The design of falsework should be based in a sound structural concept. In particular, stable load paths from top to bottom should exist. Figure 10 shows an example where this is obviously not the case.

It should be acknowledged that existing design code rules focus on local safety verification and assume that when applied to all elements of the system the global safety is achieved. However, there are

Figure 10. Example of an unstable load path

several examples that demonstrate otherwise. The global safety of a falsework system depends on the safety of members against local failure and on the system response to local failure. Buckling of a primary load-carrying member or a critical brace element in the support towers, with no alternate load paths, could trigger a chain reaction of failures causing progressive and disproportionate collapse of the entire system. Such design considerations are however not usual. In Chapter 5 a comprehensive guidance is provided regarding this important topic.

Bracing is one of the most important aspects in a falsework structure, since their performance depends greatly on the stiffness against lateral movements provided by the bracing elements.

Bracing configuration should be determined by proper structural analysis, see Chapters 4 and 6. However, producers of falsework systems often specify, in their design guidance documents, standard bracing requirements; yet, these are not always fulfilled. For example, in some projects only the exterior bays are braced, leaving the stability of internal bays resting with the (low) lateral stiffness provided by the lacing elements and by the formwork (which might be discontinuous or not designed to resist the resulting bending and membrane forces). An additional error sometimes found in support towers, is to not include sufficient bracing elements in both directions.

Critical bracing elements – i.e. those that are vital for the structural integrity, and which failure would lead to the failure of a part or the entire structure – are not always identified in the analysis and in the drawings.

Due to procedural errors, such as lack of communication and/or documentation, the layout of the structure assembled is not the same as the structure that was designed. This may occur due to assembly errors (e.g. distances between columns, position of bracing) or because of changes in the sequence of work or in the design of the permanent structure that the designer of the falsework was not made aware of.

Finally, an important topic is geometrical imperfections. The limits considered in the design should not be in any circumstance violated during assembly and operation. There are several factors that can severely influence the geometrical imperfections of falsework, such as:

- Erection procedure, influenced by the type of structural system (both of the falsework and of the permanent structure), site conditions, workers expertise, adequacy of quality assurance schemes and competence of people doing it.
- Tolerances at joints of the various elements especially at base and top jacks as well as intermediate joints such as spigot joints.
- Careful use and storage, and quality of maintenance of the various elements to correct defects such as corrosion, local damage due to impacts or out of tolerance geometrical imperfections.

Figure 11. Example of an excessive sway imperfection of a prop

7.4.2.4 Joints and Details

Joints are another very important aspect. The lateral restraint in unbraced systems is solely provided by the lacing elements and their connections to the standards. However, the bending stiffness of these joints is usually quite low, see Chapter 4, and can be even lower if they are not correctly fixed (tightening by hand is insufficient), if the joint elements show signs of damage or corrosion, if inadequate elements are used (e.g. steel rods used to replace structural pins), if different types of joints than the ones intended are used (e.g. putlog couplers instead of right-angle couplers) or if joint eccentricities larger than the ones allowed for in the design are used (see Chapter 4). Joints between brace elements and standards, or ledgers, must be checked during the erection of the system to verify if the joint eccentricities are within the tolerance limits taken into account during the design of the system.

Joints between the falsework system and the formwork system also need careful consideration during design and assembly. It is critical to not over-extend the jacks and to minimise the load eccentricities, see Figure 13.

Another important example is the gaps that may exist between vertical members of support towers, see Figure 14, which may lead to overloaded members.

7.4.2.5 Foundations

Foundations are critical elements to the safety and performance of falsework systems. For example, bridges are located in places often associated with grounds with low geo-mechanical characteristics and often bridge falsework systems are placed over the original ground without any improvement.

The foundation elements are usually concrete footings (although sole plates are also used), of reduced width, thus only mobilizing the upper layers of soil near the surface. This not only reduces the ultimate resistance of the foundation but also its stiffness: the former property is directly related with the founda-

Figure 12. Example of the use of an inappropriate element as a structural pin

Figure 13. Typical errors found in the formwork to falsework interface (CIP, 2011). ©2016 Construction Industry Publications. Used with permission

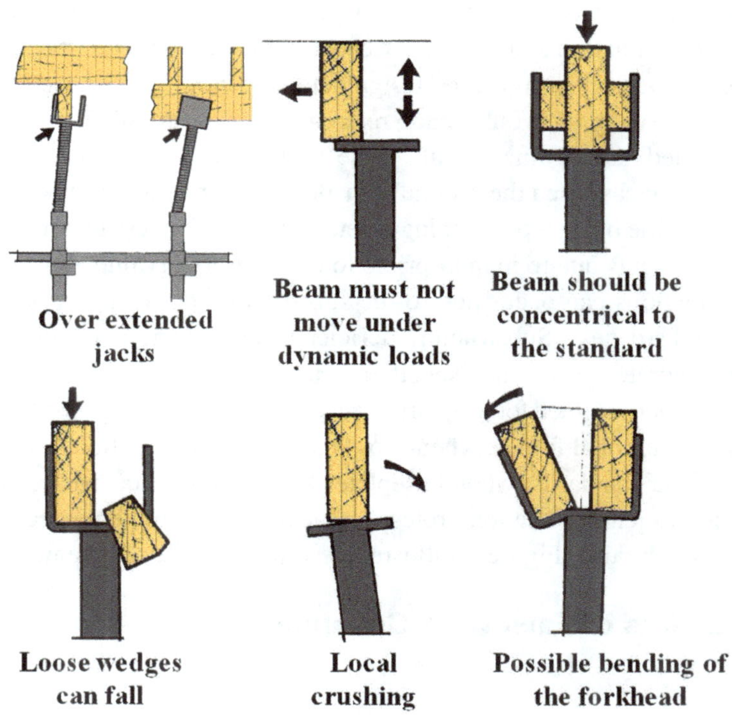

Figure 14. Example of gaps between support vertical members, adapted from fib (2009)

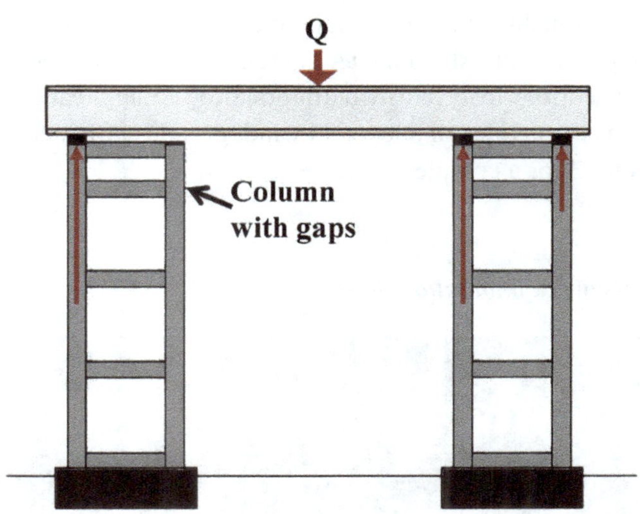

tion smaller size, being given by the soil layers within a depth approximately between one to two times the foundation smaller size (Carvalho, Pereira, & Martins, 2004).

Additionally, in several projects of falsework structures, the design of foundation elements and the safety verification of the foundation soil are often treated lightly, for example by just using the heel of a boot of an experienced inspector or engineer. Design details, control and inspection guidelines usually do not appear explicitly. Usually, it is only made reference to a permissible stress required during the construction phase, verified later against a "safe value" obtained through some simple ground testing. However, in some cases, such as when the ground is made of soil and there is a possibility for heavy rain during the construction of the bridge, performing a detailed ground investigation is justified. Not doing so can result in tragic accidents due to inappropriate foundations (Carvalho et al., 2004).

Problems with foundations can occur due to the substructure deficiencies or due to weak ground properties: resistance and stiffness. Substructure deficiencies are found when unstable foundation elements are used, when concrete weaker than specified is used in footings, when weak or damaged wood footings are (re)utilised, when inclined footings are used and the resistance against the resulting horizontal forces is relied solely on frictional forces, when gross baseplate imperfections exist which increase the instability proneness of the standards, when baseplates are severely eccentric relative to the centre of the footing or when insufficient foundation protection measures are implemented against the effects of weather events such as flash rains. Figure 15 illustrates some examples of the mentioned deficiencies.

7.4.3 Triggering Events of Falsework Collapses

7.4.3.1 Basis

Triggering events initiate the collapse but often are not the only the single cause of collapse which is hidden within the falsework system.

Triggering events are related to hazards due to permanent, variable and accidental loads applied to the structure. Permanent loads include the self-weight of the structure. Variable loads include construction loads such as the weight of the fresh concrete, the reinforcement and other materials stored in the deck, and the equipment used, the load redistribution due to the application of post-tensioning, and environmental loads such as the actions of snow and wind, ground settlements, thermal variations and seismic actions. See Chapter 3 for a complete overview.

Figure 15. Examples of falsework foundation deficiencies

Existing design philosophy requires that during the design of any structure different design situations need to be considered: persistent, transient and accidental situations. When applied to falsework systems, persistent situations can be defined as the load conditions occurring during normal operations: the permanent load, the service variable loads including the effects of post-tensioning, and the effects of wind, snow, temperature and ground movements actions. Transient situations typically correspond to the stages of assembling and dismantling of the falsework structure. Finally, accidental situations refer to exceptional scenarios such as very strong wind gusts, impacts by equipment or by vehicles and local failures for example.

In general, the most important loads that falsework structures are subjected to are the weight and pressure from concrete (the latter just while concrete is still fresh), followed by other types of construction actions. In contrary to permanent structures which only receive their full design load in rare cases (e.g. the design traffic loads on bridges are rarely reached), usually falsework structures are normally subjected for a long period of their design working life to loads whose values are close to their design values. Thus the actual safety margin of falsework structures is lower than in permanent structures, i.e. the probability of failure of temporary structures is higher than that of permanent structures (fib, 2009).

Furthermore, there is no consensus regarding the load modelling during construction, taking into account formwork/falsework interaction. Many researchers have tried to improve the available models by monitoring the construction loads during concrete pouring. The falsework-formwork interaction is extremely important, see Chapter 4, since the load distribution between standards depends on the stiffness of the formwork, e.g. the isotropic, orthotropic or anisotropic behaviour of the formwork material, the resistance of the formwork granted by the plate action and the stability of the formwork and of the formwork/falsework connection.

7.4.3.2 Vertical Loads

The sequence of loading of a falsework structure can have a major effect on the stresses in individual members of the structure. Important aspects that need proper consideration during planning and design phases include type of equipment to be used, weight and volume of storage materials, method and sequence of pouring and method and sequence of post-tensioning (Billings & Routley, 1978).

For the design of the falsework system, the most critical stage of construction is usually during pouring of concrete. However, if the falsework is designed correctly the structure should not collapse under loads smaller than the ones considered in the design. As shown in Chapter 6, in general it may require a load a lot higher than the design value to cause collapse without any other defect being present. Nevertheless, the concrete pour rate and drop height should be controlled to avoid large dynamic effects that could jeopardise the safety of the falsework.

Elements can be overloaded by inadequate post-tensioning method of the permanent structure, see Chapter 3, or by excessive temperature gradients or ground settlements.

7.4.3.3 Horizontal Loads

Horizontal forces arise from different sources such as wind, concrete pouring method, system imperfections, forces from impacts and foundation settlements.

All falsework design specifications include a requirement that the falsework must be capable of resisting a horizontal design load. However, this is sometimes neglected in design. This requirement is included to provide a criterion for bracing design and thus ensure the stability of the falsework system.

As said previously, the vertical loads generally dominate the design of falsework systems. This is because the weight of the permanent structure is very high and after hardening the bridge deck or building slab acts as a top restraint of lateral movements of the falsework system. However, before the casting of the concrete the wind load needs to be considered since the formwork may not yet exist or the formwork in-plane stiffness may not be sufficient. The use of reduced wind velocities to assess the effects on the structure must be well justified, see Chapter 5.

An anemometer (wind gauge), for example, should be required onsite and monitored continuously, and weather forecasts should be reviewed routinely. Generally, on most falsework projects, casting of concrete will not be allowed when the wind velocity exceeds the operating limits. These are usually at a Beaufort Scale 6, corresponding to a wind velocity of 14 m/s (Newman & Choo, 2003). This is known as the working wind velocity. See also Chapter 3.

7.4.3.4 Settlements

Settlements should be assessed correctly to avoid unwanted and unusual load distributions within the elements of the falsework system, and problems related to the geometry control of the permanent structure. Settlements are most often related to movements in the foundations, but elastic deformations and initial gaps between elements and within the connections can also produce settlements.

Differential settlements can occur in situations where some supports bear on bridge footings and the rest on natural ground, when different foundation types are use, or due to variations in the properties of the soil. Differential settlements translate to unbalanced loads and consequently to overloaded standards and footings. Furthermore, the occurrence of these settlements may result in the overturning of part of the structure, causing secondary stresses for which the falsework structure was not designed for. This behaviour, if neglected in the design phase may lead to the collapse of the structure.

Structures where the vertical elements are not tied up to adjacent members, i.e. without bracing elements, or with joints that do not allow the redistribution of the forces between adjacent elements, are more sensitive to the effects of differential settlement. However, elements of very stiff structures, with many bracing elements, can also be very sensitive to differential settlements as these introduce additional compression/tension loads to neighbour elements, see Chapter 6.

7.5 EXAMPLES OF FORENSIC ANALYSES

7.5.1 Past Investigations

An example of a bridge falsework system failure caused by settlements occurred on 15th April 1982, when two spans of a partially completed post-tensioned concrete bridge, being constructed at the Riley Road interchange in East Chicago, collapsed during the casting of the deck, killing 13 workers (Sikkel, 1982).

At the time of failure virtually all the forces were supported by the falsework (isolated high-capacity towers located close to the bridge piers). The failure occurred during the casting of the deck slab of the fourth span when about 100 m length of the partially finished bridge and its supporting falsework collapsed. The ensuing investigation found that the falsework as built was substantially different in several vital details from that envisaged in the design. The collapse was probably triggered by the excessive settlement of one of the temporary foundation footings of one of the falsework towers. This caused an increase in the reactions provided by the other footings which were under-designed and thus cracked. The differential settlement of the foundations caused an estimated increase in the forces in the diagonal bracing members of the tower to about 40 kN which was grossly in excess of the average value of about 28 kN for the buckling strength of the tubes, determined from later tests. This partial tower failure induced a slight sway at the top of the tower causing the main cross-members supporting the bridge to be eccentrically loaded. The welds holding these in place fractured and one cross-beam fell away imposing an eccentric force on the tower which then buckled and collapsed, precipitating the collapse of the partially-completed span.

On subsequent investigation it was found that:

1. The foundation footings of the towers had been constructed on top of about 3 m of compacted fill, but this overlaid 300-600 mm deep pockets of highly compressible black organic silt.
2. The foundation footings were only 300 mm thick, whereas the existing code required a thickness of at least 530 mm.
3. Some cracks in the foundation footings had been noted by the site surveyor a few days before the collapse, but their significance had not been appreciated.

In 1989, an inquiry research team assembled in the USA to study the causes of the collapse of Route 198 Baltimore-Washington Parkway Bridge, performed several tests of bridge falsework systems (Surdahl, Miller, & Glenn, 2010). The tests focused on establishing the failure modes of the bridge falsework systems. Vertical loads were uniformly distributed at the top sections of the system to simulate the loading conditions on the bridge falsework systems during construction. For the towers that failed, researchers found that the cross-bracing members between bridge falsework tower legs bowed out of plane, making them incapable of providing the bracing needed to restrain the lateral displacements of the bridge falsework tower legs. This loss of lateral stiffness resulted in the buckling and fracture of the bridge falsework tower legs.

An example of a collapse of a BCE occurred when a large launching gantry (180m long and about 800 tons in weight) being used in the construction of the second Thai-Lao Friendship Bridge on the Mekong River, Vietnam, collapsed making ten fatalities and ten injuries, and a cost of 4.3 million USD.

The accident took place when the gantry carrying a load of about 52 tons was reported to have bent and then suddenly buckled when moving over a temporary support. The falling pieces of equipment hit about 30 engineers and workers at the construction site. Most of who then fell into the river.

Three independent experts had been appointed by involved parties, of which two established "Faulty Design" as being the cause of the loss. Defective design of the infill trestle at the top of a temporary support was considered to be the root cause of the claim. The point of failure originated at the transom beam webs which were non-stiffened. Apparently, the design contravened several clauses of the applicable design code, the most important of which being that it omitted any consideration of the effect of horizontal loading which should have been included.

Regarding temporary grandstands and stage structures collapses, a number of accidents have occurred in the recent years:

- Earls Court, UK, 1994: temporary grandstand seating 1200 collapsed at the start of a pop concert, more than 50 injuries.
- Madonna Concert, France, 2009: stage roof collapsed during construction with two fatalities and eight injuries.
- Big Valley Jamboree, Canada, 2009: stage roof collapsed with one fatality and 75 injuries.
- Guns N' Roses Concert, Brazil, 2010: stage roof collapsed with several injuries.
- Bluesfest Concert, Canada, July 2011: main stage collapsed with three injuries.
- Indiana State Fair, the USA, 2011: main stage roof collapsed with six fatalities and 44 injuries.
- Pukkelpop Music Festival, Belgium, 2011: main stage collapsed with five fatalities and 140 injuries.
- Jovanotti Concert, Italy, 2011: ground support stage structure collapsed during construction with one fatality and 12 injuries.

Additional examples of temporary structures failure investigations are given in various bibliographic references (Andresen, 2012; El-Safty, Zinszer, & Morcous, 2008; Moore & Green, 2010; Pisheh, Shafiei, & Hatambeigi, 2009; Ratay, 2009; Surdahl et al., 2010; Thornton Tomasetti Inc, 2012). For BCE's reference is made to Béguin (2010), Mizon & Kitchener (1997) and Rosignoli (2007, 2013)

7.5.2 Access Scaffold Collapse at Milton Keynes

A 40 m high 16 m wide independent tied access scaffold attached a building in Milton Keynes collapsed on April 11th 2006. It affected the west elevation of the building. Fortunately, it happened at lunch time when most of the operatives were off the scaffold. However, there were 20 injuries, with one workman dying three days later. The initial reports in the media were that wind action was the cause of collapse. However, on the day in question tower cranes were operating locally and they can only work in winds up to 50 km/h and the local meteorological station, approximately 25 km away, recorded mean velocities of 48 km/h, thus ruling wind action out.

A local amateur photographer had taken photographs daily throughout the construction and use of the scaffold. Hence the configuration of the structure immediately before collapse was known. Three different structural engineers were independently hired to analyse the collapse. When they first appeared in court with different results the judge hearing the case required them to leave the court and return the next day with an agreed solution. The solution they arrived at is described below.

Firstly, an examination of the photographs and witness statements showed that a loading tower which had been attached at the southwest corner of the building had been removed the previous day but only after pallets of tiles had been raised and placed at the middle of platforms adjacent to the corner. The load was estimated at 2.8 tonnes. The tiles were placed in piles at the bays adjacent to the tower on three lifts and not distributed. The scaffold codes for access scaffolds typically only allow a full load of 2.0 kN/m^2 on one working level and 1.0 kN/m^2 on the working level immediately below. There was a significant overload of imposed load adjacent to the tower.

Further examination of the photographs showed that the ties were irregularly attached to the building. They were attached horizontally at spacings of between 10 to 12 m apart and vertically at irregular spacings between 2.5 to 5.0 m apart with the lowest ties 10 m above the base. These did not conform with either of the design codes BS 5973 (BSI, 1993) or BS EN 12811 (BSI, 2003) in use at the time (The UK was in transition between the earlier allowable stress design and the Eurocode based on limit state principles).

The UK standard BS 5973 allowed one standard per every 40 m^2 with a maximum separation between ties of 8.5 m and the National Access and Scaffolding Confederation design code based on BS EN 12811-TG20:05 (NASC, 2005) specified spacings of no more than 4 m horizontally or vertically for tall scaffolds. The erected scaffold was found to have ties at 60 m^2 spacings for the top lifts and approximately 100 m^2 at the bottom although in the middle the spacing was approximately at 20 m^2 spacings (Andresen, 2012). Examination of the ties showed that not only were they at incorrect spacings but that in some cases they were 300 mm above or below the transoms connecting the front to rear pairs of standards and not braced to the standard as they should have been. Additionally, the tie connections were not linked to the front standards as well. It was stated that the ties were installed by an inexperienced person without supervision and that no pull-out tests on the tie strength had been carried out (Fyall, 2012).

The scaffold had originally been applied to the north, west and south façades of the structure but in the days preceding the collapse had been partially removed from the south leaving one untied bay in the south adjacent to the tower support. Throughout the twelve month life of the scaffold, façade bracing had been removed from the bottom three levels and was missing between the fourth and fifth levels. Furthermore, ledger bracing was missing in the lower southwest corner.

The scaffold, as usual in modern UK methods, was fully boarded at all levels and had debris netting applied to the façade. Early on, during the life of the scaffold, an impermeable advertising banner 12 m × 12 m had been attached to the scaffold which could have affected wind performance but it had been removed before the collapse.

Finally, a mixture of new galvanised tubes and used rusty tubes was used. They were placed on alternate standards in an attempt to reduce the chance of a weak structural component.

The structural analysis of the scaffold showed that the factor of safety against collapse under the imposed loading was approximately 1.1 prior to the removal of the tower, and marginally over 1.0 after the tower was removed. It is not surprising that the scaffold failed.

As well as the structural faults in erection and maintenance described above, the investigation determined that no drawings or required calculations had been prepared and communicated by the main scaffolder to the sub-contractor appointed to erect and the manage the structure. The inspection regime had been carried out early in the life of the scaffold but the regime had lapsed and no documentary evidence of the scaffold throughout the remainder of its life was available.

At the resulting court case, the Coroner's verdict was that the death had been caused by the collapse of the scaffold, the main contractor was fined £132,000, a sub-contractor £64,000 and the company responsible for final erection and management went into liquidation, with the managing director now having a warrant for his arrest as he has since disappeared. Civil claims for damage are estimated between £5m and £10m.

7.5.3 Bridge Falsework Collapse at Portugal

A bridge falsework collapse occurred on a motorway construction project in Portugal, see Figure 16. Due to confidentiality the details from the location and date of the collapse cannot be publicly released.

The viaduct under construction was spanning a motorway in operation. The bridge geometry was complex with a circular curvature in plan and inclination angles both in the longitudinal and transversal elevations. Furthermore, an existing two-way road underneath the falsework system needed to remain open during the construction. Steel girders had to be assembled in order to span the road lanes with an intermediate support at the middle of the road. The supports consisted of falsework towers.

Figure 16. View of the site after the collapse. ©2016 LUSA. Used with permission

The collapse occurred while workers were pouring concrete onto the formwork (860 m^3 of a total of 1,300 m^3 were already poured). The whole length of the fresh concreted section of the deck, together with its supporting structure collapsed at once. Luckily, the collapse happened during the evening when traffic on the road below was light. Tragically, the accident caused the death of a driver and injured eight workers. The cost for the material damage section was estimated to be € 600,000.

The failure was sudden, without warning, leading to the complete collapse of the two spans over the road. After an exhaustive numerical investigation, four main enabling events were identified which could have contributed to the collapse:

1. Instability of the standards of the intermediate support (tower) of the steel girder system – no bracing was installed to prevent the buckling of these elements in the longitudinal direction.
2. Failure of some standards and top jacks of the most loaded longitudinal frames of the falsework system assembled on top of the steel girder system: the connections between the ledgers and the standards did not provide the sufficient restraint necessary to avoid high second-order effects.
3. Flexural-torsional instability of the most loaded steel girders – there were evidence that the standards were placed eccentrically to the web axis of the steel beams.
4. Failure of the base jacks of the falsework system assembled on top of the steel girder system by excessive rotation of the girders which buckled.

The system lacked bracing elements. The tower supports were braced in only one direction and the frames of the falsework system assembled on top of the steel girder system were braced only at external bays. Also, the internal forces of the system under loading were not correctly determined since the effect of a flexible support provided by the steel girders was ignored and considered as a rigid support. The connection between the girders and the supported falsework system was deficient and allowed the standards to move during assembly and operation activities.

7.6 CONCLUSION

This Chapter has overviewed the causes of temporary structures failing and concluding with two real examples of major collapses. The results of a survey of bridge falsework accidents since 1970 in 25 countries were also presented.

From the results of surveys of falsework and scaffold collapses reviewed in the Chapter, it is possible to conclude that most failures occur due to a combination of design errors (e.g. by omitting ties and not investigating ground conditions), erection failures (e.g. not erecting in accordance with the design drawings and using poor materials), and quality management flaws (related to inadequate quality assurance and quality control procedures during design, assembly and operation).

The importance of correct management of temporary structures was found to be vital and repeated collapses have been shown to occur when the management of the structures is inadequate. It is important that the role of temporary works supervisor, in the UK called a Temporary Works Controller, be seen as essential and not peripheral to the construction project. It is also essential that the designer understands the assumptions and limitations of the design methods used for temporary structures.

This is emphasized in two cases of forensic investigations to temporary structures collapses which were presented at the end of the Chapter. From them it was also shown that such events are usually very costly, constitute a high risk to people's lives and can potentially lead to the company's closure.

REFERENCES

André, J., Beale, R., & Baptista, A. (2012). A survey of failures of bridge falsework systems since 1970. *Proceedings of the ICE - Forensic Engineering, 65*(4), 161–172. doi:10.1680/feng.12.00012

Andresen, J. (2012). Investigation of a collapsed scaffold structure. *Proceedings of the Institution of Civil Engineers - Forensic Engineering, 165*(2), 95–104. doi:10.1680/feng.11.00021

Béguin, G. (2010). Catastrophic sliding of a concrete deck during its launching on the steel girders of a composite bridge. *Journal of Performance of Constructed Facilities*, *24*(2), 138–147. doi:10.1061/(ASCE)0887-3828(2010)24:2(138)

Bennett, L. (2004). Peer review of analysis of specialist group reports on causes of construction accidents (No. 218). London, UK: Health and Safety Executive (HSE).

Billings, I., & Routley, R. (1978). Falsework load measurement at Newton No. 1 bridge. *New Zealand Engineering*, *33*(6), 127–134.

Blockley, D. (2011). Engineering safety.*Proceedings of the ICE - Forensic Engineering, 164*(1), 7–13.

Bragg, S. (1975). Final report of the advisory committee on falsework. Her Majesty's Stationery Office (HMSO).

BSI. (1982). BS 5975:1982 – Code of practice for falsework (1st ed.). London, UK: British Standards Institution (BSI).

BSI. (1993). BS 5973:1993 – Code of practice for access and working scaffolds and special scaffold structures in steel (3rd ed.). London, UK: British Standards Institution (BSI).

BSI. (2003). BS EN 12811–1:2003 – Temporary works equipment – Part 1: Scaffolds – Performance requirements and general design. London, UK: British Standards Institution (BSI).

BSI. (2011). BS 5975:2008+A1:2011 – Code of practice for temporary works procedures and the permissible stress design of falsework (3rd ed.). London, UK: British Standards Institution (BSI).

Burrows, B. (1989). *Organisation and quality of falsework construction (PhD.)*. Warwick University.

Carvalho, A., Pereira, C., & Martins, R. (2004). Problems with falsework foundations (in Portuguese). In *Proceedings of the 9th Portuguese National Congress of Geotecnics* (p. 10). Aveiro: Universidade de Aveiro.

CIP. (2011). Temporary works and falsework. In *Construction health and safety manual* (p. 16). Construction Industry Publications Ltd.

CIRIA. (2003). Retention of masonry facades. Best practice guide (No. C579). London, UK: Construction Industry Research and Information Association (CIRIA).

CIRIA. (2011). Guidance on catastrophic events in construction (No. C699). London, UK: Construction Industry Research and Information Association (CIRIA).

El-Safty, A., Zinszer, M., & Morcous, G. (2008). Forensic investigation of a bridge construction scaffolding collapse. In *Proceedings of Structures Congress 2008: Crossing Borders* (pp. 1–10). Vancouver, Canada: American Society of Civil Engineers (ASCE). doi:10.1061/41016(314)198

fib. (2009). *Formwork and falsework for heavy construction. Guide to good practice*. Lausanne, Switzerland: International Federation for Structural Concrete (fib), Bulletin No. 48.

Fyall, I. (2012). Milton Keynes scaffolding collapse. Retrieved from accessindustryforurum.org.uk (2016, June 27).

Hadipriono, F., Lim, C., & Wong, K. (1986). Event tree analysis to prevent failures in temporary structures. *Journal of Construction Engineering and Management*, *112*(4), 500–513. doi:10.1061/(ASCE)0733-9364(1986)112:4(500)

Hadipriono, F., & Wang, H. (1986). Analysis of causes of falsework failures in concrete structures. *Journal of Construction Engineering and Management*, *112*(1), 112–121. doi:10.1061/(ASCE)0733-9364(1986)112:1(112)

Hadipriono, F., & Wang, H. (1987). Causes of falsework collapses during construction. *Structural Safety*, *4*(3), 179–195. doi:10.1016/0167-4730(87)90012-9

HSE. (2003). Causal factors in construction accidents. London, UK: The Health and Safety Executive (HSE).

HSE. (2011). Preventing catastrophic events in construction. London, UK: The Health and Safety Executive (HSE).

Lew, H. S. (1984). Construction failures and their lessons. *Batiment International. Building Research and Practice*, *12*(5), 272–275. doi:10.1080/09613218408545283

Maitra, A. (1994). Potential for reducing accidents by cladding scaffolds. In *Proceedings of Conference into Wind Loading on Temporary Structures* (pp. 1–15). Buxton, UK: Health and Safety Executive (HSE).

Maitra, A. (1997). Accidents associated with scaffolding in the UK. In *Proceedings of Technical Seminar, Design and use of Temporary Structures*. London, UK: Institution of Civil Engineers (ICE).

Matousek, M., & Schneider, J. (1976). Untersuchungen zur struktur des sicherheitsproblems bei bauwerken (in German) (No. 59). Zurich, Switzerland: Swiss Federal Institute of Technology Zurich (ETHZ).

Melchers, R., Baker, M., & Moses, F. (1983). Evaluation of experience. In *Proceedings of IABSE Workshop on Quality Assurance within the Building Process* (pp. 9–30). Rigi, Switzerland: International Association for Bridge and Structural Engineering (IABSE).

Milojkovic, B. (1999). *Factors of safety for standard scaffold structures* (PhD Thesis). Oxford Brookes University, Oxford, UK.

Milojkovic, B., Beale, R., & Godley, M. (2002). Determination of the factors of safety of standard scaffold structures. In *Proceedings of the 3rd International Conference on Advances in Steel Structures (ICASS 2002)* (pp. 303–310). Hong Kong, China: Elsevier. doi:10.1016/B978-008044017-0/50035-4

Mizon, B., & Kitchener, J. (1997). Second Severn crossing: Viaduct superstructure and piers. *Proceedings of the ICE -. Civil Engineering (New York, N.Y.), 120*(2), 35–48.

Moore, M., & Green, T. (2010). Investigation of falsework failures during construction of a cast-in-place bridge structure. In *Proceedings of the 2010 Structures Congress* (pp. 22–31). American Society of Civil Engineers (ASCE). doi:10.1061/41130(369)3

NASC. (2005). *NASC TG20:05– Guide to good practice for scaffolding with tube and fittings*. London, UK: National Access and Scaffolding Confederation (NASC).

NASC. (2013). *Guide to good practice for scaffolding with tube and fittings*. London, UK: National Access and Scaffolding Confederation.

Newman, J., & Choo, B. (2003). *Advanced concrete technology: Processes*. Butterworth-Heinemann.

Pisheh, Y., Shafiei, H., & Hatambeigi, M. (2009). A case study of failure due to inappropriate usage of forming scaffold system. In *Proceedings of 5th Forensic Engineering Congress 2009* (pp. 536–545). Washington, DC: American Society of Civil Engineers (ASCE). doi:10.1061/41082(362)55

Ratay, R. (2009). *Forensic structural engineering handbook*. McGraw Hill Professional.

Reason, J. (1990). *Human error*. Cambridge University Press. doi:10.1017/CBO9781139062367

Rosignoli, M. (2007). Robustness and stability of launching gantries and movable shuttering systems - Lessons learned. *Structural Engineering International, 17*(2), 133–140. doi:10.2749/101686607780680673

Rosignoli, M. (2013). *Bridge construction equipment*. London, UK: ICE Publishing. doi:10.1680/bce.58088

Scheer, J. (2010). *Failed bridges: Case studies, causes and consequences*. London, UK: Wiley VCH. doi:10.1002/9783433600634

Schneider, J. (2006). Structural & Construction Safety. Specially recorded for IABSE E-Learning 2006, Switzerland.

SCOSS. (2002). Falsework: Full circle? (No. Topic Paper SC/T/02/01). London, UK: Standing Committee on Structural Safety (SCOSS).

SCOSS. (2005). 15th Biennial report from SCOSS. London, UK: Standing Committee on Structural Safety (SCOSS).

Sexsmith, R. (1998). Reliability during temporary erection phases. *Engineering Structures, 20*(11), 999–1003. doi:10.1016/S0141-0296(97)00195-8

Sikkel, R. (1982). Health and safety in construction. In *Proceedings of IABSE Workshop on Health and Safety in Construction* (pp. 37–56). Tokyo, Japan: International Association for Bridge and Structural Engineering (IABSE).

Starossek, U. (2009). *Progressive collapse of structures*. Thomas Telford Ltd. doi:10.1680/pcos.36109

Steven, S. (2010). The different phases in construction – Design in health and safety to the project life cycle. In *ICE Manual of health and safety in construction* (pp. 51–69). London: Thomas Telford Ltd.

Surdahl, R., Miller, D., & Glenn, V. (2010). The positive legacy of a bridge collapse. Public Roads, Federal Highway Administration, US Department of Transportation, 73(5).

Thornton Tomasetti Inc. (2012). *Indiana state fair commission. August 13, 2011 Collapse incident. Investigative report*. Indiana State Fair Commission.

Walker, A. (1981). Study and analysis of the first 120 failure cases. In Proceedings of Structural Failures in Buildings (pp. 15–40). London, UK: The Institution of Structural Engineers (IStructE).

Wardhana, K., & Hadipriono, F. (2003). Analysis of recent bridge failures in the United States. *Journal of Performance of Constructed Facilities, 17*(3), 144–150. doi:10.1061/(ASCE)0887-3828(2003)17:3(144)

Wong, S., Onof, C., & Hobbs, R. (2005). Models for evaluating the costs of bridge failure.*Proceedings of the ICE - Bridge Engineering, 158*(3), 117–128. doi:10.1680/bren.2005.158.3.117

Chapter 8
Quality Management

ABSTRACT

Ensuring that temporary structures projects are managed well so that budgets are maintained and safety is ensured throughout the project is the objective of this chapter. The chapter presents recommendations for the quality control, quality assurance and quality improvement processes to be undertaken in the design, assembly, use and maintenance of the temporary structures to reduce the associated risks. The chapter starts by discussing the importance of having clear management structures for projects involving collaboration between the client, designer and the construction engineer. This leads to the importance of an overall project supervisor, sometimes called a Temporary Works Controller, who has ultimate authority for the safe execution of the project. The use of Structural Health Monitoring is described with particular reference to its ability to make the erection of temporary structures projects safer.

8.1 INTRODUCTION

Temporary structures have a major role in the execution time, cost, quality, durability, safety, efficiency, utility and aesthetics of any construction project. Therefore, it is not surprising that a correct choice, good planning, designing and operation of temporary structures are keys for the success of every construction project. In order to achieve a successful project, continuous knowledge exchange and synchronised planning between the permanent structure designer, the project contractor, the temporary structures designer, the temporary structures sub-contractor and others must exist, see Figure 1 for example.

Unfortunately this is not always a reality. As so clearly demonstrated in Chapter 7, poor project management is often a cause of temporary structures collapses. For example, Milojkovic (1999) demonstrated that a simple domestic scaffold under conditions of poor design and erection even if with good site management, could exhibit a global safety factor of just above unity, whereas the same scaffold under good design, erection and site management should have a global safety factor of approximately ten. Even more importantly, it was found that with poor site management, regardless of the quality of design, the maximum value of the global safety factor was not larger than unity. This implies that the

DOI: 10.4018/978-1-5225-2199-0.ch008

Copyright © 2017, IGI Global. Copying or distributing in print or electronic forms without written permission of IGI Global is prohibited.

structure would be on the verge of collapse at all times and any sources of inadequate design or improper erection would cause collapse. See Tables 7.1 to 7.3 for the details.

While recognising that nature of things is ultimately to fail, the focus of efforts should be in ensuring that the risk level of the latter occurring is acceptable during the planned design working life of structures. Of course, it is not possible to know exactly if failure will occur in the future. As a result, there are large uncertainty levels and complexities involved in developing plans for the future. In this context, decisions tend to be short-term intended: regulatory authorities may be reluctant to impose more severe requirements to legal documents, investments focus more on replacing and renewing as needed rather than modernising infrastructure and expenditure takes place in response to a crisis rather than proactively planning and managing infrastructure assets such as temporary structures. In addition, the mainstream overarching management objective has been to operate infrastructure systems at near maximum capacity. This, however, causes systems to be less resilient against anticipated or unknown hazards during the systems lifespan; optimisation for one set of conditions creates vulnerabilities to changes in those conditions. One should always consider that "failure is inevitable, the question is when".

The economic risk from natural or man-made causes perceived by stakeholders often represents a small percentage of the capital at risk. This might be often the case, but does not take fully into consideration the follow-up consequences, e.g. loss of life, very high economic losses when compared with potential investment costs, and impacts on GDP due to disruption of service, reputational damage, contractual penalties and the potential for litigation (Guthrie & Konaris, 2012).

Quality is defined in ISO 9000 as the degree to which a set of inherent characteristics of an object fulfils a need or expectation that is stated, generally implied or obligatory (ISO, 2015). In the latter definition, "object" means "anything perceivable or conceivable" (ISO, 2015). Quality management can include establishing quality policies and quality objectives, and processes to achieve these quality objectives through quality planning, quality assurance, quality control, and quality improvement (ISO, 2015).

The definition of Quality Assurance (QA) is defined to be those activities which are focused on providing confidence that the quality requirements will be fulfilled. It focuses on ensuring that defects are prevented and that potential defects in new products or processes are eliminated. It requires regular audits of performance and good quality management producing high quality documentation at all stages.

Quality Control (QC), on the other hand, encompasses the activities which are focused on fulfilling the quality requirements of the product or process. It therefore aims, identify and remedy any defects which might occur within the process or project.

In this Chapter, the most important quality management topics related to temporary structures are presented and discussed. The information presented can be used as measures to improve the management of temporary structures and to ensure that QA/QC is maintained. Additional information is provided in various bibliographic references (BSI, 2011, 2016; CIP, 2011; CIRIA, 2015; fib, 2009; HSE, 2010, 2013, 2015, Knoll, 2011, 2013; Rosignoli, 2013; The Concrete Society, 2012; TWf, 2012b, 2014). It is expected that the reader will acquire knowledge on the following topics:

1. Control of design.
2. Control of site activities.
3. Management of the project.

Figure 1. Activities and responsibilities in falsework design and construction for standard (top) and special (bottom) projects recommended in CIP (2011). ©2016 Construction Industry Publications. Used with permission

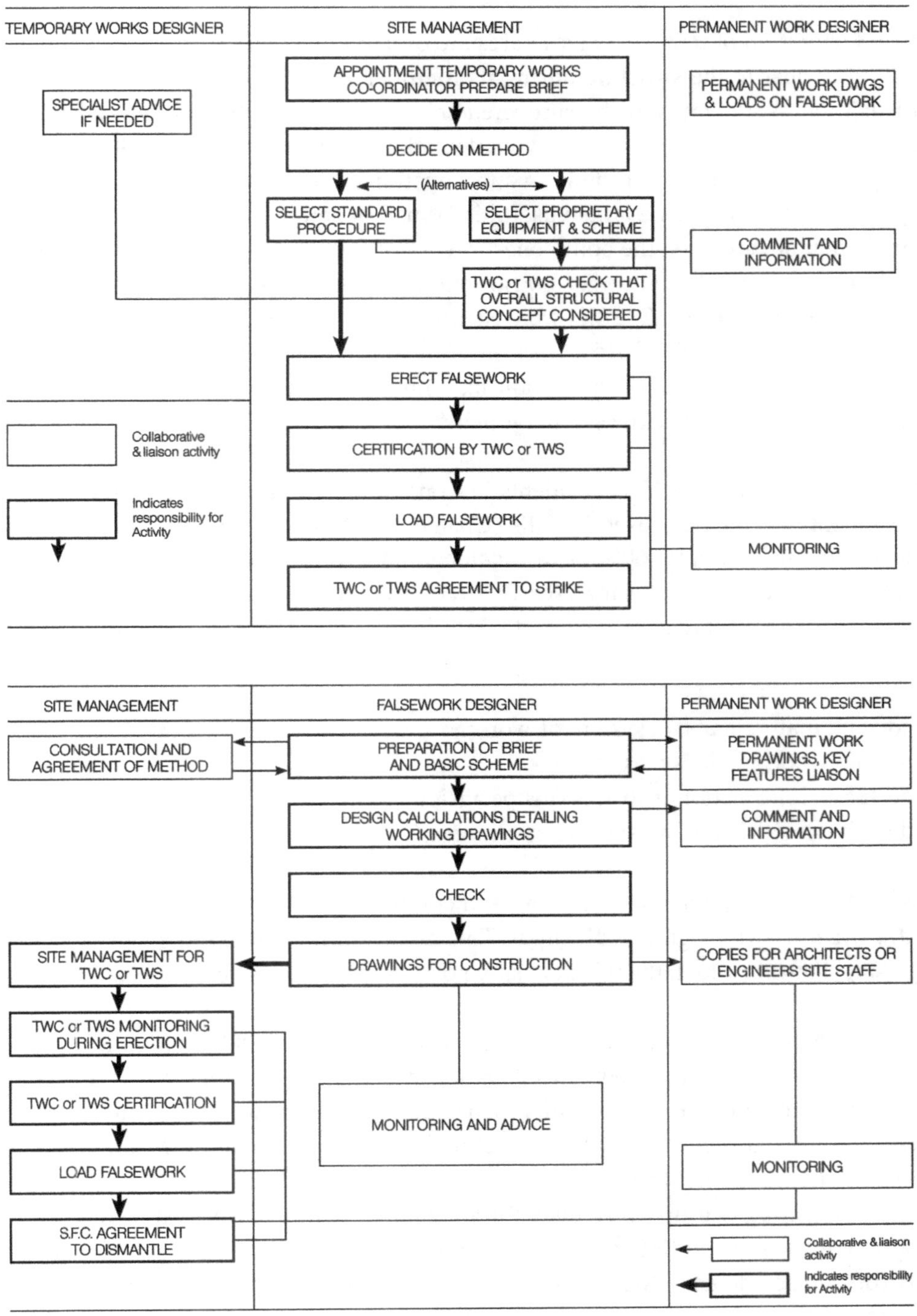

8.2 MANAGEMENT OF TEMPORARY STRUCTURES PROJECTS

8.2.1 Health and Safety

Safe design and operation of temporary structures are of paramount concern to the health and safety of operatives using or near to the structure during its lifetime. The websites of the European Agency for Safety and Health at Work (https://osha.europa.eu/en), the UK's Health and Safety Executive (http://www.hse.gov.uk/), Safe Work Australia (www.safeworkaustralia.gov.au), Hong Kong Labour Depart (www.labour.gov.hk/eng/osh) and of the USA's Occupational Safety & Health Administration (https://www.osha.gov/index.html) provide ample useful information and guidance.

Falls from height are amongst the commonest causes of fatality and injury. For example, within the period 2003-13 there were 401 fatalities on construction sites in Australia, this being 15% of all fatalities at work in the country. Of these 112 (28%) were caused by falls from ladders, mobile stairs or ramps and scaffolding (Safe Work Australia, 2015). Back injuries followed by injuries to fingers, thumbs, knees and shoulders were the commonest form of non-fatal injury and nearly half of the injuries occurred to tradesmen such as carpenters, bricklayers and joiners, 37% of the injuries were caused mobile pant and transport

Similarly, in 2004 there were 3833 accidents in Construction in the Hong Kong SAR of which 47% of the fatal accidents (22) were caused by falls (Chan et al., 2008). It is notable that this research also showed that of the small sample, 60% of the accidents involved workers who had less than 10 years experience in the Construction industry. Nearly 30% of the fatal accidents occurred in falls from less than 5 m in height. The researchers suggested the following procedures in order to reduce the risk of falls from height:

- "Provide and maintain a safe system of work;
- Provide safe working platforms;
- Provide safety information/instruction/supervision;
- Provide suitable fall arresting systems/anchorages".

An algorithm for the automated monitoring and control of fall hazards was developed by Navon & Kolton (2007), together with a graphical output. This model allows fall hazards to be identified as early as the design stage, during which there is normally no reference to the safety aspects of the construction stage.

In Finland, an interesting approach to safety was used as they found that just monitoring sites for work and safety risk factors did not engender full safe practice. Therefore, over a three year period the safety authorities made a competition for best quality management practices with the results published nationally. The results were that there was a 40% reduction in falls from height, 50% fewer site deficiencies noted, 60% less deficiencies in electricity and lighting and 46% reduction in the non-use of protective equipment and risk taking (European Agency for Safety and Health at Work, 2004).

In 2003, the UK Health and Safety Executive produced a report describing the results of an investigation into 100 site accidents (HSE, 2003). Nearly 45% were caused by falls from height, 17% being trapped by something overturning or collapsing, 17% by being struck by a moving vehicle and the rest by several different causes such as being struck by a moving object. The report produced recommendations, still valid for today. Namely that:

- Safety is the responsibility from everybody who is associated with the design, execution or management of a project.
- Good communications are vital in ensuring safe work.
- Attention should be given to the selection and use of equipment, tools and materials with safety always being considered.
- A great improvement in standards of site layout required with contractors and sub-contractors raising expectations about what constitutes good and acceptable practice.
- Personal protective equipment should be appropriate for the job being undertaken. If not, it risks being not used.
- The construction industry should be benchmarked against other industries to bring its safety up to equivalent levels (at the time of the report, 33% of all work related fatalities occurred in the construction industry).
- Risk assessment and management must be incorporated in all management procedures.
- "Bureaucracy" and "safety" are different. Improved safety does not imply extra bureaucracy.
- Safety does not cost normally but if it does, regulatory bodies should impose requirements so that everyone is on an equal basis.

The report led to a guide which is available to anyone on www.hse.gov.uk/ (HSE, 2006). It is designed for small contractors who may not be fully aware of all the principles of good quality management practice. Emphasis is made on the importance of proper preparation in planning and organising the work ensuring that health and a safety is considered from the job's beginning. The guide emphasises that the most frequent causes of accidental death and injury are:

- Falls from height - prevented by toeboards and guardrails which are often missing or incorrectly positioned and ensuring platforms are easy to access and in safe positions.
- Mobile plant - typically, injuries occur by reversing vehicles or plant overturning (the first author observed a forklift truck overturn outside a university department of construction!).
- Falling material and collapses - demolition and disassembly of falsework and scaffolds may cause collapses if instability is not analysed and prevented (see the Milton Keynes scaffold collapse in Chapter 7, Section 7.5.2). Injuries also occur when material either rolls or is knocked off platforms or falls when being raised (a head of a university construction department told the first author that he prevented a major material fall from a lifting process when the operatives to speed the process up had made a hook from a piece of reinforcing rod which he observed becoming straight when the material was being lifted).
- Electrical accidents - these occur when operatives use unsafe and untested equipment or when overhead or buried cables are cut.
- Trips – the commonest cause of injury with (in 2006) over 1000 major injuries reported and caused by ineffective management of access routes such as footpaths, stairways and corridors.

A similar set of guidelines was produced in Australia (Safe Work Australia, 2008).

Guidelines were also produced in Poland for the safe use of scaffolds (Błazik-Borowa & Szer, 2014). In particular, they emphasized:

- Loading of platforms above their fixed capacity is not allowed;
- Workers must not be allowed to gather on platforms;
- Climbing trestles, stringers, transoms and scaffold rails is forbidden;
- Tools must not be left at the edge of a platform;
- People on platforms were forbidden from making sudden moves, bending over guardrails, placing materials and tools on one side of a scaffold;
- Workers must use appropriate safety equipment when operating at heights;
- Employees must have medical examinations to enable them to work satisfactorily;
- Inspections of scaffolds must occur after adverse weather conditions such as a storm or wind in excess of 10 m/sec.

8.2.2 Planning and Procurement

Every project should start with a clear definition of the objectives and the requirements needed to achieve them. Relevant stakeholders should bring to the table their expertise and together decide the methods needed to meet the project objectives (SIA, 2003).

There are several stakeholders directly or indirectly responsible for temporary structures: researchers, designers, producers, clients, consultants, insurers, contractors, sub-contractors and finally workers. The installation or assemblage, operation and dismantling of temporary structures are usually performed by a specialised sub-contractor, selected during the construction phase, which normally follows the construction method specified in the tender documents.

The temporary structures project must fulfil the design philosophy and requirements indicated in the detailed design project of the permanent structure. To this end and taking into account that every bridge and building construction project has inherent specificities, the temporary structures project is generally a specially developed project performed by the sub-contractor, often expert design companies, using proprietary equipment. Therefore each project is different; and although there usually are similarities from one project to another, each project must be carried out on an individual basis. Many aspects of a new project will turn out to be different from those of previous projects, and there are often requirements and other considerations which are different from any previous job.

It is on the planning and design phases where correct and rational decision-making is more efficiently undertaken. During these phases, risks can be identified, mitigated or reduced, and decisions are easier to implement as the cost of changes and error corrections is smaller than the corresponding cost during construction or use of the infrastructure. Appropriate plan and design procedures are critical to achieve lean and safe construction and use.

Responsible contractors appraise a set of feasible construction methods and evaluate costs and risks before making a selection. The criteria for choosing the type of temporary structures are manifold. For example for bridge construction it involves: the geometrical characteristics of the superstructure, namely the layout of the bridge (plan and elevations), deck type and its material as well as the height of the piers, the length of the bridge and of each span and the spans uniformity, to the ground properties, bridge context (deep valleys, crossing a waterway or a road, open field or urban area, ease of access, size of space available, etc), labour costs, logistic issues such as availability of materials and equipment, designer and contractor expertise, etc. Pre-design aids for concrete bridges are presented in fib (2000) and The Concrete Centre (2008). Ayaho, Hideyuki, & Hideaki (1997) proposed a system for selecting the erection method for steel bridges.

In this context, different types of BCE are tailored to be used in the construction of complex bridges such as cable-stayed bridges, suspension bridges and also girder bridges (the majority) when the spans exceed 40 to 60 m or when the number of spans is large, or the height of the piers exceeds 30 m, when the ground properties are weak or when the bridge is built in urban areas or in difficult access areas. In Chapter 2, a possible range of applications of bridge temporary structures based on the material of bridge decks: steel, composite and concrete is presented.

The detailed design of any structure can only be undertaken when all the data concerning the physical requirements of the location are collected. A visit to the future location of the structure by the engineer and the designer is an essential stage of the project. The main information to be collected on the spot includes (Chen & Duan, 2013):

- "**Topography**: It is useful to have a topographic account and a plan view of the site indicating the possibilities of access, as well as the available areas for the installation of the construction site, storages, and so on.
- **Hydrology**: In the case of a bridge crossing of a river, the frequency and importance of floods, solid debit, and possible carriage of floating objects susceptible to striking piers must be known. Apart from impacts, the greatest danger is scour effects. It is advisable to estimate the potential height of scour in the neighbourhood of the supports and to limit as much as possible the number of supports in the aquatic site.
- **Geological and Geotechnical Data**: These data, which concern the nature and properties of the ground and the foundation, without forgetting the knowledge of the level of the groundwater, are very important. Their collection constitutes a decisive stage for the choice of the type of foundation. An insufficient study can mean modifications of the project or a very expensive extension of the already executed foundations will be required if the ground does not have the expected properties. Geotechnical tests are generally rather expensive and the designer has to organize the tests according to the size and the importance of the construction works. The designer has to make them at first for the envisaged location of the supports and collect the test results, which would already have been made in the neighbourhood.".

8.2.3 Design

Design is divided in four phases: design brief and statement, conceptual design, detailed design and site support. Quality during all these four phases should be guaranteed by the use of rules specified in design codes, standards, and by engineering good practices. In Europe, the reference design framework is made of the Eurocodes, execution and test standards, material and product specifications (procedures such as those involved in CE marking). The Eurocodes strictly cover design and not site quality management, unlike other countries such the USA (ANSI, 2011) where construction procedures are emphasised. Indeed, in the UK, an old design code BS 5975 (BSI, 2011) which is based on Allowable Stress Design is still in use with a latest appendix dated 2011 as it covers site quality management, such as the appointment of a Temporary Works Coordinator.

Quality assurance and quality control (QA/QC) during design is essential. Fröderberg & Thelandersson (2013) carried out a round robin experiment where 14 different designers were asked to determine the forces at some particular elements of a reinforced concrete building. Analysing the results, they found a very large variability (approximately 20% COV) between results. They attributed it primarily to a lack of knowledge. They also observed that structural design is often based on past experience rather than on the actual calculation model results. They concluded that modern design codes in general cover the design uncertainties of complex structural systems inadequately. Improvement through collaboration between two or more design firms in producing a design is recommended by Knoll (2013) as the different independent reviewers may find each other's errors more easily.

The influence of resistance related, model and material, design assumptions in temporary structures has quite a large range of degrees. Errors in the numerical model (e.g. mesh size, element types, loading sequence and many other variables in nonlinear analyses) need to be estimated and evaluated using proper criteria. If they fail to pass the acceptance criteria, they must be corrected.

It is usual to perform error sensitivity analyses on structural performance considering only one type of error each time. Based on these analyses, it may be concluded that a single type of error has no significant influence on the structural performance. However, this conclusion misses an obvious fact: it is not each individual importance that really matters, but the interaction between multiple types of errors. For example: two types of errors may not lead separately to a large deterioration of the structural performance but when both are present they may interact and the combined effect can be catastrophic. Therefore, one must interpret the results of such kind of analyses with care (see for example, the description of the effects on global safety factors caused by defect combinations for a simple access scaffold in Chapter 7, Section 7.3).

Therefore, simulation governance is essential in any design. This includes verification and validation of design, uncertainty quantification, and independent design reviews. Ideally, the designer should be able to hand the analysis results accompanied with a statement similar to "Here is the analysis model results with the associated bias of X and confidence band of Y applied. I am 90% confident that if we ran a test, the result would lie between these two values".

Then, there is the subject of actions related design assumptions. In particular, concerning actions that were unknown or not accounted for due to design errors or because code rules are wrong/incomplete or simply because the design code accepted the risks. Depending on the sensitivity of structures against design assumptions, different levels of complexity could be assigned to design analysis. See Chapter 5 for guidance on providing adequate structural robustness against unidentified hazard events.

Here the meaning of "less sensitive" deserves a comment. Under the design codes framework, the meaning of "less sensitive" relates to the structural performance considered as the difference between the imposed actions and structural resistance, in general associated with the first structural element/component failure. Therefore, a structure that is "less sensitive" to design assumptions only implies that it will still satisfy design code requirements (resistance > action effect) despite errors during design and execution, although the follow-up consequences after first failure (including the type of collapse mode) may be different than the one assumed (or "chosen") during design.

Under existing design codes framework, there are no provisions to assess the sensitivity of the structure to damage progression against design assumptions and execution quality levels. Therefore, rules to accommodate this situation should be developed and introduced, e.g. design rules including structural robustness and execution/inspection classes linked with consequence classes. In Chapter 5 such a scheme is presented.

8.2.4 Assembly, Operation and Disassembly

Assuming the design is done correctly, translating the behaviour considered during design to the behaviour of the built structure is an even a more evolving task. The latter requires the coordination between the designer, owner of the work, contractor, material supplier, workers, QA/QC staff, etc. This is difficult, but steps can be made towards it by attaching supervision/inspection classes to consequence classes. See Chapter 5 for a proposal of such a scheme.

The assembly method, agreed by the designer and the contractor prior to commencing the works, should guarantee that the intended layout of the structure specified in the design documentation is verified.

It must not be forgotten that temporary structures are by-and-large one of the most common components of construction projects. They are reused many times, including during the construction of a given bridge, requiring multiple assembly, operation and dismantling cycles. These repetitive and routine activities can cause loss of attention and contribute to accidents. Additionally, some construction workers are chance makers, i.e. accident makers. One should not forget that the primary responsibility for ensuring health and safety should not only lie with those who create risks but also with those who work with them. Finally, due to being reused several times, deterioration processes such as corrosion, local damage or accumulation of geometrical imperfections will occur and limit the behaviour and strength of the system. It is essential that during assembly, damaged element are identified which must be reinforced or replaced.

During operation of temporary structures, a structural health monitoring (SHM), which alerts the contractor/designer/owner of potential problems, should be implemented. The designer should assist in the development of the monitoring program. This is particularly relevant for the wind action during all phases of the project.

SHM is the process of determining and implementing a strategy to determine the faults that occur (Farrar & Worden, 2007, 2012). It typically involves the installation of different types of sensors at predefined locations of the temporary structure to measure physical variables considered relevant to monitor and understand the structural behaviour. Loads, internal forces, material stresses and strains, element's displacements are variables commonly measured in a SHM programme. SHM is not limited to structural variables but also environmental variables can be monitored such as wind velocity and air temperature. Recent sensor technologies offer the possibility of wireless monitoring using self-powered equipment, ideal for remote real-time long-term monitoring.

Examples of sensors to measure displacements are inclinometers, tilt meters, linear variable differential transformers (LVDTs), fibre-optic Bragg grating (FBG) sensors, land survey methods and GPS systems. Examples of sensors to measure material's stresses and strains are strain gauges, FBG sensors and piezoelectric sensors. Examples of sensors to measure loads are load cells and piezoelectric sensors. Finally, wind velocity can be measured using anemometers.

Farrar & Worden (2007, 2012) list the "axioms" of structural health monitoring. They are:

- "All materials have inherent faults or defects". Faults cause the material to fail but the material may have defects which allow the material to perform but in a sub-optimal manner.
- "The assessment of damage requires a comparison between two system states". Different procedures may be used to determine the extent of damage. Farrar & Worden (2007, 2012) describe the process of using an FE system to determine the damage. Firstly, a model is created of a perfect

structure and then imperfections introduced until a match with a tested structure achieved, thereby enabling the damage to be ascertained.

- "Identifying the existence and location of damage can be done in an unsupervised learning mode, but identifying the type of damage present and the severity can generally only be done in a supervised learning mode". Quoting the example above with an FE model, it is possible that different damage patterns may simulate the same approximate test and only close examination may determine which state models the true damage.
- "Sensors cannot measure damage. Feature extraction through signal processing and statistical classification is necessary to covert sensor information into damage information". The authors point out that for example sensors can never determine stress directly but only strain and if stress is required the stress-strain relationship must be used to find the stress.
- "Without intelligent feature extraction, the more sensitive a measurement is to damage, the more sensitive it is to changing operational and environmental conditions". For example, measurements were made on a damaged plate at two different temperatures. The temperature difference produced different stress patterns which affected the results.
- "The length and timescales associated with damage initiation and evolution dictate the required properties of the Structural Health Monitoring System (SHM)". For example, the monitoring system determining the effects of a concrete pour on a formwork/falsework structure will be different depending upon the rate of pour and the time over which the pour takes place.
- "There is a trade-off between the sensitivity to damage of an algorithm and its noise rejection capacity".
- "The size of damage that can be detected from changes in system dynamics is inversely proportional to the frequency range of excitation".

These axioms define the principles required to enable an appropriate SHM system to be designed. Farrar & Worden (2007, 2012) present four reasons for using SHM. They are:

- Evaluating the life-safety and economic justification for monitoring;
- Determining the cases of damage which are most concern in the system being investigated;
- Determining the conditions, both environmental and operational, under which the system functions;
- Determining the limitations under which data can be acquired.

SHM also offers the possibility to obtain information concerning potential structural vulnerabilities ahead of time enabling engineers to make informed and timely decisions with respect to preventive repair, retrofit, replacement or limitation of use of a temporary structure.

The above set of reasons implies that monitoring should be used for cases where there is a severe risk of injury/death to operatives and others in the vicinity of the construction and the economic consequences of it.

Farrar & Worden (2007, 2012) go on to point out challenges to the use of SHM. Firstly, local damage may not be detected in a global monitoring system which could affect the overall performance of a system. Secondly, the electronic system set up to monitor the behaviour of the structure must be defined before placed in situ. This requires a full understanding of the potential causes of damage so that the monitoring is effective. Thirdly, these systems may be required to operate for a considerable period of time in an unsupervised learning mode as data from damage studies on the system is not available. Finally, there is a difficulty in persuading management that SHM provides an economic saving over current procedures.

Further information about health-monitoring systems can be found in the books by Farrar & Worden (2012) and Webster & Eren (2014).

SHM systems are relevant to all types of temporary structures, in particular during the operation phase. For example, had a SHM system been installed at Milton Keynes the potential for collapse would have been determined and correct procedures adopted to prevent the collapse. See Chapter 7.

A monitoring method of the axial forces and lateral displacements of building shoring systems was suggested by Huang et al. (2000). The allowable limit values for these two variables as well as the locations of the monitoring equipment were proposed. Strain gauges were recommended for measuring the axial forces and LVDTs for measuring the lateral displacements. The strain gauges and LVDTs should be installed at all the critical locations. If the axial forces exceed the design values obtained from a calculation model then the actual behaviour of the system deviates from the one considered in the design phase and the principles of the operation must be reassessed, both in terms of design and use of the temporary structures. The lateral displacements constitute a finer control variable since they can form the basis of an early warning system of the performance of the system.

Zhang, Wang, & Song (2015) developed an SHM system to determine the looseness in Cuplok® falsework systems. They used three stress PZTs (Piezoelectric lead zirconate titanate transducers) and three shear PZTs mounted on a standard and a ledger, respectively. Using wavelet analysis they showed that the looseness in the cuplok joint (connecting the ledger to the standard) reduced as the connection was tightened. This procedure could be used on a site where it is relevant to monitor the looseness of critical joints.

The disassembly operations of temporary structures should be planned with equal care to those used for the design of the permanent works. Notice that disassembly of temporary structures may require possibly assembly of other temporary structures.

8.2.5 Inspection, Supervision and Maintenance

The purpose of inspection and supervision is to ensure that temporary structures are assembled in accordance with the specifications and drawings provided by the designer, and that this continues throughout operation. The importance of early supervision and checking cannot be overemphasized as basic faults can usually be remedied at a low overall cost.

Inspection should typically occur at the following stages:

- Before temporary structures being taken into use for the first time;
- At regular intervals not exceeding seven days since the last inspection, with a report being prepared for site management after each inspection;
- Before loading of the temporary structures, and for special structures (elements), with intermediate checks during the loading sequence;
- After substantial addition, dismantling or any other alteration;
- In advance of a critical event (e.g. some hours before a strong gale wind is forecasted);
- After adverse weather conditions, e.g. high winds, etc.;
- after any event likely to have affected its strength or stability.

For all types of temporary structures, it is advantageous to appoint a specialist assigned by the equipment manufacturer or a specialised company to carry out both the inspection and supervision.

It is Recommended the establishment of inspection permit documents which should be signed by all parties involved in the project, and contain the main findings of the inspection, the assessment criteria used and the decision. For example: "Permit to Load" (before loading), "Permit to Proceed" (after loading) and "Permit to Dismantle" (before disassembly). This process will be repeated as required until completion. Note that this procedure is a requirement in ANSI/ASSE A10.8:2011 (ANSI, 2011).

A non-exhaustive list of inspection points is listed below (additional points are listed in BS 5975):

- Check if temporary structures are assembled correctly: check for general layout, jacks' extension lengths, number and spacing of vertical and of horizontal members, amount and configuration of diagonal bracing, loose connections, use of correct accessories (e.g. structural pins), type of foundation ground, alignment, level and type of foundation elements, load eccentricities, type and positioning of bearings, type of hydraulic jacks. Perform a comparison with the information provided in the drawings and in other design documentation;
- Check if the elements specified in the design documentation are used (e.g. type and geometry of main elements, steel grades, type and dimension of bolts, etc.);
- Check for the use of "copycat" elements usually of inferior mechanical properties, reject if found or alternatively send to test properties and assess if they comply with the design requirements (the first author assisted in the analysis of "look-alikes" components of a well-known proprietary scaffold system manufactured in the Middle East. It was found that the "look-alikes" components were up to 10% weaker than the genuine components);
- Check for damages in elements (tubes, girders, couplers, baseplates, bearings, bolt threads, welds, etc.), namely regarding corrosion, distortions, eccentricities, cracks, dents, etc.;
- Check that welds have not been "missed" or have not cracked due to brittle fracture. Couplers, baseplates should be checked to ensure the integrity of the welds;
- Check if rejected elements are not reused;
- Check the presumed bearing value of soil against the design value. Factors such as estimated settlement and changes in ground water level must also be assessed;
- Check gaps between temporary structures and permanent structure and between temporary structures and supporting structures;
- Check if steel beams subject to point loads should have web stiffeners, and if they are accurately positioned under point loads;
- Check if hydraulic jacks have the required capacity in both pressure and stroke extension;
- Check if electrical equipment (e.g. power generator, lifting equipment) is adequate and is working properly;
- Check if SHM equipment is positioned correctly and is working properly;
- Check if corrective measures are done fully and correctly;
- Check abnormal deformations of elements, ground settlements during loading;
- Check if temporary structures are disassembled in the correct sequence, requiring possibly assembly of auxiliary temporary structures.

At the end of an inspection, a tag indicating status and capacity of the temporary structure, name and signature of the qualified person and date of inspection must be present on the structure.

Concerning maintenance, general guidance with respect to the most common storage and handling systems are specified in The Concrete Society (2005). In particular, provisions are specified to control risks that may arise if, for example, products are heavily oiled and/or bundled. It is recommended that steel and alloy elements be separated so that they are not used inappropriately. All loose material in corroded steel elements (e.g. tubes, girders, etc.) should be cleaned. The dimensions of the non-corroded geometry of the elements should then be measured and checked against the minimum requirements set in product standards. For example, the external diameter of scaffold/falsework steel tubes should conform to the requirements of BS EN 39 (BSI, 2001). In all cases, maintenance should ensure that the continuous reuse of materials and products allows the adequate margins of structural safety to be preserved and the design objectives to be attained.

8.2.6 Temporary Works Coordinator

In Great Britain every construction project which is likely to involve more than 30 days or 500 person days of construction work is legally bound to fulfil the requirements set in The Construction (Design and Management) Regulations 2015 (CDM2015) (HSE, 2015). The CDM2015 Regulations specifies the duties of all stakeholders involved in a construction project with respect to planning, management and monitoring of health, safety and welfare in construction projects. This document also specifies the responsibilities of the duty-holders. In particular, the clients shall appoint a CDM coordinator as their key adviser who will assist them with their duties during the construction project. Additionally, the construction phase cannot start until the principal contractor has prepared a construction phase plan (a document recording the health and safety arrangements, site rules and any special measures for construction work) (HSE, 2015). A temporary structure must be of such design and so installed and maintained as to withstand any foreseeable loads which may be imposed on it, and must only be used for the purposes for which it is so designed, installed and maintained (HSE, 2015).

BS 5975 (BSI, 2011), recommends the appointment of a Temporary Works Coordinator (TWC) to coordinate and supervise the activities of all concerned and to ensure the works are brought to a safe conclusion. Similarly, ANSI/ASSE A10.8 (ANSI, 2011) requires the appointment of an experienced supervisor to have overall control of scaffold structures. Additionally, Temporary structures Supervisors (TWSs) can be appointed to assist the TWC with their duties. Checking and inspection by competent TWSs should be a continuous process, starting with the materials to be used, the foundations, and progressive inspection and checks as the structure is erected. Leaving such checks until the temporary structure is complete is useless. Errors in the materials used, in the foundations and in the assembly procedure will be impossible to correct without dismantling. Possible checklists have been presented above and are also provided in CIP (2011).

BS 5975 (BSI, 2011) requires that (see also Section 8.2.8):

- "It is essential for the TWC to be competent and to have relevant up-to-date training and both the qualifications and the experience appropriate to the complexity of the project. The appointment of the TWC should be made known to all concerned." (Clause 7.1.2).
- "It is essential for the TWSs to be competent and to have relevant up-to-date training and both the qualifications and the experience appropriate to the complexity of the project." (Clause 7.3.2).

BS 5975 (Clause 7.2) also states that the TWC must (BSI, 2011):

- Be the first point of contact between designer and all site staff;
- Be responsible for ensuring that the designer's requirements for control of temporary structures is implemented on the site;
- Be responsible to a designated individual for seeing that all the temporary structure is erected in conformance with drawings and specifications;
- Have the authority to stop work if it is not in conformance with specifications and have overall responsibility for ensuring that loadings are as specified;
- Record the responsibilities delegated to TWSs when appointed;
- Ensure that any risks which were identified at the design stage or in the assumed methods of construction, or any loading constraints identified by the designer of the permanent works are included within the design brief;
- Ensure that a safe temporary structures design has been produced;
- Ensure that a design check is carried out by a second person, not previously involved in the design for competency, structural adequacy and compliance with the brief;
- Register and record all drawings, calculations or any other relevant documents for the final design. It must also include records of any agreed changes to the original drawings including checks that the changes do not affect the safety of the structure;
- When it has been confirmed that the permanent structure has gained adequate strength and/or stability, ensuring that a permit to unload (take out of use) the temporary structures is issued by either the TWC or the TWS;
- Ensuring that safe dismantling procedures are in place before authorising disassembly;
- Ensure that any relevant information for the health and safety file is transmitted to the CDM coordinator.

8.2.7 Documentation and Communication

It is critical that the information is well documented, communicated, understood and followed by those responsible and working on the site. The scope and detail of the documentation should be commensurate with the complexity of the project, the nature of the hazards and the severity of risks.

A design brief should be prepared to serve as the starting point for subsequent decisions, design work, calculations and drawings. All concerned with the construction should contribute towards the preparation of the brief. The brief should include all data relevant to the design of the temporary structure. Below, the type of information that might be required for the preparation of the design brief is indicated (TWf, 2014):

- Appropriate drawings of the permanent works;
- Appropriate clauses from the specification for the permanent works;
- Statement of any requirement to design the temporary structures in accordance with a particular standard or guidance document, for example BS 5975 or BS EN 12812;
- Information on any significant risk associated with the design of the permanent works;
- The programme for the construction of the permanent works;
- The programme for the various phases of the design, design check, any external approvals required, and procurement and erection of the temporary structures;

- The timing for the removal of the temporary structures in relation to the ability of the permanent works to be self-supporting;
- Equipment and materials available for use in the temporary structures;
- Environmental information such as the location, altitude and topography of the site, the distance from the nearest sea, rainfall, water levels and current velocities;
- Site investigation data and reports relating to the areas under and adjacent to the foundations of the temporary structures; this should include information on all underground and over-head services;
- Loads that may be induced in the temporary structures by permanent works that have been completed, such as the application of staged post-tensioning, load re-distribution and any movements of significance including any settlements or deflections that can be anticipated from the permanent works as load is progressively increased;
- Any limitations stated by the designer of the permanent works on the position and extent of loads imposed by the temporary structures onto elements of the permanent works which have been constructed such as loads imposed by successive floors of multi-storey construction onto lower floors or loading of permanent foundations required to support the temporary structures.

The detailed design report should be a well structured document, containing an overview of the structure and how the structure was designed to be assembled and used, along with other design related assumptions and limitations. This document should also contain a section with detailed design calculations starting from the design codes used, the structural system definition, loads and load cases considered, reactions and internal forces obtained, design methods used and proofs evidence. The appendices should contain general and detailed drawings and the method statement. As a minimum the detailed design report should contain sufficient and transparent information concerning:

- Detailed drawings and specifications;
- Design loading with details of partial factors;
- Results of site surveys and soil investigations;
- Stability criteria, where relevant;
- Design load capacity of all members;
- Limits on positions of construction joints (if joints are to be in the final structure, otherwise may be left to the contractor) ;
- Lifting positions on members to be erected as single pieces and specification of maximum jack extensions to ensure stability of the lifting operation;
- Influence of post-tensioning on adjacent members of the temporary structures (e.g. falsework and BCE);
- Tying pattern required for temporary structures (e.g. scaffolds) and details of required tie capacity (if the designer is specifying these, otherwise contractor must supply) ;
- The safe maximum rate of placing of the concrete and a maximum safe height of local heaping;
- Specification of inspection levels for types of elements and criteria to be used in the quality control;
- Specification of critical phases that need prior approval;
- Specification of critical phases to monitor, which elements and quantities to monitor (internal to the system, e.g. forces and displacements, and external to the system, e.g. wind velocity), providing tolerance values and contingency response measures to be taken if the latter are violated;

- Measures to mitigate or prevent accidental actions, e.g. concrete jersey barriers placed protecting the structure against vehicular impacts;
- Recommendation of a maintenance scheme.

Method statements should explain the procedural steps to be followed in detail using both schematic and descriptive elements. They should address the particular needs of the site (foundation testing for example) and detail the planned sequences and methods of work relative to temporary structures including the erection, operation and monitoring manuals, highlighting the critical stages and key operations and verifications (e.g. maximum point loads, working wind velocities and geometrical tolerances). Method statements should be prepared in such a way that they enable supervisors and managers to ensure that persons on site are made aware of how the work is to be carried out, including the sequence of operations, the plant and types of equipment to be used and the precautions to be taken, as appropriate. Each method statement should be succinct and should form a single document, including site plans, annotated diagrams and a detailed programme for the work, in order to clearly communicate to those carrying out the work on site what is required. A logical order should be followed. It should be easy to understand, and agreed by and known to all levels of management and supervision. The use of photographic sequences has been found to be effective. Checklists for method statements of examples of temporary structures are provided in CIP (2011).

As well as construction plans, the design drawings should include details of the temporary structures removal operations, methods and sequences of removal, and equipment to be used. The drawings must show the size of all load-supporting members, connections and joints, and bracing systems. All design-controlling dimensions, including elements length and spacing must be shown (Blank, Blank, & Kondazi, 2014). Unfortunately, it is common for construction drawings to be insufficiently clear and instructions to be either missing or not definitive (Durkee, 2014).

The designer should communicate clearly to other stakeholders that the design process is not risk free and acceptable risks are involved.

A clear and efficient communication channel needs to be in place so that when any design changes of the temporary structure or the permanent structure occur, the interested parties are well informed and in advance so that they are able to carry out the necessary actions with time. If there are several contractors working a site, there is need for close cooperation to avoid the operations of one contractor jeopardize the safety and work of the others. It is recommended that in these cases, a single entity takes the responsibility for each task.

Any change that may affect the basis of the design should be communicated to the entity responsible for the design. Examples of such changes are:

- Change of sequence of loading;
- Change of permanent structure proportions;
- Placing net or full sheeted cladding that will affect the magnitude and distribution of wind load and need to be included in temporary structures design calculations;
- Stacking of building materials on temporary structures or on completed permanent works such that stability of the partially completed permanent work or the temporary structure could be affected (see the effects on the Milton Keynes Collapse, Chapter 7, Section 7.5).

8.2.8 Competence and Training

Competence is about being able to do the work safely, without endangering others, and meeting the legal health and safety requirements. A competent person is someone with sufficient knowledge of the specific tasks to be undertaken and the risks which the work will entail, with sufficient training, experience and ability to enable them to carry out their duties in relation to the project, to recognize their limitations, and to take appropriate action to prevent harm to those carrying out or affected by the work.

There are no definitive textbooks or manuals which define the correct procedure for all construction jobs. The construction particulars are usually dependent on each firm's experience, expertise, policies, and practices: i.e. their competence. The teams involved in the planning, design, operation, inspection and supervision should be competent in the area of temporary structures, with sufficient experience and appropriate knowledge, and should establish direct communication channels between them, together with clear and well defined requirements and responsibility of each task, and a list of the most relevant hazards and a specification of the appropriate measures to control them.

BS 5975 (BSI, 2011), gives recommendations and guidance on the procedural controls to be applied to all aspects of temporary structures. In particular, it recommends the appointment of a competent Temporary Works Coordinator (TWC) to coordinate and supervise the activities of all concerned, to ensure the works are brought to a safe conclusion. Guidance to assess individual competency is provided in Carpenter (2010), HSE (2010) and TWf (2012a):

- Sufficient engineering knowledge and understanding so that the individual is able to read, understand and implement all the requirements of specifications and drawings;
- Sufficient management and leadership ability to plan and manage people and resources;
- Independent judgement to be able to identify the limits of his personnel and the team's knowledge and skills;
- A sound knowledge of legislation, the hazards and safe systems of work and the ability to manage health, safety and welfare within the individual area of responsibility;
- The ability to communicate well with others at all levels, ability to discuss ideas and plans competently and with confidence that his judgement is correct;
- Experience appropriate to the complexity of the project;
- Knowledge of temporary structures procedures and the issues associated with the type of structures being used.

It is expected that all of those involved with temporary structures have sufficient training to enable them to perform. They should be required to attend training sessions at least once in every three years. Training should focus on the following objectives:

- Understanding design intent: construction sequence, monitoring of various parameters during construction;
- Understanding assembly, operation and maintenance requirements of the equipment;
- Understanding structure- equipment interaction to identify loading combinations to which structure will be subjected during each stage of construction.

Methods for assessment of competency include, but are not limited to:

- Minimum periods of "observed" experience - taken to be indicative of competent performance;
- Qualifications and training - used as an indication of the level of underpinning knowledge;
- Verbal or written examination of a person's knowledge and/or attitudes.

8.2.9 Quality Improvement

Table 1 provides a sequence of activities that improve the quality management of temporary structures and ensure that QA/QC is maintained. These should be implemented in the Plan, Do, Check, Act cycle as illustrated in Figure 2 (HSE, 2013). Whilst management procedures, method statements and risk registers are all important tools for mitigating safety related risks, it is also desirable to create a safety-aware culture whereby everyone on construction sites or other hazardous environments is looking out for risks and looking out for each other. Also, documented procedures are necessary, but not sufficient, for safety-aware behaviour. They are of no value if not enforced on site by experienced supervisors who understand the inherent risks and have the authority to take responsive action. The supervisor should have not just the right experience, but also the relevant authority (Sorensen, 2002).

Table 1. Various quality issues that may arise during the life of a structure and methods of preventing them or reducing/eliminating their effects, adapted from Canisius (2011)

Activity during the life of the structure	Type of quality issues that can affect structural performance	Methods of preventing occurrences and reducing or eliminating their effects
Conceptual design	Poor concept, giving a structural type sensitive to errors and poor quality	Employment of knowledgeable Engineers, preferably with experience
Detailed design	Erroneous calculations	Capable staff. Check calculations. Use verified software.
	Not consider or erroneously consider an important safety aspect	Capable staff who understand structural behaviour. Staff to be conversant with Codes to be employed. Staff to be aware when to seek help. Independent checking. Use validated software.
	Wrong assumptions on structural or material behaviour	
	Incorrect detailing	Capable staff. Checking.
	Erroneous notes in drawings	Capable staff. Checking.
Procurement	Poor control measures	Contracts to have adequate quality control measures and acceptance testing.
	Priority of costs over quality	Adequate funds.
	Inconsistent quality	Choose suppliers with adequate QC measures.
Assembly	Poor material quality on site	Test for quality. Reject poor quality. Adequate suppliers.
	Damaged elements	Inspect prior to and after construction. Reject/Repair.
	Incorrect setting out	Well trained staff. Checking prior to start of construction. Early discovery would reduce complexities of correction.

continued on following page

Table 1. Continued

Operation	Incorrect structural members	Check.
	Poor operation phasing and sequence	Adequate supervision and procedures. Well trained staff.
	Poor inspection	Adequate supervision and procedures. Well trained staff.
	Poor monitoring and control	Adequate supervision and procedures. Well trained staff.
	Overloading elements.	Adequate supervision and procedures. Load limit signs. Well trained staff.
	Contractor changing details without permission of designer. Designer not checking a contractor's changes.	Adequate supervision and procedures. Safety conscious staff.
	Inadequate site investigations	Adequate specifications. Adequate supervision and procedures. Well trained staff.
	Poor communication	Adequate specifications. Adequate supervision and procedures. Well trained staff.
	Changes that affect the structure	Awareness of workers. Awareness of contractors. Availability of as constructed drawings.
Maintenance	Damage to elements or their protective measures	Adequate periodic inspection and maintenance.
	Not maintaining regularly	Owner/Manager commitment to maintain.
	Improper repair	Adequate specifications. Adequate supervision and procedures. Well trained staff.
Disassembly	Bad demolition sequence, caused by defective understanding of structural behaviour.	Awareness of contractors. Adequate contractors. Availability of as-constructed drawings.

Figure 2. The Plan, Do, Check, Act cycle. Adapted from HSE (2013)

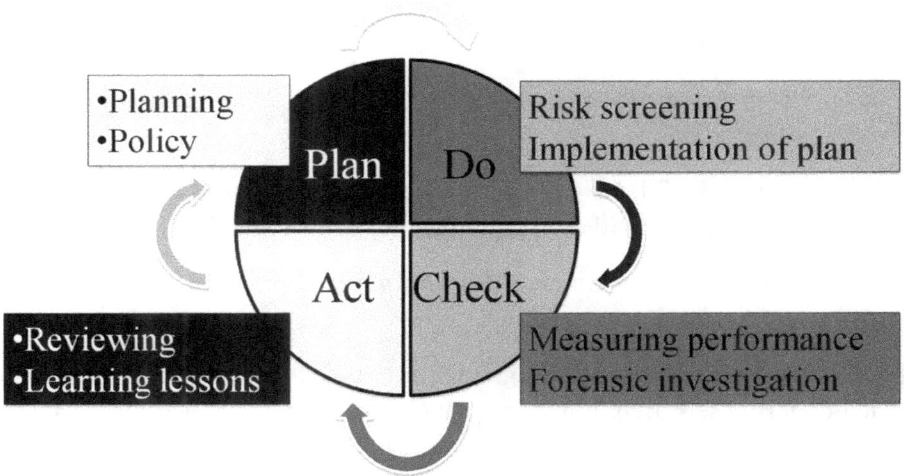

8.3 CONCLUSION

Good quality and safe construction practices can only be achieved if all the participants in a project work together. This Chapter has presented a set of methods and procedures to enable this to occur. In particular, emphasis was given to the following:

- The appointment of a supervisor or supervisors to ensure that temporary structures are designed, erected, use and dismantled safely;
- That complete records are made of all stages in the execution of a project, from design through to completion, and these are readily available on-site for inspection by any competent person;
- That Structural Health Monitoring should be considered for all large temporary structures projects;
- Checks must be made on designs by an independent second person before any erection is undertaken;
- All operatives involved with a project should be competent with respect to the assigned responsibilities and be adequately trained, and re-trained regularly, in their specialisms with an emphasis on the safety of themselves and others.

8.4 REFERENCES

ANSI. (2011). ANSI/ASSE A10.8:2011 – Scaffolding safety requirements. American National Standards Institute (ANSI).

Ayaho, M., Hideyuki, K., & Hideaki, N. (1997). Development of a knowledge-based expert system for selection of steel bridge erection method.*Proceedings of NEW TECH Lisbon 97: New Technologies in Structural Engineering*, 2, 203–211.

Blank, S., Blank, M., & Kondazi, H. (2014). Concrete bridge construction. In *Bridge engineering handbook: Construction and maintenance* (2nd ed.; Vol. 2, p. 18). CRC Press.

Błazik-Borowa, E., & Szer, J. (2014). The analysis of the stages of scaffolding life with regard to the decrease in the hazard at building works. *Archives of Civil and Mechanical Engineering*, *15*(2), 516–524. doi:10.1016/j.acme.2014.09.009

BSI. (2001). *Loose steel tubes for tube and coupler scaffolds – Technical delivery conditions*. London, UK: British Standards Institution.

BSI. (2011). BS 5975:2008+A1:2011 – Code of practice for temporary works procedures and the permissible stress design of falsework (3rd ed.). London, UK: British Standards Institution (BSI).

BSI. (2016).*Temporary works – Application of European Standards in design – Guide*. London, UK: British Standards Institution.

Canisius, G. (2011). Structural robustness design for practising engineers (COST Action TU0601 – Robustness of Structures). Brussels, Belgium: European Cooperation in Science and Technology (COST).

Carpenter, J. (2010). Falsework. In *ICE Manual of health and safety in construction* (pp. 205–213). London: Thomas Telford Ltd.

Chan, A., Wong, F., Chan, D., Yam, M., Kwok, A., Lam, E., & Cheung, E. (2008). Work at height fatalities in the repair, maintenance, alteration, and addition works. *Journal of Construction Engineering and Management, 134*(7), 527–535. doi:10.1061/(ASCE)0733-9364(2008)134:7(527)

Chen, W., & Duan, L. (2013). *Handbook of international bridge engineering*. CRC Press. doi:10.1201/b15520

CIP. (2011). Temporary works and falsework. In *Construction health and safety manual* (p. 16). Construction Industry Publications Ltd.

CIRIA. (2015). CDM2015 – Construction work sector guidance for designers (No. C755). London, UK: Construction Industry Research and Information Association (CIRIA).

Durkee, W. (2014). Steel bridge construction. In *Bridge engineering handbook: Construction and maintenance* (2nd ed.; Vol. 2, p. 50). CRC Press.

European Agency for Safety and Health at Work. (2004). *Building in safety. Prevention of risks in construction in practice*. Brussels, Belgium: European Agency for Safety and Health at Work.

Farrar, C., & Worden, K. (2007). An introduction to structural health monitoring. *Philosophical Transactions of the Royal Society of London A: Mathematical, Physical and Engineering Sciences, 365*(1851), 303–315.

Farrar, C., & Worden, K. (2012). *Structural health monitoring: A machine learning perspective*. Chichester, UK: John Wiley & Sons. doi:10.1002/9781118443118

fib. (2000). *Guidance for good bridge design*. Lausanne, Switzerland: International Federation for Structural Concrete (fib), Bulletin No. 9.

fib. (2009). *Formwork and falsework for heavy construction. Guide to good practice*. Lausanne, Switzerland: International Federation for Structural Concrete (fib), Bulletin No. 48.

Fröderberg, M., & Thelandersson, S. (2013). Modelling uncertainty and gross errors in preliminary design of buildings. In *Proceedings of 11th International Conference on Structural Safety and Reliability (ICOSSAR 2013)* (pp. 439–444). New York: Taylor & Francis Group.

Guthrie, P., & Konaris, T. (2012). *Infrastructure and resilience*. London, UK: Foresight, Government Office for Science.

HSE. (2003). Causal factors in construction accidents. London, UK: The Health and Safety Executive (HSE).

HSE. (2006). Health and safety in construction (No. HSG150). London, UK: The Health and Safety Executive (HSE).

HSE. (2010). The management of temporary works in the construction industry (No. SIM 02/2010/04). London, UK: The Health and Safety Executive (HSE).

HSE. (2013). Managing for health and safety (No. HSG65). London, UK: The Health and Safety Executive (HSE).

HSE. (2015). The construction (Design and management) regulations 2015 (CDM2015). London, UK: Health and Safety Executive (HSE).

Huang, Y., Chen, W., Chen, H., Yen, T., Kao, Y., & Lin, C. (2000). A monitoring method for scaffold-frame shoring systems for elevated concrete formwork. *Computers & Structures*, *78*(5), 681–690. doi:10.1016/S0045-7949(00)00051-1

ISO. (2015). ISO 9000:2015 – Quality management systems – Fundamentals and vocabulary (4th ed.). Geneva, Switzerland: International Organization for Standardization (ISO).

Knoll, F. (2011). Error proneness in construction. In *Applications of statistics and probability in civil engineering* (pp. 2745–2752). London: Taylor & Francis Group. doi:10.1201/b11332-405

Knoll, F. (2013). Avoiding errors through collaboration. In *Proceedings of 11th International Conference on Structural Safety and Reliability (ICOSSAR 2013)* (pp. 111–117). New York: Taylor & Francis Group.

Milojkovic, B. (1999). *Factors of safety for standard scaffold structures* (PhD Thesis). Oxford Brookes University, Oxford, UK.

Navon, R., & Kolton, O. (2007). Algorithms for automated monitoring and control of fall hazards. *Journal of Computing in Civil Engineering*, *21*(1), 21–28. doi:10.1061/(ASCE)0887-3801(2007)21:1(21)

Rosignoli, M. (2013). *Bridge construction equipment*. London, UK: ICE Publishing. doi:10.1680/bce.58088

Safe Work Australia. (2008). *National code of practice for the prevention of falls in general construction*. Commonwealth of Australia.

Safe Work Australia. (2015). *Work related injuries and fatalities in Construction, Australia, 2003 to 2013*. Safe Work Australia.

SIA. (2003). *SIA 260 – Bases pour l'élaboration des projets de structures porteuses*. Société Suisse des Ingénieurs et des Architectes.

Sorensen, J. (2002). Safety culture: A survey of the state-of-the-art. *Reliability Engineering & System Safety*, *76*(2), 189–204. doi:10.1016/S0951-8320(02)00005-4

The Concrete Centre. (2008). *Concrete bridges: The benefits of concrete in bridge design and construction*. London, UK: The Concrete Centre.

The Concrete Society. (2005). *A guide to the safe transportation of formwork and falsework equipment*. London, UK: The Concrete Society.

The Concrete Society. (2012). *Formwork: A guide to good practice* (3rd ed.). London, UK: The Concrete Society, Report CS30.

TWf. (2012a). Competencies of the TWC (No. TW/11/027). London, UK: Temporary Works forum.

TWf. (2012b). Design management of temporary works (No. TW/12/006). London, UK: Temporary Works forum.

TWf. (2014). Clients' guide to temporary works (No. TWf2014:02). London, UK: Temporary Works forum.

Webster, J., & Eren, H. (2014). *Measurement, instrumentation, and sensors handbook* (2nd ed.). London, UK: Taylor & Francis Group.

Zhang, L., Wang, C., & Song, G. (2015). Health status monitoring of cuplok scaffold joint connection based on wavelet packet analysis. *Shock and Vibration*, *2015*, 7. doi:10.1155/2015/695845

Nomenclature

NOTATION

In the book, matrices are denoted with upper case bold letters and vectors are denoted with lower case bold letters. Italic upper or lower case letters denote variables. Functions are written as *h*(). The use of *f*() is reserved for the probability density function, and *F*() denotes the cumulative distribution function. Roman capital P denotes the probability of an event.

UNITS

Except where specifically noted, this book uses the SI units of kilograms, metres, seconds, Pascals, Newtons, degrees and hertz (i.e. kg, m, s, Pa, N and Hz, respectively).

To Convert From	To	Divide by
	Length	
metre (m)	foot (ft)	0.304 8
millimetre (mm)	inch (in)	25.4
metre (m)	yard (yd)	0.914 4
	Area	
square millimetre (mm^2)	square inch	645.16
square metre (m^2)	square foot	0.092 903 04
square metre (m^2)	square yard	0.836 127 36
	Volume	
cubic metre (m^3)	cubic yard	0.764 555
cubic metre (m^3)	cubic foot	0.028 316 85
	Mass	
kilogram (kg)	pound	0.453 592 37
	Density	
kilogram metre (kg/m^3)	pound foot	0.138 255
	Force	
Newton (N)	pound-force	4.448 222

continues on following page

Units Continued

To Convert From	To	Divide by
Moment		
Newton metre (N.m)	pound-force foot	1.355 818
Newton metre (N.m)	pound-force inch	0.112 984 8
Pressure, stress		
kilopascal (kPa)	pound-force per square inch (psi)	6.894 757
kilopascal (kPa)	pound-force per square foot	0.047 880 26

ABBREVIATIONS

AASHTO: American Association of State Highway and Transportation Officials
ACI: American Concrete Institute
AGS: Association of Geotechnical and Geoenvironmental Specialists
ALARP: As Low As Reasonably Practicable
ANCOLD: Australian National Committee on Large Dams
AS: Australian Standard
ASCE: American Society of Civil Engineers
ASD: Allowable Stress Design
BCE: Bridge Construction Equipment
BCSA: British Constructional Steelwork Association
BPN: Bayesian Probabilistic Networks
BS: British Standard
BSI: British Standards Institute
CBA: Cost Benefit Analysis
CDF: Cumulative Distribution Function
CEA: Cost Effectiveness Analysis
CEN: European Committee for Standardization
CIP: Construction Industry Publications
CIRIA: Construction Industry Research and Information Association
CFD: Computational Fluid Dynamics
CPF: Cost of Preventing a Fatality
CPT: Cumulative Prospect Theory
CSA: Canadian Standards Association
EN: Euronorm (European standard)
ETA: Event Tree Analysis
FEA: Finite Element Analysis
FMEA: Failure Mode and Effect Analysis
FORM: First Order Reliability Method
FSM: Full Span Method
FTA: Fault Tree Analysis
GMNIA: Geometrical and Material Nonlinear Imperfect Analysis

HM: Her Majesty's Government (UK)
HAZOP: Hazard and Operability analysis
HSE: Health and Safety Executive
ICE: Institution of Civil Engineers
INBAR: International Network for Bamboo and Rattan
ISO: International Standards Organisation
IStructE: Institute of Structural Engineers
JCSS: Joint Committee on Structural Safety
LQI: Life Quality Index
LRFD: Load Resistance Factor Design
LSD: Limit State Design
MC: Monte Carlo Methods
MSS: Moveable Scaffold Systems
NA: National Annex to a European code
NASC: National Access and Scaffolding Confederation
NPV: Net Present Value
NWP: Numerical Weather Prediction
PASMA: Prefabricated Access Suppliers' and Manufacturers' Association
SAA: Standards Association Australia
SCOSS: Standing Committee on Structural Safety
SFEM: Stochastic Finite Element Methods
SLS: Serviceability Limit State
SORM: Second Order Reliability Method
TIA: Telecommunications Industry Association
TWC: Temporary Works Coordinator
TWf: Temporary Works Forum
TWS: Temporary Works Supervisor
UNISDR: United Nations Office for Disaster Risk Reduction
UK DfT: UK Department for Trade
UKOOA: UK Offshore Operators Association
ULS: Ultimate Limit State
USDOD: United States of America Defence Department
VPF: Value of Preventing a Fatality
WEF: World Economic Forum
WTA: Willingness to accept
WTP: Willingness to pay
fib: International federation for structural concrete

Compilation of References

AASHTO. (2003). AASHTO GSCB–2–M – Guide specifications for design and construction of segmental concrete bridges, with 2003 interim revisions (2nd ed.). Washington, DC: American Association of State Highway and Transportation Officials (AASHTO).

AASHTO. (2008). AASHTO GSBTW–1–M – Guide design specifications for bridge temporary works, with 2008 interim revisions (1st ed.). Washington, DC: American Association of State Highway and Transportation Officials (AASHTO).

AASHTO. (2008a). AASHTO CHBTW–1–M – Construction handbook for bridge temporary works, with 2008 interim revisions (1st ed.). Washington, DC: American Association of State Highway and Transportation Officials (AASHTO).

AASHTO. (2008b). AASHTO GSBTW–1–M – Guide design specifications for bridge temporary works, with 2008 interim revisions (1st ed.). Washington, DC: American Association of State Highway and Transportation Officials (AASHTO).

AASHTO. (2016). AASHTO LRFDUS–7 – LRFD Bridge design specifications, with 2015 and 2016 interim revisions (7th ed.). Washington, DC: American Association of State Highway and Transportation Officials (AASHTO).

AASHTO. (2016a). AASHTO LRFDCONS–3 – LRFD Bridge construction specifications, with 2016 interim revisions (3rd ed.). Washington, DC: American Association of State Highway and Transportation Officials (AASHTO).

AASHTO. (2016b). AASHTO LRFDUS–7 – LRFD Bridge design specifications, with 2015 and 2016 interim revisions (7th ed.). Washington, DC: American Association of State Highway and Transportation Officials (AASHTO).

Abdel-Jaber, M., Beale, R. G., & Godley, M. (2006). A theoretical and experimental investigation of pallet rack structures under sway. *Journal of Constructional Steel Research*, *62*(1–2), 68–80. doi:10.1016/j.jcsr.2005.04.008

Abdel-Jaber, M., Beale, R., Godley, M., & Abdel-Jaber, M. (2009). Rotational strength and stiffness of tubular scaffold connectors.*Proceedings of the ICE - Structures and Buildings, 162*(6), 391–403. doi:10.1680/stbu.2009.162.6.391

Acheson, D. (1990). *Elementary fluid dynamics*. Oxford, UK: Oxford University Press.

ACI. (2004). ACI 347–04 – Guide to formwork for concrete. Washington, DC: American Concrete Institute (ACI).

ACI. (2011). ACI 355.3R–11 – Guide for design of anchorage to concrete: Examples using ACI 318 Appendix D. Washington, DC: American Concrete Institute (ACI).

ACI. (2014). ACI 318–14 – Building code requirements for structural concrete and commentary. Washington, DC: American Concrete Institute (ACI).

ACI. (2014). ACI 347R–14 – Guide to formwork for concrete. Washington, DC: American Concrete Institute (ACI).

ACI. (2014a). ACI 318–14 – Building code requirements for structural concrete and commentary. Washington, DC: American Concrete Institute (ACI).

ACI. (2014b). ACI 347R–14 – Guide to formwork for concrete. Washington, DC: American Concrete Institute (ACI).

AGS. (2011). *UK specification for ground investigation* (2nd ed.). London, UK: ICE Publishing.

Ahmadi-Kashani, K., Brun, C., & Papanikolas, P. (2007). Design and construction of Megalorema Viaduct, Greece. *Proceedings of the ICE - Bridge Engineering, 160*(3), 117–127. doi:10.1680/bren.2007.160.3.117

AISC. (2010). AISC 360–10 – Specification for structural steel buildings. American Institute of Steel Construction (AISC).

Al-Qarra, B. (1999). The Dee Estuary Bridge: Control of geometry during construction. *Proceedings of the ICE -. Civil Engineering (New York, N.Y.), 132*(1), 31–39.

ANCOLD. (2003). Guidelines on risk assessment. The Australian National Committee on Large Dams (ANCOLD).

Ando, T. (2010). *Bayesian model selection and statistical modeling*. Boca Raton, FL: Chapman & Hall/CRC. doi:10.1201/EBK1439836149

André, J. (2008). *Behaviour and quality of steel telescopic props. Analytical, numerical and experimental investigation* (M.Sc. Thesis). Instituto Superior Técnico, Technical University of Lisbon, Lisbon, Portugal. (in Portuguese)

André, J. (2014). *Determination of the main parameters affecting the performance of bridge falsework systems* (PhD Thesis). Oxford Brookes University, Oxford, UK.

André, J., Baptista, A., & Camotim, D. (2009). Numerical investigation of adjustable telescopic steel props. In *Proceedings of the 12th International Conference on Civil, Structural and Environmental Engineering Computing* (p. 25). Madeira, Portugal: Civil-Comp Press.

André, J., Baptista, A., & Camotim, D. (2009b). Numerical investigation of adjustable telescopic steel props. In *Proceedings of the 12th International Conference on Civil, Structural and Environmental Engineering Computing* (p. 25). Madeira, Portugal: Civil-Comp Press.

André, J., Baptista, A., & Camotim, D. (2007). Experimental investigation of steel telescopic props (in Portuguese). In *Proceedings of the VI Portuguese National Conference of Steel and Composite Construction* (p. 12). Porto, Portugal: The Portuguese Steelwork Association (cmm).

André, J., Baptista, A., & Camotim, D. (2009a). Design buckling curve for steel telescopic props under compression (in Portuguese). In *Proceedings of the VII Portuguese National Conference of Steel and Composite Construction* (p. 10). Lisbon, Portugal: The Portuguese Steelwork Association (cmm).

André, J., Beale, R., & Baptista, A. (2012a). A survey of failures of bridge falsework systems since 1970.*Proceedings of the ICE - Forensic Engineering, 65*(4), 161–172. doi:10.1680/feng.12.00012

André, J., Beale, R., & Baptista, A. (2012b). Bridge construction equipment: An overview of the existing design guidance. *Structural Engineering International, 22*(3), 15. doi:10.2749/101686612X13363869853419

André, J., Beale, R., & Baptista, A. (2013a). Experimental behaviour of bridge falsework joints. In *Proceedings of the IX Portuguese National Conference of Steel and Composite Construction* (p. 10). Guimarães, Portugal: The Portuguese Steelwork Association (cmm).

André, J., Beale, R., & Baptista, A. (2013b). Recent advances and existing challenges in the design of bridge falsework systems. *Civil Engineering and Environmental Systems, 30*(2), 130–145. doi:10.1080/10286608.2012.733374

André, J., Beale, R., & Baptista, A. (2014). Numerical investigation of bridge falsework structures. In *Proceedings of the 5th Portuguese Structural Engineering Conference (JPEE 2014)* (p. 15). Lisbon, Portugal: Portuguese National Laboratory for Civil Engineering (LNEC).

André, J., Beale, R., & Baptista, A. (2014b). Robustness and risk evaluation of bridge falsework structures. In *Proceedings of the 5th Portuguese Structural Engineering Conference (JPEE 2014)* (p. 15). Lisbon, Portugal: Portuguese National Laboratory for Civil Engineering (LNEC).

André, J., Beale, R., & Baptista, A. (2015). New indices of structural robustness and structural fragility. *Structural Engineering & Mechanics*, *56*(6), 1063–1093. doi:10.12989/sem.2015.56.6.1063

André, J., Beale, R., & Baptista, A. (2015a). Numerical analysis of bridge falsework. In *Proceedings of the 8th International Conference on Advances in Steel Structures (ICASS 2015)* (p. 20). Lisbon, Portugal: Elsevier.

André, J., Beale, R., & Baptista, A. (2015b). Performance of bridge falsework structures under external hazards. In *Proceedings of the X Portuguese National Conference of Steel and Composite Construction* (p. 10). Coimbra, Portugal: The Portuguese Steelwork Association (cmm).

André, J., Beale, R., & Baptista, A. (2017). *Risk analysis of bridge falsework Cuplok systems*. Accepted for Publication in Structure and Infrastructure Engineering.

Andresen, J. (2012). Investigation of a collapsed scaffold structure. *Proceedings of the Institution of Civil Engineers - Forensic Engineering, 165*(2), 95–104. doi:10.1680/feng.11.00021

Ang, A., & Tang, W. (1975). *Probability concepts in engineering planning and design: Basic principles* (1st ed.; Vol. 1). Wiley.

Ang, A., & Tang, W. (2007). *Probability concepts in engineering: Emphasis on applications in civil and environmental engineering* (2nd ed.). London, UK: Wiley.

ANSI. (2011). ANSI/ASSE A10.8:2011 – Scaffolding safety requirements. American National Standards Institute (ANSI).

ANSYS. (2016). *ANSYS Fluent user's guide*. Pennsylvania, USA: ANSYS, Inc.

Apeland, S., Aven, T., & Nilsen, T. (2002). Quantifying uncertainty under a predictive, epistemic approach to risk analysis. *Reliability Engineering & System Safety*, *75*(1), 93–102. doi:10.1016/S0951-8320(01)00122-3

Artamonov, A., Bendel, C., Blueckert, C., Landrin, H., Spencer, M., & Wallace, P., … Wannick, H. (2008). Bridges – Construction, insurance and risk management. In *Proceedings of the 41st annual Conference of The International Association of Engineering Insurers (IMIA)* (p. 44). Gleneagles, Scotland: The International Association of Engineering Insurers (IMIA).

ASCE. (2010). ASCE 7–10 – Minimum design loads for buildings and other structures. American Society of Civil Engineers (ASCE).

ASCE. (2014). ASCE/SEI 37–14 – Design loads on structures during construction. American Society of Civil Engineers (ASCE).

ASTM. (2014). ASTM D143–14 – Standard test methods for small clear specimens of timber. The American Society for Testing and Materials (ASTM).

Aven, T. (2004). On how to approach risk and uncertainty to support decision-making. *Risk Management*, *6*(4), 27–39. doi:10.1057/palgrave.rm.8240196

Ayaho, M., Hideyuki, K., & Hideaki, N. (1997). Development of a knowledge-based expert system for selection of steel bridge erection method. In *Proceedings of NEWTECH Lisbon 97: New Technologies in Structural Engineering*, 2, 203–211.

Ayyub, B., & McCuen, R. (2011). *Probability, statistics, and reliability for engineers and scientists*. London, UK: Taylor and Francis.

Baker, C. (2007). Wind engineering: Past, present and future. *Journal of Wind Engineering and Industrial Aerodynamics, 95*(9–11), 843–870. doi:10.1016/j.jweia.2007.01.011

Baker, J., Schubert, M., & Faber, M. (2008). On the assessment of robustness. *Structural Safety, 30*(3), 253–267. doi:10.1016/j.strusafe.2006.11.004

Bakhoum, M. (2014). Bridge construction methods. In *Bridge engineering handbook: Construction and maintenance* (2nd ed.; Vol. 2, p. 60). CRC Press.

Baptista, A., & Silva, S. (2002). Promotion of quality in the management of construction temporary structures (in Portuguese). In *Proceedings of Portuguese National Conference of Structural Engineering* (pp. 583–594). Lisbon, Portugal: Portuguese National Laboratory for Civil Engineering (LNEC).

Bashor, R., & Kareem, A. (2009). Comparative study of major international standards.*Proceedings of the 7th Asia-Pacific Conference on Wind Engineering (APCWE-VII)*, 8.

Bathe, K. (2006). *Finite element procedures*. Klaus-Jurgen Bathe.

Bažant, Z., Le, J., Greening, F., & Benson, D. (2008). What did and did not cause collapse of World Trade Center Twin Towers in New York? *Journal of Engineering Mechanics, 134*(10), 892–906. doi:10.1061/(ASCE)0733-9399(2008)134:10(892)

Bažant, Z., & Verdure, M. (2007). Mechanics of progressive collapse: Learning from World Trade Center and building demolitions. *Journal of Engineering Mechanics, 133*(3), 308–319. doi:10.1061/(ASCE)0733-9399(2007)133:3(308)

BCA. (2001). *Early striking and improved backpropping for efficient flat slab construction*. Crowthorne, UK: British Cement Association.

BCSA. (2005). BCSA Guide to steel erection in windy conditions (No. 39/05). London, UK: The British Constructional Steelwork Association Ltd. (BCSA).

Beale, R., & Godley, M. (2003). Common causes of failure of metal scaffolds in the UK and stability checks on scaffolds. In *Proceedings of the Technical Seminar on Metal Scaffolding (Falsework) – Design, Construction and Safety* (pp. 62–79). Hong Kong, China: Hong Kong Institute of Steel Construction.

Beale, R. (2014). Scaffold research — A review. *Journal of Constructional Steel Research, 98*, 188–200. doi:10.1016/j.jcsr.2014.01.016

Beale, R., & Godley, M. (1995). The analysis of scaffold structures using LUSAS. *Proceedings of LUSAS, 95*, 10–24.

Beale, R., & Godley, M. (2006). Numerical modelling of tube and fitting access scaffold systems. *Advanced Steel Construction, 2*(3), 199–223.

Beavor, N., & Cai, J. (2006). Design and construction of two launched bridges.*Proceedings of the ICE - Bridge Engineering, 159*(3), 137–145. doi:10.1680/bren.2006.159.3.137

Bedford, T., & Cooke, R. (2001). *Probabilistic risk analysis: Foundations and methods*. Cambridge, UK: Cambridge University Press. doi:10.1017/CBO9780511813597

Béguin, G. (2010). Catastrophic sliding of a concrete deck during its launching on the steel girders of a composite bridge. *Journal of Performance of Constructed Facilities, 24*(2), 138–147. doi:10.1061/(ASCE)0887-3828(2010)24:2(138)

Bellinger, W. (2007). *The economic analysis of public policy*. Routledge. doi:10.4324/9780203946480

Benjamin, J., & Cornell, C. (1970). *Probability, statistics, and decision–making for civil engineers* (3rd ed.). New York: McGraw-Hill.

Bennett, L. (2004). Peer review of analysis of specialist group reports on causes of construction accidents (No. 218). London, UK: Health and Safety Executive (HSE).

Berry, L., Lindsay, J., Salomon, E., & Veale, S. (2006). Risk, responsibility and regulation – Whose risk is it anyway?. London, UK: UK's Better Regulation Commission.

Billings, I., & Routley, R. (1978). Falsework load measurement at Newton No. 1 bridge. *New Zealand Engineering*, *33*(6), 127–134.

Birch, N., Booth, J., & Walker, M. (1971). Effect of site factors on the load capacities of adjustable steel props (No. R27). London, UK: Construction Industry Research and Information Association (CIRIA).

Birch, N., Walker, M., & Lee, C. (1977). Safe working loads for adjustable steel props; the influence of prop conditions and site workmanship (No. TN 79). London, UK: Construction Industry Research and Information Association (CIRIA).

Birnbaum, M. (2008). New paradoxes of risky decision making. *Psychological Review*, *115*(2), 463–501. doi:10.1037/0033-295X.115.2.463 PMID:18426300

Blackmore, P. (2004). Wind loads on unclad structures (No. Special Digest 5). Bracknell, UK: Building Research Establishment (BRE).

Blackmore, P., & Freathy, P. (2004). Wind loads on temporary stage decks (No. Special Digest 483). Bracknell, UK: Building Research Establishment (BRE).

Blair, T., & Woods, K. (1990). Seismic scaffolding concerns? Let it 'walk'. *Nuclear Plant Journal*, *8*(3), 76–77.

Blank, S., Blank, M., & Kondazi, H. (2014). Concrete bridge construction. In *Bridge engineering handbook: Construction and maintenance* (2nd ed.; Vol. 2, p. 18). CRC Press.

Błazik-Borowa, E., & Gontarz, J. (2016). The influence of the dimension and configuration of geometric imperfections on the static strength of a typical façade scaffolding. *Archives of Civil and Mechanical Engineering*, *16*(3), 269–281. doi:10.1016/j.acme.2015.11.003

Błazik-Borowa, E., & Szer, J. (2014). The analysis of the stages of scaffolding life with regard to the decrease in the hazard at building works. *Archives of Civil and Mechanical Engineering*, *15*(2), 516–524. doi:10.1016/j.acme.2014.09.009

Blocken, B. (2011). Computational wind engineering: Theory and applications. In C. Baniotopoulos, C. Borri, & T. Stathopoulos (Eds.), *Environmental wind engineering and design of wind energy structures* (pp. 55–93). Udine, Italy: Springer. doi:10.1007/978-3-7091-0953-3_3

Blockley, D. (1992). *Engineering safety*. London, UK: McGraw-Hill Book Company.

Blockley, D. (2011). Engineering safety.*Proceedings of the ICE - Forensic Engineering*, *164*(1), 7–13.

Boggs, D., & Peterka, J. (1992). Wind speeds for design of temporary structures. In *Proceedings of the 10th ASCE Structures Congress*. San Antonio, TX: American Society of Civil Engineers (ASCE).

Bokan, H., Janjic, D., & Heiden, M. (2006). Optimisation of the tensioning schedule for cable-stayed bridges using dynamic software. In *Proceedings of the IABSE Symposium 'Responding to Tomorrow's Challenges in Structural Engineering'* (p. 9). Budapest, Hungary: International Association for Bridge and Structural Engineering (IABSE). doi:10.2749/222137806796169182

Bowles, D. (2003). ALARP evaluation: Using cost effectiveness and disproportionality to justify risk reduction. In *Proceedings of ANCOLD 2003 Conference on Dams* (p. 17). Hobart, Australia: The Australian National Committee on Large Dams (ANCOLD).

Bowles, D. (2007). Tolerable risk for dams: How safe is safe enough? In *Proceedings of the US Society on Dams Annual Conference* (p. 24). Philadelphia: United States Society on Dams (USSD).

Bowles, J. (1997). *Foundation analysis and design* (5th ed.). New York: McGraw-Hill.

Box, G., & Draper, N. (1987). *Empirical model–building and response surfaces*. Wiley.

Bragg, S. (1975). Final report of the advisory committee on falsework. Her Majesty's Stationery Office (HMSO).

Bragg, S. (1975). Final report of the advisory committee on falsework. London: Her Majesty's Stationery Office (HMSO).

Brand, R. (1975). *Falsework and access scaffolds in tubular steel*. London, UK: McGraw-Hill.

Browning, G. (2011). *Human perception of vibrations due to synchronised crowd loading in grandstands* (PhD Thesis). University of Bath, Bath, UK.

Brownjohn, J., Racic, V., & Chen, J. (2016). Universal response spectrum procedure for predicting walking-induced floor vibration. *Mechanical Systems and Signal Processing*, *70–71*, 741–755. doi:10.1016/j.ymssp.2015.09.010

BSI. (1982). BS 5975:1982 – Code of practice for falsework (1st ed.). London, UK: British Standards Institution (BSI).

BSI. (1982a). BS 4074:1982 – Specification for metal props and struts. London, UK: British Standards Institution (BSI).

BSI. (1982b). BS 5507–3:1982 – Methods of test for falsework equipment – Part 3: Props. London, UK: British Standards Institution (BSI).

BSI. (1982c). BS 5975:1982 – Code of practice for falsework (1st ed.). London, UK: British Standards Institution (BSI).

BSI. (1990a). BS 1139–5:1990, HD 1000:1988 – Metal scaffolding – Part 5: Specification for materials, dimensions, design loads and safety requirements for service and working scaffolds made of prefabricated elements. London, UK: British Standards Institution (BSI).

BSI. (1990b). BS 4360:1990 – Specification for weldable structural steels. London, UK: British Standards Institution (BSI).

BSI. (1993). BS 5973:1993 – Code of practice for access and working scaffolds and special scaffold structures in steel (3rd ed.). London, UK: British Standards Institution (BSI).

BSI. (1993a). BS 5973:1993 – Code of practice for access and working scaffolds and special scaffold structures in steel (3rd ed.). London, UK: British Standards Institution (BSI).

BSI. (1993b). *Structural steel I and H sections. Tolerances on shape and dimensions*. London, UK: British Standards Institution.

BSI. (1996). BS 5975:1996 – Code of practice for falsework (2nd ed.). London, UK: British Standards Institution (BSI).

BSI. (1997). BS 6399–2:1997 – Loading for buildings. Code of practice for wind loads. London, UK: British Standards Institution (BSI).

BSI. (1999). *Adjustable telescopic steel props – Product specifications, design and assessment by calculation and tests*. London, UK: British Standards Institution.

BSI. (2000). BS EN 1337–1:2000 – Structural bearings – Part 1: General design rules. London, UK: British Standards Institution (BSI).

BSI. (2001). *Loose steel tubes for tube and coupler scaffolds – Technical delivery conditions*. London, UK: British Standards Institution.

BSI. (2001).*Loose steel tubes for tube and coupler scaffolds – Technical delivery conditions*. London, UK: British Standards Institution.

BSI. (2002). BS EN 12811–3:2002 – Temporary works equipment – Part 3: Load testing. London, UK: British Standards Institution (BSI).

BSI. (2002a). BS EN 1991–1–1:2002 – Actions on structures – Part 1–1: General actions – Densities, self–weight, imposed loads for buildings. London, UK: British Standards Institution (BSI).

BSI. (2002b). UK National annex to BS EN 1991–1–1:2002. London, UK: British Standards Institution (BSI).

BSI. (2003). BS EN 12811–1:2003 – Temporary works equipment – Part 1: Scaffolds – Performance requirements and general design. London, UK: British Standards Institution (BSI).

BSI. (2003a). BS EN 12810 – Façade scaffolds made of prefabricated components – Parts 1–2. London, UK: British Standards Institution (BSI).

BSI. (2003a). BS EN 12810–1:2003 – Facade scaffolds made of prefabricated components – Part 1: Product specifications. London, UK: British Standards Institution (BSI).

BSI. (2003b). BS EN 12810–2:2003 – Facade scaffolds made of prefabricated components – Part 2: Particular methods of structural design. London, UK: British Standards Institution (BSI).

BSI. (2003b). BS EN 12811–1:2003 – Temporary works equipment – Part 1: Scaffolds – Performance requirements and general design. London, UK: British Standards Institution (BSI).

BSI. (2003c). BS EN 12811–1:2003 – Temporary works equipment – Part 1: Scaffolds – Performance requirements and general design. London, UK: British Standards Institution (BSI).

BSI. (2004). BS EN 10025–2:2004 – Hot rolled products of structural steels - Part 2: Technical delivery conditions for non-alloy structural steels. London, UK: British Standards Institution (BSI).

BSI. (2004a). BS EN 1991–1–3:2003 – Actions on structures – Part 1–3: General actions – Snow loads. London, UK: British Standards Institution (BSI).

BSI. (2004a).*Mobile access and working towers made of prefabricated elements. Materials, dimensions, design loads, safety and performance requirements*. London, UK: British Standards Institution.

BSI. (2004b). BS EN 12811 – Temporary works equipment – Parts 1–3. London, UK: British Standards Institution (BSI).

BSI. (2004b). BS EN 1337–2:2004 – Structural bearings – Part 2: Sliding elements. London, UK: British Standards Institution (BSI).

BSI. (2004c). BS EN 1337–4:2004 – Structural bearings – Part 4: Roller bearings. London, UK: British Standards Institution (BSI).

BSI. (2004d). BS EN 1337–7:2004 – Structural bearings – Part 7: Spherical and cylindrical ptfe bearings. London, UK: British Standards Institution (BSI).

BSI. (2004e). BS EN 1998–1:2004 – Design of structures for earthquake resistance – Part 1: General rules, seismic actions and rules for buildings. London, UK: British Standards Institution (BSI).

BSI. (2004f). BS EN 10025–2:2004 – Hot rolled products of structural steels - Part 2: Technical delivery conditions for non-alloy structural steels. London, UK: British Standards Institution (BSI).

BSI. (2004g). BS EN 12811–2:2004 – Temporary works equipment – Part 2: Information on materials. London, UK: British Standards Institution (BSI).

BSI. (2004h). *Temporary works equipment. Load bearing towers of prefabricated components. Particular methods of structural design.* London, UK: British Standards Institution.

BSI. (2005). BS EN 1990:2002+A1:2005 – Basis of structural design. London, UK: British Standards Institution (BSI).

BSI. (2005). BS EN 74–1:2005 – Couplers, spigot pins and baseplates for use in falsework and scaffolds – Part 1: Couplers for tubes – Requirements and test procedures. London, UK: British Standards Institution (BSI).

BSI. (2005a). BS EN 1990:2002+A1:2005 – Basis of structural design. London, UK: British Standards Institution (BSI).

BSI. (2005a). BS EN 74–1:2005 – Couplers, spigot pins and baseplates for use in falsework and scaffolds – Part 1: Couplers for tubes – Requirements and test procedures. London, UK: British Standards Institution (BSI).

BSI. (2005b). BS EN 1990:2002+A1:2005 – Basis of structural design. London, UK: British Standards Institution (BSI).

BSI. (2005b). BS EN 1991–1–6:2005 – Actions on structures – Part 1–6: General actions – Actions during execution. London, UK: British Standards Institution (BSI).

BSI. (2005b). BS EN 1993–1–1:2005 – Design of steel structures – Part 1–1: General rules and rules for buildings. London, UK: British Standards Institution (BSI).

BSI. (2005c). BS EN 1991–1–6:2005 – Actions on structures – Part 1–6: General actions – Actions during execution. London, UK: British Standards Institution (BSI).

BSI. (2005d). BS EN 1992–2:2005 – Design of concrete structures – Part 2: Concrete bridges – Design and detailing rules. London, UK: British Standards Institution (BSI).

BSI. (2005e). BS EN 1993–1–1:2005 – Design of steel structures – Part 1–1: General rules and rules for buildings. London, UK: British Standards Institution (BSI).

BSI. (2005f). BS EN 1993–1–8:2005 – Design of steel structures – Part 1–8: Design of joints. London, UK: British Standards Institution (BSI).

BSI. (2006a). BS EN 10210–1:2006 – Hot finished structural hollow sections of non–alloy and fine grain steels – Part 1: Technical delivery conditions. London, UK: British Standards Institution (BSI).

BSI. (2006a). BS EN 1991–3:2006 – Actions on structures – Part 3: Actions induced by cranes and machines. London, UK: British Standards Institution (BSI).

BSI. (2006a). BS EN 1993–1–5:2006 – Design of steel structures – Part 1–5: Plated structural elements. London, UK: British Standards Institution (BSI).

BSI. (2006b). BS EN 10219–1:2006 – Cold formed welded structural hollow sections of non–alloy and fine grain steels – Part 1: Technical delivery requirements. London, UK: British Standards Institution (BSI).

BSI. (2006b). BS EN 1993–1–11:2006 – Design of steel structures – Part 1–11: Design of structures with tension components. London, UK: British Standards Institution (BSI).

BSI. (2006b). BS EN 1993–3–1:2006 – Design of steel structures – Part 3–1: Towers, masts and chimneys – Towers and masts. London, UK: British Standards Institution (BSI).

BSI. (2006c). BS EN 1993–2:2006 – Design of steel structures – Part 2: Steel bridges. London, UK: British Standards Institution (BSI).

BSI. (2006d). BS EN 10210–2:2006 – Hot finished structural hollow sections of non–alloy and fine grain steels – Part 2: Tolerances, dimensions and sectional properties. London, UK: British Standards Institution (BSI).

BSI. (2006e). BS EN 10219–2:2006 – Cold formed welded structural hollow sections of non–alloy and fine grain steels – Part 2 : Tolerances, dimensions and sectional properties. London, UK: British Standards Institution (BSI).

BSI. (2007). BS EN 1997–2:2007 – Geotechnical design – Part 2: Ground investigation and testing. London, UK: British Standards Institution (BSI).

BSI. (2007a). BS EN 1993–1–6:2007 – Design of steel structures – Part 1–5: Strength and stability of shell structures. London, UK: British Standards Institution (BSI).

BSI. (2007a). BS EN 1997–2:2007 – Geotechnical design – Part 2: Ground investigation and testing. London, UK: British Standards Institution (BSI).

BSI. (2007b). BS EN 1993–6:2007 – Design of steel structures – Part 6: Crane supporting structures. London, UK: British Standards Institution (BSI).

BSI. (2007b). UK National annex to BS EN 1991–1–3:2003. London, UK: British Standards Institution (BSI).

BSI. (2007c). UK National annex to BS EN 1992–2:2005. London, UK: British Standards Institution (BSI).

BSI. (2008). UK National annex to BS EN 1993–1–1:2005. London, UK: British Standards Institution (BSI).

BSI. (2008a). Report on the calibration of Eurocode for wind loading (BS EN 1991-1-4) and its UK National Annex against the current UK wind code (BS 6399: part 2). London, UK: British Standards Institution (BSI).

BSI. (2008a). UK National annex to BS EN 1993–1–1:2005. London, UK: British Standards Institution (BSI).

BSI. (2008b). UK National annex to BS EN 1991–1–6:2005. London, UK: British Standards Institution (BSI).

BSI. (2008b). UK National annex to BS EN 1993–1–5:2006. London, UK: British Standards Institution (BSI).

BSI. (2008c). UK National annex to BS EN 1993–1–8:2005. London, UK: British Standards Institution (BSI).

BSI. (2008c). UK National annex to BS EN 1998–6:2005. London, UK: British Standards Institution (BSI).

BSI. (2008d). UK National annex to BS EN 1993–1–11:2006. London, UK: British Standards Institution (BSI).

BSI. (2008e). UK National annex to BS EN 1993–2:2006. London, UK: British Standards Institution (BSI).

BSI. (2008f). UK National annex to BS EN 1998–1:2004. London, UK: British Standards Institution (BSI).

BSI. (2009). *Steel static storage systems – Adjustable pallet racking systems – Principles for structural design*. London, UK: British Standards Institution.

BSI. (2009a). *Execution of concrete structures*. London, UK: British Standards Institution.

BSI. (2009a). *Steel static storage systems – Adjustable pallet racking systems – Principles for structural design*. London, UK: British Standards Institution.

BSI. (2009a). *Structural timber. Strength classes*. London, UK: British Standards Institution.

BSI. (2009b). BS EN 1998–2:2005+A1:2009 – Design of structures for earthquake resistance – Part 2: Bridges. London, UK: British Standards Institution (BSI).

BSI. (2009b). PD 6698:2009 – Recommendations for the design of structures for earthquake resistance to BS EN 1998. London, UK: British Standards Institution (BSI).

BSI. (2009b). UK National annex to BS EN 1997–2:2007. London, UK: British Standards Institution (BSI).

BSI. (2009c). PD 6688–1–4:2009 – Background information to the national annex to BS EN 1991–1–4 and additional guidance. London, UK: British Standards Institution (BSI).

BSI. (2009c). PD 6703:2009 – Structural bearings. Guidance on the use of structural bearings. London, UK: British Standards Institution (BSI).

BSI. (2009d). PD 6698:2009 – Recommendations for the design of structures for earthquake resistance to BS EN 1998. London, UK: British Standards Institution (BSI).

BSI. (2009d). UK National annex to BS EN 1990:2002+A1:2005. London, UK: British Standards Institution (BSI).

BSI. (2009e). UK National annex to BS EN 1991–3:2006. London, UK: British Standards Institution (BSI).

BSI. (2009e). UK National annex to BS EN 1993–6:2007. London, UK: British Standards Institution (BSI).

BSI. (2009f). UK National annex to BS EN 1997–2:2007. London, UK: British Standards Institution (BSI).

BSI. (2010). BS EN 1991–1–4:2005+A1:2010 – Actions on structures – General actions – Part 1–4: Wind actions. London, UK: British Standards Institution (BSI).

BSI. (2010a). BS EN 1991–1–4:2005+A1:2010 – Actions on structures – General actions – Part 1–4: Wind actions. London, UK: British Standards Institution (BSI).

BSI. (2010a). BS EN 74–3:2007 – Couplers, spigot pins and baseplates for use in falsework and scaffolds – Part 3: Plain base plates and spigot pins – Requirements and test procedures. London, UK: British Standards Institution (BSI).

BSI. (2010b). BS EN 1991–1–4:2005+A1:2010 – Actions on structures – General actions – Part 1–4: Wind actions. London, UK: British Standards Institution (BSI).

BSI. (2010b). UK National annex to BS EN 1993–3–1:2006. London, UK: British Standards Institution (BSI).

BSI. (2011). BS 5975:2008+A1:2011 – Code of practice for temporary works procedures and the permissible stress design of falsework (3rd ed.). London, UK: British Standards Institution (BSI).

BSI. (2011a). BS 5975:2008+A1:2011 – Code of practice for temporary works procedures and the permissible stress design of falsework (3rd ed.). London, UK: British Standards Institution (BSI).

BSI. (2011b). BS EN 197-1:2011 – Cement. Composition, specifications and conformity criteria for common cements. London, UK: British Standards Institution (BSI).

BSI. (2011b). *Falsework. Performance requirements and general design.* London, UK: British Standards Institution.

BSI. (2011b).*Falsework. Performance requirements and general design.* London, UK: British Standards Institution.

BSI. (2011c). BS EN 1090–2:2008+A1:2011 – Execution of steel structures and aluminium structures – Part 2: Technical requirements for steel structures. London, UK: British Standards Institution (BSI).

BSI. (2011c). UK National annex to BS EN 1991–1–4:2005+A1:2010. London, UK: British Standards Institution (BSI).

BSI. (2011d). *Falsework. Performance requirements and general design.* London, UK: British Standards Institution.

BSI. (2011e). UK National annex to BS EN 1991–1–4:2005+A1:2010. London, UK: British Standards Institution (BSI).

BSI. (2012). *Adjustable telescopic aluminium props – Product specifications, design and assessment by calculation and tests*. London, UK: British Standards Institution.

BSI. (2012).*Adjustable telescopic aluminium props – Product specifications, design and assessment by calculation and tests*. London, UK: British Standards Institution.

BSI. (2013). BS EN 1997–1:2004+A1:2013 – Geotechnical design – Part 1: General rules. London, UK: British Standards Institution (BSI).

BSI. (2013). BS EN 1999–1–1:2007+A2:2013 – Design of aluminium structures – Part 1–1: General structural rules. London, UK: British Standards Institution (BSI).

BSI. (2013a). *Concrete. Specification, performance, production and conformity*. London, UK: British Standards Institution.

BSI. (2013b). BS EN 1997–1:2004+A1:2013 – Geotechnical design – Part 1: General rules. London, UK: British Standards Institution (BSI).

BSI. (2014). BS EN 1991–1–7:2006+A1:2014 – Actions on structures – Part 1–7: General actions – Accidental actions. London, UK: British Standards Institution (BSI).

BSI. (2014a). BS EN 13001–2:2014 – Crane safety. General design – Part 2: Load actions. London, UK: British Standards Institution (BSI).

BSI. (2014a). BS EN 1991–1–7:2006+A1:2014 – Actions on structures – Part 1–7: General actions – Accidental actions. London, UK: British Standards Institution (BSI).

BSI. (2014b). BS EN 13001–2:2014 – Crane safety. General design – Part 2: Load actions. London, UK: British Standards Institution (BSI).

BSI. (2014b). UK National annex to BS EN 1997–1:2004+A1:2013. London, UK: British Standards Institution (BSI).

BSI. (2014c). UK National annex to BS EN 1991–1–7:2006+A1:2014. London, UK: British Standards Institution (BSI).

BSI. (2014d). UK National annex to BS EN 1997–1:2004+A1:2013. London, UK: British Standards Institution (BSI).

BSI. (2015). BS 1139–2.2:2009+A1:2015 – Metal scaffolding – Part 2: Couplers – Section 2.2: Aluminium couplers and special couplers in steel. London, UK: British Standards Institution (BSI).

BSI. (2015a). BS 5930:2015 – Code of practice for ground investigations. London, UK: British Standards Institution (BSI).

BSI. (2015b). BS 8004:2015 – Code of practice for foundations. London, UK: British Standards Institution (BSI).

BSI. (2016). BS 7121–1:2016 – Code of practice for safe use of cranes – Part 1: General. London, UK: British Standards Institution (BSI).

BSI. (2016).*Temporary works – Application of European Standards in design – Guide*. London, UK: British Standards Institution.

Buller, P. (1993). Gales of January and February 1990: Damage to buildings and structures (No. BR 248) (p. 24). Bracknell, UK: Building and Research Establishment (BRE).

Bunni, N. (2003). *Risk and insurance in construction* (2nd ed.). London, UK: Taylor & Francis.

Burrows, B. (1989). *Organisation and quality of falsework construction (PhD)*. Warwick University.

Burrows, B. (1989). *Organisation and quality of falsework construction (PhD.)*. Warwick University.

Caetano, E. (2007). Cable vibrations in cable–stayed bridges. Zurich, Switzerland: International Association for Bridge and Structural Engineering (IABSE).

Cajot, L., Haller, M., Conan, Y., Sedlacek, G., Kraus, O., Rondla, J., … Johansson, B. (2003). Probabilistic quantification of safety of a steel structure highlighting the potential of steel versus other materials. Brussels, Belgium: European Commission.

Calgaro, J., Tschumi, M., & Gulvanessian, H. (2010). *Designers' Guide to Eurocode 1: Actions on bridges*. London, UK: Thomas Telford Ltd. doi:10.1680/dgeab.31586

Canisius, G. (2011). Structural robustness design for practising engineers (COST Action TU0601 – Robustness of Structures). Brussels, Belgium: European Cooperation in Science and Technology (COST).

Canisius, G., & Maitra, A. (1999). Strength of adjustable props - effect of shear-pin position.*Proceedings of the ICE - Structures and Buildings, 134*(2), 139–146. doi:10.1680/istbu.1999.31380

Cao, S. (2013). CFD applications in structural wind engineering. In Y. Tamura & A. Kareem (Eds.), *Advanced structural wind engineering* (pp. 301–324). Tokyo, Japan: Springer Japan. doi:10.1007/978-4-431-54337-4_11

Cao, S. (2013). Strong winds and their characteristics. In Y. Tamura & A. Kareem (Eds.), *Advanced structural wind engineering* (pp. 1–25). Tokyo, Japan: Springer. doi:10.1007/978-4-431-54337-4_1

Cardwell, S. (2010). *Bridge deck types and construction methods.* Presented at the IABSE WG 6 Seminar: State-of-the-art Bridge Deck Erection; Safe and Efficient use of Special Equipment, Singapore.

Carlton, R., & Llurba, X. (2007). Modelling prop loads during concrete slab construction. *Structures & Buildings, 160*(4), 221–229.

Carpenter, J. (2010). Falsework. In *ICE Manual of health and safety in construction* (pp. 205–213). London: Thomas Telford Ltd.

Carvalho, A., Pereira, C., & Martins, R. (2004). Problems with falsework foundations (in Portuguese). In *Proceedings of the 9th Portuguese National Congress of Geotecnics* (p. 10). Aveiro: Universidade de Aveiro.

Casey, M., & Wintergerste, T. (2000). *Best practices guidelines: ERCOFTAC special interest group on 'Quality and Trust in Industrial CFD.* Brussels, Belgium: ERCOFTAC.

Castillo, E., Hadi, A., Balakrishnan, N., & Sarabia, J. (2004). *Extreme value and related models with applications in engineering and science*. Wiley.

CEGB. (1966). Report of the Committee of Inquiry into collapse of cooling towers at Ferrybridge, Monday 1 November, 1965. London, UK: Central Electricity Generating Board (CEGB).

CEN. (1996). Background document to ENV 1991–2–6:1996 – Basis of design and actions on structures – Part 2–6: Actions during execution. Brussels, Belgium: European Committee for Standardization (CEN).

CEN. (2009). CEN/TS 1992–4:2009 – Design of fastenings for use in concrete – Parts 1–5. Brussels, Belgium: European Committee for Standardization (CEN).

Chan, A., Wong, F., Chan, D., Yam, M., Kwok, A., Lam, E., & Cheung, E. (2008). Work at height fatalities in the repair, maintenance, alteration, and addition works. *Journal of Construction Engineering and Management, 134*(7), 527–535. doi:10.1061/(ASCE)0733-9364(2008)134:7(527)

Chandrangsu, T. (2010). *Advanced analysis and probabilistic-based design of support scaffold systems* (PhD Thesis). University of Sydney, Sydney, Australia.

Chandrangsu, T., & Rasmussen, K. (2009). Full-scale tests and advanced structural analysis of formwork subassemblies. In *Proceedings of the 6th International Conference on Advances in Steel Structures* (pp. 1083–1090). Hong Kong, China: Hong Kong Institute of Steel Construction.

Chandrangsu, T., & Rasmussen, K. (2011a). Investigation of geometric imperfections and joint stiffness of support scaffold systems. *Journal of Constructional Steel Research*, *67*(4), 576–584. doi:10.1016/j.jcsr.2010.12.004

Chandrangsu, T., & Rasmussen, K. (2011b). Structural modelling of support scaffold systems. *Journal of Constructional Steel Research*, *67*(5), 866–875. doi:10.1016/j.jcsr.2010.12.007

Chan, S., Dymiotis-Wellington, C., & Zhou, Z. (2002). Second-order analysis and design of steel scaffold using multiple eigen-imperfection modes. In *Proceedings of the 3rd International Conference on Advances in Steel Structures (ICASS 2002)* (pp. 312–327). Hong Kong, China: Elsevier. doi:10.1016/B978-008044017-0/50037-8

Chan, S., & Xian, X. (2004). *Engineering and mechanical properties of structural bamboo*. Hong Kong, China: Hong Kong Polytechnic University.

Chan, S., Zhou, Z., Chen, W., Peng, J., & Pan, A. (1995). Stability analysis of semirigid steel scaffolding. *Engineering Structures*, *17*(8), 568–574. doi:10.1016/0141-0296(95)00011-U

Charuvisit, S., Hino, Y., Ohdo, K., Maruta, E., & Kanda, M. (2007). Wind tunnel experiment on wind pressures acting on the scaffolds in strong winds. *Journal of Wind Engineering*, *32*(1), 1–10. doi:10.5359/jwe.32.1

Chen, J., Li, G., & Racic, V. (2016). Acceleration response spectrum for predicting floor vibration due to occupants jumping. *Engineering Structures*, *112*, 71–80. doi:10.1016/j.engstruct.2016.01.013

Chen, W., & Duan, L. (2013). *Handbook of international bridge engineering*. CRC Press. doi:10.1201/b15520

Chen, W., & Han, D. (2007). *Plasticity for structural engineers* (2nd ed.). J. Ross Publishing.

Chilton, S., Jones-Lee, M., Loomes, G., Robinson, A., Hopkins, L., & Cookson, R. … Beattie, J. (1998). Valuing health and safety controls. A literature review (No. CRR 171). London, UK: Health and Safety Executive (HSE).

Chu, A., Zhou, Z., Koon, C., Chan, S., Peng, J., & Pan, A. (1996). Design of steel scaffolding using an integrated design and analysis approach. In *Proceedings of International Conference on Advances in Steel Structures* (pp. 245–250). Hong Kong, China: Pergamon. doi:10.1016/B978-008042830-7/50039-5

Chung, K., & Chan, S. (2002). *Proceedings of the International Seminar 'Bamboo Scaffolds in Building Construction' - An Alternative and Supplement to Metal Scaffolds*. Hong Kong, China: International Network for Bamboo and Rattan (INBAR).

Chung, K., Chan, S., & Yu, W. (2002). Design of bamboo scaffolds. In *Proceedings of the International Seminar 'Bamboo Scaffolds in Building Construction' - An Alternative and Supplement to Metal Scaffolds* (pp. 65–88). Hong Kong, China: Hong Kong Polytechnic University.

Chung, K., Yu, W., & Chan, S. (2002). Mechanical properties and engineering data of structural bamboo. In *Proceedings of the International Seminar 'Bamboo Scaffolds in Building Construction' - An Alternative and Supplement to Metal Scaffolds* (pp. 1–23). Hong Kong, China: Hong Kong Polytechnic University.

Chung, K. F., & Yu, W. (2002). Mechanical properties of structural bamboo for bamboo scaffoldings. *Engineering Structures*, *24*(4), 429–442. doi:10.1016/S0141-0296(01)00110-9

Chung, K., & Chan, S. (2002). *Design of bamboo scaffolds*. Hong Kong, China: INBAR and Hong Kong Polytechnic University.

Chung, K., & Siu, Y. (2002). *Erection of bamboo scaffolds*. Hong Kong, China: INBAR and Hong Kong Polytechnic University.

CIP. (2011). Temporary works and falsework. In *Construction health and safety manual* (p. 16). Construction Industry Publications Ltd.

CIRIA. (1971). Effect of site factors on the Load capacities of adjustable steel props (No. 27). London, UK: Construction Industry Research and Information Association (CIRIA).

CIRIA. (1977). Rationalization of safety and serviceability factors in structural codes (No. 63). London, UK: Construction Industry Research and Information Association (CIRIA).

CIRIA. (1977). Safe working loads for adjustable steel props: The influence of prop conditions and site workmanship (No. TN79). London, UK: Construction Industry Research and Information Association (CIRIA).

CIRIA. (1978). A mathematical model of the structure of strong winds (No. 76). London, UK: Construction Industry Research and Information Association (CIRIA).

CIRIA. (1996). Control of risk: A Guide to the systematic management of risk from construction (No. SP125). London, UK: Construction Industry Research and Information Association (CIRIA).

CIRIA. (2000). Civil engineering design and construct - A guide to integrating design into the construction process (No. C534). London, UK: Construction Industry Research and Information Association (CIRIA).

CIRIA. (2003). Retention of masonry facades. Best practice guide (No. C579). London, UK: Construction Industry Research and Information Association (CIRIA).

CIRIA. (2009). Whole-life infrastructure asset management: Good practice guide for civil infrastructure (No. C677). London, UK: Construction Industry Research and Information Association (CIRIA).

CIRIA. (2011). Guidance on catastrophic events in construction (No. C699). London, UK: Construction Industry Research and Information Association (CIRIA).

CIRIA. (2015). CDM2015 – Construction work sector guidance for designers (No. C755). London, UK: Construction Industry Research and Information Association (CIRIA).

Claeskens, G., & Hjort, N. (2008). *Model selection and model averaging*. Cambridge, UK: Cambridge University Press. doi:10.1017/CBO9780511790485

Coles, S. (2001). *An introduction to statistical modeling of extreme values*. London, UK: Springer-Verlag. doi:10.1007/978-1-4471-3675-0

Collings, D. (2001). The A13 viaduct: construction of a large monolithic concrete bridge deck. *Proceedings of the ICE - Structures and Buildings, 146*(1), 85–91. doi:10.1680/stbu.2001.146.1.85

Comité Euro-International du Béton. (1993). *CEB–FIP Model code 90*. Thomas Telford Ltd.

Congdon, P. (2006). *Bayesian statistical modelling* (2nd ed.). Chichester, UK: John Wiley & Sons. doi:10.1002/9780470035948

Cook, H. (2007). Designer's guide to EN 1991–1–4 – Eurocode 1: Actions on structures. Part 1–4: General actions. Wind actions. London, UK: Thomas Telford Ltd.

Cooke, R. (1991). Experts in Uncertainty. Opinion and Subjective Probability in Science. Oxford, UK: Oxford University Press. Retrieved from https://global.oup.com/academic/product/experts-in-uncertainty-9780195064650

Cooke, R., & Goossens, L. (2004). Expert judgement elicitation for risk assessments of critical infrastructures. *Journal of Risk Research*, *7*(6), 643–656. doi:10.1080/1366987042000192237

Cook, N. (1999). *Wind loading: A practical guide to BS 6399–2*. London, UK: Thomas Telford Ltd. doi:10.1680/wl.27558

Costa, A., Crespo, A., Navarro, J., Lizcano, G., Madsen, H., & Feitosa, E. (2008). A review on the young history of the wind power short-term prediction. *Renewable & Sustainable Energy Reviews*, *12*(6), 1725–1744. doi:10.1016/j.rser.2007.01.015

Crémer, J. (2003). *Conception des ponts*. Liège, Belgium: Liège University.

Crisfield, M. (1997). *Nonlinear finite element analysis of solids and structures* (Vol. 2). New York, USA: Wiley.

Cropper, M., & Sussman, F. (1990). Valuing future risks to life. *Journal of Environmental Economics and Management*, *19*(2), 160–174. doi:10.1016/0095-0696(90)90066-8

CSA. (1975). S269.1–1975 R2003 – Falsework for construction purposes – Part 1: Design. Canada: Canadian Standards Association (CSA).

CSA. (2014). S6–14 – Canadian highway bridge design code. Canada: Canadian Standards Association (CSA).

Dagnew, A., & Bitsuamlak, G. (2014). Computational evaluation of wind loads on a standard tall building using LES. *Wind and Structures An International Journal*, *18*(5), 567–598. doi:10.12989/was.2014.18.5.567

Dallard, P., Fitzpatrick, T., & Flint, A. (2001). The Millennium bridge, London: Problems and solutions. *The Structural Engineer*, *79*(8), 15–17.

Das, P. (1997). *Safety of bridges*. Thomas Telford Ltd.

Davenport, A. (1960). Rationale for determining design wind velocities. *Journal of the Structural Division*, *86*(5), 39–68.

Davenport, A. (1967). Gust loading factors. *Journal of the Structural Division*, *93*(ST3), 11–34.

Deaves, D. (1981). Computations of wind flow over changes in surface roughness. *Journal of Wind Engineering and Industrial Aerodynamics*, *7*(1), 65–94. doi:10.1016/0167-6105(81)90068-4

Department of Transportation, Division of Structures. (2001). *California falsework manual* (32nd ed.). State of California.

Det Norske Veritas. (2002). Marine risk asessment. London, UK: The Health and Safety Executive (HSE).

DIN. (1987). *Telescopic steel props – Safety requirements, testing*. Berlin, Germany: Deutsches Institut für Normung.

DIN. (1987).*Telescopic steel props – Safety requirements, testing*. Berlin, Germany: Deutsches Institut für Normung.

Ditlevsen, O., & Friis-Hansen, P. (2005). Life quality time allocation index – an equilibrium economy consistent version of the current life quality index. *Structural Safety*, *27*(3), 262–275. doi:10.1016/j.strusafe.2004.12.001

Done, J., Davis, C., & Weisman, M. (2004). The next generation of NWP: Explicit forecasts of convection using the weather research and forecasting (WRF) model. *Atmospheric Science Letters*, *5*(6), 110–117. doi:10.1002/asl.72

Dowell, J., Weiss, S., Hill, S., & Infield, D. (2014). Short-term spatio-temporal prediction of wind speed and direction. *Wind Energy (Chichester, England)*, *17*(12), 1945–1955. doi:10.1002/we.1682

Dowrick, D. (2009). *Earthquake resistant design and risk reduction* (2nd ed.). John Wiley & Sons. doi:10.1002/9780470747018

Drew, D., Barlow, J., & Lane, S. (2013). Observations of wind speed profiles over Greater London, UK, using a Doppler lidar. *Journal of Wind Engineering and Industrial Aerodynamics*, *121*, 98–105. doi:10.1016/j.jweia.2013.07.019

Durkee, W. (2014). Steel bridge construction. In *Bridge engineering handbook: Construction and maintenance* (2nd ed.; Vol. 2, p. 50). CRC Press.

Dusenberry, D., & Hamburger, R. (2006). Practical means for energy-based analyses of disproportionate collapse potential. *Journal of Performance of Constructed Facilities*, *20*(4), 336–348. doi:10.1061/(ASCE)0887-3828(2006)20:4(336)

Dyrbye, C., & Hansen, S. (1999). *Wind loads on structures*. Chichester, UK: John Wiley & Sons.

Eales, R., Smith, S., Twigger-Ross, C., Sheate, W., Özdemiroglu, E., Fry, C., & Tomlinson, P. (2003). *Integrated appraisal methods (No. E2–044/TR)*. Bristol, UK: UK Environment Agency.

Edwards, R., LaDue, J., Ferree, J., Scharfenberg, K., Maier, C., & Coulbourne, W. (2013). Tornado intensity estimation: Past, present, and future. *Bulletin of the American Meteorological Society*, *94*(5), 641–653. doi:10.1175/BAMS-D-11-00006.1

Efron, B. (1979). Bootstrap methods: Another look at the jackknife. *Annals of Statistics*, *7*(1), 1–26. doi:10.1214/aos/1176344552

Eh Huang, N., & Shen, S. (2005). *Hilbert-Huang transform and its applications* (Vol. 5). Singapore: World Scientific.

Eligehausen, R., Mallée, R., & Silva, J. (2006). *Anchorage in concrete construction*. Berlin, Germany: Ernst & Sohn.

Ellingwood, B. (2008). Reliability based codes in the United States. Challenges to professional engineering practice. *Proceedings of the 4th International ASRANet Colloquium on Integrating for Structural Analysis, Risk & Reliability.*

Ellingwood, B., Smilowitz, R., Dusenberry, D., Duthinh, D., Lew, H., & Carino, N. (2007). Best practices for reducing the potential for progressive collapse in buildings. National Institute of Standards and Technology (NIST). doi:10.6028/NIST.IR.7396

Ellis, B., & Ji, T. (2004). The response of structures to dynamic crowd loads (No. Digest 426). Bracknell, UK: Building Research Establishment (BRE).

Ellis, B., & Ji, T. (1997). Human-structure interaction in vertical vibrations.*Proceedings of the ICE - Structures and Buildings, 122*, 1–9.

Ellis, B., Ji, T., & Littler, J. (2000). The response of grandstands to dynamic crowd loads.*Proceedings of the ICE - Structures and Buildings, 140*, 355–365. doi:10.1680/stbu.2000.140.4.355

El-Safty, A., Zinszer, M., & Morcous, G. (2008). Forensic investigation of a bridge construction scaffolding collapse. In *Proceedings of Structures Congress 2008: Crossing Borders* (pp. 1–10). Vancouver, Canada: American Society of Civil Engineers (ASCE). doi:10.1061/41016(314)198

El-Sheikh, M., & Chen, W. (1989a). *Behaviour and strength of telescopic steel shores (No. CE-STR-88-7).* West Lafayette, IN: Purdue University.

El-Sheikh, M., & Chen, W. (1989b). Maximum probabilistic shore load in multistorey R/C buildings. *Computers & Structures*, *32*(6), 1347–1357. doi:10.1016/0045-7949(89)90311-8

Enright, J., Harris, R., & Hancock, G. (2000). Structural stability of braced scaffolding and formwork with spigot joints. In *Proceedings of 15th International Specialty Conference in Cold-Formed Steel Design and Construction*, 357–376.

EOTA. (2013a). ETAG 001: Metal anchors for use in concrete – Parts 1–6. Brussels, Belgium: European Organisation for Technical Assessment (EOTA).

EOTA. (2013b). ETAG 029: Metal injection anchors for use in masonry. Brussels, Belgium: European Organisation for Technical Assessment (EOTA).

European Agency for Safety and Health at Work. (2004). *Building in safety. Prevention of risks in construction in practice*. Brussels, Belgium: European Agency for Safety and Health at Work.

European Commission. (2013a). An EU Strategy on adaptation to climate change. Brussels, Belgium: European Commision (EC).

European Commission. (2013b). An EU Strategy on adaptation to climate change. Adapting infrastructure to climate change. Brussels, Belgium: European Commision (EC).

Faber, M. (2009). Risk and safety in engineering course lecture notes. Swiss Federal Institute of Technology Zurich (ETHZ).

Faber, M., & Stewart, M. (2003). Risk assessment for civil engineering facilities: Critical overview and discussion. *Reliability Engineering & System Safety*, *80*(2), 173–184. doi:10.1016/S0951-8320(03)00027-9

Faella, C., Piluso, V., & Rizzano, G. (1999). *Structural steel semirigid connections: Theory, design, and software*. Boca Raton, FL: CRC Press.

Farrar, C., & Worden, K. (2007). An introduction to structural health monitoring. *Philosophical Transactions of the Royal Society of London A: Mathematical, Physical and Engineering Sciences, 365*(1851), 303–315.

Farrar, C., & Worden, K. (2012). *Structural health monitoring: A machine learning perspective*. Chichester, UK: John Wiley & Sons. doi:10.1002/9781118443118

Fawcett, L. (2005). *Statistical methodology for the estimation of environmental extremes* (PhD Thesis). University of Newcastle, Newcastle upon Tyne, UK.

Fawcett, L., & Walshaw, D. (2006). Markov chain models for extreme wind speeds. *Environmetrics*, *17*(8), 795–809. doi:10.1002/env.794

Fawcett, L., & Walshaw, D. (2008). Bayesian inference for clustered extremes. *Extremes*, *11*(3), 217–233. doi:10.1007/s10687-007-0054-y

Fawcett, L., & Walshaw, D. (2012). Estimating return levels from serially dependent extremes. *Environmetrics*, *23*(3), 272–283. doi:10.1002/env.2133

Feng, X. (1994). *The influence of semi-rigid connections on the behaviour of slender structures* (PhD Thesis). Oxford Brookes University, Oxford, UK.

Feng, X., Godley, M., & Beale, R. (1998). Rotational stiffnesses of semi-rigid baseplates.*Proceedings of 14th International Speciality Conference on Cold-Formed Steel Structures*, 323–334.

Fernāndez, J., Hermanns, L., Alarcón, E., & Fraile, A. (2013). A review of load models for the prediction of footfall-induced vibration in structures. In *Proceedings of 11th International Conference on Vibration Problems* (p. 9). Lisbon, Portugal: DEC/FCT/UNL and IDMEC/IST.

Ferry Borges, J., & Castanheta, M. (1968). Structural safety (1st ed.). Lisbon, Portugal: Laboratorio National de Engenharia Civil (LNEC).

Ferry Borges, J., & Castanheta, M. (1985). Structural safety (3rd ed.). Lisbon, Portugal: Laboratorio National de Engenharia Civil (LNEC).

fib. (2000). *Guidance for good bridge design*. Lausanne, Switzerland: International Federation for Structural Concrete (fib), Bulletin No. 9.

fib. (2009). *Formwork and falsework for heavy construction. Guide to good practice*. Lausanne, Switzerland: International Federation for Structural Concrete (fib), Bulletin No. 48.

Field, C., Barros, V., Stocker, T., Qin, D., Dokken, D., Ebi, K., … Midgley, P. (2012). Managing the risks of extreme events and disasters to advance climate change adaptation. A special report of Working Groups I and II of the Intergovernmental Panel on Climate Change (IPCC). Cambridge University Press, Cambridge, UK: Intergovernmental Panel on Climate Change (IPCC).

Fischhoff, B., Lichtenstein, S., Slovic, P., Derby, S., & Keeney, R. (1981). *Acceptable risk*. Cambridge University Press.

Flay, R. (2013). Bluff body aerodynamics. In Y. Tamura & A. Kareem (Eds.), *Advanced structural wind engineering* (pp. 59–84). Tokyo, Japan: Springer Japan. doi:10.1007/978-4-431-54337-4_3

Foley, A., Leahy, P., Marvuglia, A., & McKeogh, E. (2012). Current methods and advances in forecasting of wind power generation. *Renewable Energy*, *37*(1), 1–8. doi:10.1016/j.renene.2011.05.033

Frangopol, D., & Curley, J. (1987). Effects of damage and redundancy on structural reliability. *Journal of Structural Engineering*, *113*(7), 1533–1549. doi:10.1061/(ASCE)0733-9445(1987)113:7(1533)

Franke, J., Hirsch, C., Jensen, A., Krüs, H., Schatzmann, M., & Westbury, P., … Wright, N. (2004). Recommendations on the use of CFD in Wind Engineering. *Proceedings of the International Conference on Urban Wind Engineering and Building Aerodynamics: COST C14 – Impact of Wind and Storm on City life and Built Environment*, 11.

Frank, R., Bauduini, C., Driscoll, R., Kavvadas, M., Ovesen, N., Orr, T., & Schuppener, B. (2013). *Designers' Guide to Eurocode 7: Geotechnical design – General rules*. London, UK: Thomas Telford Ltd.

Fröderberg, M., & Thelandersson, S. (2013). Modelling uncertainty and gross errors in preliminary design of buildings. In *Proceedings of 11th International Conference on Structural Safety and Reliability (ICOSSAR 2013)* (pp. 439–444). New York: Taylor & Francis Group.

Frye, M., & Morris, G. (1975). Analysis of Flexibly connected steel frames. *Canadian Journal of Civil Engineering*, *2*(3), 280–291. doi:10.1139/l75-026

Fu, G., & Frangopol, D. (1990). Balancing weight, system reliability and redundancy in a multiobjective optimization framework. *Structural Safety*, *7*(2–4), 165–175. doi:10.1016/0167-4730(90)90066-X

Fujita, T. (1985). *The downburst: Microburst and macroburst: report of projects NIMROD and JAWS (Satellite and Mesometeorology Research Project)*. Chicago: University of Chicago.

Fyall, I. (2012). *Milton Keynes scaffolding collapse*. Retrieved from accessindustryforurum.org.uk (2016, June 27).

Gao, H., Gao, W., Li, C., & Huo, X. (2013). Influence of the structural performance of steel tubular scaffold based on measured imperfection. *Applied Mechanics and Materials*, *256–259*, 754–757.

Gerhardt, H., & Janser, F. (1994). Wind loads on permeable façades. *Journal of Wind Engineering and Industrial Aerodynamics*, *53*(1-2), 37–48. doi:10.1016/0167-6105(94)90017-5

Gerrand, C. (1987). The equivalent orthotropic elastic properties of plywood. *Wood Science and Technology*, *21*(4), 335–348. doi:10.1007/BF00380201

Giebel, G., Brownsword, R., Kariniotakis, G., Denhard, M., & Draxl, C. (2011). The state–of–the–art in short–term prediction of windpower (2nd ed.). Copenhagen, Denmark: ANEMOS.plus.

Goda, K., & Hong, H. (2008). Application of cumulative prospect theory: Implied seismic design preference. *Structural Safety*, *30*(6), 506–516. doi:10.1016/j.strusafe.2007.09.007

Godley, M. (1990). *The performance of SGB Cuplok connectors (No. R221)*. Oxford, UK: Oxford Polytechnic, Department of Civil Engineering, Building & Cartography.

Godley, M., & Beale, R. (1997). Sway stiffness of scaffold structures. *The Structural Engineer*, *75*(1), 4–12.

Godley, M., & Beale, R. (2001). Analysis of large proprietary access scaffold structures.*Proceedings of the ICE - Structures and Buildings*, *146*(1), 31–39. doi:10.1680/stbu.2001.146.1.31

Godley, M., Beale, R., & Feng, X. (1998). Rotational stiffness of semi-rigid baseplates.*Proceedings of the 14th Specialty Conference on Cold-Formed Steel Structures*, 323–335.

Goodwin, P., & Wright, G. (2010). *Decision analysis for management judgment* (4th ed.). Wiley.

Gorlin, W. (2009). Wind loads for temporary structures: Making the case for industrywide standards. *Journal of Architectural Engineering*, *15*(2), 35–36. doi:10.1061/(ASCE)1076-0431(2009)15:2(35)

Graubner, C., & Proske, T. (2005). Formwork pressure - A new concept for the calculation.*Proceedings of the 2nd North American Conference on the Design and Use of Self-Consolidating Concrete and the Fourth RILEM International Symposium on Self-Compacting Concrete*, 605–612.

Graubner, C., & Proske, T. (2010). Formwork pressure of highly workable concretes– Experiments focused on setting and vibration and design approach. In *Proceedings of the 6th RILEM Conference on Self-Compacting Concrete*. Montreal: Springer Verlag.

Gulvanessian, H., Calgaro, J., & Holický, M. (2012). *Designer's guide to Eurocode: Basis of structural design* (2nd ed.). London, UK: Thomas Telford Ltd. doi:10.1680/bsd.41714

Gutenberg, B., & Richter, C. (1954). *Seismicity of the earth and associated phenomena* (2nd ed.). Princeton, NJ: Princeton University Press.

Guthrie, P., & Konaris, T. (2012). *Infrastructure and resilience*. London, UK: Foresight, Government Office for Science.

Gylitoft, K., & Mroz, K. (1995). Load-carrying capacity of scaffolds. *Structural Engineering International*, *5*(1), 37–42. doi:10.2749/101686695780601619

Hadipriono, F., Lim, C., & Wong, K. (1986). Event tree analysis to prevent failures in temporary structures. *Journal of Construction Engineering and Management*, *112*(4), 500–513. doi:10.1061/(ASCE)0733-9364(1986)112:4(500)

Hadipriono, F., & Wang, H. (1986). Analysis of causes of falsework failures in concrete structures. *Journal of Construction Engineering and Management*, *112*(1), 112–121. doi:10.1061/(ASCE)0733-9364(1986)112:1(112)

Hadipriono, F., & Wang, H. (1987). Causes of falsework collapses during construction. *Structural Safety*, *4*(3), 179–195. doi:10.1016/0167-4730(87)90012-9

Haimes, Y. (2009). *Risk modeling, assessment and management* (3rd ed.). New York: John Wiley & Sons.

Haldar, A., & Mahadevan, S. (2000). *Probability, reliability, and statistical methods in engineering design*. New York: John Wiley & Sons.

Halperin, K., & McCann, M. (2004). An evaluation of scaffold safety at construction sites. *Journal of Safety Research*, *35*(2), 141–150. doi:10.1016/j.jsr.2003.11.004 PMID:15178232

Hartford, D., & Baecher, G. (2004). *Risk and uncertainty in dam safety*. London, UK: Thomas Telford Ltd. doi:10.1680/rauids.32705

Harung, H., Lightfoot, E., & Duggan, D. (1975). The strength of scaffold towers under vertical loading. *The Structural Engineer*, *53*(1), 23–30.

Hastie, T., Tibshirani, R., & Friedman, J. (2009). *The elements of statistical learning – Data mining, inference, and prediction* (2nd ed.). New York: Springer.

Hastings, J., Zhao, Q., & Burdette, E. (2010). Steel girder stability during bridge erection: AASHTO LRFD check on L/b ratios. *Journal of Bridge Engineering*, *15*(6), 759–762. doi:10.1061/(ASCE)BE.1943-5592.0000102

Hayes, K. (2011). *Uncertainty and uncertainty analysis methods (No. EP102467)*. Dickson, Australia: CSIRO Mathematics, Informatics and Statistics.

Helwany, C. (2007). *Applied soil mechanics with ABAQUS aplications*. John Wiley & Sons. doi:10.1002/9780470168097

Hewson, N. (2003). *Prestressed concrete bridges: Design and construction*. London, UK: Thomas Telford. doi:10.1680/pcbdac.32231

Hida, S., Ibrahim, F., Capers, H., Bailey, G., Friedland, I., Kapur, J., … Sivakumar, B. (2010). Assuring bridge safety and serviceability in Europe (No. FHWA-PL-10-01). Washington, DC: U.S. Department of Transportation.

Hill, H. (2004). Rational and irrational design loads for temporary' structures. *Practice Periodical on Structural Design and Construction*, *9*(3), 125–129. doi:10.1061/(ASCE)1084-0680(2004)9:3(125)

HM Treasury. (2003). *Appraisal and evaluation in central government (The green book)*. Controller of Her Majesty's Stationery Office.

HM Treasury. (2004). *Management of risk – Principles and concepts (The orange book)*. Controller of Her Majesty's Stationery Office.

Holmes, J. (2014). Along- and cross-wind response of a generic tall building: Comparison of wind-tunnel data with codes and standards. *Journal of Wind Engineering and Industrial Aerodynamics*, *132*, 136–141. doi:10.1016/j.jweia.2014.06.022

Holmes, J. (2015). *Wind loading of structures* (3rd ed.). CRC Press.

Holmes, J., Allsop, A., & Ginger, J. (2014). Gust durations, gust factors and gust response factors in wind codes and standards. *Wind and Structures An International Journal*, *19*(3), 339–352. doi:10.12989/was.2014.19.3.339

Holmes, J., Kwok, K., & Ginger, J. (2012). *Wind loading handbook for Australia and New Zealand: Background to AS/NZS1170.2 wind actions*. Sydney, Australia: Australasian Wind Engineering Society.

Holmes, M., & Hindson, D. (1979). Structural behaviour of load bearing falsework. *Proceedings of the ICE*, *67*(3), 721–741.

Hong Kong Buildings Department. (2006). *Guidelines for the design and construction of bamboo scaffolds*. Hong Kong, China: Occupational Safety and Health Branch.

Hong Kong Labour Department. (2014). *Code of practice for bamboo scaffolding safety* (3rd ed.). Hong Kong, China: Occupational Safety and Health Branch.

House of Lords. (2006). *Government policy on the management of risk* (No. 5th Report of Session 2005–06). London, UK: The Stationery Office Limited.

Hoxey, R. (1994). Full-scale studies of wind-loading on light-weight structures, windbreaks and walls at Silsoe Research Institute. In *Proceedings of Conference into Wind Loading on Temporary Structures* (pp. 83–98). Buxton, UK: Health and Safety Executive (HSE).

HSE. (1992). The tolerability of risk from nuclear power stations. London, UK: The Health and Safety Executive (HSE).

HSE. (1994). *Proceedings of the conference on wind loading on temporary structures*. London, UK: Health and Safety Executive (HSE).

HSE. (1995). Generic terms and concepts in the assessment and regulation of industrial risks. London, UK: The Health and Safety Executive (HSE).

HSE. (2001). Reducing risks, protecting people: HSE's decision making process (No. R2P2). London, UK: The Health and Safety Executive (HSE).

HSE. (2003). Causal factors in construction accidents. London, UK: The Health and Safety Executive (HSE).

HSE. (2006). Health and safety in construction (No. HSG150). London, UK: The Health and Safety Executive (HSE).

HSE. (2009a). Construction intelligence report: Analysis of construction injury and ill health intelligence 1997/98 - 2008/09. London, UK: The Health and Safety Executive (HSE).

HSE. (2009b). ND guidance on the demonstration of ALARP (as low as reasonably practicable) (No. T/AST/005-Issue 4-Rev 1). London, UK: The Health and Safety Executive (HSE).

HSE. (2010). The management of temporary works in the construction industry (No. SIM 02/2010/04). London, UK: The Health and Safety Executive (HSE).

HSE. (2011). Preventing catastrophic events in construction. London, UK: The Health and Safety Executive (HSE).

HSE. (2013). Managing for health and safety (No. HSG65). London, UK: The Health and Safety Executive (HSE).

HSE. (2015). The construction (Design and management) regulations 2015 (CDM2015). London, UK: Health and Safety Executive (HSE).

Huang, S., Li, Q., & Xu, S. (2007). Numerical evaluation of wind effects on a tall steel building by CFD. *Journal of Constructional Steel Research*, *63*(5), 612–627. doi:10.1016/j.jcsr.2006.06.033

Huang, Y., Chen, H., Rosowsky, D., & Kao, Y. (2000). Load carrying capacities and failure modes of scaffold-shoring systems. Part 1: Modelling and experiments. *Structural Engineering & Mechanics*, *10*(1), 53–66. doi:10.12989/sem.2000.10.1.053

Huang, Y., Chen, W., Chen, H., Yen, T., Kao, Y., & Lin, C. (2000). A monitoring method for scaffold-frame shoring systems for elevated concrete formwork. *Computers & Structures*, *78*(5), 681–690. doi:10.1016/S0045-7949(00)00051-1

Huang, Y., Kao, Y., & Rosowsky, D. (2000). Load carrying capacities and failure modes of scaffold-shoring systems. Part 2: An analytical model and its closed-form solution. *Structural Engineering & Mechanics*, *10*(1), 67–79. doi:10.12989/sem.2000.10.1.067

Hu, C., Ge, Z., & Jing, W. (2011). Research on ultimate bearing capacity of coupler steel tube falsework with initial defect. *Advanced Materials Research*, *163–167*, 32–35.

Huizing, J., Blakeley, R., & Ramsay, G. (1977). Falsework. *New Zealand Engineering*, *32*(1), 2–9.

Humar, J. (2012). *Dynamics of structures* (3rd ed.). Leiden, The Netherlands: CRC Press.

Hurd, M. (1995). Formwork for concrete (6th ed.). American Concrete Institute (ACI), ACI SP-4.

Ikäheimonen, J. (1997). *Construction loads on shores and stability of horizontal formworks* (PhD Thesis). Royal Institute of Technology (KTH), Stockholm, Sweden.

Ilick, J., Arora, V., & Dolejs, J. (2015). Validation of finite element updated models of the developed facade scaffold anchor.*IOP Conference Series: Materials Science and Engineering*, *96*, 9.

Irtaza, H. (2009). *Experimental and computational determination of wind loads on netted/sheeted scaffolds* (PhD Thesis). Oxford Brookes University, Oxford, UK.

Irtaza, H., Beale, R., & Godley, M. (2007). Determination of the effects of wind load on bare tube access scaffold structures using computational fluid dynamics. In *Proceedings of the 5th International Conference on Advances in Steel Structures (ICASS 2007)*, 637–642.

Irtaza, H., Beale, R., & Godley, M. (2007). Determination of the effects of wind load on bare tube access scaffold structures using computational fluid dynamics.*Proceedings of the 5th International Conference on Advances in Steel Structures (ICASS 2007)*,637–642.

Irtaza, H., Beale, R., & Godley, M. (2012). A wind-tunnel investigation into pressure distribution around sheet clad scaffolds. *Journal of Wind Engineering and Industrial Aerodynamics*, *103*, 86–95. doi:10.1016/j.jweia.2012.03.004

Irtaza, H., Beale, R., Godley, M., & Jameel, A. (2013). Comparison of wind pressure measurements on Silsoe experimental building from full-scale observation, wind-tunnel experiments and various CFD techniques. *International Journal of Engineering Science and Technology*, *5*(1), 28–41.

Irtaza, H., Beale, R., Godley, M., & Jameel, A. (2014). Pressure distribution around net clad scaffolds using computation fluid dynamics techniques. *IOSR Journal of Mechanical and Civil Engineering*, *4*, 54–62.

ISO. (1998). ISO 2394:1998 – General principles on reliability for structures (3rd ed.). Geneva, Switzerland: International Organization for Standardization (ISO).

ISO. (2001). ISO 12494:2001 – Atmospheric icing of structures. Geneva, Switzerland: International Organization for Standardization (ISO).

ISO. (2005). ISO 6507-1:2005 – Metallic materials – Vickers hardness test – Test method. Geneva, Switzerland: International Organization for Standardization (ISO).

ISO. (2007). ISO 10137:2007 – Bases for design of structures. Serviceability of buildings and walkways against vibrations (2nd ed.). Geneva, Switzerland: International Organization for Standardization (ISO).

ISO. (2009). ISO 4354:2009 – Wind actions on structures. Geneva, Switzerland: International Organization for Standardization (ISO).

ISO. (2009a). ISO 13824:2009 – Bases for design of structures – General principles on risk assessment of systems involving structures. Geneva, Switzerland: International Organization for Standardization (ISO).

ISO. (2009a). ISO 6892–1:2009 – Metallic materials – Tensile testing – Part 1: Method of test at room temperature. Geneva, Switzerland: International Organization for Standardization (ISO).

ISO. (2009b). ISO 31000:2009 – Risk management – Principles and guidelines. Geneva, Switzerland: International Organization for Standardization (ISO).

ISO. (2009b). ISO 8375:2009 Ed 2 – Timber structures. Glued laminated timber. Test methods for determination of physical and mechanical properties (2nd ed.). Geneva, Switzerland: International Organization for Standardization (ISO).

ISO. (2009c). ISO 31010:2009 – Risk management – Risk assessment techniques. Geneva, Switzerland: International Organization for Standardization (ISO).

ISO. (2012). ISO 8686–1:2012 – Cranes – Design principles for loads and load combinations – Part 1: General. Geneva, Switzerland: International Organization for Standardization (ISO).

ISO. (2013). ISO 8492:2013 – Metallic materials – Tube – Flattening test. Geneva, Switzerland: International Organization for Standardization (ISO).

ISO. (2014). ISO 6506-1:2014 – Metallic materials – Brinell hardness test – Test method. Geneva, Switzerland: International Organization for Standardization (ISO).

ISO. (2015). ISO 2394 :2015 – General principles on reliability for structures (4th ed.). Geneva, Switzerland: International Organization for Standardization (ISO).

ISO. (2015). ISO 9000:2015 – Quality management systems – Fundamentals and vocabulary (4th ed.). Geneva, Switzerland: International Organization for Standardization (ISO).

IStructE. (2007). Temporary demountable structures: Guidance on procurement, design and use (3rd ed.). London, UK: The Institution of Structural Engineers.

Ivanyi, M., & Baniotopoulos, C. (2014). *Semi-rigid joints in structural steelwork*. Vienna, Austria: Springer Verlag Wien.

Jagger, T., & Elsner, J. (2006). Climatology models for extreme hurricane winds near the United States. *Journal of Climate*, *19*(13), 3220–3236. doi:10.1175/JCLI3913.1

Jaspart, J., & Weynard, K. (2016). *Design of joints in steel and composite structures*. Berlin, Germany: John Wiley & Sons. doi:10.1002/9783433604762

JCSS. (2001). *JCSS Probabilistic model code*. Joint Committee on Structural Safety.

Johnston, D. (2014). Formwork for concrete (8th ed.). American Concrete Institute (ACI), ACI SP-4.

Jones-Lee, M., Loomes, G., Yates, D., Mitchell, E., Spackman, M., McQuaid, J., & Thomson, T. (2008). T440: Fatalities and weighted injuries. London, UK: Rail Safety and Standards Board (RSSB).

Kahneman, D., & Tversky, A. (1979). Prospect theory: An analysis of decision under risk. *Econometrica*, *47*(2), 263–291. doi:10.2307/1914185

Kaplan, S., & Garrick, B. (1981). On the quantitative definition of risk. *Risk Analysis*, *1*(1), 11–27. doi:10.1111/j.1539-6924.1981.tb01350.x PMID:11798118

Kareem, A., & McCullough, M. (2013). Numerical simulation of wind effects. In Y. Tamura & A. Kareem (Eds.), *Advanced structural wind engineering* (pp. 261–299). Tokyo, Japan: Springer Japan. doi:10.1007/978-4-431-54337-4_10

Kasperski, M. (2013). Estimation of the design wind speed. In Y. Tamura & A. Kareem (Eds.), *Advanced structural wind engineering* (pp. 27–58). Tokyo, Japan: Springer. doi:10.1007/978-4-431-54337-4_2

Khun, M., & Johnson, K. (2013). *Applied predictive modeling*. New York: Springer. doi:10.1007/978-1-4614-6849-3

Kindmann, R., & Kraus, M. (2012). *Steel structures: Design using FEM*. Berlin, Germany: Ernst & Sohn Verlag. doi:10.1002/9783433600771

Kishi, N., & Chen, W. (1990). Moment-rotation relations of semirigid connections with angles. *Journal of Structural Engineering*, *116*(7), 1813–1834. doi:10.1061/(ASCE)0733-9445(1990)116:7(1813)

Knoll, F. (2013). Avoiding errors through collaboration. In *Proceedings of 11th International Conference on Structural Safety and Reliability (ICOSSAR 2013)* (pp. 111–117). New York: Taylor & Francis Group.

Knoll, F., & Vogel, T. (2009). Design for robustness. Zurich: International Association for Bridge and Structural Engineering (IABSE).

Knoll, F. (2011). Error proneness in construction. In *Applications of statistics and probability in civil engineering* (pp. 2745–2752). London: Taylor & Francis Group. doi:10.1201/b11332-405

Kolmogoroff, A. (1933). *Grundbegriffe der wahrscheinlichkeitsrechnung*. Berlin, Germany: Springer Berlin Heidelberg. doi:10.1007/978-3-642-49888-6

Konakli, K., & Faber, M. (2014). Value of information analysis in structural safety. In *Proceedings of 2nd International Conference on Vulnerability and Risk Analysis and Management (ICVRAM) and the 6th International Symposium on Uncertainty, Modeling, and Analysis (ISUMA)* (pp. 1605–1614). Liverpool, UK: American Society of Civil Engineers (ASCE). doi:10.1061/9780784413609.161

Korol, R., & Sivakumaran, K. (2014). Reassessing the plastic hinge model for energy dissipation of axially loaded columns. *Journal of Structures, 2014*(1), 7.

Kumamoto, H. (2007). *Satisfying safety goals by probabilistic risk assessment*. London, UK: Springer-Verlag.

Kwok, K. (2013). Wind-induced vibrations of structures. In Y. Tamura & A. Kareem (Eds.), *Advanced structural wind engineering* (pp. 121–155). Tokyo, Japan: Springer Japan. doi:10.1007/978-4-431-54337-4_5

LaViolette, M., Wipf, T., Lee, Y., Bigelow, J., & Phares, B. (2007). Bridge construction practices using incremental launching. Washington, DC: American Association of State Highway and Transportation Officials (AASHTO).

Le Guen, J. (2008). *Tolerability of risk: UK principles & practice for controlling work activities*. Presented at the Workshop on Tolerable Risk Evaluation sponsored by US Army Corps of Engineers. U.S. Department of the Interior: Bureau of Reclamation.

Leadbetter, M., Bailey, G., & Rootzén, H. (1983). *Extremes and related properties of random sequences and processes*. New York: Springer-Verlag. doi:10.1007/978-1-4612-5449-2

Lee, M., & Daebritz, M. (2010). *Moveable scaffolding systems - MSS*. Presented at the IABSE WG 6 Seminar: State-of-the-art Bridge Deck Erection; Safe and Efficient use of Special Equipment, Singapore.

Leonhardt, F. (1979). *Vorlesungen über massivbau – Teil 6 – Grundlagen des massivbrückenbaues*. Berlin, Germany: Springer-Verlag. doi:10.1007/978-3-642-61863-5

Levy, H. (2006). *Stochastic dominance: Investment decision making under uncertainty* (2nd ed.). New York: Springer.

Lew, H. S. (1984). Construction failures and their lessons. *Batiment International Building Research and Practice*, *12*(5), 272–275. doi:10.1080/09613218408545283

Liddle, R. (2010). The launching of the River Esk bridge, UK.*Proceedings of the ICE - Bridge Engineering, 163*(3), 137–146. doi:10.1680/bren.2010.163.3.137

Lightfoot, E., & Bhula, D. (1977a). A test rig for scaffold couplers. *Matériaux et Construction*, *10*(3), 168–173. doi:10.1007/BF02474850

Lightfoot, E., & Bhula, D. (1977b). The idealization of scaffold couplers for performance tests and scaffold analysis. *Matériaux et Construction*, *10*(3), 159–168. doi:10.1007/BF02474849

Lightfoot, E., & Le Messurier, A. (1974). Elastic analysis of frameworks with elastic connections. *Journal of the Structural Division*, *100*(ST6), 1297–1309.

Lightfoot, E., & Lemessurier, A. (1975). Instability of space frames having elastically connected and offset members. In *Proceedings of the 2nd International Conference on Space Structures*, 143–149.

Lightfoot, E., Olivetto, G., & Merchant, W. (1977). The collapse strength of tubular steel scaffold assemblies. *Proceedings of the ICE*, *63*(2), 311–329.

Lindley, D., Pandya, D., & Khanpour, A. (2001). Estimating seismic response of freestanding scaffolding. In *Transactions of the 16th International Conference on Structural Mechanics in Reactor Technology*. Washington, DC: International Association for Structural Mechanics in Reactor Technology (IASMRT).

Lindner, J., & Magnitzke, P. (1990). Stability of scaffold structures with sheeting[in German]. *Stahlbau*, *59*(2), 39–44.

Liu, H., Chen, Z., Wang, X., & Zhou, T. (2010). Theoretical analysis and experimental research on stability behavior of structural steel tube and coupler falsework with X-bracing. *Advanced Steel Construction*, *6*(4), 949–962.

Liu, H., Zhao, Q., Wang, X., Zhou, T., Wang, D., Liu, J., & Chen, Z. (2010). Experimental and analytical studies on the stability of structural steel tube and coupler scaffolds without X-bracing. *Engineering Structures*, *32*(4), 1003–1015. doi:10.1016/j.engstruct.2009.12.027

Lucko, G., & Garza, J. (2003). Constructability considerations for balanced cantilever construction. *Practice Periodical on Structural Design and Construction*, *8*(1), 47–56. doi:10.1061/(ASCE)1084-0680(2003)8:1(47)

Maitra, A. (1994). Potential for reducing accidents by cladding scaffolds. In *Proceedings of Conference into Wind Loading on Temporary Structures* (pp. 1–15). Buxton, UK: Health and Safety Executive (HSE).

Maitra, A. (1997). Accidents associated with scaffolding in the UK. In *Proceedings of Technical Seminar, Design and use of Temporary Structures*. London, UK: Institution of Civil Engineers (ICE).

Mallée, R., Fuchs, W., & Eligehausen, R. (2013). *Design of fastenings for use in concrete - The CEN/TS 1992-4 provisions*. Weinheim, Germany: Wiley. doi:10.1002/9783433602737

Maquoi, R., & Rondal, J. (1978). Mise en équations des nouvelles courbes européennes de flambement. *Revue Construction Métallique*, *1*, 17–29.

Marsden, J., McDonald, L., Bowles, D., Davidson, R., & Nathan, R. (2007). Dam safety, economic regulation and society's need to prioritise health and safety expenditures.*Proceedings of the NZSOLD/ANCOLD Workshop on 'Promoting and Ensuring the Culture of Dam Safety'*, 12.

Masumoto, H., Hara, K., & Yamashita, M. (1994). Design plan of super multi-span continuous Menshin bridge with deck length of 725m. In *Proceedings of the 3rd U.S.- Japan Workshop on Earthquake Protective Systems for Bridges* (Vol. 4, pp. 69–84). U.S. National Center for Earthquake Engineering Research.

Matousek, M., & Schneider, J. (1976). Untersuchungen zur struktur des sicherheitsproblems bei bauwerken (in German) (No. 59). Zurich, Switzerland: Swiss Federal Institute of Technology Zurich (ETHZ).

Matousek, M., & Schneider, J. (1976). Untersuchungen zur struktur des sicherheitsproblems bei bauwerken. Zurich, Switzerland: Swiss Federal Institute of Technology Zurich (ETHZ). (in German)

McDonald, L., Bowles, D., Hartford, D., van der Meer, J., & Vrijling, J. (2005). Risk assessment in dam safety management. International Commission of Large Dams (ICOLD), Bulletin 130.

McGill, W., & Ayyub, B. (2007). The meaning of vulnerability in the context of critical infrastructure protection. In *Critical infrastructure protection: Elements of risk* (pp. 25–48). George Mason University School of Law.

Melchers, R., Baker, M., & Moses, F. (1983). Evaluation of experience. In *Proceedings of IABSE Workshop on Quality Assurance within the Building Process* (pp. 9–30). Rigi, Switzerland: International Association for Bridge and Structural Engineering (IABSE).

Melchers, R. (1999). *Structural reliability analysis and prediction* (2nd ed.). Wiley.

Metheringham, S., & Townshend, J. (2005). A1 River Tyne bridge, UK-design and construction. *Bridge Engineering, 158*(2), 45–52.

Milojkovic, B. (1999). *Factors of safety for standard scaffold structures* (PhD Thesis). Oxford Brookes University, Oxford, UK.

Milojkovic, B., Beale, R., & Godley, M. (1996). Modelling scaffold connections. In *Proceedings of 4th ACME UK Annual Conference* (pp. 85–88). Glasgow, UK: Association of Computational Mechanics in Engineering (ACME).

Milojkovic, B., Beale, R., & Godley, M. (2002). Determination of the factors of safety of standard scaffold structures. In *Proceedings of the 3rd International Conference on Advances in Steel Structures (ICASS 2002)* (pp. 303–310). Hong Kong, China: Elsevier. doi:10.1016/B978-008044017-0/50035-4

Mizon, B., & Kitchener, J. (1997). Second Severn crossing: Viaduct superstructure and piers. *Proceedings of the ICE -. Civil Engineering (New York, N.Y.), 120*(2), 35–48.

Mohammadi, J., & Heydari, A. (2008). Seismic and wind load considerations for temporary structures. *Practice Periodical on Structural Design and Construction, 13*(3), 128–134. doi:10.1061/(ASCE)1084-0680(2008)13:3(128)

Moore, M., & Green, T. (2010). Investigation of falsework failures during construction of a cast-in-place bridge structure. In *Proceedings of the 2010 Structures Congress* (pp. 22–31). American Society of Civil Engineers (ASCE). doi:10.1061/41130(369)3

Morgenthal, G., Sham, R., & Schwarz, A. (2008). The 4000 tonne lift of Stonecutters bridge steel back-spans, H.K.*Proceedings of the ICE - Bridge Engineering, 161*(4), 197–208. doi:10.1680/bren.2008.161.4.197

Muhanna, R., Rao, M., & Mullen, R. (2013). Advances in interval finite element modelling of structures. *Life Cycle Reliability and Safety Engineering, 2*(3), 15–22.

Munich Re. (2015). *NatCatSERVICE. Loss events worldwide 1980 – 2014*. Munich Re.

Narasimhan, H., & Faber, M. (2009). Categorisation and assessment of robustness related provisions in European standards.*Proceedings of the Joint Workshop of COST Actions TU0601 and E55*, 147–164.

NASC. (2005). *NASC TG20:05– Guide to good practice for scaffolding with tube and fittings*. London, UK: National Access and Scaffolding Confederation (NASC).

NASC. (2011). *Anchorage systems for scaffolding*. London, UK: National Access and Scaffolding Confederation.

NASC. (2013). *Guide to good practice for scaffolding with tube and fittings*. London, UK: National Access and Scaffolding Confederation.

Nathwani, J., Lind, N., & Pandey, M. (1997). *Affordable safety by choice: The life quality method*. Waterloo, Canada: Institute of Risk Research, University of Waterloo.

Navon, R., & Kolton, O. (2007). Algorithms for automated monitoring and control of fall hazards. *Journal of Computing in Civil Engineering*, *21*(1), 21–28. doi:10.1061/(ASCE)0887-3801(2007)21:1(21)

Newman, J., & Choo, B. (2003). *Advanced concrete technology: Processes*. Butterworth-Heinemann.

Nieto, F., Hernández, S., Jurado, J., & Romera, L. (2010). Code provisions for wind Loads on short road bridges: Spanish IAP code, UNE-EN 1991-1-4 & 2007 AASHTO LRFD bridge design specifications. In *Proceedings of the 2010 Structures Congress* (pp. 2122–2133). American Society of Civil Engineers (ASCE). doi:10.1061/41130(369)192

Nishijima, K. (2009). *Issues of sustainability in engineering decision analysis* (PhD Thesis). ETH Zurich, Zurich, Switzerland.

Nitschke, J. (2010). *Form-traveller systems*. Presented at the IABSE WG 6 Seminar: State-of-the-art Bridge Deck Erection; Safe and Efficient use of Special Equipment, Singapore: International Association for Bridge and Structural Engineering (IABSE).

Nwogugu, M. (2006). A further critique of cumulative prospect theory and related approaches. *Applied Mathematics and Computation*, *179*(2), 451–465. doi:10.1016/j.amc.2005.11.102

Oakley, J., & OHagan, A. (2004). Probabilistic sensitivity analysis of complex models: A Bayesian approach. *Journal of the Royal Statistical Society. Series B, Statistical Methodology*, *66*(3), 751–769. doi:10.1111/j.1467-9868.2004.05304.x

Oberkampf, W., & Roy, C. (2010). *Verification and validation in scientific computing*. Cambridge, UK: Cambridge University Press. doi:10.1017/CBO9780511760396

Okubo, K., Hongo, T., & Kondo, K. (2010). Evaluation of wind force coefficients of the self-climbing scaffold (in Japanese). In *Proceedings of the 21st National Symposium on Wind Engineering*. Japan Association for Wind Engineering.

Ota, M. (1950). Studies on the properties of bamboo stem. Part 4: Study on the form of specimen in the test of compressive strength. *Journal Japanese Forestry Society*, *32*, 65–69.

Ottosen, N., & Ristinmaa, M. (2005). *The mechanics of constitutive modeling*. Amsterdam, The Netherlands: Elsevier.

Ouchi, F. (2004). *A literature review on the use of expert opinion in probabilistic risk analysis*. Washington, DC: The World Bank. doi:10.1596/1813-9450-3201

Pallett, P., Burrow, M., Clark, L., & Ward, R. (2001). *Investigation into aspects offalsework (No. crr01394)*. Birmingham, UK: The University of Birmingham.

Palutikof, J., Brabson, B., Lister, D., & Adcock, S. (1999). A review of methods to calculate extreme wind speeds. *Meteorological Applications*, *6*(2), 119–132. doi:10.1017/S1350482799001103

PASMA. (2000). Operators code of practice for PASMA aluminium alloy towers (7th ed.). Glasgow, UK: Prefabricated Access Suppliers' and Manufacturers' Association (PASMA).

Peng, J. (2002). Stability analyses and design recommendations for practical shoring systems during construction. *Journal of Construction Engineering and Management*, *128*(6), 536–544. doi:10.1061/(ASCE)0733-9364(2002)128:6(536)

Peng, J. (2004). Structural modeling and design considerations for double-layer shoring systems. *Journal of Construction Engineering and Management*, *130*(3), 368–377. doi:10.1061/(ASCE)0733-9364(2004)130:3(368)

Peng, J., Ho, C., Lin, C., & Chen, W. (2015). Load-carrying capacity of single-row steel scaffolds with various setups. *Advanced Steel Construction*, *11*(2), 185–210.

Peng, J., Pan, A., & Chan, S. (1998). Simplified models for analysis and design of modular falsework. *Journal of Constructional Steel Research*, *48*(2–3), 189–209. doi:10.1016/S0143-974X(98)00198-9

Peng, J., Pan, A., Chen, W., Yen, T., & Chan, S. (1997). Structural modeling and analysis of modular falsework systems. *Journal of Structural Engineering*, *123*(9), 1245–1251. doi:10.1061/(ASCE)0733-9445(1997)123:9(1245)

Peng, J., Pan, A., Rosowsky, D., Chen, W., Yen, T., & Chan, S. (1996a). High clearance scaffold systems during construction - I. Structural modelling and modes of failure. *Engineering Structures*, *18*(3), 247–257. doi:10.1016/0141-0296(95)00144-1

Peng, J., Pan, A., Rosowsky, D., Chen, W., Yen, T., & Chan, S. (1996b). High clearance scaffold systems during construction - II. Structural analysis and development of design guidelines. *Engineering Structures*, *18*(3), 258–267. doi:10.1016/0141-0296(95)00145-X

Peng, J., Rosowsky, D., Pan, A., Chen, W., Chan, S., & Yen, T. (1994). *Concrete placement load effects during construction - Part I: Structural modeling and analysis (No. CE-STR-94-15)*. West Lafayette, IN: Purdue University.

Peng, J., Wang, P., Chan, S., & Huang, C. (2012). Load capacities of single-layer shoring systems: An experimental study. *Advances in Structural Engineering*, *15*(8), 1389–1410. doi:10.1260/1369-4332.15.8.1389

Peng, J., Wu, C., Chan, S., & Huang, C. (2013). Experimental and numerical studies of practical system scaffolds. *Journal of Constructional Steel Research*, *91*, 64–75. doi:10.1016/j.jcsr.2013.07.028

Peng, J., Wu, C., Shih, M., & Yang, Y. (2013). Experimental study of load capacities of tubular steel adjustable shores used in construction. *International Journal of Structural Stability and Dynamics*, *13*(04), 32. doi:10.1142/S0219455412500630

Peng, J., Yen, T., Pan, A., Chen, W., & Chan, S. (1996). Structural modeling and analysis of scaffold systems. In *Proceedings of International Conference on Advances in Steel Structures* (pp. 251–256). Hong Kong, China: Pergamon. doi:10.1016/B978-008042830-7/50040-1

Peurifoy, R., & Oberlender, G. (2010). *Formwork for Concrete Structures* (4th ed.). New York: McGraw-Hill Education.

Pieńko, M., & Błazik-Borowa, E. (2013). Numerical analysis of load-bearing capacity of modular scaffolding nodes. *Engineering Structures*, *48*, 1–9. doi:10.1016/j.engstruct.2012.08.028

Pierre, L., Kopp, G., Surry, D., & Ho, T. (2005). The UWO contribution to the NIST aerodynamic database for wind loads on low buildings: Part 2. Comparison of data with wind load provisions. *Journal of Wind Engineering and Industrial Aerodynamics*, *93*(1), 31–59. doi:10.1016/j.jweia.2004.07.007

Pisheh, Y., Shafiei, H., & Hatambeigi, M. (2009). A case study of failure due to inappropriate usage of forming scaffold system. In *Proceedings of 5th Forensic Engineering Congress 2009* (pp. 536–545). Washington, DC: American Society of Civil Engineers (ASCE). doi:10.1061/41082(362)55

Podolny, W., & Muller, J. (1982). *Construction and design of prestressed concrete segmental bridges*. New York: Wiley.

Poon, C., & Yip, R. (2008). Comparison of timber and metal formwork systems. *Water Resources Management*, *161*(1), 29–36.

Potts, D. M., & Zdravkovic, L. (1999). Finite element analysis in geotechnical engineering.: Vol. 1. *Theory*. London, UK: Thomas Telford Publishing.

Potts, D. M., & Zdravkovic, L. (2001). Finite element analysis in geotechnical engineering.: Vol. 2. *Application*. London, UK: Thomas Telford Publishing.

Póvoas, A. (2012). The utilization of movable scaffolding systems in large spans. *Structural Engineering International, 22*(3), 395–400. doi:10.2749/101686612X13363869853374

Prabhakaran, U. (2009). *Nonlinear analysis of scaffolds with semirigid connections* (PhD Thesis). Oxford Brookes University, Oxford, UK.

Prabhakaran, U., Beale, R., & Godley, M. (2011). Analysis of scaffolds with connections containing looseness. *Computers & Structures, 89*(21–22), 1944–1955. doi:10.1016/j.compstruc.2011.03.016

Prestney, L. (2012). Quake lessons for scaffolding safety. *Build Magazine, New Zealand, 130*, 1.

Proske, T., & Graubner, C. (2007). Pressure on formwork using SCC - experimental studies and modelling. In *Proceedings of 5th International RILEM Symposium on Self-Compacting Concrete* (pp. 473–478). Ghent, Belgium: RILEM Publications SARL.

Proske, T., & Graubner, C. (2008). Formwork pressure asserted by highly workable concretes. *Darmstadt Concrete - Annual Journal on Concrete and Concrete Structures, 23*.

Proske, D. (2008). *Catalogue of risks. Natural, technical, social and health risks*. Berlin, Germany: Springer Berlin Heidelberg.

Quinion, D. (1984). Load variations in bridge falsework. In *Proceedings of IABSE 12th Congress* (p. 1104). Vancouver, Canada: International Association for Bridge and Structural Engineering (IABSE).

Racic, V., Pavic, A., & Brownjohn, J. (2009). Experimental identification and analytical modelling of human walking forces: Literature review. *Journal of Sound and Vibration, 326*(1–2), 1–49. doi:10.1016/j.jsv.2009.04.020

Rackwitz, R. (2004). Discounting for optimal and acceptable technical facilities involving risks. *HERON, 49*(2), 139–170.

Rademaekers, K., Laan, J., Boeve, S., & Lise, W. (2011). Investment needs for future adaptation measures in EU nuclear power plants and other electricity generation technologies due to effects of climate change. Brussels, Belgium: Directorate-General for Energy, European Commision (EC).

RAIB. (2009). Collision with debris from bridge GE19 near London Liverpool Street 28 May 2008. London, UK: Rail Accident Investigation Branch (RAIB), Department for Transport.

Rao, M., Mullen, R., & Muhanna, R. (2011). A new interval finite element formulation with the same accuracy in primary and derived variables. *International Journal of Reliability and Safety, 5*(3/4), 336. doi:10.1504/IJRS.2011.041184

Rasmussen, K. (2003). Full-range stress–strain curves for stainless steel alloys. *Journal of Constructional Steel Research, 59*(1), 47–61. doi:10.1016/S0143-974X(02)00018-4

Ratay, R. (2009). *Forensic structural engineering handbook*. McGraw Hill Professional.

Reason, J. (1990). *Human error*. Cambridge University Press. doi:10.1017/CBO9781139062367

Richards, P., & Hoxey, R. (1993). Appropriate boundary conditions for computational wind engineering models using the k-ϵ turbulence model. *Journal of Wind Engineering and Industrial Aerodynamics, 46–47*, 145–153. doi:10.1016/0167-6105(93)90124-7

Richards, P., & Hoxey, R. (2008). Wind loads on the roof of a 6 m cube. *Journal of Wind Engineering and Industrial Aerodynamics, 96*(6–7), 984–993. doi:10.1016/j.jweia.2007.06.032

Richards, P., Hoxey, R., Connell, B., & Lander, D. (2007). Wind-tunnel modelling of the Silsoe Cube. *Journal of Wind Engineering and Industrial Aerodynamics, 95*(9–11), 1384–1399. doi:10.1016/j.jweia.2007.02.005

Richards, P., & Norris, S. (2011). Appropriate boundary conditions for computational wind engineering models revisited. *Journal of Wind Engineering and Industrial Aerodynamics*, *99*(4), 257–266. doi:10.1016/j.jweia.2010.12.008

Richards, P., & Robinson, M. (1999). Wind loads on porous structures. *Journal of Wind Engineering and Industrial Aerodynamics*, *83*(1–3), 455–465. doi:10.1016/S0167-6105(99)00093-8

Rimington, J., McQuaid, J., & Trbojevic, V. (2003). *Application of risk-based strategies to workers' health and safety protection: UK experience*. Ministry of Social Affairs and Employment.

Rosignoli, M. (2000). Thrust and guide devices for launched bridges. *Journal of Bridge Engineering*, *5*(1), 75–83. doi:10.1061/(ASCE)1084-0702(2000)5:1(75)

Rosignoli, M. (2001). Deck segmentation and yard optimisation for launched bridges. *Concrete International*, *23*(2), 64–73.

Rosignoli, M. (2002). *Bridge launching*. London, UK: Thomas Telford Ltd.

Rosignoli, M. (2007). Robustness and stability of launching gantries and movable shuttering systems - Lessons learned. *Structural Engineering International*, *17*(2), 133–140. doi:10.2749/101686607780680673

Rosignoli, M. (2010). Self-launching erection machines for precast concrete bridges. *PCI Journal*, *55*(1), 36–57. doi:10.15554/pcij.01012010.36.57

Rosignoli, M. (2013). *Bridge construction equipment*. London, UK: ICE Publishing. doi:10.1680/bce.58088

Rosowsky, D. (1995). Estimation of design loads for reduced reference periods. *Structural Safety*, *17*(1), 17–32. doi:10.1016/0167-4730(94)00037-Q

Rosowsky, D., Huang, Y., Chen, W., & Yen, T. (1994). Modeling concrete placement loads during construction. *Structural Engineering Review*, *6*(2), 71–84.

Rosowsky, D., Huston, D., Fuhr, P., & Chen, W. (1994). Measuring formwork loads during construction. *Concrete International*, *16*(11), 21–25.

SAA. (1995). *3610:1995 – Formwork for concrete. Standards Association of Australia.*

SAA. (2002). *1170.0:2002 – Structural design actions – Part 0: General principles. Standards Association of Australia.*

SAA. (2007). AS 5100 – Bridge design – Parts 1–7. Standards Association of Australia (SAA).

SAA. (2011). *1170.2:2011 – Structural design actions – Part 2: Wind actions. Australia: Standards Association of Australia.*

Safe Work Australia. (2008). *National code of practice for the prevention of falls in general construction*. Commonwealth of Australia.

Safe Work Australia. (2015). *Work related injuries and fatalities in Construction, Australia, 2003 to 2013*. Safe Work Australia.

Sahin, S. (2009). *A comparative study of AISC-360 and Eurocode 3 strength limit states (M.Sc.). Middle East Technical University.*

Salvadori, A. (2009). Ultimate strength of adjustable telescopic steel props according to standard EN 1065. *Journal of Constructional Steel Research*, *65*(10–11), 1964–1970. doi:10.1016/j.jcsr.2009.06.006

Savage, S. (2009). *The flaw of averages: Why we underestimate risk in the face of uncertainty*. John Wiley & Sons.

Schardt, R. (1994). Generalized beam theory - An adequate method for coupled stability problems. *Thin-walled Structures, 19*(2), 161–180. doi:10.1016/0263-8231(94)90027-2

Scheer, J. (2010). *Failed bridges: Case studies, causes and consequences*. London, UK: Wiley VCH. doi:10.1002/9783433600634

Schneider, J. (2006). Structural & Construction Safety. Specially recorded for IABSE E-Learning 2006, Switzerland.

Schneider, J. (2006b). Structural & Construction Safety. Specially recorded for IABSE E-Learning 2006, Switzerland.

Schneider, J. (2006a). *Introduction to safety and reliability of structures* (2nd ed.). IABSE.

Schweiger, H., & Peschl, G. (2005). Reliability analysis in geotechnics with the random set finite element method. *Computers and Geotechnics, 32*(6), 422–435. doi:10.1016/j.compgeo.2005.07.002

SCOSS. (2002). Falsework: Full circle? (No. Topic Paper SC/T/02/01). London, UK: Standing Committee on Structural Safety (SCOSS).

SCOSS. (2005). 15th Biennial report from SCOSS. London, UK: Standing Committee on Structural Safety (SCOSS).

Sétra. (2007). Prestressed concrete bridges built using the cantilever method. Design guide. Paris, France: Service d'études sur les transports, les routes et leurs aménagements (Sétra).

Sexsmith, R. (1998). Reliability during temporary erection phases. *Engineering Structures, 20*(11), 999–1003. doi:10.1016/S0141-0296(97)00195-8

Sexsmith, R., & Reid, S. (2003). Safety factors for bridge falsework by risk management. *Structural Safety, 25*(2), 227–243. doi:10.1016/S0167-4730(02)00058-9

SGB. (2006). Cuplok: User's manual. SGB Group (now HARSCO).

SGB. (2009). *Cuplok: Datasheet*. SGB Group (now HARSCO).

SIA. (2003). *SIA 260 – Bases pour l'élaboration des projets de structures porteuses*. Société Suisse des Ingénieurs et des Architectes.

Sikkel, R. (1982). Health and safety in construction. In *Proceedings of IABSE Workshop on Health and Safety in Construction* (pp. 37–56). Tokyo, Japan: International Association for Bridge and Structural Engineering (IABSE).

Sim, J. (2006). *Human-structure interaction in cantilever grandstands* (PhD Thesis). University of Oxford, Oxford, UK.

Simiu, E., & Scanlan, R. (1996). *Wind effects on structures: Fundamentals and applications to design* (3rd ed.). New York: John Wiley and Sons.

Simões da Silva, L., Rebelo, C., Nethercot, D., Marques, L., Simões, R., & Vila Real, P. (2009). Statistical evaluation of the lateral–torsional buckling resistance of steel I-beams, Part 2: Variability of steel properties. *Journal of Constructional Steel Research, 65*(4), 832–849. doi:10.1016/j.jcsr.2008.07.017

Simulia. (2012). *Abaqus theory manual. Section 1.3.1: Rotation variables*. Simulia.

Smith, J. (2006). Structural robustness analysis and the fast fracture analogy. *Structural Engineering International, 16*(2), 118–123. doi:10.2749/101686606777962521

Solari, G. (2016). Thunderstorm response spectrum technique: Theory and applications. *Engineering Structures, 108*, 28–46. doi:10.1016/j.engstruct.2015.11.012

Sommer, H., Seethaler, R., Chanel, O., Herry, M., Masson, S., & Vergnaud, J. (1999). Health costs due to road traffic-related air pollution. An impact assessment project of Austria, France and Switzerland. Economic evaluation. Bern, Switzerland: World Health Organization (WHO).

Sorensen, J. (2002). Safety culture: A survey of the state-of-the-art. *Reliability Engineering & System Safety*, *76*(2), 189–204. doi:10.1016/S0951-8320(02)00005-4

Spackman, M. (2009). Review of the J–value literature – Final report. London, UK: The Health and Safety Executive (HSE).

Spackman, M., Evans, A., Jones-Lee, M., Loomes, G., Holder, S., & Webb, H. (2011). *Updating the VPF and VPIs: Phase 1: Final report*. UK Department for Transport, Local Government and the Regions.

Spackman, M., & Holder, S. (2007). *Discount rates for rail safety scheme appraisals. Final report for the office of rail regulation*. NERA Economic Consulting.

Starossek, U. (2006). Progressive collapse of structures: Nomenclature and procedures. *Structural Engineering International*, *16*(2), 113–117. doi:10.2749/101686606777962477

Starossek, U. (2009). *Progressive collapse of structures*. Thomas Telford Ltd. doi:10.1680/pcos.36109

Starossek, U., & Haberland, M. (2008). Measures of structural robustness — Requirements & applications.*Proceedings of the ASCE SEI 2008 Structures Congress – Crossing Borders*, 10.

Starossek, U., & Wolff, M. (2005). Progressive collapse: design strategies. In *Proceedings of IABSE Symposium, Structures and Extreme Events* (p. 8). Lisbon, Portugal: International Association for Bridge and Structural Engineering (IABSE).

Stathopoulos, T., & Baniotopoulos, C. (2007). *Wind effects on buildings and design of wind–sensitive structures* (Vol. 493). Springer. doi:10.1007/978-3-211-73076-8_1

Steven, S. (2010). The different phases in construction – Design in health and safety to the project life cycle. In *ICE Manual of health and safety in construction* (pp. 51–69). London: Thomas Telford Ltd.

Sudret, B. (2007). *Uncertainty propagation and sensitivity analysis in mechanical models. Contributions to structural reliability and stochastic spectral methods*. Paris, France: Ecole Nationale des Ponts et Chauss'ees.

Surdahl, R., Miller, D., & Glenn, V. (2010). The positive legacy of a bridge collapse. Public Roads, Federal Highway Administration, US Department of Transportation, 73(5).

Szabó, B., & Babuška, I. (2011). *Introduction to finite element analysis: Formulation, verification and validation*. Chichester, UK: John Wiley & Sons, Ltd. doi:10.1002/9781119993834

Tamboli, A. (2009). *Handbook of steel connection. Design and details*. McGraw Hill Professional.

Tamura, Y., Wu, D., & Yang, Q. (2015). Relation between design load level and lifetime of individual building and its elements. In *Proceedings of the 2nd International Conference on Performance-based and Lifecycle Structural Engineering (PLSE 2015)* (p. 10). Brisbane, Australia: The University of Queensland.

Tesfamariam, S., & Goda, K. (2013). *Handbook of seismic risk analysis and management of civil infrastructure systems*. Cambridge, UK: Woodhead Publishing Limited. doi:10.1533/9780857098986

The Concrete Centre. (2008). *Concrete bridges: The benefits of concrete in bridge design and construction*. London, UK: The Concrete Centre.

The Concrete Society. (1971). *Falsework - Report of the joint committee, The Concrete Society and the Institution of Structural Engineers (No. TRCS 4)*. London, UK: The Concrete Society.

The Concrete Society. (1986). *Formwork: A guide to good practice* (1st ed.). London, UK: The Concrete Society, Report CS30.

The Concrete Society. (1995). *Formwork: A guide to good practice* (2nd ed.). London, UK: The Concrete Society, Report CS30.

The Concrete Society. (2003). *Guide to flat slab formwork and falsework*. London, UK: The Concrete Society, Report CS140.

The Concrete Society. (2005). *A guide to the safe transportation of formwork and falsework equipment*. London, UK: The Concrete Society.

The Concrete Society. (2012). *Formwork: A guide to good practice* (3rd ed.). London, UK: The Concrete Society, Report CS30.

Thoft-Christensen, P., & Baker, M. (1982). *Structural reliability theory and its applications*. Springer. doi:10.1007/978-3-642-68697-9

Thornton Tomasetti Inc. (2012). *Indiana state fair commission. August 13, 2011 Collapse incident. Investigative report*. Indiana State Fair Commission.

TIA. (2016). TIA–222–G – Structural standard for antenna supporting structures and antennas. Arlington, VA: Telecommunications Industry Association (TIA).

Tischer, T., & Kuprenas, J. (2003). Bridge falsework productivity - Measurement and influences. *Journal of Construction Engineering and Management*, *129*(3), 243–250. doi:10.1061/(ASCE)0733-9364(2003)129:3(243)

Todinov, M. (2007). *Risk–based reliability analysis and generic principles for risk reduction*. Amsterdam, The Netherlands: Elsevier.

Turkstra, C., & Madsen, H. (1980). Load combinations in codified structural design. *Journal of the Structural Division*, *106*(12), 2527–2543.

Tversky, A., & Kahneman, D. (1992). Advances in prospect theory: Cumulative representation of uncertainty. *Journal of Risk and Uncertainty*, *5*(4), 297–323. doi:10.1007/BF00122574

TWf. (2012a). Competencies of the TWC (No. TW/11/027). London, UK: Temporary Works forum.

TWf. (2012b). Design management of temporary works (No. TW/12/006). London, UK: Temporary Works forum.

TWf. (2014). Clients' guide to temporary works (No. TWf2014:02). London, UK: Temporary Works forum.

U.S. DOT. (2009). *Treatment of the economic value of a statistical life in departmental analyses – 2009 annual revision*. Washington, DC: U.S. Department of Transportation.

Uemura, A., Kanda, K., Sakamoto, T., & Ito, S. (2000). Design, fabrication and erection of the stiffening girder of 3rd Kurushima Kaikyo bridge. *Kawasaki Steel Giho*, *32*(2), 95–102.

UK DfT. (1999). *Guidance on robustness and provision against accidental actions. The current application of requirements A3 of the building regulations 1991. Final report including background documentation*. London, UK: UK Department for Transport, Local Government and the Regions.

UK DfT. (2011). *Guidance documents – Expert. Tag unit 3.4: The safety objective*. London, UK: UK Department for Transport.

UKOOA. (1999). Industry guidelines on a framework for risk related decision support (No. HS006). London, UK: UK Offshore Operators Association (UKOOA).

UNISDR. (2012). *Impacts of disasters since the 1992 Rio de Janeiro Earth Summit. United Nations Office for Disaster Risk Reduction*. UNISDR.

UNISDR. (2015). Making development sustainable: The future of disaster risk management. Global assessment report on disaster risk reduction. Geneva, Switzerland: United Nations Office for Disaster Risk Reduction (UNISDR).

USDOD. (2010). Unified facilities criteria (UFC) – Design of buildings to resist progressive collapse. Washington, DC: United States of America Defence Department, Report UFC 4-023-03.

Van der Hoven, I. (1957). Power spectrum of horizontal wind speed in the frequency range from 0.0007 to 900 cycles per hour. *Journal of Meteorology*, *14*(2), 160–164. doi:10.1175/1520-0469(1957)014<0160:PSOHWS>2.0.CO;2

van Gelder, P. (2000). *Statistical methods for the risk-based design of civil structures* (PhD Thesis). Delft University of Technology, Delft, The Netherlands.

Versteeg, H., & Malalasekera, W. (2007). *An introduction to computational fluid dynamics: The finite volume method* (2nd ed.). New York: Pearson.

Vieira, R., Virtuoso, F., & Pereira, E. (2014). A higher order model for thin-walled structures with deformable cross-sections. *International Journal of Solids and Structures*, *51*(3–4), 575–598. doi:10.1016/j.ijsolstr.2013.10.023

Virlogeux, M. (1993). Normandie bridge: Design and construction.*Proceedings of the ICE - Structures and Buildings, 99*(3), 281–302.

Viscusi, W. (1993). The value of risks to life and health. *Journal of Economic Literature*, *31*(4), 1912–1946.

Viscusi, W., & Aldy, J. (2003). The value of a statistical life: A critical review of market estimates throughout the world. *Journal of Risk and Uncertainty*, *27*(1), 5–76. doi:10.1023/A:1025598106257

Viscusi, W., & Gayer, T. (2002). Safety at any price? *Regulation*, *25*(3), 54–63.

Voelkel, W. (1990). German assessment for the Cuplok scaffold system within the framework of the comparison of the procedures in different European countries - Part 1: Test report (No. T 2381/1). Berlin, Germany: Institut fur Bautechnik (IfB).

von Neumann, J., & Morgenstern, O. (1944). *Theory of games and economic behavior*. Princeton University Press.

Von Storch, H., & Zwiers, F. (1999). *Statistical analysis in climate research*. Cambridge, UK: Cambridge University Press. doi:10.1017/CBO9780511612336

Vrijling, J., van Gelder, P., Goossens, L., Voortman, H., & Mandey, P. (2004). A framework for risk criteria for critical infrastructures: Fundamentals and case studies in the Netherlands. *Journal of Risk Research*, *7*(6), 569–579. doi:10.1080/1366987032000081178

Vrijling, J., van Hengel, W., & Houben, R. (1998). Acceptable risk as a basis for design. *Reliability Engineering & System Safety*, *59*(1), 141–150. doi:10.1016/S0951-8320(97)00135-X

Walker, A. (1981). Study and analysis of the first 120 failure cases. In Proceedings of Structural Failures in Buildings (pp. 15–40). London, UK: The Institution of Structural Engineers (IStructE).

Wang, L., McCullough, M., & Kareem, A. (2013). A data-driven approach for simulation of full-scale downburst wind speeds. *Journal of Wind Engineering and Industrial Aerodynamics, 123,* 171–190.

Wang, F., Tamura, Y., & Yoshida, A. (2013). Wind loads on clad scaffolding with different geometries and building opening ratios. *Journal of Wind Engineering and Industrial Aerodynamics*, *120*, 37–50. doi:10.1016/j.jweia.2013.06.015

Wang, F., Tamura, Y., & Yoshida, A. (2014). Interference effects of a neighboring building on wind loads on scaffolding. *Journal of Wind Engineering and Industrial Aerodynamics*, *125*, 1–12. doi:10.1016/j.jweia.2013.11.009

Wang, P., Tang, T., & Zheng, H. (2004). Analysis of cable-stayed bridges during construction by cantilever methods. *Computers & Structures*, *82*(4–5), 329–346. doi:10.1016/j.compstruc.2003.11.003

Wang, X., Guo, P., & Huang, X. (2011). A review of wind power forecasting models. *Energy Procedia*, *12*, 770–778. doi:10.1016/j.egypro.2011.10.103

Wardhana, K., & Hadipriono, F. (2003). Analysis of recent bridge failures in the United States. *Journal of Performance of Constructed Facilities*, *17*(3), 144–150. doi:10.1061/(ASCE)0887-3828(2003)17:3(144)

Webster, J., & Eren, H. (2014). *Measurement, instrumentation, and sensors handbook* (2nd ed.). London, UK: Taylor & Francis Group.

Weesner, L., & Jones, H. (2001). Experimental and analytical capacity of frame scaffolding. *Engineering Structures*, *23*(6), 592–599. doi:10.1016/S0141-0296(00)00087-0

WEF. (2015). *Global risks 2015*. Geneva, Switzerland: World Economic Forum.

Whitaker, S., Graves, R., James, M., & McCann, P. (2003). Safety with access scaffolds: Development of a prototype decision aid based on accident analysis. *Journal of Safety Research*, *34*(3), 249–261. doi:10.1016/S0022-4375(03)00025-2 PMID:12963071

Willford, M., & Allsop, A. (1990). Design guide for wind loads on unclad framed building structures during construction (No. BR 173). Bracknell, UK: Building and Research Establishment (BRE).

Willford, M., & Young, P. (2006). *A design guide for footfall induced vibration of structures (No. CCIP-016)*. Surrey, UK: The Concrete Centre.

Williams, C. (1994). The structural design of fabric structures to resist wind loading. In *Proceedings of Conference into Wind Loading on Temporary Structures* (pp. 74–82). Buxton, UK: Health and Safety Executive (HSE).

Wilshere, C. (1983). *Falsework. ICE works construction guide*. London, UK: Thomas Telford Ltd.

Wilson, J., & Holmes, C. (2007). Seismic vulnerability and mitigation during construction of cable-stayed bridges. *Journal of Bridge Engineering*, *12*(3), 364–372. doi:10.1061/(ASCE)1084-0702(2007)12:3(364)

Wong, S., Onof, C., & Hobbs, R. (2005). Models for evaluating the costs of bridge failure.*Proceedings of the ICE - Bridge Engineering*, *158*(3), 117–128. doi:10.1680/bren.2005.158.3.117

Wright, N., & Easom, G. (2003). Non-linear k–ε turbulence model results for flow over a building at full-scale. *Applied Mathematical Modelling*, *27*(12), 1013–1033. doi:10.1016/S0307-904X(03)00123-9

Xiao, N., Huang, H., Wang, Z., Pang, Y., & He, L. (2011). Reliability sensitivity analysis for structural systems in interval probability form. *Structural and Multidisciplinary Optimization*, *44*(5), 691–705. doi:10.1007/s00158-011-0652-9

Xie, N., & Wang, G. (2009). Test analysis on hidden defect in high falsework and its effect on structural reliability. In *Proceedings of the 8th International Conference on Reliability, Maintainability and Safety, ICRMS 2009* (pp. 1077–1080). Chengdu, China: Institute of Electrical and Electronics Engineers (IEEE). doi:10.1109/ICRMS.2009.5270000

Xie, Q., Lu, S., Cóstola, D., & Hensen, J. (2014). An arbitrary polynomial chaos-based approach to analyzing the impacts of design parameters on evacuation time under uncertainty. *Fire Safety Science*, *11*, 14. doi:10.3801/IAFSS.FSS.11-1077

Yakhot, V., Orszag, S., Thangam, S., Gatski, T., & Speziale, C. (1992). Development of turbulence models for shear flows by a double expansion technique. *Physics of Fluids A: Fluid Dynamics (1989-1993)*, *4*(7), 1510–1520.

Yen, T., Huang, Y., Chen, W., & Lin, Y. (1995). *Design of scaffold systems for concrete buildings during construction (No. CE-STR-95-4)*. West Lafayette, IN: Purdue University.

Yue, F., Li, G., & Yuan, Y. (2012). Design methods of integral-lift tubular steel scaffolds for high-rise building construction. *Structural Design of Tall and Special Buildings*, *21*(1), 46–56. doi:10.1002/tal.635

Yue, F., Yuan, Y., Li, G., Ye, K., Chen, Z., & Wang, Z. (2005). Wind load on integral-lift scaffolds for tall building construction. *Journal of Structural Engineering*, *131*(5), 816–824. doi:10.1061/(ASCE)0733-9445(2005)131:5(816)

Yuen, T. (1994). *Bamboo scaffolding in Hong Kong. Part II* (Dissertation submitted for the partial fulfilment of the Bachelor Degree of Engineering). Hong Kong Polytechnic University, Hong Kong, China.

Yu, W., & Chung, K. (2000). *Qualification tests on Mao Jue under compression and bending*. Hong Kong, China: Hong Kong Polytechnic University.

Yu, W., Chung, K., & Chan, S. (2003). Column buckling of structural bamboo. *Engineering Structures*, *25*(6), 755–768. doi:10.1016/S0141-0296(02)00219-5

Zhang, H., Mullen, R., & Muhanna, R. (2010). Finite element structural analysis using imprecise probabilities based on p-box representation. In *Proceedings of the 4th International Workshop on Reliable Engineering Computing (REC 2010)* (pp. 211–225). University of Singapore: Research Publishing Services.

Zhang, H., Chandrangsu, T., & Rasmussen, K. (2010). Probabilistic study of the strength of steel scaffold systems. *Structural Safety*, *32*(6), 393–401. doi:10.1016/j.strusafe.2010.02.005

Zhang, H., Mullen, R., & Muhanna, R. (2012). Structural analysis with probability boxes. *International Journal of Reliability and Safety*, *6*(1), 110–129.

Zhang, L., Wang, C., & Song, G. (2015). Health status monitoring of cuplok scaffold joint connection based on wavelet packet analysis. *Shock and Vibration*, *2015*, 7. doi:10.1155/2015/695845

Zienkiewicz, O., Taylor, R., & Fox, D. (2014). *The finite element method for solid and structural mechanics* (7th ed.). Oxford, UK: Butterworth-Heinemann.

Zienkiewicz, O., Taylor, R., & Zhu, J. (2013). *The finite element method: Its basis and fundamentals* (7th ed.). Oxford, UK: Butterworth-Heinemann.

About the Authors

Robert Beale obtained his BSc in Mathematics from Southampton University and a PhD in the analysis of semi-rigid connections for racking and scaffold structures from Oxford Brookes University where he worked as a Lecture in Engineering Mathematics and a researcher into slender structures. He has published over 120 papers over a wide range of subjects ranging from fluid mechanics to mono-tube dampers to engineering structures, with an emphasis on pallet rack and scaffold structures. He has undertaken consultancy into the behaviour of rack and scaffold connections and supervised and externally examined research degrees both within the UK and internationally. Robert Beale has been invited to give seminars in Hong Kong, Malaysia and China and to attend and present his work at conferences throughout the world.

João André obtained his Diploma in Civil Engineering and MSc in Structural Engineering from "Instituto Superior Técnico" (part of University of Lisbon), and a PhD in Structural Engineering from Oxford Brookes University. He worked as a Professor for two years in "Universidade Lusófona" teaching courses on steel and reinforced concrete structures. He has been working in the Structures Department of the Portuguese National Laboratory Civil Engineering (LNEC) since 2005 where he currently serves as a Postdoctoral Research Fellow. He has published over 30 papers over a wide range of subjects ranging from numerical and experimental analyses, robustness and risk analyses, with emphasis on temporary structures. He was appointed member of the project team responsible for defining the "Robustness Framework" for the revision of the European Structural Eurocodes and he is the National Expert of the Working Group WG6 of CEN/TC250. He is currently working in two European COST Action research projects concerning communication and bridge structures.

Index

I

M

P

R

S

T

U

W

Y

Purchase Print + Free E-Book or E-Book Only*

Purchase a print book through the IGI Global Online Bookstore and receive the e-book for free or purchase the e-book only! Shipping fees apply.

www.igi-global.com

Recommended Reference Books

ISBN: 978-1-4666-7387-8
© 2015; 994 pp.
List Price: $412

ISBN: 978-1-4666-5824-0
© 2014; 617 pp.
List Price: $292

ISBN: 978-1-4666-8130-9
© 2015; 343 pp.
List Price: $172

ISBN: 978-1-4666-7248-2
© 2015; 794 pp.
List Price: $268

ISBN: 978-1-4666-7320-5
© 2015; 343 pp.
List Price: $172

ISBN: 978-1-4666-5800-4
© 2014; 556 pp.
List Price: $188

*IGI Global now offers the exclusive opportunity to receive a free e-book with the purchase of the publication in print, or purchase any e-book publication only. You choose the format that best suits your needs. This offer is only valid on purchases made directly through IGI Global's Online Bookstore and not intended for use by book distributors or wholesalers. Shipping fees will be applied for hardcover purchases during checkout if this option is selected.

Should a new edition of any given publication become available, access will not be extended on the new edition and will only be available for the purchased publication. If a new edition becomes available, you will not lose access, but you would no longer receive new content for that publication (i.e. updates). The free e-book is only available to single institutions that purchase printed publications through IGI Global. Sharing the free e-book is prohibited and will result in the termination of e-access.

Publishing Information Science and Technology Research Since 1988

IGI GLOBAL
DISSEMINATOR OF KNOWLEDGE

www.igi-global.com | Sign up at www.igi-global.com/newsletters | facebook.com/igiglobal | twitter.com/igiglobal

Stay Current on the Latest Emerging Research Developments

Become an IGI Global Reviewer for Authored Book Projects

The overall success of an authored book project is dependent on quality and timely reviews.

In this competitive age of scholarly publishing, constructive and timely feedback significantly decreases the turnaround time of manuscripts from submission to acceptance, allowing the publication and discovery of progressive research at a much more expeditious rate. Several IGI Global authored book projects are currently seeking highly qualified experts in the field to fill vacancies on their respective editorial review boards:

Applications may be sent to:
development@igi-global.com

Applicants must have a doctorate (or an equivalent degree) as well as publishing and reviewing experience. Reviewers are asked to write reviews in a timely, collegial, and constructive manner. All reviewers will begin their role on an ad-hoc basis for a period of one year, and upon successful completion of this term can be considered for full editorial review board status, with the potential for a subsequent promotion to Associate Editor.

If you have a colleague that may be interested in this opportunity, we encourage you to share this information with them.

www.igi-global.com

InfoSci®-Books

A Database for Progressive Information Science and Technology Research

Maximize Your Library's Book Collection!

Invest in IGI Global's InfoSci®-Books database and gain access to hundreds of reference books at a fraction of their individual list price.

The InfoSci®-Books database offers unlimited simultaneous users the ability to precisely return search results through more than 68,000 full-text chapters from nearly 3,000 reference books in the following academic research areas:

Business & Management Information Science & Technology • Computer Science & Information Technology
Educational Science & Technology • Engineering Science & Technology • Environmental Science & Technology
Government Science & Technology • Library Information Science & Technology • Media & Communication Science & Technology
Medical, Healthcare & Life Science & Technology • Security & Forensic Science & Technology • Social Sciences & Online Behavior

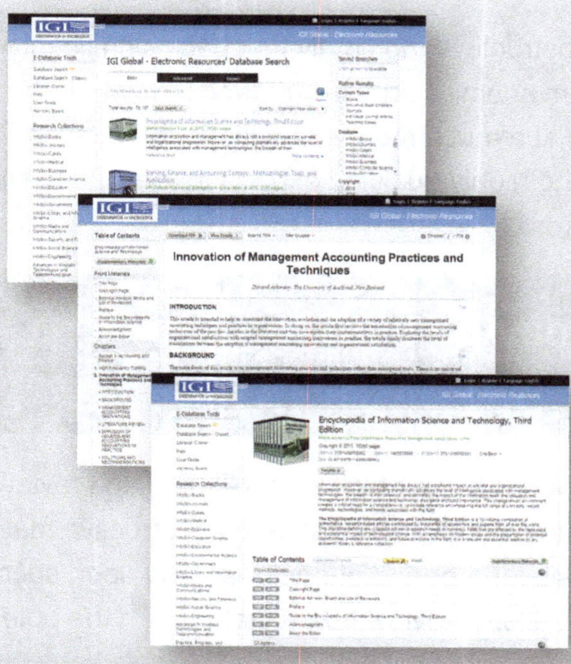

Peer-Reviewed Content:
- Cutting-edge research
- No embargoes
- Scholarly and professional
- Interdisciplinary

Award-Winning Platform:
- Unlimited simultaneous users
- Full-text in XML and PDF
- Advanced search engine
- No DRM

Librarian-Friendly:
- Free MARC records
- Discovery services
- COUNTER4/SUSHI compliant
- Training available

To find out more or request a free trial, visit: www.igi-global.com/eresources

IGI GLOBAL DISSEMINATOR OF KNOWLEDGE
www.igi-global.com

IGI Global
Proudly Partners with

eContent Pro

eContent Pro specializes in the following areas:

Academic Copy Editing
Our expert copy editors will conduct a full copy editing procedure on your manuscript and will also address your preferred reference style to make sure your paper meets the standards of the style of your choice.

Expert Translation
Our expert translators will work to ensure a clear cut and accurate translation of your document, ensuring that your research is flawlessly communicated to your audience.

Professional Proofreading
Our editors will conduct a comprehensive assessment of your content and address all shortcomings of the paper in terms of grammar, language structures, spelling, and formatting.

IGI Global Authors, Save 10% on eContent Pro's Services!

Scan the QR Code to Receive Your 10% Discount

The 10% discount is applied directly to your eContent Pro shopping cart when placing an order through IGI Global's referral link. Use the QR code to access this referral link. eContent Pro has the right to end or modify any promotion at any time.

Email: customerservice@econtentpro.com

econtentpro.com

IRMA INTERNATIONAL

Information Resources Management Association

Become an IRMA Member

Members of the **Information Resources Management Association (IRMA)** understand the importance of community within their field of study. The Information Resources Management Association is an ideal venue through which professionals, students, and academicians can convene and share the latest industry innovations and scholarly research that is changing the field of information science and technology. Become a member today and enjoy the benefits of membership as well as the opportunity to collaborate and network with fellow experts in the field.

IRMA Membership Benefits:

- **One FREE Journal Subscription**
- **30% Off Additional Journal Subscriptions**
- **20% Off Book Purchases**
- Updates on the latest events and research on Information Resources Management through the IRMA-L listserv.
- Updates on new open access and downloadable content added to Research IRM.
- A copy of the Information Technology Management Newsletter twice a year.
- A certificate of membership.

IRMA Membership $195

Scan code or visit **irma-international.org** and begin by selecting your free journal subscription.

Membership is good for one full year.

www.irma-international.org

Encyclopedia of Information Science and Technology, Third Edition (10 Vols.)

Mehdi Khosrow-Pour, D.B.A. (Information Resources Management Association, USA)
ISBN: 978-1-4666-5888-2; **EISBN:** 978-1-4666-5889-9; © 2015; 10,384 pages.

The **Encyclopedia of Information Science and Technology, Third Edition** is a 10-volume compilation of authoritative, previously unpublished research-based articles contributed by thousands of researchers and experts from all over the world. This discipline-defining encyclopedia will serve research needs in numerous fields that are affected by the rapid pace and substantial impact of technological change. With an emphasis on modern issues and the presentation of potential opportunities, prospective solutions, and future directions in the field, it is a relevant and essential addition to any academic library's reference collection.

[1] 30% discount offer cannot be combined with any other discount and is only valid on purchases made directly through IGI Global's Online Bookstore (www.igi-global.com/books), not intended for use by distributors or wholesalers. Offer expires December 31, 2016.

Free Lifetime E-Access with Print Purchase

Take 30% Off Retail Price:

Hardcover with Free E-Access:[2] $2,765
~~List Price: $3,950~~

E-Access with Free Hardcover:[2] $2,765
~~List Price: $3,950~~

Recommend this Title to Your Institution's Library: www.igi-global.com/books

[2] IGI Global now offers the exclusive opportunity to receive free lifetime e-access with the purchase of the publication in print, or purchase any e-access publication and receive a free print copy of the publication. You choose the format that best suits your needs. This offer is only valid on purchases made directly through IGI Global's Online Bookstore and not intended for use by book distributors or wholesalers. Shipping fees will be applied for hardcover purchases during checkout if this option is selected.

The lifetime of a publication refers to its status as the current edition. Should a new edition of any given publication become available, access will not be extended on the new edition and will only be available for the purchased publication. If a new edition becomes available, you will not lose access, but you would no longer receive new content for that publication (i.e. updates). Free Lifetime E-Access is only available to single institutions that purchase printed publications through IGI Global. Sharing the Free Lifetime E-Access is prohibited and will result in the termination of e-access.

www.ingramcontent.com/pod-product-compliance
Lightning Source LLC
LaVergne TN
LVHW080531300926
849162LV00007B/1581